ADVANCES IN ELECTRIC POWER AND ENERGY

IEEE Press
445 Hoes Lane
Piscataway, NJ 08854

ADVANCES IN ELECTRIC POWER AND ENERGY

Static State Estimation

EDITED BY

MOHAMED E. EL-HAWARY
Dalhousie University

Published by John Wiley & Sons, Inc., Hoboken, New Jersey.

Published simultaneously in Canada.

Library of Congress Cataloging-in-Publication Data:

Names: El-Hawary, M. E., editor. | John Wiley & Sons, Inc., publisher.
Title: Advances in electric power and energy : static state estimation / edited by Mohamed E El-Hawary, Dalhousie University.
Description: Hoboken, New Jersey : John Wiley & Sons, Inc., [2021] | Series: IEEE Press | Includes bibliographical references and index.
Identifiers: LCCN 2020028681 (print) | LCCN 2020028682 (ebook) | ISBN 9781119480464 (hardback) | ISBN 9781119584506 (hardback) | ISBN 9781119480365 (adobe pdf) | ISBN 9781119480440 (epub)
Subjects: LCSH: Electric power systems–State estimation.
Classification: LCC TK1005 .A35 2020 (print) | LCC TK1005 (ebook) | DDC 621.319/1–dc23
LC record available at https://lccn.loc.gov/2020028681
LC ebook record available at https://lccn.loc.gov/2020028682

Cover Design: Wiley
Cover Images: (top) © Sam Robinson/Getty Images, (middle) © Steve Ramplin/EyeEm/Getty Images

Set in 10/12pt Times by SPi Global, Pondicherry, India

TO FRED C. SCHWEPPE, 1933–1988

He brought state estimation to electric power, and later led the development of the theoretical bases of competitive electric power markets. A teacher and innovator of great understanding and creativity, "solving the problem" was secondary to bringing out the best in those he worked with.

CONTENTS

ABOUT THE EDITOR

Mohamed "Mo" El-Aref El-Hawary, age 76, of Halifax passed away on Friday, 26 July 2019. Born in Sohag, Egypt, he was predeceased by his parents Mahmood and Amina El-Hawary of Alexandria, Egypt. He is survived by his wife, Ferial (El-Bibany) El-Hawary, Halifax; daughter, Elizabeth "Bette" El-Hawary, Halifax; sons, Robert "Bob" El-Hawary, London, UK, and Rany "Ron" (Tricia Lane) El-Hawary, Halifax; sisters, Mervat El-Hawary, Mona El-Hawary, and Mawaheb (Heba) El-Hawary (all located in Alexandria, Egypt); and grandchildren, Alexa, Ben, Grace, Ryan, Eoin, Kegan, Duncan, and Liam.

Dr. El-Hawary was a Professor of Electrical and Computer Engineering at Dalhousie University in Halifax, Nova Scotia, Canada. He had a BSc in Electrical Engineering, Distinction and First-Class Honors, University of Alexandria, Egypt, 1965, and a PhD in Electrical Engineering, University of Alberta, Edmonton, 1972, where he was an Izaak Walton Killam Memorial Fellow from 1970 to 1972. He was Associate Professor of Electrical Engineering at the Federal University of Rio de Janeiro for two years and subsequently served for eight years on faculty at Memorial University of Newfoundland since 1974. He was appointed Chairman of Electrical Engineering Program in 1976. In 1981, he joined the Technical University of Nova Scotia (TUNS) as Professor of Electrical Engineering. In 1997, TUNS was amalgamated with Dalhousie University. Dr. El-Hawary has been Associate Dean of Engineering at Dalhousie between 1995 and 2007, Director of International and External Relations for the Faculty of Engineering in 2008–2009, and Chair of the Senate of Dalhousie University in 2001–2007. He cherished having had the opportunity to be part of educating, mentoring, and touching the lives and careers of countless students in the field of Electrical Engineering over his long and distinguished career.

Throughout his career, Mo authored over 10 textbooks and almost 200 full journal papers. He was the Institute of Electrical and Electronics Engineering (IEEE) Press Power Engineering Series Editor and Founding Editor in Chief of the IEEE Systems, Man and Cybernetics Magazine, and Power Letters of PES.

He was Associate Editor for the three major Electric Machines and Power Systems' Journals and Editor of Electrical Power Engineering, McGraw-Hill Encyclopedia of Science and Technology. He was a Fellow of IEEE, Canadian Academy of Engineering, Engineers Canada, and the Engineering Institute of Canada. He was a Distinguished Lecturer of the IEEE Power and Energy Society.

He served as a member of the Board of Directors and Secretary of IEEE and as President of IEEE Canada. He served on the IEEE Publication Services and Products Board, Fellows Committee, IEEE Press Board Chairman, Power Engineering Society (PES): System Operations Committee Chair and, member of HKN Board, and Vice President, Development, IEEE Canada Foundation. He has been recipient of IEEE Canada, W. S. Read Service Award, 2010. In 1999 IEEE awarded him the EAB Meritorious Achievement, Power Engineering Educator of the Year, and IEEE Canada General A.G.L. McNaughton Gold Medal.

ABOUT THE CONTRIBUTORS

Venkataramana Ajjarapu currently holds the David Nicholas Professor of Electrical and Computer Engineering at Iowa State University. His area of expertise includes power system stability, reactive power control, and optimization. He is a Fellow of the IE.

Aditya Ashok is a senior research engineer in the Electricity Infrastructure and Buildings division at the Pacific Northwest National Laboratory (PNNL) and has been with PNNL since February 2016. Aditya received his doctoral degree in Electrical Engineering from Iowa State University in May 2017. Aditya's research interests include analyzing cyber vulnerabilities in energy delivery systems, assessing potential impacts to system operations, reliability, and economics, and developing novel algorithms to mitigate cyber vulnerabilities and help enhance the overall security and resilience of energy delivery systems.

Bernd Brinkmann received his Bachelor's degree in Electrical Engineering from the University of Applied Sciences Bielefeld, Germany, in 2011. After gaining experience as a design engineer and software developer, he is currently pursuing the PhD degree in Electrical Power Engineering at the University of Tasmania, Australia. His research interests include state estimation uncertainty, distribution network observability, and optimal meter placement.

Eduardo Caro received the Electrical Engineering degree from the Technical University of Catalonia, Barcelona, Spain, 2007, and the PhD degree in the University of Castilla-La Mancha, Spain, 2011. He is currently an Assistant Professor at the Universidad Politécnica de Madrid, Madrid, Spain. His research interests include power system estimation, optimization, and electricity load forecasting.

Sungyun Choi received the BE degree in Electrical Engineering from Korea University, Seoul, South Korea, in 2002 and the MS and PhD degrees in Electrical and Computer Engineering from Georgia Institute of Technology, Atlanta, GA, USA, in 2009 and 2013, respectively. From 2002 to 2005, he was a Network and System Engineer, and from 2007 to 2013, he was a Research Assistant with the Power System Control and Automation Laboratory, Atlanta, GA, USA. Since 2014, he has been a Senior Researcher with Smart Power Grid Research Center, Korea Electrotechnology Research Institute, Uiwang, South Korea. His research interests include smart grid technology, autonomous operation of microgrids, power system protection, distributed dynamic state estimation, and communication networks and systems in power industries.

George J. Cokkinides was born in Athens, Greece, in 1955. He received the BS, MS, and PhD degrees from the Georgia Institute of Technology, Atlanta, GA, USA, in 1978, 1980, and 1985, respectively. From 1983 to 1985, he was a Research Engineer at the Georgia Tech Research Institute. Since 1985, he has been with the University of South Carolina, Columbia, SC, USA, where he is currently an Associate Professor of Electrical Engineering. His research interests include power system modeling and simulation, power electronics applications, power system harmonics, and measurement instrumentation. Professor Cokkinides is a member of the IEEE Power and Energy Society.

Venkata Dinavahi received the BEng. degree in electrical engineering from the Visveswaraya National Institute of Technology (VNIT), Nagpur, India, in 1993, the MTech. degree in Electrical Engineering from the Indian Institute of Technology (IIT) Kanpur, India, in 1996, and the PhD degree in Electrical and Computer engineering from the University of Toronto, Ontario, Canada, in 2000. Presently he is a Professor with the Department of Electrical and Computer Engineering, University of Alberta, Edmonton, Alberta, Canada. His research interests include real-time simulation of power systems and power electronic systems, electromagnetic transients, device-level modeling, large-scale systems, and parallel and distributed computing. He is a Fellow of IEEE.

James W. Feltes received his BS degree with honors in Electrical Engineering from Iowa State University and his MS degree in Electrical Engineering from Union College.

He joined Power Technologies, Inc. (PTI), now part of Siemens Power Transmission and Distribution Inc., in 1979 and is currently a senior manager. At PTI, he has participated in many studies involving planning, analysis, and design of transmission and distribution systems. He has also been involved in many projects involving the development of models for studies of power system dynamics, testing to record equipment response, and model parameter derivation.

He is a registered professional engineer in the state of New York and a Fellow of the IEEE. He is a member of the IEEE Power Engineering Society and Industry Applications Society and is active on several IEEE committees and task forces.

Georgios B. Giannakis (Fellow'97) received his Diploma in Electrical Engineering from the National Technical University of Athens, Greece, 1981. From 1982 to 1986 he was with the University of Southern California (USC), where he received his MSc in Electrical Engineering, 1983, MSc in Mathematics, 1986, and PhD in Electrical Engineering, 1986. He was a faculty member with the University of Virginia from 1987 to 1998, and since 1999 he has been a professor with the University of Minnesota, where he holds an ADC Endowed Chair, a University of Minnesota McKnight Presidential Chair in ECE, and serves as director of the Digital Technology Center.

His general interests span the areas of statistical learning, communications, and networking – subjects on which he has published more than 470 journal papers, 770 conference papers, 25 book chapters, two edited books, and two research monographs. His current research focuses on Data Science, and Network Science with applications to the Internet of Things, and power networks with renewables. He is the (co-) inventor of 34 issued patents, and the (co-) recipient of 9 best journal paper awards from the IEEE Signal Processing (SP) and Communications Societies, including the G. Marconi Prize Paper Award in Wireless Communications. He also received the IEEE-SPS Norbert Wiener Society Award (2019); EURASIP's A. Papoulis Society Award (2020); Technical Achievement Awards from the IEEE-SPS (2000) and from EURASIP (2005); the IEEE ComSoc Education Award (2019); the G. W. Taylor Award for Distinguished Research from the University of Minnesota, and the IEEE Fourier Technical Field Award (2015). He is a foreign member of the Academia Europaea, and Fellow of the National Academy of Inventors, the European Academy of Sciences, IEEE and EURASIP. He has served the IEEE in a number of posts, including that of a Distinguished Lecturer for the IEEE-SPS.

Manimaran Govindarasu currently holds the Mehl Professor of Computer Engineering at Iowa State University. His area of expertise includes CPS for the smart grid, cyber security, and real-time systems and networks. He is a Fellow of the IEEE.

Ye Guo is an Assistant Professor at Tsinghua-Berkeley Shenzhen Institute, Tsinghua University. He received his bachelor degree in 2008 and doctoral degree in 2013, both from the Department of Electrical Engineering, Tsinghua University. He was a Postdoctoral Associate at Cornell University between 2014 and 2018. His research interests include distributed optimization, game and market theory, state estimation, and their applications in power and energy systems. He has received the Best-of-the-Best paper award and another Best Paper Award at IEEE PES General Meeting, 2019, and another Best Paper Award at IEEE PES General Meeting 2020. He also received the Best Poster Award at PSERC IAB Meeting 2018.

Ibrahim Omar Habiballah is an Associate Professor of EE Department at King Fahd University of Petroleum and Minerals, Saudi Arabia. In his area he taught several undergraduate and graduate courses in electrical, power systems, power transmission, and electrical machines. His research interests include power systems in general, power system state estimation, power system optimization, HV insulators, and energy conservation.

Araceli Hernández received the PhD degree in Electrical Engineering from the Universidad Politécnica de Madrid (UPM), Madrid, Spain, in 2000. Currently, she works at the Department of Control, Electrical and Electronic Engineering and Computing at UPM, where she is an Associate Professor. Her fields of interest include power system analysis and power quality estimation and measurement.

Hadis Karimipour received the PhD degree from the University of Alberta in 2016. She is currently a Postdoctoral Fellow at the Department of Electrical and Computer Engineering at the University of Calgary. Her research interests include large-scale power system state estimation, cyber-physical modeling, cybersecurity of the smart grids, and parallel and distributed computing.

Vassilis Kekatos is an Assistant Professor at the Bradley Department of ECE at Virginia Tech. He obtained his Diploma, MSc, and PhD in Computer Science and Engineering from the University of Patras, Greece, in 2001, 2003, and 2007, respectively. He was a recipient of a Marie Curie Fellowship during 2009–2012 and a research associate with the ECE Department at the University of Minnesota, where he received the postdoctoral career development award (honorable mention). During 2014, he stayed with the University of Texas at Austin and the Ohio State University as a visiting researcher. His research focus is on optimization and learning for future energy systems. He is currently serving in the editorial board of the IEEE Trans. on Smart Grid.

Mert Korkali received his MS and PhD degrees in Electrical Engineering from Northeastern University, Boston, MA, USA, in 2010 and 2013, respectively. He is currently a Research Staff Member at the Computational Engineering Division at Lawrence Livermore National Laboratory, Livermore, CA, USA. Previously, he was a Postdoctoral Research Associate at the University of Vermont, Burlington, VT, USA. His current research interests lie at the broad interface of robust state estimation and fault location in power systems, extreme event modeling, cascading failures, uncertainty quantification, and probabilistic grid planning. He is the Co-chair of the IEEE Task Force on Standard Test Cases for Power System State Estimation. He is currently serving as an Editor of the *IEEE Open Access Journal of Power and Energy* and of the *IEEE Power Engineering Letters*, and an Associate Editor of *Journal of Modern Power Systems and Clean Energy*. Dr. Korkali is a Senior Member of IEEE.

Massimo La Scala is Professor of Electrical Energy Systems at Politecnico di Bari and IEEE Fellow. He has been Principal Investigator of numerous research projects in smart grids and smart cities and scientific consultant of the Ministry of the Economic Development in Italy and of AEEGSI the Italian Regulatory Authority of Electricity, Gas and Water. He is the director of the "Laboratory for the development of renewables and energy efficiency: Lab ZERO" at Politecnico di Bari.

Yu Liu was born in Hefei, China, in 1990. He received the BS and MS degrees in Electric Power Engineering from Shanghai Jiao Tong University, Shanghai, China, in 2011 and 2013, respectively, and the MS degree in electrical and computer engineering in 2013 from Georgia Institute of Technology, Atlanta, GA, USA, where he is currently working toward the PhD degree in electrical and computer engineering. His research interests include power system protection, parameter estimation, and circuit fault locating.

A.P. Sakis Meliopoulos was born in Katerini, Greece, in 1949. He received the ME and EE Diploma in Electrical Engineering from the National Technical University of Athens, Athens, Greece, in 1972 and the MSEE and PhD degrees in electrical engineering from the Georgia Institute of Technology, Atlanta, GA, USA, in 1974 and 1976, respectively. He is presently a Georgia Power Distinguished Professor. He has published three books, holds three patents, and has published more than 300 technical papers. Professor Meliopoulos received the IEEE Richard Kaufman Award in 2005, and in 2010, he received the George Montefiore Award from the Montefiore Institute, Belgium. He is the Chairman of the Georgia Tech Protective Relaying Conference and a member of Sigma Xi.

Hyde M. Merrill received the BA degree in Mathematics and MS degree in Electrical Engineering from the University of Utah and the PhD degree in Electrical Engineering from the Massachusetts Institute of Technology. He is a registered professional engineer in New York and a Fellow of the IEEE.

He has worked for the American Electric Power Service Corp, the MIT Energy Lab, Power Technologies, Inc., the Rensselaer Polytechnic Institute, and Merrill Energy LLC. In 2015, he joined the University of Utah as Adjunct Professor. He teaches power systems and leads research on blackouts.

Lamine Mili is a Professor of Electrical and Computer Engineering at Virginia Tech. He is an IEEE Fellow and a member of Institute of Mathematical Statistics and the American Statistical Association. His research interests include power system analysis and control, power system dynamics and stability, and robust statistics as applied to engineering problems.

Michael Negnevitsky received his BE (Hons.) and PhD degrees from the Byelorussian University of Technology, Belarus, in 1978 and 1983, respectively. Currently, he is a Professor in Power Engineering and Computational Intelligence and Director of the Centre for Renewable Energy and Power Systems, University of Tasmania, Australia. He is a Chartered Professional Engineer, Fellow of Engineers Australia, and Member of the National ITEE College Board. His research interests include power system security, renewable energy, and state estimation.

Marco Pau received the MS degree (cum laude) in Electrical Engineering and the PhD degree in Electronic Engineering and Computer Science from the University of Cagliari, Italy, in 2011 and 2015, respectively. Currently, he is research associate at the Institute for Automation of Complex Power Systems at the E.ON Energy Research Center, RWTH Aachen University, Germany, where he leads the team for Distribution Grid Monitoring and Automation. His research activities mainly concern the design of solutions for the monitoring and automation of distribution systems as well as techniques for the smart management of active distribution grids.

Paolo Attilio Pegoraro received the MS (summa cum laude) degree in Telecommunications engineering and the PhD degree in Electronic and Telecommunication Engineering from the University of Padova, Padua, Italy, in 2001 and 2005, respectively. From 2015 to 2018 he was an Assistant Professor with the Department of Electrical and Electronic Engineering, University of Cagliari, Cagliari, Italy, where he is currently Associate Professor. He has authored or coauthored over 110 scientific papers. His current research interests include the development of new measurement techniques for modern power networks, with attention to synchronized measurements and state estimation.

Dr. Pegoraro is a Senior Member of IEEE Instrumentation and Measurement Society, member of TC 39 (Measurements in Power Systems) and of IEC TC 38/WG 47. He is an Associate Editor of the IEEE Transactions on Instrumentation and Measurement.

Ferdinanda Ponci graduated with PhD in Electrical Engineering from the Politecnico di Milano, in 2002. She joined the Department of Electrical Engineering, University of South Carolina, as an Assistant Professor in 2003 and became Associate Professor in 2008. In 2009, she joined the Institute for Automation of Complex Power Systems, RWTH Aachen University, where she is currently Professor for "Monitoring and distributed control for power systems."

She is Senior Member of IEEE and of the AdCom of the IEEE Instrumentation and Measurement Society.

Md. Ashfaqur Rahman is a PhD candidate in the Department of Electrical and Computer Engineering in Clemson University, Clemson, SC, USA. He received his BS from Bangladesh University of Engineering and Technology in 2009 and MS from Texas Tech University in 2012. He has a total of 7 technical papers with 98 citations with h-index and i-index be 3. His current research interests include the development of a distributed dynamic state estimator. He also worked on false data injection attack, parallel and distributed computation, state prediction, contingency analysis, optimal power flow, etc. He has served as a reviewer of IEEE journals and conference papers.

Sara Sulis received the MS degree in Electrical Engineering and the PhD degree in Industrial Engineering from the University of Cagliari, Cagliari, Italy, in 2002 and 2006, respectively. She is currently Associate Professor of Instrumentation and Measurements with the University of Cagliari. Dr. Sulis is a Senior Member of the IEEE, member of the Instrumentation and Measurement Society, of the IEEE TC 39 "Measurements in Power Systems," and of the CENELEC TC 38 "Instrument Transformers." She has authored or coauthored more than 100 scientific papers. Her current research interests include distributed measurement systems designed to perform state estimation and harmonic sources estimation of distribution networks.

Hongbin Sun is a Professor in the Department of Electrical Engineering, Tsinghua University, Beijing, China, the Changjiang Chair Professor of Education Ministry of China, and an IEEE Fellow. He received double BS degrees in 1992 and PhD in 1997, respectively, both from Electrical Engineering, Tsinghua University. His research interests include automatic voltage control (AVC), smart grid, renewable energy and electrical vehicle integration, and power system operation and control.

Lang Tong is the Irwin and Joan Jacobs Professor of Engineering at Cornell University and the Cornell site Director of Power Systems Engineering Research Center (PSERC). He received a BE degree from Tsinghua University and a PhD degree in Electrical Engineering from the University of Notre Dame. He held visiting positions at Stanford University, the University of California at Berkeley, the Delft University of Technology, and the Chalmers University of Technology in Sweden.

Lang Tong's current research focuses on data analytics, optimization, and economic problems in energy and power systems. A Fellow of IEEE and the 2018 Fulbright Distinguished Chair in Alternative Energy, he received paper awards from the IEEE Circuit and Systems, Signal Processing, Communications, and Power and Energy Systems societies.

Ganesh Kumar Venayagamoorthy is the Duke Energy Distinguished Professor of Power Engineering and Professor of Electrical and Computer Engineering at Clemson University. Dr. Venayagamoorthy is the Founder (2004) and Director of the Real-Time Power and Intelligent Systems Laboratory (http://rtpis.org). He holds an Honorary Professor position in the School of Engineering at the University of Kwazulu-Natal, Durban, South Africa. Dr. Venayagamoorthy received his PhD and MSc (Eng.) degrees in Electrical Engineering from the University of Natal, Durban, South Africa, in February 2002 and April 1999, respectively. He received his B.Eng. (Honors) degree with a First Class from Abubakar Tafawa Balewa University, Bauchi, Nigeria, in March 1994. He holds a MBA degree in Entrepreneurship and Innovation from Clemson University, SC (2016). Dr. Venayagamoorthy's interests are in the research, development, and innovation of smart grid technologies and operations, including computational intelligence, intelligent sensing and monitoring, intelligent systems, integration of renewable energy sources, power system optimization, stability and control, and signal processing. He is an inventor of technologies for scalable computational intelligence for complex systems and dynamic stochastic optimal power flow. He led the brain2grid project funded by US NSF. He has published over 500 refereed technical articles. His publications are cited >18,000 times with a h-index of 64. Dr. Venayagamoorthy has been involved in over 75 sponsored projects in excess of US $12 million. Dr. Venayagamoorthy has given over 500 invited keynotes, plenaries, presentations, tutorials, and lectures in over 40 countries to date. He has several international educational and research collaborations. Dr. Venayagamoorthy is a Senior Member of the IEEE, and a Fellow of the IET, UK, and the SAIEE.

Gang Wang received the BEng. degree in Automatic Control from the Beijing Institute of Technology, Beijing, China, in 2011, and the PhD degree in Electrical Engineering from the University of Minnesota, Minneapolis, USA, in 2018, where he stayed as a postdoctoral researcher until 2020. Since August 2020, he has been a professor with the School of Automation, Beijing Institute of Technology. His research interests focus on the areas of signal processing, deep learning, and reinforcement learning with applications to cyber-physical systems and data science. He was the recipient of the Excellent Doctoral Dissertation Award from the Chinese Association of Automation in 2019, the Best Student Paper Award from the 2017 European Signal Processing Conference, and the Best Conference Paper at the 2019 IEEE Power & Energy Society General Meeting.

Wenchuan Wu is a Professor in the Department of Electrical Engineering, Tsinghua University, Beijing, China. He received his BS in 1996, MS in 1999, and PhD degrees in 2003 all from the Electrical Engineering Department, Tsinghua University. His research interests include Energy Management System, active distribution system operation and control, and EMTP-TSA hybrid real-time simulation. He is an Associate Editor of *IEE Proceedings – Generation, Transmission and Distribution and Journal of Electric Power Components and Systems*.

Yuanhai Xia is an Electrical Engineer with China State Construction Engineering Corporation (Middle East). He has one and half years' experience in building electric and half year in high voltage power transmission. He got his MSc from

KFUPM, Saudi Arabia, in electrical and power system in 2014. He is familiar with international and domestic codes/standards, AutoCAD drawing, master excel skills with VB programming, and other programming language such as Matlab, python, and Linux shell.

Boming Zhang is a Professor in the Department of Electrical Engineering, Tsinghua University, Beijing, China. He received MEng. from Harbin Institute of Technology in 1982 and PhD from Tsinghua University in 1985, both in Electrical Engineering. He has been serving for Tsinghua University since 1985. His research area includes power system analysis, computer application in power system control center, etc. He won IEEE PES/CSEE Yu-Hsiu Ku Electrical Engineering Award in 2015.

Junbo Zhao (SM'19) received the PhD degree from the Bradley Department of Electrical and Computer Engineering, Virginia Tech, Blacksburg, VA, USA, in 2018. He was an Assistant Professor (Research) with Virginia Tech from May 2018 to August 2019. He did the summer internship at the Pacific Northwest National Laboratory from May 2017 to August 2017. He is currently an Assistant Professor with Mississippi State University, Starkville, MS, USA. He has written three book chapters and published more than 70 peer-reviewed journal and conference papers, among which there are three ESI papers. His research interests are power system modeling, state estimation, dynamics and cybersecurity, synchrophasor applications, renewable energy integration and control, and robust statistical signal processing and machine learning.

Dr. Zhao is a co-recipient of the best paper award of 2019 IEEE PES ISGT Asia, and the best reviewer of the IEEE TRANSACTIONS ON POWER SYSTEMS 2018 and the IEEE TRANSACTIONS ON SMART GRID 2019. He is currently the Chair of the IEEE Task Force on Power System Dynamic State and Parameter Estimation, and the Secretary of the IEEE Working Group on State Estimation Algorithms and the IEEE Task Force on Synchrophasor Applications in Power System Operation and Control. He serves as the Associate Editor of the IEEE TRANSACTIONS ON POWER SYSTEMS, the IEEE TRANSACTIONS ON SMART GRID, and *International Journal of Electrical Power and Energy Systems*, and the Subject Editor of *IET Generation, Transmission and Distribution*.

Hao Zhu is an Assistant Professor of ECE at University of Texas at Austin. She received a BE degree from Tsinghua University in 2006 and MSc and PhD degrees from the University of Minnesota in 2009 and 2012, all in Electrical Engineering. Her current research interests include power grid monitoring, distribution system operations and control, and energy data analytics. She received the NSF CAREER Award in 2017, the Siebel Energy Institute Seed Grant Award and the US AFRL Summer Faculty Fellowship in 2016.

CHAPTER 1

GENERAL CONSIDERATIONS

Mohamed E. El-Hawary

Dalhousie University in Halifax, Nova Scotia, Canada

In this introductory chapter, we introduce the concept of state estimation (SE) in electric power system and trace its evolution from a historical perspective. SE emerged as an indispensable real-time tool that is part of a suite of applications designed to support and enable electric power operators' "situational awareness." The term "situational awareness" in the context of power grid operation is "understanding the present environment and being able to accurately anticipate future problems to enable effective actions."

This chapter offers a detailed discussion of the role of SE in practice. A guide to the chapters included in this volume is offered to conclude the chapter.

1.1 PRELUDE

At the IEEE Power Industry Computer Applications (PICA) conference held on 18–21 May 1969 in Denver, Colorado, Professor Fred C. Schweppe and his associates presented a three-part paper on static state estimation and related detection and identification problems in electric power systems. The papers were subsequently published in the *IEEE Transactions on Power Apparatus and Systems* [1–3]. The first paper [1] introduced the overall problem statement, mathematical modeling, and general algorithms for state estimation, detection, and identification (SEDI) using weighted least squares (WLS) approximations. The second paper [2] discussed an approximate mathematical model and the resulting simplifications in SEDI. The third paper [3] dealt with implementation problems, considerations of dimensionality, execution speed and storage, and the time-varying nature of actual power systems.

A year later, Merrill and Schweppe [4] introduced a bad data suppression (BDS) estimator, which is computationally very similar to WLS approximation.

Advances in Electric Power and Energy: Static State Estimation, First Edition.
Edited by Mohamed E. El-Hawary.

Published 2021 by John Wiley & Sons, Inc.

The concept is no more complex, and bad data detection and identification can be performed "for free," since BDS requires no more computer time or complexity than does WLS, and in the absence of bad data, BDS reduces to WLS.

1.2 DEFINING SSE

In 1974, Schweppe and Handschin [5] described state estimation (SE) using the following metaphor: "The life blood of the control system is a base of clean pure data defining the system state and status (voltages, network configuration). This life blood is obtained from the nourishment provided by the measurements gathered from around the system (data acquisition). A static state estimator is the digestive system which removes the impurities from the measurements and converts them into a form which the brain (man or computer) of the central control system can readily use to make 'action' decisions on system economy, quality, and security."

Reference [1] formally defines the static state of an electric power system as the vector of voltage magnitudes and angles at all network buses. The static state estimator (SSE) is a data processing algorithm for converting imperfect redundant meter readings and other available information to an estimate of the static state.

Item 603-02-09 of the International Electrotechnical Commission (IEC) Electropedia [6] offers the following definition of "state estimation" as "the computation of the most probable currents and voltages within the network at a given instant by solving a system of mostly nonlinear equations whose parameters are obtained by means of redundant measurements."

The North American Electric Reliability Corporation (NERC) Real-Time Tools Best Practices Task Force (RTBPTF) 2008 final report [7] offers the following definition: "A state estimator is an application that performs statistical analysis using a set of imperfect, redundant, telemetered power-system data to determine the system's current condition. The system condition or state is a function of several variables: bus voltages, relative phase angles, and tap changing transformer positions. A state estimator can typically identify bad analog telemetry, estimate non-telemetered flows and voltages, and determine actual voltage and thermal violations in observable areas."

According to [5], SSE has evolved rapidly to online implementations beginning with the Norwegian Tokke installation [8] followed by the larger AEP installation [9] soon after. Not long later, T. E. Dy Liacco [10] stated: "Although the number of control centers with State Estimation is still rather small, the number is increasing at a rapid rate. The requirement for State Estimation at a modern control center has become the rule, rather than the exception."

The fundamental problem of state estimation can be defined as an over determined system of nonlinear equations solved as an unconstrained weighted least squares (WLS) minimization problem. The WLS estimator minimizes the weighted sum of the squares of the residuals. Residuals are the error or difference between the estimated and the actual values [11]. Many papers and books treat the broad generic area of "state estimation" in system theory [12–14]. State estimation concepts can be applied in other power systems areas [15–20].

1.3 THE NEED FOR STATE ESTIMATION

Security control is the main strategy used in the operation of electric power systems, where actions are taken to prevent an impending emergency, to correct an existing emergency, or to recover from an emergency. Knowing the state of the system under steady-state conditions is the key to security control.

Control centers may be classified into two types, according to the information base available. In one type of control center, the raw power system data as obtained in real time is an adequate information base for operation. The other type of control center goes beyond the mere acquisition of data. By applying state estimation, a far better and a more comprehensive information base than raw data is obtained.

The importance of the real-time load flow fed by state estimation lies in its use as basis for security analysis. With the load flow as a base, reference allows analyzing the effects on the system of any contingency event. In contrast, without state estimation, there is not much to be done with raw data except to check it for abnormal values.

A further important feature of state estimation is the ability to detect the presence of bad data (outliers) and to identify which data is in error. Corrections can then be expedited in the field on the faulty instrumentation. Without state estimation, there is no effective, systematic way of finding measurement errors. Some sort of data validation has been attempted wherein power measurements around a bus are summed up and flows at both ends of a branch checked against each other, but these checks apart from being inconclusive end up being too complicated as it has to take into account the topology of the network. Now network topology is handled systematically and correctly by state estimation. Hence for all the checking done by so-called data validation programs, it is best to go directly to state estimation.

In the modeling of power systems for security control functions, there are usually external networks, i.e. networks or subnetworks, which are not being telemetered by the control center and which are not observable. There are two approaches for estimating the state of these external networks. One approach is to use pseudo-measurements, based on statistics and forecasts, of the injections at the nodes of the external network. The pseudo-measurements are then assigned relatively low weights and included as part of the measurement set in the state estimation routine. The second approach is to perform the state estimation only on the observable part. The state of the external network is then obtained by finding a load flow solution using the pseudo-measurements as inputs with the boundary node voltages held at the values determined by the state estimation of the observable part.

For bad data identification, early state estimators singled out measurements with the highest values of the weighted residual. Newer bad data identification techniques use both the weighted residual and the normalized residual. Either the values of the normalized residual or the ratios of the normalized ones from the weighted residual are used to identify bad data. Bad data rejection is a time-consuming procedure at control centers especially if there are more than one measurement in error.

The concept of security control cannot be fully realized without a complete information base that is derived from the voltage magnitudes and phase angles of all buses in a power system obtained via state estimation [1–3].

1.4 STATIC STATE ESTIMATION IN PRACTICE

The material in this section is based on a NERC Task Force report [7]. To quote the Task Force:

> This report presents the findings and recommendations of the North American Electric Reliability Corporation (NERC) Real-Time Tools Best Practices Task Force (RTBPTF) concerning minimum acceptable capabilities and best practices for real-time tools necessary to ensure reliable electric system operation and reliability coordination. RTBPTF's undertaking is based on the U.S.-Canada Power System Outage Task Force findings that key causes of the August 14, 2003 northeast blackout included absence of situational awareness and inadequate reliability tools. That report also notes the need for visualization display systems to monitor system reliability.
>
> RTBPTF's recommendations result from an extensive, three-year process of fact-finding and analysis supported by the results of the Real-Time Tools Survey, the most comprehensive survey ever conducted of current electric industry practices.
>
> RTBPTF's findings and recommendations are firmly grounded in the results of the Real-Time Tools Survey, a more than 300-page, web-based document with nearly 2,000 questions on a broad scope of current industry practices and plans for using real-time tools.

While [21] referred to RTUs as the eyes and ears and hands of the master station, the phrase came to be commonly used to refer to the state estimator as the eyes and ears of the real-time operator. Indeed, in current practice the state estimator prevails as an "essential" tool for power system operators' "situational awareness." Existing NERC reliability standards assume the use of state estimators to aid RCs and TOPs in maintaining situational awareness for the bulk electric system. The state estimator must be available and able to produce an accurate solution because many applications rely on the state estimator solution as base case.

State estimators are commercially available allowing SCADA/EMS vendors to provide viable state estimators off the shelf with some customization and fully integrated with users' production SCADA/EMS systems. State estimators are used as input to monitor MVA/ampere loadings and low and high bus voltages, voltage drop, voltage node angle separation, SCADA, and visualization. Therefore, it is important that it be available and produce an accurate solution.

Single-pass methods execute one estimation that simultaneously includes internal and external networks, observable/internal network, and nonobservable/

external network solved together. The two-pass method deals with two state estimates, one for the internal system and another for the external system or for the entire system (observable/internal network and nonobservable/external network solved separately). Many in the power industry prefer using a single-pass over a two-pass solution.

According to [22], the single-pass method suffers from numerical instability. An enhancement to the one-pass method uses a set of critical external pseudo-measurements. Some alternative two-pass state estimators require a load flow study for the external system. Both two-pass methods reduce the effects of boundary errors in the internal system solution by properly weighing the external pseudo-measurements, but they may result in very high or negative loads and generations in the external system. Zero-injection buses are more commonly treated as high-confidence bus injection measurements than as hard constraints.

For an overwhelming majority of users, the state estimator solution is used as a base case for reliability-analysis applications such as contingency analysis (CA), power flow (PF), and as input to system analysis tools such as:

1. Online/operator PF
2. Offline PF
3. Locational marginal pricing (LMP)
4. Voltage stability analysis
5. Security-constrained economic dispatch

In some cases, the state estimator is used primarily as the basis for information communicated to operators regarding power system status; e.g. the state estimator drives the alarm application that alerts operators to impending power system events.

1.4.1 SE Performance Issues

It is common practice to rely on periodic triggers to run state estimators every two minutes. Manual and SCADA events such as breaker trips and analog rates of change are also used. Moreover, the average state estimator execution time ranges from one second to two minutes (with an average of about 20 seconds).

It is difficult to recommend specific state estimator voltage and angle convergence tolerances because of the different algorithms employed by different state estimators and the way specific convergence parameters are used in these algorithms. For example, some state estimators check convergence based on changes of the absolute values of voltage magnitudes and voltage phase angles (relative to ground) between successive iterations.

Common industry practice for the voltage-magnitude convergence-tolerance criteria (per unit) is a maximum of 0.1 (0.01 kV per unit) for both internal/observable and external/unobservable systems. For the angle difference in radians, the tolerance is 0.0100.

1.4.2 Weights Assigned to Measurements

The state estimator requires measurement weights (confidences) that affect its solution. The weights for telemetered and non-telemetered measurements are selected according to the following:

1. Use individually defined weights for at least some of the telemetered measurements used by the state estimators.
2. Use globally defined weights for at least some of the telemetered measurements used by the state estimators.

The basis for weights applied to at least some analog values used by the state estimator is either a generic percentage metering error or specific meter accuracies.

1.4.3 SE Availability Considerations

The state estimator must be highly available and must also be able to provide a reasonable, accurate, and robust solution that meets the purposes for which it is intended. Practitioners report that the average time during which state estimator solutions are unavailable is 15 minutes or less per outage for almost all users. In addition, unavailability of the state estimator for up to 30 minutes is considered as having no significant impact on system operations.

Having state estimator failures less than 30 minutes apart is perceived as having a "significant" impact on system operations. This however varies according to internal policies and market considerations.

1.4.4 SE Solution Quality (Accuracy)

State estimator availability requirements are complemented by solution-quality requirements to ensure that operators are given accurate information allowing them to be fully aware of the system situation in a timely manner.

Operators report that they can detect and identify bad analog measurements and remove them from the state estimator measurement set. Users quantified the real/reactive power mismatch tolerance criteria for their internal/observable systems is in the 0.05 MW (per unit) – 170 MW real power mismatch tolerance range and a 0.001 Mvar (per unit) – 500 Mvar reactive power mismatch tolerance range. The average real and reactive mismatch tolerance criteria reported were 35 MW and 69.5 Mvar, respectively.

Macedo [23] states that state estimator MVA mismatch should be less than 10 MVA. He does not distinguish between internal and external systems.

1.4.4.1 Metrics to Evaluate SE Solution Quality

More than one metric is used to evaluate the accuracy of the results of the state estimator solution:

- Cost index is also referred to as "performance index" or "quadratic cost." In general, it measures the sum of the squares of the normalized estimate errors (residuals). Increasing cost index values could indicate deteriorating state estimator solution quality. This is the most commonly used indicator, whose values range between 45 and 58%.
- Chi-squared criterion is the second most used, and its value ranges between 36 and 42%.
- Measurement error/bias analysis is used as a performance indicator.
- Average residual value is used as a performance indicator.

The reliability entity should track the selected metric over time to establish the pattern and determine what indicates a problem with state estimator solution quality. Deviation from the "normal range" of these metrics should trigger state estimator maintenance and support. These metrics are important because they could affect the CA solution.

Many factors affect SE solution-quality metrics such as:

1. Electrical device modeling, connectivity, and telemetry data mapping. If the topology is incorrect, the state estimator may not converge or may yield grossly incorrect results. A topology error may be caused by either inaccurate status of breakers and switching devices or errors in the network model.
2. Availability and quality of telemetry data. Telemetry data are essential components of the state estimation process.
3. Inadequate observability. State estimation is extended to the unobservable parts of the network through the addition of pseudo-measurements that are computed based on load prediction using load distribution factors, or they can represent non-telemetered generation assumed to operate at a base-case output level. The quality of pseudo-measurements may be bad if they are not updated regularly to reflect current conditions.
4. Measurement redundancy of the network is defined as the ratio of the number of measurements to the number of state variables in the observable area of the network.

1.4.4.2 Methods for Evaluating SE Solution Quality (Accuracy)

The following methods are used to evaluate the accuracy of the state estimator results:

- Continually monitor and minimize the amount of bad data detected by correcting model, telemetry, and bad status.
- Compare critical telemetry with the state estimator solution (ties, major lines, large units, etc.).
- Use measurement error/bias analysis to detect and resolve telemetry and model problems.

- Periodically review all stations to correct high residuals and minimize all residuals as much as reasonably possible.
- Compare CA results to actual system.
- Compare power flow results with actual system.
- Compare state estimator actual violations to see if they closely match actual SCADA violations.
- Compare state estimator total company load/generation/interchange integrated over time to see if it closely matches billing metering.

1.4.5 Using SE to Monitor External Facilities

The state estimator solution quality when it is used to monitor external facilities depends on the accuracy of their models. The external network models could affect the quality of state estimator solutions by:

1. Propagation of errors into the internal model solution from the external model solution. This applies to one-pass state estimators if the external network model solution is mainly based on forecasted and/or pseudo-measurements rather than telemetered data. The external network model equivalencing methods could also cause errors to propagate. For two-pass state estimator, there could be boundary problems (between the internal/observable solution and the external/unobservable solution) that could cause the total network solution to not converge.
2. Measurement density in the external system. Many buses in external models are measurement unobservable. The low values for the external-status-point-to-external-bus ratios for many respondents (i.e. less than one status point per bus) indicate that many external buses do not have telemetered breaker/switch information, which implies a bus-branch-type external model (i.e. a planning model) for many buses.
3. Convergence issues related to external models and/or telemetry data for external model. Measurements for the external network model usually originate from data links. As a result, data availability depends on data link availability.
4. The impact of interchange transactions, especially for the external portion of the model, could influence the state estimator solution.
5. Adding detail or expanding the external network model could affect the throughput (execution time) of the state estimator application.

External network model improvements are expected to enhance the accuracy of the results.

- Adding breaker/switch detail to the external and internal models
- Adding extensive telemetry to the external and internal models
- Adding lower-voltage detail to the external and internal models

- Adding one or more control areas to the external model
- Creating a new external model

1.4.6 SE Maintenance/Troubleshooting and Support Practices

Many users have state estimator support personnel available continuously. Most users monitor state estimator status on a continuous basis (24 × 7 × 365) and maintain their state estimators with in-house staff, and some use vendor staff in addition to in-house staff for support.

Most users notify operators and control room staff of a state estimator failure. State estimator status is presented primarily via alarm tools and physical displays. Some users page and send email notifying of a state estimator failure.

Operators attempt to resolve state estimator problems prior to notifying support personnel.

Many users have a process to investigate and debug unsolved/non-converged and bad/inaccurate state estimator solutions.

The operator receives an alarm notification of state estimator problems and then calls for support personnel as needed to solve the problem. Alternatively support personnel are on call and connect remotely after business hours to fix reported problems. Support personnel may be paged automatically by the application(s) to troubleshoot the problem.

For example, the Electric Reliability Council of Texas (ERCOT) requires the following state estimator performance measures [24]:

1. State Estimator to converge 97% of runs during a one-month period.
2. On transmission elements identified as causing 80% of congestion cost in the latest year for which data is available, the residual difference between State Estimator results and Power Flow results for critically monitored transmission element MW flows are required to be less than 3% of the associated element emergency rating on at least 95% of samples measured in a one-month period.
3. On transmission elements identified as causing 80% of congestion cost in the latest year for which data is available, the difference between the MW telemetry value and the MW State Estimator value shall be less than 3% of the associated element emergency rating on at least 95% of samples measured in a one-month period.
4. On 20 most important station voltages designated by ERCOT and approved by ROS; the telemetered voltage minus State Estimator voltage shall be within 2% of the telemetered voltage measurement involved for at least 95% of samples measured during a one-month period.
5. On all transmission elements greater than 100kV; the difference between State Estimator MW solution and the SCADA measurement will be less than

10 MW or 10% of the associated emergency rating (whichever is greater) on 99.5% of all samples during a one-month period. All equipment failing this test will be reported to the associated TSP for repair within 10 days of detection.

1.5 APPLICATIONS THAT USE SE SOLUTION

The state estimator solution is a base case for CA and PF for almost all users. Some users employ the state estimator solution in security-constrained economic dispatch and LMP and offline power flow applications [25]. In summary, the following applications use the state estimator solution as a base case:

Contingency analysis

Online/operator PF

Locational marginal pricing

Security-constrained economic dispatch

Voltage stability analysis

Dynamic stability analysis

A brief discussion of some of these functions is given next.

1.5.1 Contingency Analysis

As a real-time application, CA uses current SE system conditions to determine the effects of specific, simulated outages (lines, generators, or other equipment) on power system security or higher load, flow, or generation levels. In addition, CA considers unexpected failure or outage of a system component (transmission lines, generators, circuit breaker, switch, or any other electrical equipment) and naturally line overloads or voltage violations or higher load, flow, or generation level [26–27].

Failed or nonfunctional CA application has been identified as a key cause of many significant blackouts. Therefore, it must be highly available and redundant. The information produced by CA allows the operator to implement mitigation actions ahead of a contingency and maintain the reliability of the electric power system.

1.5.2 Power Flow (Online/Operator)

PF calculates the state of the electric power system in the form of flows, voltages, and angles based on load, generation, net interchange, and facility status data. PF determines system state. PF are available in both online and offline versions.

Online load flow programs are a standard part of all energy management system specification [28].

Online PF is widely used to assess system conditions or perform look-ahead analysis. It is also used in "$n-1$" CA and to identify potential future voltage collapse or reliability problems.

Real-time reliability tools can only provide results that accurately represent current and potential reliability problems if these tools have real-time PF and voltage values and status data for other elements included in their models. The accuracy of the information that real-time reliability tools provide depends on the accuracy of the data supplied to the tools.

1.5.3 Locational Marginal Pricing

The equal incremental cost (system lambda) rule arises in conventional economic dispatch of a system of fossil fuel thermal generating units serving an active power load (demand) neglecting transmission losses. This case uses one active power balance equation (APBE) to model the electric network physical constraints. Accounting for transmission losses in the APBE leads to the well-known loss penalty factors that are used to penalize the incremental cost of generation for each unit. Loss penalty factors whose values are greater than one correspond to units whose losses increase with the load demand and are further away from the load center. In optimal PF, the electric network is modeled using the PF equations and results in two lambdas (one for the active power equation and the second for the reactive equation) for each node in the system. This is the basis of LMP.

LMP reflects the wholesale value of electric energy at different pricing nodes (locations) considering the operating characteristics, physical constraints, and losses caused by the physical constraints and limits of the transmission system and patterns of load generation. Pricing nodes include individual points on the transmission system, load zones (i.e. aggregations of pricing nodes), external nodes, and nodes where the independent system operator interconnects with a neighboring region and the Hub. The Hub is a collection of locations that represent an uncongested price for electric energy, facilitate electric energy trading, and enhance transparency and liquidity in the marketplace [29–32].

Different locations in the system have different LMPs since transmission and reserved constraints prevent the next least expensive megawatt (MW) of electric energy from reaching all locations of the grid. Even during periods when the least expensive megawatt can reach all locations, the marginal cost of physical losses will result in different LMPs at different locations.

Typically, the LMPs are calculated every five minutes. The LMP at a load zone is used to:

1. Establish the price for electric energy purchases and sales at specific locations throughout the wholesale electricity market for compensating generators and charging loads.
2. Collect transmission congestion charges.

3. Determine compensation for holders of financial transmission rights.

It is common practice to evaluate the LMP at a load zone as the weighted average of all the nodes within that load zone.

In practice, an LMP consists of three components:

1. Energy component of all LMPs that is the price for electric energy at the "reference point," which is the load-weighted average of the system node prices.
2. Congestion component related to the marginal cost of congestion at a given node or external node relative to the load-weighted average of the system node prices. In a manner like that of the load zone LMP, the congestion component of a zonal price is the weighted average of the congestion components of the nodal prices that comprise the zonal price. Moreover, the congestion component of the Hub price is the average of the congestion components of the nodes belonging to the Hub.
3. Loss component at a given node or external node reflects the cost of losses at that location relative to the load-weighted average of the system node prices. The loss component of a zonal price is the weighted average of the loss components of the nodal prices that constitute the zonal price. The loss component of the Hub price is the average of the loss components of the nodes that belong to the Hub.

1.5.4 Security-Constrained Economic Dispatch

Conventional economic dispatch (ED) provides to the load frequency control (LFC) both economic base points and participation factors to control system frequency and net interchange in an economic fashion. The constrained economic dispatch (CED) application aims to meet the following objectives:

- Provide economic base points to LFC
- Promote reliability of service by respecting network transmission limitations
- Provide constrained participation factors
- Contribute no major impact on speed or storage requirements over present EDC

Security-constrained economic dispatch determines the level at which each committed resource should be operated while enforcing the "security" aspects of the system [33–34].

1.5.5 Voltage Stability Assessment

The voltage stability assessment (VSA) application uses a current state estimator model of the real-time system to determine additional load or transfer the system can sustain in a given direction before it encounters voltage instability [35–36]. VSA runs in near real time and aids in the determination of system operating limits based on a recent snapshot of the real-time system. VSA may derive minimum voltages at key buses below which voltage collapse might occur if the system experiences additional stresses. It may also provide information on minimum dynamic reactive reserves required in local areas.

VSA is different from offline voltage stability analysis tools used by planners for medium-term or long-term studies. But if such tools are used for studying near real-time snapshots to answer operator's voltage stability questions, they would be included in the VSA real-time tool suite.

1.5.6 Dynamic Stability Assessment

The dynamic stability assessment (DSA) application (or a suite of applications) executes in near real time to assist in determining stability-related system operating limits using a snapshot of the real-time system (i.e. current state estimator output). It may also provide an indication of dynamic stability margin for the most critical fault/contingency condition.

DSA is different from offline stability analysis tools usually used for medium- or long-term studies [37–39]. However, if such tools are used by operators studying near real-time snapshots to aid in voltage stability evaluation, these tools should be included as part of the DSA suite.

For more background material, the reader is referred to [40–42].

1.6 OVERVIEW OF CHAPTERS

Apart from this introductory chapter, there are 13 chapters devoted to state of the art in this vibrant area.

In Chapter 2, Eduardo Caro and Araceli Hernández discuss a mathematical programming approach to state estimation in power systems. They focus on WLS, least absolute value (LAV), quadratic constant, and quadratic linear criterions, among others. Additionally, the statistical correlation among measurements is analyzed and included, enhancing both estimation accuracy and the bad data identification capabilities. All procedures are illustrated by simple but insightful examples.

From a computational perspective, quadratic constant and LAV techniques perform faster than the conventional WLS estimator, saving up to 75% CPU time (compared with the WLS method directly solved as an optimization problem). On the other hand, mathematical programming formulation of some estimators (such as least median of squares and least trimmed of squares approaches) encounter non-convexities and a significant number of binary variables, resulting in higher computational burdens.

If the measurement set is corrupted with errors that the χ^2 test cannot detect, WLS estimation results quality deteriorates, providing the worst estimation quality (if LMS and LTS approaches are not considered). It is observed that QL and QC techniques outperform the rest of estimators, followed by LMR procedure. As expected, the authors report that their computational experiments reported in this chapter suggest that alternative estimators are potential substitutes for traditional WLS method. The state of the art of current nonlinear optimization solvers and recent advances in computational equipment allows using robust estimators in real electric energy systems.

Chapter 3 draws on two bodies of knowledge – electric power engineering and network (graph) theory – to develop and apply a new failure network, an application of line outage distribution factors. Here Hyde M. Merrill and James W. Feltes present an approach to measure how susceptible is an electric power system to cascading outages (stress) that lead to blackouts. The problem is defined in the context of a new perspective of the electric power system. A failure network is defined, based on well-known line outage distribution factors. Following the practice of network (graph) theory, the structure and properties of this network are analyzed with metrics that measure stress (susceptibility to cascading outages). The metrics can be applied in real-time operations or in planning to identify vulnerability to cascading. Three studies are described, two on very large North American systems, the other on the smaller national system of Peru. New insights are presented, and a new class of power system options is identified, to reduce susceptibility to cascading rather than to increase transfer capability.

The ideas have application in planning as well as in real time. For operations, it depends, as Schweppe expected, on data from a state estimator. But the additional computations go far beyond simply calculating flows and injections using Ohm's and Kirchhoff's laws.

The authors of Chapter 4, "Model-Based Anomaly Detection for Power System State Estimation", Aditya Ashok, Manimaran Govindarasu, and Venkataramana Ajjarapu, recognize that state estimators depend on SCADA measurements from the various remote substations, which introduces several vulnerabilities due to malicious cyberattacks. The security and resiliency of the power system state estimator are important since its output is used by several other network applications in the EMS such as real-time CA, power markets, etc. While SE is designed to detect and recover from some degree of bad data injected due to measurement errors, or even measurement loss due to telemetry issues, they could be impacted by malicious cyberattacks causing loss of observability, operational, and market impacts. A holistic approach to attack-resilient SE should involve a combination of attack-resilient planning approaches to improve attack prevention capabilities in conjunction with attack-resilient anomaly detection and robust SE formulations to improve attack detection and mitigation resulting in a defense-in-depth architecture.

This chapter starts with a broad survey of relevant state-of-the-art literature that addresses the vulnerability of SE to stealthy false data injection attacks and topology-based attacks. This is followed by a review of some offline attack prevention approaches to enhance the redundancy of SE against those stealthy cyberattacks and online techniques for attack detection and mitigation that address anomaly detection, bad data detection, and other formulations of SE. A model-based anomaly detection approach uses short-term load forecasts, generation schedules, and available secure PMU data to detect anomalies due to stealthy cyberattacks. This approach is complementary to traditional bad data detection methods in SE to detect stealthy cyberattacks. This chapter offers some insights into the performance of the proposed anomaly detection approach using a case study on the IEEE 14-bus system. Finally, this chapter provides a summary of the contributions and promising directions for emerging research topics that show promise for the future including PMU-based linear state estimator, integrated hybrid SE formulations with PMU data and SCADA, robust and dynamic SE formulations, and MTD-based approaches for SE that leverage redundancy and randomization of measurements.

A. P. Sakis Meliopoulos, Yu Liu, Sungyun Choi, and George J. Cokkinides propose in Chapter 5 a scheme for real-time operation and protection of microgrids based on distributed dynamic state estimation (DDSE). First, the DDSE can be used for setting-less component protection that applies dynamic state estimation on a component under protection with real-time measurements and dynamic models of the component. Based on the results, the well-known χ^2 test yields the confidence level that quantifies the goodness of fit of models to measurements, indicating the health status of the component. With this approach, renewable DERs in microgrids can be protected on an autonomous and adaptive basis. Meanwhile, the estimated state variables of each component are converted to phasor data with time tags and then collected to the DERMS of microgrids. These aggregated phasor data that are once filtered by the DDSE are input to the static state estimator in the DERMS along with unfiltered data sent from conventional meters, relays, and digital fault recorders, ultimately generating real-time operating conditions of microgrids. This chapter also provides numerical simulations to compare the DDSE-based approach with conventional centralized state estimation in terms of data accuracy and computational speeds.

Any component in microgrids can be protected by the proposed setting-less protection method, capable of tracking full dynamic characteristics of a device under protection. This method can provide adaptive protection in microgrids, where unpredictable fault conditions or abnormal states may arise. It is important to point out that the setting-less protection is fundamentally based on the physical characteristics, thus requiring no additional settings for grid conditions. The authors suggest that the approach also facilitates real-time operation by reducing the state estimation computation time as well as by enhancing the accuracy of estimation results. In this sense, the DDSE can be of great importance to the real-time operation and management of microgrids in which the penetration of renewable DERs has recently increased.

In Chapter 6, "Distributed Robust Power System State Estimation" by Vassilis Kekatos, H. Zhu, G. Wang, and Georgios B. Giannakis discuss some of the recent advances in power system state estimation (PSSE). The Cramer–Rao lower bound (CRLB) on the covariance of any unbiased estimator is first derived for the PSSE setup. Following a review of conventual Gauss–Newton iterations, contemporary PSSE solvers leveraging relaxations to convex programs and successive convex approximations are explored. To overcome the high complexity involved, a scheme named "feasible point pursuit", relying on successive convex approximations is advocated. A decentralized PSSE paradigm is presented to provide the means for coping with the computationally intensive SDP formulations, which is tailored for the interconnected nature of modern grids, while it can also afford processing PMU data in a timely fashion. Novel bad data processing models and fresh perspectives linking critical measurements to cyberattacks on the state estimator are presented. Motivated by advances in online convex optimization, model-free, and model-based state trackers, the authors offer a fresh perspective on state tracking under model-free and model-based estimators. With the current focus on low- and medium-voltage distribution grids, solvers for unbalanced and multiphase operating conditions are desirable. Smart meters and synchrophasor data from distribution grids (also known as micro-PMUs) call for new data processing solutions. Advances in machine learning and statistical signal processing, such as sparse and low-rank models, missing and incomplete data, tensor decompositions, deep learning, nonconvex and stochastic optimization tools, and (multi)kernel-based learning to name a few, are currently providing novel paths to grid monitoring tasks while realizing the vision of smarter energy systems.

Mert Korkali in Chapter 7, "Robust Wide-Area Fault Visibility and Structural Observability in Power Systems with Synchronized Measurement Units," presents work merging robust state estimation and optimal sensor deployment with the objective to achieve system-wide fault visibility and structural observability in modern power systems equipped with wide-area measurement systems (WAMSs). The first part of this chapter introduces a method that enables synchronized measurement-based fault visibility in large-scale power systems. The approach uses the traveling waves that propagate throughout the network after fault conditions and requires capturing arrival times of fault-initiated traveling waves using synchronized sensors so as to localize the fault with the aid of the recorded times of arrival (ToAs) of these waves. The second part of this chapter is devoted to optimization model for the deployment (placement) of PMUs paving the way for complete topological (structural) observability in power systems under various considerations, including PMU channel limits, zero-injection buses, and a single PMU failure.

In Chapter 8, authors Junbo Zhao, Lamine Mili, and Massimo La Scala recall that in the power system environment, the distribution of the measurement noise is usually unknown and frequently deviates from the assumed Gaussian distribution model, yielding outliers. Under these conditions, the performance of the current state estimators that rely on Gaussian assumption can deteriorate significantly. In addition, the sampling rates of SCADA and PMU measurements are quite different, causing a time skewness problem. Under the title "A Two-Stage Robust Power System State Estimation Method with Unknown Measurement Noise," the authors propose a robust state estimation framework to address the unknown non-Gaussian noise and the measurement time skewness issue. In the framework, the Schweppe-type Huber generalized maximum-likelihood (SHGM) estimator is advocated for SCADA measurement-based robust state estimation. They show that the state estimates provided by the SHGM estimator follow roughly a Gaussian distribution. This effectively allows combining it with the buffered PMU measurements for final state estimation. Robust Mahalanobis distances are proposed to detect outliers and assign appropriate weights to each buffered PMU measurement. Those weights are further utilized by the SHGM estimator to filter out non-Gaussian PMU measurement noise and help suppress outliers. Extensive simulation results carried out on the IEEE-30 bus test system demonstrate the effectiveness and robustness of the proposed method.

Chapter 9 by Ibrahim Omar Habiballah and Yuanhai Xia: "Least-Trimmed-Absolute-Value State Estimator" is intended to improve the accuracy of estimation results considering complex situations induced by multiple types of bad data. In addition to conventional state estimators such as WLS and LAV, other robust estimators are used to detect and filter out bad data. This includes, among many, least median squares and least-trimmed square estimators. The authors introduce an efficient robust estimator known as least-trimmed-absolute-value estimator. The algorithm arises from the two estimators: LAV and LTS and benefits the merits of both. It can detect and eliminate both single and multiple bad data more efficiently. DC estimation is conducted on 6-bus system and IEEE 14-bus system first; then these two systems and the IEEE 30-bus system are used to conduct AC estimation experiments. Various types of bad data are simulated to evaluate the performance of the proposed robust estimator.

A new probabilistic approach to state estimation in distribution networks based on confidence levels is introduced in Chapter 10. Here, Bernd Brinkmann and Michael Negnevitsky state that their proposal uses the confidence that the estimated parameters are within their constraints as a primary output of the estimator. By using the confidence value, it is possible to combine information about the estimated value as well as the accuracy of the estimate into a single number. Their motivation is that the traditional approach to state estimation only provides the estimated values to the network operator without any information about the accuracy of the estimates. This works well in transmission networks where a large number of redundant measurements are generally available. However, due to economic constraints, the number of available real-time measurements in distribution networks is usually low. This can lead to a significant amount of uncertainty in the state estimation result. This makes it difficult to adapt the traditional state estimation approach to distribution networks.

A probabilistic observability assessment is also presented in this chapter using a similar probabilistic approach. The traditional approach to observability in distribution networks is limited because even if a network is classified as observable, the state estimation result could be completely decoupled from reality. The presented method on the other hand determines if the state of a distribution network can be estimated with a degree of accuracy that is sufficient to evaluate if the true value of the estimated parameters is within their respective constraints.

This approach has been demonstrated in case studies using real 13-bus and 145-bus feeders. The results show that even if a large amount of uncertainty is present in the state estimation result, the proposed approach can provide practical information about the network state in a form that is easy to interpret.

The premise of "Advanced Distribution System State Estimation in Multi-Area Architectures" is that distribution grids are characterized by a very large number of nodes and different voltage levels. Moreover, different portions of the system can be operated by different distribution system operators. In this context, multi-area approaches can be indispensable key tools to perform DSSE efficiently. Chapter 11 by Marco Pau, Paolo Attilio Pegoraro, Ferdinanda Ponci, and Sara Sulis presents state of the art, challenges, and novel approaches for multi-area state estimation (MASE) in distribution systems. A new methodology, based on a two-step procedure, is presented in detail. This procedure is designed to accurately estimate the status of a large-scale distribution network, relying on a distributed measurement system in a multi-area framework. Criteria for the sub-area's division are presented along with the issues, requirements, and challenges of estimation steps. Benefits of the use of synchronized measurements obtained by phasor measurement units (PMUs) in terms of the accuracy and efficiency enhancements in the estimation results are presented and discussed.

Chapter 12 by Ye Guo, Lang Tong, Wenchuan Wu, Hongbin Sun, and Boming Zhang is under the title "Hierarchical Multi-Area State Estimation" and is motivated by the need for a coordinated state estimator for multi-area power systems. Of course, the proposed method should provide the same state estimate as a centralized estimator but solved in a distributed manner. In this chapter, the authors review earlier relevant work in the field, including two-level single-iteration estimators, inter-area Gauss–Newton methods and intra-area Gauss–Newton methods. In particular, the authors focus on recently published work where local system operators communicate their sensitivity functions to the coordinator. These sensitivity functions fully represent local optimal conditions, and consequently, this method has improved rate of convergence.

The application of parallel processing for static/dynamic state estimation is motivated by the desire for faster computation for online monitoring of the system behavior. In Chapter 13, Hadis Karimipour and Venkata Dinavahi investigate the process of accelerating static/dynamic estimation for large-scale networks.

In the first part, using an additive Schwarz method, the solution of each subsystem is carried out by using the conventional numerical techniques and exchanging the boundary data among subsystems. To increase the accuracy a slow coherency method was used to decide the domain decomposition. In addition, load balancing by distributing equal workload among processors is utilized to minimize inter-processor communication. The advantages of the proposed approach over existing approaches include reducing execution time by splitting equal amount of work among several processors, minimizing the effect of boundary buses in accuracy and not requiring major changes in existing power system state estimation paradigm. Next, the proposed method is implemented in massively parallel architecture of GPU. As shown in the results, the advantage of utilizing GPU for parallelization is significant when the size of the system is increased.

The proposed method is general and can be extended to any number of GPUs connected in a cluster. Results show that more GPUs can reduce expected computation time. Result comparisons verified the accuracy and efficiency of the proposed method. In addition, the performance of the slow coherency method as the partitioning tool was analyzed, and it was concluded that for different fault locations in the system, results derived from this method had lower amounts of error.

Chapter 14 is "Dishonest Gauss Newton Method-Based Power System State Estimation on a GPU", by Md. Ashfaqur Rahman and Ganesh Kumar Venayagamoorthy. The authors acknowledge that real-time power system control requires accelerating the computation processes. While many methods to speed up the computational process are available, it is worthwhile to explore current parallel computation technology to develop faster estimators. The authors use the term "dishonest Gauss Newton method," but the technique is based on the PARTAN (short for Parallel tangent). Their study concerns a graphics processing unit (GPU) implementation. As the method is not explored extensively in the literature, its accuracy is investigated first. Then different aspects of the parallel implementation are explained. It takes a few hundreds of microseconds for IEEE 118-bus systems, which are found to be the fastest in the existing reported times. For very large systems, the required configuration of a GPU and the corresponding time are also estimated. Finally, the distributed method-based parallelization is also implemented.

REFERENCES

1. Schweppe, F.C. and Wildes, J. (Jan./Feb. 1970). Power system static state estimation, part I: exact model. *IEEE Transactions on Power Apparatus and Systems* PAS-89 (1): 120–125.
2. Schweppe, F.C. and Rom, D.B. (Jan./Feb. 1970). Power system static state estimation part II: approximate model. *IEEE Transactions on Power Apparatus and Systems* PAS-89 (1): 125–130.
3. Schweppe, F.C. (Jan./Feb. 1970). Power system static state estimation, part III: implementation. *IEEE Transactions on Power Apparatus and Systems* PAS-89 (1): 130–135.
4. Merrill, H.M. and Schweppe, F.C. (Nov./Dec. 1971). Bad data suppression in power system static state estimation. *IEEE Transactions on Power Apparatus and Systems* PAS-90 (6): 2718–2725.
5. Schweppe, F.C. and Handschin, E.J. (Jul. 1974). Static state estimation in electric power systems. *Proceedings of the IEEE* PAS-62 (7): 972–982.
6. International Electrotechnical Commission (1986). International Electrotechnical Commission, Geneva, Switzerland. http://www.electropedia.org/iev/iev.nsf/display?openform&ievref=603-02-09 (accessed 2 November 2020).
7. Real-Time Tools Best Practices Task Force. (Mar. 2008). Real-Time Tools Survey Analysis and Recommendations, 569pp. https://www.nerc.com/pa/rrm/ea/August%2014%202003%20Blackout%20Investigation%20DL/Real-Time_Tools_Survey_Analysis_and_Recommendations_March_2008.pdf (accessed 16 November 2020).
8. Svoen, J., Fismen, S.A., Faanes, H.H., and Johannessen, A. (1972). The online closed-loop approach for control of generation and overall protection at Tokke power plants. *International Conference on Large High-Tension Electric Systems*, CIGRE, Paris, France. Paper 32-06.

9. Dopazo, J.F., Ehrmann, S.T., Klitin, O.A., and Sasson, A.M. (Sep./Oct. 1973). Justification of the AEP real time load flow project. *IEEE Transactions on Power Apparatus and Systems* PAS-92: 1501–1509.
10. Dy Liacco, T.E. (1982). The role of state estimation in power system operation. Implementation of state estimation techniques in real time control of power systems, IFAC, Identification and System Parameter Estimation, Washington, DC.
11. Wood, A.J., Wollenberg, B.F., and Sheblé, G.B. (Nov. 2013). *Power Generation, Operation, and Control*, 3e. Wiley.
12. Schweppe, F.C. (1973). *Uncertain Dynamic Systems*. Englewood Cliffs, NJ: Prentice-Hall.
13. Gelb, A. (1974). *Applied Optimal Estimation*. MIT Press.
14. Crassidis, J.L. and Junkins, J.L. (2004). *Optimal Estimation of Dynamic Systems*. CRC Press.
15. Handschin, E. and Galiana, F.D. (Jun. 1973). Hierarchical state estimation for real-time monitoring of electric power systems. *Proceedings of 8th PICA Conference*, Minneapolis, MN, pp. 304–312.
16. Guo, Y., Tong, L., Wu, W. et al. (Jan./Feb. 2017). Hierarchical multi-area state estimation via sensitivity function exchanges. *IEEE Transactions on Power Systems* 32 (1): 442–453.
17. Kashyap, N., Werner, S., and Huang, Y.-F. (2018). Decentralized PMU-assisted power system state estimation with reduced interarea communication. *IEEE Journal of Selected Topics in Signal Processing* 12 (4): 607–616.
18. Galiana, F. and Schweppe, F. (1972). A weather dependent probabilistic model for short term forecasting. *IEEE Winter Power Meeting*, New York. Paper C72 171-2.
19. Chang, C.S. and Yi, M. (1998). Real-time pricing related short-term load forecasting. *Proceedings of EMPD '98. 1998 International Conference on Energy Management and Power Delivery*, Singapore, Vol. 2, pp. 411–416.
20. Moore, R.L. and Schweppe, F.C. (Jun. 1973). Adaptive coordinated control for nuclear power plant load changes. *Proceedings of 8th PICA Conference*, Minneapolis, MN, pp. 180–186.
21. Smith, H.L. and Block, W.R. (Jan. 1993). RTUs slave for supervisory systems (power systems). *IEEE Computer Applications in Power* 6 (1): 27–32.
22. Korres, G.N. (June/July 2002). A partitioned state estimator for external network modeling. *IEEE Transactions on Power Systems* 17 (3): 834–842.
23. Macedo, F. (2004). Reliability software minimum requirements & best practices. *FERC Technical Conference*, July 14. www.slideserve.com/hallie/reliability-software-minimum-requirements-best-practices (accessed 7 November 2020).
24. ERCOT, State Estimator Standards, TAC Approved: May 2, 2013. http://www.ercot.com/content/mktrules/obd/documents/State-Estimator-Standards-Approved-TAC-050213.doc (accessed 10 November 2020).
25. Wu, F.F., Moslehi, K., and Bose, A. (Nov. 2005). Power system control centers: past, present, and future. *Proceedings of the IEEE* 93 (11): 1890–1908.
26. Stott, B., Alsac, O., and Monticelli, A.J. (1987). Security analysis and optimization. *Proceedings of the IEEE* 75 (12): 1623–1644.
27. Debs, A.S. (1988). *Modern Power Systems Control and Operation*. Springer.
28. Johnson, W.A., Potts, G.W., Wrubel, J.N., and Schulte, R.P. (Nov./Dec. 1983). On-line load flows from a system operators viewpoint. *IEEE Transactions on Power Apparatus and Systems* PAS-102 (6): 1818–1822.

29. Ilic, M., Galiana, F., and Fink, L. (1998). *Power Systems Restructuring: Engineering and Economics*. Norwell, MA: Kluwer.
30. Frame, J. (2001). Locational marginal pricing. *2001 IEEE Power Engineering Society Winter Meeting*. Conference Proceedings, vol. 1, p. 377.
31. Li, F., Pan, J., and Chao, H. (Apr. 2004). Marginal loss calculation in competitive electrical energy markets. *Proceedings of the 2004 IEEE International Conference on Electric Utility Deregulation, Restructuring and Power Technologies (DRPT2004)*. Hong Kong, China, Vol. 1, pp. 205–209.
32. Kirschen, D. and Strbac, G. (2004). *Fundamentals of Power System Economics*. New York: Wiley.
33. Adler, R.B. and Fischl, R. (Mar./Apr. 1977). Security constrained economic dispatch with participation factors based on worst case bus load variations. *IEEE Transactions on Power Apparatus and Systems* 96 (2): 347–356.
34. Sanders, C.W. and Monroe, C.A. (Jul./Aug. 1987). An algorithm for real-time security constrained economic dispatch. *IEEE Transactions on Power Systems* 2 (4): 1068–1074.
35. Taylor, C. (1994). *Power System Voltage Stability*. McGraw-Hill.
36. van Cutsem, T. and Vournas, C. (1998). *Voltage Stability of Electric Power Systems*. Springer.
37. Kundur, P. (1994). *Power System Stability and Control*. McGraw-Hill.
38. Sauer, P.W. and Pai, M.A. (2006). *Power System Dynamics and Stability*. Stipes Publishing L.L.C.
39. Grigsby, L.L. (2012). *Power System Stability and Control*, 3e. CRC Press.
40. Heydt, G.T. (1986). *Computer Analysis Methods for Power Systems*. Macmillan Publishing Company.
41. Crow, M.L. (2009). *Computational Methods for Electric Power Systems*, 2e. CRC Press.
42. Momoh, J.A. and Mili, L. (2010). *Operation and Control of Electric Energy Processing Systems*. Wiley/IEEE Press.

CHAPTER 2

STATE ESTIMATION IN POWER SYSTEMS BASED ON A MATHEMATICAL PROGRAMMING APPROACH

Eduardo Caro and Araceli Hernández
Universidad Politécnica de Madrid, Madrid, Spain

2.1 INTRODUCTION

This chapter revisits the state estimation problem in power system and develops a direct approach of a mathematical programming solution for the most relevant estimators. Traditionally, power system state estimation is tackled by solving the system of nonlinear equations corresponding to the first-order necessary optimality conditions of the estimation problem.

Besides the traditional approach, the estimation problem can be solved directly, which has a number or advantages such as (i) including inequality constraints representing physical hard limits, (ii) taking advantage of currently available state-of-the-art nonlinear programming solvers, (iii) treating sparsity in an efficient and implicit manner, etc.

For the interested reader, pioneering state estimation works include [1–6]. A detailed description of state estimation in power system is provided in [7], which includes an appropriate literature review.

Advances in Electric Power and Energy: Static State Estimation, First Edition.
Edited by Mohamed E. El-Hawary.

Published 2021 by John Wiley & Sons, Inc.

2.2 FORMULATION

The traditional weighted least of squares (WLS) state estimation problem has the form

$$\underset{x}{\text{minimize}}\, J(\boldsymbol{x}) = \sum_{i=1}^{m} w_i (h_i(\boldsymbol{x}) - z_i)^2 \tag{2.1}$$

subject to

$$\boldsymbol{f}(\boldsymbol{x}) = \boldsymbol{0} \tag{2.2}$$

$$\boldsymbol{g}(\boldsymbol{x}) \leq \boldsymbol{0} \tag{2.3}$$

where $\boldsymbol{x}$ is the $n \times 1$ state variable vector, $J(\boldsymbol{x})$ the objective function (weighted quadratic error), $\boldsymbol{h}(\boldsymbol{x})$ a $m \times 1$ functional vector expressing the measurements as a function of the state variables, $\boldsymbol{w}$ a $m \times 1$ weighting factor vector, $\boldsymbol{z}$ the $m \times 1$ measurement vector, $\boldsymbol{f}(\boldsymbol{x})$ the $p \times 1$ equality constraint vector mainly enforcing conditions at transit buses (no generation and no demand), and $\boldsymbol{g}(\boldsymbol{x})$ the $q \times 1$ inequality constraint vector enforcing physical limits of the system.

Among the components of vector $\boldsymbol{h}(\boldsymbol{x})$, note that active and reactive and power injections at bus i are computed as

$$P_i = v_i \sum_{j \in \Xi} v_j \left(G_{ij} \cos(\theta_i - \theta_j) + B_{ij} \sin(\theta_i - \theta_j) \right) \tag{2.4}$$

$$Q_i = v_i \sum_{j \in \Xi} v_j \left(G_{ij} \sin(\theta_i - \theta_j) - B_{ij} \cos(\theta_i - \theta_j) \right) \tag{2.5}$$

where P_i and Q_i are the active and reactive power injection at bus i, respectively; v_i and θ_i are the voltage magnitude and angle at bus i, respectively; G_{ij} and B_{ij} are the real and imaginary part of the bus admittance matrix, respectively; and Ξ is the set of all buses. G_{ii} and B_{ii} can be computed as indicated in [7].

The active and reactive power flow ij are computed as

$$P_{ij} = v_i v_j \left(G_{ij} \cos(\theta_i - \theta_j) + B_{ij} \sin(\theta_i - \theta_j) \right) - G_{ij} v_i^2 \tag{2.6}$$

$$Q_{ij} = v_i v_j \left(G_{ij} \sin(\theta_i - \theta_j) - B_{ij} \cos(\theta_i - \theta_j) \right) + v_i^2 \left(B_{ij} - b_{ij}^{\mathrm{S}}/2 \right) \tag{2.7}$$

where P_{ij} and Q_{ij} are the active and reactive power flow from bus i to bus j, respectively, and b_{ij}^{S} the shunt susceptance of line ij.

As it is customary, measurements are considered to be independent Gaussian-distributed random variables.

Example 2.1 Traditional Formulation

The 4-bus power system depicted in Figure 2.1 is considered throughout this chapter for illustrative purposes. The network data is provided in Table 2.1. This example includes a generating bus, two demand buses, and a transit

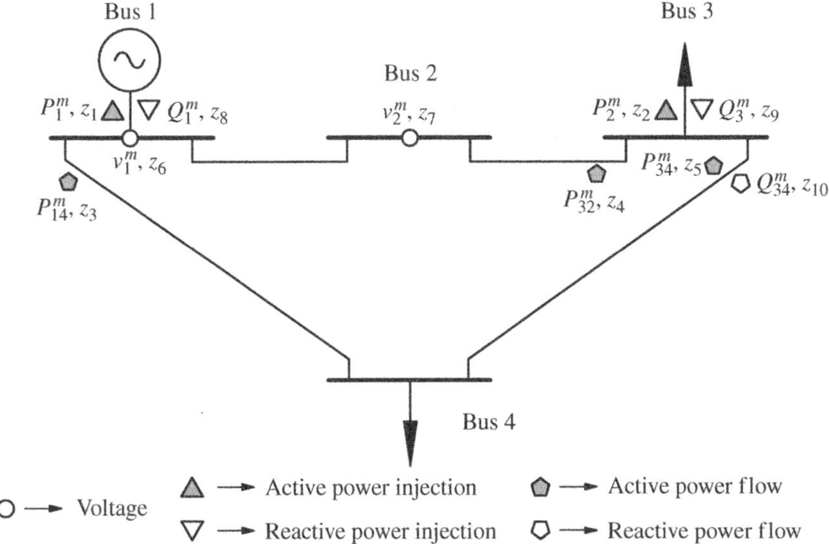

Figure 2.1 One-line diagram and measurement configuration of a 4-bus power system.

TABLE 2.1 Line parameters.

Bus i	Bus j	B_{ij} (p.u.)
1	2	10
1	4	10
2	3	10
3	4	10

bus. Considering bus 4 as the reference bus, the state variable vector $\boldsymbol{x}$ has the form

$$\boldsymbol{x} = (\theta_1, \theta_2, \theta_3, v_1, v_2, v_3, v_4)^T$$

Ten measurements are considered as shown in Figure 2.1: two voltage measurements at bus 1 and 2; two active/reactive power injection measurements at bus 1 and 3; three active power flow measurements at lines 1–4, 3–2, and 3–4; and one reactive power flow measurement at line 3–4. The standard deviation of any measurement other than voltage is 0.02, while the standard deviation of voltage measurements is 0.01. The considered measurement configuration provides a redundancy ratio of $(10 + 2)/7 = 1.71$. Note that the measurement set includes 10 actual measurements plus two exact *measurements* corresponding to active and reactive power injections at the transit bus.

The state estimation problem for this example has the form

minimize$_x$

$$
\begin{aligned}
J(\boldsymbol{x}) = {} & w_1\left(v_1\left(v_2 B_{12}\sin\left(\theta_1-\theta_2\right)+v_4 B_{14}\sin\left(\theta_1-0\right)\right)-z_1\right)^2 \\
& + w_2\left(v_3\left(v_2 B_{32}\sin\left(\theta_3-\theta_2\right)+v_4 B_{34}\sin\left(\theta_3-0\right)\right)-z_2\right)^2 \\
& + w_3\left(v_1 v_4 B_{14}\sin\left(\theta_1-0\right)-z_3\right)^2 \\
& + w_4\left(v_3 v_2 B_{32}\sin\left(\theta_3-\theta_2\right)-z_4\right)^2 \\
& + w_5\left(v_3 v_4 B_{34}\sin\left(\theta_3-0\right)-z_5\right)^2 \\
& + w_6\left(v_1-z_6\right)^2 \\
& + w_7\left(v_2-z_7\right)^2 \\
& + w_8\left(v_1\left(-v_1 B_{11}-v_2 B_{12}\cos\left(\theta_1-\theta_2\right)-v_4 B_{14}\cos\left(\theta_1-0\right)\right)-z_8\right)^2 \\
& + w_9\left(v_3\left(-v_3 B_{33}-v_2 B_{32}\cos\left(\theta_3-\theta_2\right)-v_4 B_{34}\cos\left(\theta_3-0\right)\right)-z_9\right)^2 \\
& + w_{10}\left(-v_3 v_4 B_{34}\cos\left(\theta_3-0\right)+v_3^2 B_{34}-z_{10}\right)^2
\end{aligned}
$$

subject to 2 equality constraints for the transit bus and 7 inequality constraints enforcing physical limits:

$$
\begin{aligned}
& v_2\left(v_1 B_{21}\sin\left(\theta_2-\theta_1\right)+v_3 B_{23}\sin\left(\theta_2-\theta_3\right)\right)=0 \\
& v_2\left(-v_1 B_{21}\cos\left(\theta_2-\theta_1\right)-v_2 B_{22}-v_3 B_{23}\cos\left(\theta_2-\theta_3\right)\right)=0
\end{aligned}
$$

$$
\begin{aligned}
& P_1^{\min} \le v_1\left(v_2 B_{12}\sin\left(\theta_1-\theta_2\right)+v_4 B_{14}\sin\left(\theta_1-0\right)\right) \le P_1^{\max} \\
& Q_1^{\min} \le v_1\left(-v_1 B_{11}-v_2 B_{12}\cos\left(\theta_1-\theta_2\right)-v_4 B_{14}\cos\left(\theta_1-0\right)\right) \le Q_1^{\max} \\
& -\pi \le \theta_i \le \pi, \quad i=1,\ldots,3
\end{aligned}
$$

2.3 CLASSICAL STATE ESTIMATION PROCEDURE

Traditionally, problem (2.1)–(2.3) is simplified by ignoring inequality constraints (2.3) and then solving the system of nonlinear equations constituted by the first-order optimality conditions of (2.1)–(2.2), i.e.

$$\sum_{i=1}^{m} \nabla_x\left[w_i(h_i(\boldsymbol{x})-z_i)^2\right] + \sum_{i=1}^{p} \lambda_i \nabla_x f_i(\boldsymbol{x}) = \boldsymbol{0} \tag{2.8}$$

$$f_i(\boldsymbol{x}) = 0 \quad i = 1, \ldots, p \tag{2.9}$$

or

$$\boldsymbol{H}^T \boldsymbol{W}[z-\boldsymbol{h}(\boldsymbol{x})] + \boldsymbol{F}^T\boldsymbol{\lambda} = \boldsymbol{0} \tag{2.10}$$

$$\boldsymbol{f}(\boldsymbol{x}) = \boldsymbol{0} \tag{2.11}$$

where $\boldsymbol{W}$ is a $m \times m$ diagonal matrix of the measurement weights w_i, $\boldsymbol{F} = \nabla_x \boldsymbol{f}(\boldsymbol{x})$ is the $p \times n$ equality constraint Jacobian, and $\boldsymbol{\lambda}$ is the $p \times 1$ Lagrangian multiplier vector associated with equality constraints (2.2).

The nonlinear system of Eqs. (2.10)–(2.11) can be solved by Newton through the iteration below:

$$\begin{bmatrix} H^T W H & F^T \\ F & 0 \end{bmatrix} \begin{bmatrix} \Delta x^{(\nu+1)} \\ -\lambda^{(\nu+1)} \end{bmatrix} = \begin{bmatrix} H^T W \Delta z^{(\nu)} \\ -f\left(x^{(\nu)}\right) \end{bmatrix} \tag{2.12}$$

where $\Delta z^{(\nu)} = z - h\,(x^{(\nu)})$.

The linear system of Eqs. (2.12) is iteratively solved until Δx is sufficiently small. Further details can be found in [7].

It should be noted that the solution approach based on (2.12) was developed at the time that no efficient mathematical programming solvers (in terms of accuracy, required computing time, and sparsity treatment) were available. However, such solvers are nowadays available.

Example 2.2 Classical Solution Example

The system in Figure 2.1 is considered in this example. In order to solve the nonlinear system (2.10)–(2.11) by the Newton method, matrices H and F and vectors Δz and $f(x)$ should be computed at each iteration. The diagonal terms of the measurement weight matrix W correspond to the inverse of the variance of the measurement errors. Expressions for these matrices are

$$H(x) = \begin{pmatrix} \dfrac{\partial h_1(x)}{\partial x_1} & \cdots & \dfrac{\partial h_1(x)}{\partial x_7} \\ \vdots & \ddots & \vdots \\ \dfrac{\partial h_{10}(x)}{\partial x_1} & \cdots & \dfrac{\partial h_{10}(x)}{\partial x_7} \end{pmatrix}$$

$$F(x) = \begin{pmatrix} \dfrac{\partial f_1(x)}{\partial x_1} & \cdots & \dfrac{\partial f_1(x)}{\partial x_7} \\ \dfrac{\partial f_2(x)}{\partial x_1} & \cdots & \dfrac{\partial f_2(x)}{\partial x_7} \end{pmatrix}$$

$$W = \text{Diag}(2500, 2500, 2500, 2500, 2500, 10\,000, 10\,000, 2500, 2500, 2500)$$

The minimum and maximum active/reactive power injections in puMW/puMVar are $P_1^{\max} = 3$, $P_1^{\min} = 0.8$, $Q_1^{\max} = 3.5$, and $Q_1^{\min} = -3.5$. Measurements are generated by solving the power flow with a height accuracy and then adding independent Gaussian-distributed errors to the *exact* values of the measurements. These measurements and the exact values are provided in Tables 2.2–2.4. In these tables, superscript "true" indicates true value, superscript "m" measurement, and a "∧" estimated value.

The initial solution considered for state variables is the flat voltage level, and a convergence tolerance of 10^{-5} on the largest Δx_i is considered. The convergence, illustrated in Table 2.5, is obtained after four iterations.

TABLE 2.2 Voltages: measured, true and estimated values.

Bus no.	v_i^{m} (p.u.)	v_i^{true} (p.u.)	$\hat{v}_i$ (p.u.)	θ_i^{true} (rad)	$\hat{\theta}_i$ (rad)
1	1.0934	1.0965	1.0960	0.1819	0.1820
2	1.0252	1.0220	1.0222	0.1122	0.1128
3	—	0.9531	0.9540	0.0320	0.0334
4	—	0.9174	0.9175	0.0000	0.0000

TABLE 2.3 Power injections: measured, true and estimated values.

	P_i^{m}	P_i^{true}	$\hat{P}_i$	Q_i^{m}	Q_i^{true}	$\hat{Q}_i$
Bus no.		(MW p.u.)			(MVAr p.u.)	
1	2.6093	2.6000	2.5940	2.9597	2.9743	2.9578
3	−0.4634	−0.5000	−0.4819	−0.2606	−0.2800	−0.2674

TABLE 2.4 Power flows: measured, true and estimated values.

	P_{ij}^{m}	P_{ij}^{true}	$\hat{P}_{ij}$	Q_{ij}^{m}	Q_{ij}^{true}	$\hat{Q}_{ij}$
Line no.		(MW p.u.)			(MVAr p.u.)	
1–4	1.8120	1.8198	1.8200	—	—	—
3–2	−0.7615	−0.7802	−0.7740	—	—	—
3–4	0.2650	0.2802	0.2921	0.3448	0.3447	0.3524

TABLE 2.5 State-variable updates and convergency summary.

	Iterations				
State variable	0	1	2	3	4
θ_1	0.0000	0.1824	0.1821	0.1820	0.1820
θ_2	0.0000	0.1056	0.1130	0.1128	0.1128
θ_3	0.0000	0.0287	0.0336	0.0334	0.0334
v_1	1.0000	1.1003	1.0959	1.0960	1.0960
v_2	1.0000	1.0183	1.0221	1.0222	1.0222
v_3	1.0000	0.9363	0.9538	0.9540	0.9540
v_4	1.0000	0.8895	0.9174	0.9175	0.9175
$J(\hat{x})$	49 850.5529	562.7272	4.3639	4.2674	4.2674

2.3.1 Bad Measurement Detection

If measurements are Gaussian distributed and independent, and if the weighting factors w_i correspond with the inverse of the measurement variances ($w_i = 1/\sigma_i^2$), the distribution of $J(x)$ is a χ^2 with $m+r-n$ degrees of freedom [8], where r is twice the number of transit buses (exact *measurements*). Thus, we can write

$$\text{Prob}\left(J(\hat{x}) \leq \chi^2(1-\alpha, m+r-n)\right) = 1-\alpha \tag{2.13}$$

where $1-\alpha$ is the confidence level.

Therefore, for a given α (e.g. 0.01), the value $\chi^2(1-\alpha, m+r-n)$ can be computed and the χ^2 test applied, i.e. if $J(\hat{x}) < \chi^2(1-\alpha, m+r-n)$ there is no bad measurement at the $1-\alpha$ confidence level; otherwise there is. Further details can be found in [7].

Example 2.3 Bad Measurement Detection Example

The voltage measurement at bus 1 (Figure 2.1) is considered to be 1.05 instead of 1.0933, i.e. a bad measurement is introduced. Solving the estimation problem, the optimal objective function value is $J(\hat{x}) = 16.6008$.

The test threshold at 0.99 confidence level ($\alpha = 0.01$) with $10+2-7=5$ degrees of freedom is $\chi^2(0.99{,}5) = 15.0863$. Therefore, since $\chi^2(0.99{,}5) <$ 16.6008, we conclude that bad data plague the measurement set with a 0.99 confidence level.

For the initial case and since $J(\hat{x}) = 4.2674 < \chi^2(0.99, 5)$, no bad data affect the measurement set with at 0.99 confidence level.

2.3.2 Identification of Erroneous Measurements

If the bad measurement test detects bad measurements, these measurements should be identified. This is accomplished below.

Consider the nonlinear measurement model

$$\boldsymbol{e} = \boldsymbol{z} - \boldsymbol{h}(\boldsymbol{x}^{\text{true}}) \tag{2.14}$$

where $\boldsymbol{x}^{\text{true}}$ is the unknown true state vector and $\boldsymbol{e}$ is the measurement error vector. Note that

$$\begin{aligned} E[\boldsymbol{e}] &= E[\boldsymbol{z} - \boldsymbol{h}(\boldsymbol{x}^{\text{true}})] = E[\boldsymbol{z}] - E[\boldsymbol{h}(\boldsymbol{x}^{\text{true}})] \\ &= E[\boldsymbol{z}] - \boldsymbol{h}(\boldsymbol{x}^{\text{true}}) = \boldsymbol{0} \end{aligned} \tag{2.15}$$

which is a typical assumption in state estimation.

Consider the differential measurement equation:

$$\text{d}\boldsymbol{e} = \text{d}\boldsymbol{z} - \boldsymbol{H}\text{d}\boldsymbol{x} \tag{2.16}$$

If measurements are Gaussian distributed and independent, the least squares estimator $\text{d}\hat{\boldsymbol{x}}$ can be obtained by minimizing the weighted sum of square deviations of the differential errors:

$$\underset{\text{d}\boldsymbol{x}}{\text{minimize}}(\text{d}\boldsymbol{z} - \boldsymbol{H}\text{d}\boldsymbol{x})^T \boldsymbol{W}(\text{d}\boldsymbol{z} - \boldsymbol{H}\text{d}\boldsymbol{x}) \tag{2.17}$$

This minimization problem leads to the system of equations:

$$\left(\boldsymbol{H}^T\boldsymbol{W}\boldsymbol{H}\right)\text{d}\boldsymbol{x} = \boldsymbol{H}^T\boldsymbol{W}\text{d}\boldsymbol{z} \tag{2.18}$$

known as the system of *normal equations* [8]. Assuming that the gain matrix

$$\boldsymbol{G} = \boldsymbol{H}^T \boldsymbol{W} \boldsymbol{H} \tag{2.19}$$

has an inverse, the least squares estimates $\mathrm{d}\hat{x}$ can be written explicitly as

$$\mathrm{d}\hat{x} = \boldsymbol{G}^{-1} \boldsymbol{H}^T \boldsymbol{W} \mathrm{d}z \tag{2.20}$$

from which we conclude that $\mathrm{d}\hat{x}$ is a linear function of $\mathrm{d}z$.

The residual and the differential residual are defined as

$$\boldsymbol{r} = z - \boldsymbol{h}(\hat{x}) \tag{2.21}$$

$$\mathrm{d}\boldsymbol{r} = \mathrm{d}z - \boldsymbol{H}\mathrm{d}\hat{x} \tag{2.22}$$

Thus

$$\mathrm{d}\boldsymbol{r} = \mathrm{d}z - \boldsymbol{H}\mathrm{d}\hat{x} = \mathrm{d}z - \boldsymbol{P}\mathrm{d}z = (\boldsymbol{I}_m - \boldsymbol{P})\mathrm{d}z \tag{2.23}$$

where $\boldsymbol{P} = \boldsymbol{H}\boldsymbol{G}^{-1}\boldsymbol{H}^T\boldsymbol{W}$ is known as *hat* or *projection* matrix.

Integrating (2.23), the linear transformation from z to $\boldsymbol{r}$ at the optimum is

$$\boldsymbol{r} = [\boldsymbol{I}_m - \boldsymbol{P}]z + \boldsymbol{k} = \boldsymbol{S}z + \boldsymbol{k} \tag{2.24}$$

where $\boldsymbol{k}$ is a constant vector and $\boldsymbol{S}$ is known as the residual sensitivity matrix.

If measurements z are Gaussian distributed and independent with zero mean and covariance matrix $\boldsymbol{R}$ ($\boldsymbol{R} = \boldsymbol{W}^{-1}$) and $\boldsymbol{h}(\hat{x})$ is an unbiased estimator of $\boldsymbol{h}(\boldsymbol{x}^{\mathrm{true}})$, i.e. $E[\boldsymbol{h}(\hat{x})] = \boldsymbol{h}(\boldsymbol{x}^{\mathrm{true}})$, the residual vector $\boldsymbol{r}$ provided by the linear transformation (2.24) is Gaussian distributed with parameters given by

$$\begin{aligned} E[\boldsymbol{r}] &= E[z - \boldsymbol{h}(\hat{x})] = E[z] - E[\boldsymbol{h}(\hat{x})] \\ &= E[z] - \boldsymbol{h}(\boldsymbol{x}^{\mathrm{true}}) = E[\boldsymbol{e}] = \boldsymbol{0} \end{aligned} \tag{2.25}$$

$$\begin{aligned} \boldsymbol{\Omega} &= E\left[(\boldsymbol{r} - E[\boldsymbol{r}])(\boldsymbol{r} - E[\boldsymbol{r}])^T\right] \\ &= E\left[(\boldsymbol{S}z + \boldsymbol{k} - \boldsymbol{S}[z] - \boldsymbol{k})(\boldsymbol{S}z + \boldsymbol{k} - \boldsymbol{S}E[z] - \boldsymbol{k})^T\right] \\ &= E\left[\boldsymbol{S}(z - E[z])(z - E[z])^T\boldsymbol{S}^T\right] \end{aligned} \tag{2.26}$$

and due to (2.14) and (2.25):

$$\begin{aligned} \boldsymbol{\Omega} &= E\left[\boldsymbol{S}\boldsymbol{e}\boldsymbol{e}^T\boldsymbol{S}^T\right] \\ &= \boldsymbol{S}E\left[\boldsymbol{e}\boldsymbol{e}^T\right]\boldsymbol{S}^T \\ &= \boldsymbol{S}\boldsymbol{R}\boldsymbol{S}^T = \boldsymbol{S}\boldsymbol{R} \end{aligned} \tag{2.27}$$

The normalized residual ($N(0, 1^2)$) of measurement i is

$$r_i^N = \frac{1}{\sqrt{\Omega_{ii}}}(z_i - h_i(\hat{x})) \tag{2.28}$$

The largest residual identifies a bad measurement with a $1 - \alpha$ confidence level (e.g. 0.99) if $|r_i^N| > \Phi^{-1}(1 - \alpha/2)$. Further details can be found in [7].

TABLE 2.6 Residuals and normalized residuals.

Measurement	r_i	r_i^N
P_1	–0.0154	2.2757
P_3	–0.0183	1.5836
P_{14}	0.0078	–0.5386
P_{32}	–0.0127	0.8996
P_{34}	0.0275	–1.9219
v_1	0.0258	–3.6530
v_2	–0.0245	3.5700
Q_1	–0.0032	2.6631
Q_3	–0.0018	1.0033
Q_{34}	0.0041	–1.0586

Example 2.4 Bad Measurement Identification Example
The considered covariance matrix $\boldsymbol{R}$ is

$$\boldsymbol{R} = 10^{-4}\text{Diag}(4, 4, 4, 4, 4, 1, 1, 4, 4, 4)$$

Matrices $\boldsymbol{P}$ and $\boldsymbol{\Omega}$ are computed using expressions (2.19) and (2.27), respectively. The residual (from (2.21)) and the normalized residual of measurements (from (2.28)) are also calculated and provided in Table 2.6.

The threshold for detection at a 0.99 confidence level ($\alpha = 0.01$) is $\Phi^{-1}(1 - 0.01/2) = 2.5758$. The normalized residual with highest absolute value in Table 2.6 (column 3) corresponds to v_1, and, because its value is larger than 2.5758, it is identified as a bad measurement, and, therefore, it is removed from z.

After removing this bad measurement, observability is checked again. In this case, the system remains observable, and the state estimation is carried out again. After estimating the state, the objective function $J(\hat{x}) = 4.1388 < \chi^2(0.99, 4)$, and therefore no additional bad measurement is identified.

2.4 MATHEMATICAL PROGRAMMING SOLUTION

Problem (2.1)–(2.3) can be directly solved by using mathematical programming techniques through a nonlinear solver. This approach is considered because current available mathematical programming solvers treat sparsity efficiently and are robust. Moreover, they are computationally efficient and provide highly accurate results. In addition, these solvers allow incorporating easily inequality constraints representing physical limits [9].

For instance, problem (2.1)–(2.3) can be solved by using solver CONOPT [10] or MINOS [11] under the General Algebraic Modeling System (GAMS) [12], which is a high-level modeling system for mathematical programming. It consists of a language compiler and a set of integrated high-performance solvers.

GAMS is tailored for complex, large-scale modeling applications and allows building large maintainable models that can be adapted quickly to new situations. Modeling systems similar to GAMS are AMPL [13] and AIMMS [14].

Example 2.5 Mathematical Programming Problem
In order to solve the example in Figure 2.1 through mathematical programming techniques, the solver CONOPT under GAMS is used [10]. The solution is provided in Tables 2.2–2.4. The objective function evaluated at the estimated state $\hat{x}$ is $J(\hat{x}) = 4.2674$. Needless to say, the solutions provided by the classical approach and the mathematical programming one are the same.

2.5 ALTERNATIVE STATE ESTIMATORS

The most common state estimation method within electric energy systems is WLS technique. This well-known procedure was first developed by Schweppe et al. [4–6] in the 1970s and was formulated as an optimization problem.

State of the art of current available nonlinear optimization solvers, recent advances in computational speed, and emerging multi-core Hyper-Threading Technology (HTT) processors allow the estimation problem to be solved directly [15]. Two main advantages can be led by proceeding in this manner: (i) alternative estimators (such as least absolute value, quadratic-constant, or least median of squares, among others) can be easily used, and (ii) decomposition techniques can be applied, resulting in decentralized estimators.

Bearing in mind that recent progress in nonlinear and mixed integer nonlinear optimization techniques, the aims of this chapter are (i) to formulate the most common state estimation algorithms as mathematical programming problems and (ii) to compare them from the numerical and computational perspectives.

From the optimization perspective, the following sections provide formulations as mathematical programming problems for the most common state estimation procedures, comparing them from both the numerical and computational perspectives.

For the sake of generality, the general mathematical programming formulation of most state estimators can be expressed as

$$\underset{x}{\text{minimize}}\, J(x) \tag{2.29a}$$

subject to

$$f(x) = \mathbf{0} \tag{2.29b}$$

$$g(x) \leq \mathbf{0} \tag{2.29c}$$

The actual expression of the scalar function $J(x)$ depends on the estimator employed. It can generally be expressed as a function of vector $y(x)$ defined as

$$y_i(x) = w_i(h_i(x) - z_i) \tag{2.30}$$

2.5.1 Weighted Least of Squares

WLS technique is a non-robust method that has been well studied in the technical literature [1, 4–7, 16, 17]. This estimator is non-robust since a single bad measurement can significantly distort the estimation results. The objective function of the WLS estimator is computed as the sum of all squared weighted measurement errors.

2.5.1.1 WLS General Formulation

WLS general formulation is

$$\underset{\boldsymbol{x}}{\text{minimize}}\, J(\boldsymbol{x}) = \sum_{i=1}^{m} y_i^2(\boldsymbol{x}) \tag{2.31a}$$

subject to

$$\boldsymbol{f}(\boldsymbol{x}) = \boldsymbol{0} \tag{2.31b}$$

$$\boldsymbol{g}(\boldsymbol{x}) \leq \boldsymbol{0} \tag{2.31c}$$

where m is the number of measurements.

Figure 2.2 depicts the objective function for the WLS estimator in the case of one single error. The violet and green curves represent a high and low weighting factor ω_i, respectively. Note that the weighted measurement error $y_i(\boldsymbol{x}) = w_i \cdot (h_i(\boldsymbol{x}) - z_i)$ appears as $y_i(\boldsymbol{x})^2$ in the objective function $J(\boldsymbol{x})$.

WLS estimation is typically obtained by solving the first-order optimality conditions of problem (2.31) via the Newton–Raphson method [18]. Note that nonlinear problem (2.31) is already a mathematical programming problem that is ready to be solved by an appropriate solver, e.g. MINOS [11], within an optimization environment such as AMPL [13] or GAMS [12].

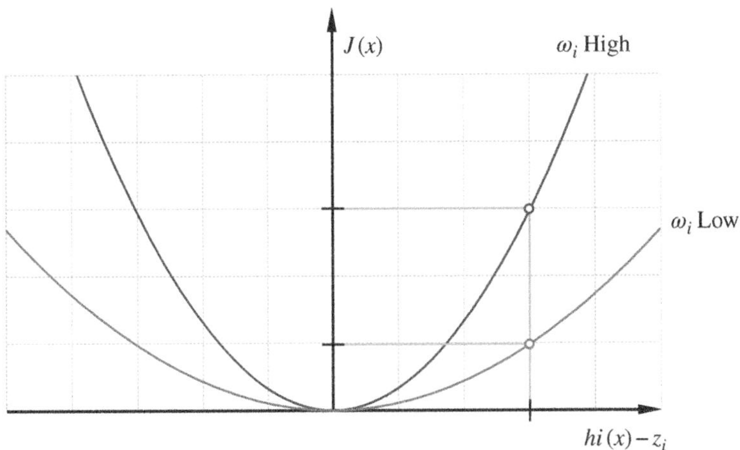

Figure 2.2 Objective function for the WLS estimator as a function of the error (one measurement).

2.5.2 Weighted Least Absolute Value

Since a single bad measurement can severely distort WLS estimation results, other estimators have been proposed with the purpose of improving the robustness of the estimation. Least absolute value (LAV) belongs to the category of *robust estimators*, i.e. procedures that are less sensitive to bad measurements than WLS technique. This estimator has also been exhaustively analyzed in technical literature [19–22].

The objective function of the LAV estimator is computed as the sum of the absolute values of all weighted measurement errors.

2.5.2.1 LAV General Formulation

The LAV general formulation is

$$\underset{\boldsymbol{x}}{\text{minimize}}\, J(\boldsymbol{x}) = \sum_{i=1}^{m} |y_i(\boldsymbol{x})| \tag{2.32a}$$

subject to

$$\boldsymbol{f}(\boldsymbol{x}) = \boldsymbol{0} \tag{2.32b}$$

$$\boldsymbol{g}(\boldsymbol{x}) \leq \boldsymbol{0} \tag{2.32c}$$

Figure 2.3 depicts the objective function for LAV estimator in the case of one single error. The violet and green curves correspond to a high and low weighting factor ω_i, respectively. Note that the weighted measurement error $y_i(\boldsymbol{x}) = w_i \cdot (h_i(\boldsymbol{x}) - z_i)$ appears as $|y_i(\boldsymbol{x})|$ in the objective function $J(\boldsymbol{x})$.

For the sake of generality, problem (2.32) is formulated by considering weighted measurement errors (see (2.30)). Note that some references label this estimator as WLAV (weighted least absolute value).

The objective function (2.32a) comprises the absolute value of vector $\boldsymbol{y}(\boldsymbol{x})$. This function is nonlinear but can be linearized straightforwardly, resulting in the mathematical programming formulation in Section 2.5.2.2.

2.5.2.2 LAV Mathematical Programming Formulation

LAV mathematical programming formulation is

$$\underset{\boldsymbol{x},\boldsymbol{s}}{\text{minimize}}\, J(\boldsymbol{x}) = \sum_{i=1}^{m} s_i \tag{2.33a}$$

subject to

$$-s_i \leq y_i(\boldsymbol{x}) \leq s_i \quad \forall i \tag{2.33b}$$

$$\boldsymbol{f}(\boldsymbol{x}) = 0 \tag{2.33c}$$

$$\boldsymbol{g}(\boldsymbol{x}) \leq \boldsymbol{0} \tag{2.33d}$$

In (2.33), note that the required linearization involves the addition of (i) a continuous variable vector $\boldsymbol{s}$ and (ii) two sets of constraints (2.33b).

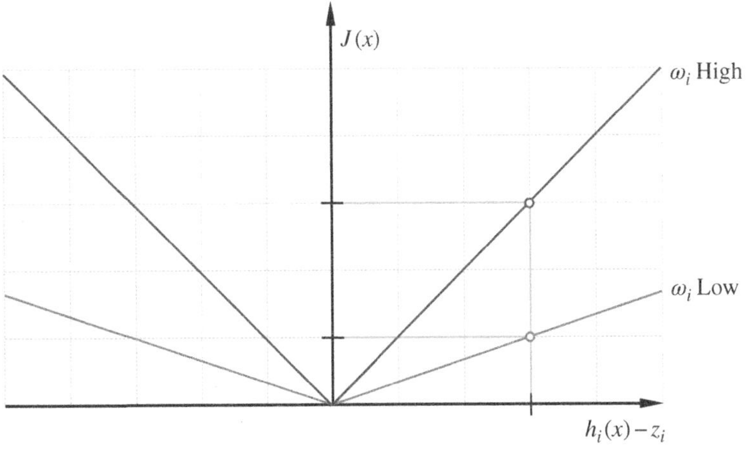

Figure 2.3 Objective function for the LAV estimator as a function of the error (one measurement).

2.5.3 Quadratic-Constant Criterion

As previously mentioned, one significant drawback of WLS estimator is the lack of robustness to bad data. Other non-quadratic estimators have been thereby developed to overcome this disadvantage, such as the aforementioned LAV method.

Quadratic-constant and quadratic-linear algorithms (QC and QL) combine the benefits of maximum likelihood least squares estimation and the bad data rejection properties of the LAV estimator.

Figure 2.4 allows the objective function for the QC estimator in the case of one single error. The violet and green curves correspond to a high and low weighting factor ω_i, respectively. Note that the behavior of function $J(\boldsymbol{x})$ for the QC estimator is the same as the WLS when the weighted measurement error $y_i(\boldsymbol{x})$ is within the given bounds.

2.5.3.1 QC General Formulation

The general formulation for QC estimator is

$$\underset{\boldsymbol{x}}{\text{minimize}}\, J(\boldsymbol{x}) = \sum_{i=1}^{m} s_i(\boldsymbol{x}) \tag{2.34a}$$

subject to

$$\boldsymbol{f}(\boldsymbol{x}) = \boldsymbol{0} \tag{2.34b}$$

$$\boldsymbol{g}(\boldsymbol{x}) \leq \boldsymbol{0} \tag{2.34c}$$

$$s_i(\boldsymbol{x}) = \left\{ \begin{array}{ll} y_i^2(\boldsymbol{x}) & \text{if} |y_i(\boldsymbol{x})| \leq T \\ T^2 & \text{if} |y_i(\boldsymbol{x})| \geq T \end{array} \right\} \forall i \tag{2.34d}$$

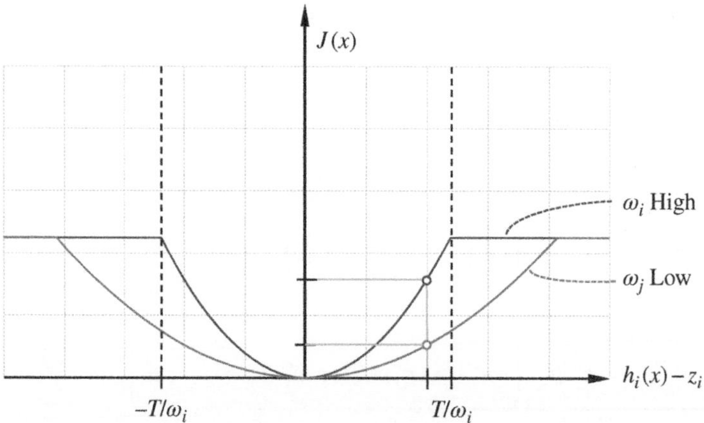

Figure 2.4 Objective function for the QC estimator as a function of the error (one measurement).

The objective function of problem (2.34) has a quadratic shape for each of those measurement error values that are within a tolerance T; otherwise, the objective function component has a constant value (see Figure 2.4). Note that a tolerance T must be selected in advance.

2.5.3.2 QC Mathematical Programming Formulation

QC mathematical programming formulation is

$$\underset{\boldsymbol{x},\boldsymbol{b}}{\text{minimize}}\ J(\boldsymbol{x}) = \sum_{i=1}^{m}\left[(1-b_i)y_i^2(\boldsymbol{x}) + b_iT^2\right] \tag{2.35a}$$

subject to

$$\boldsymbol{f}(\boldsymbol{x}) = \boldsymbol{0} \tag{2.35b}$$

$$\boldsymbol{g}(\boldsymbol{x}) \leq \boldsymbol{0} \tag{2.35c}$$

$$b_i = \{0, 1\} \quad \forall i \tag{2.35d}$$

If problem (2.34) is to be expressed as a standard mathematical programming problem, then a binary variable vector $\boldsymbol{b}$ must be added to the optimization variable set. The resulting formulation (2.35) is a mixed integer nonlinear problem.

Although problem (2.35) is a mixed integer nonlinear programming problem, the set of constraints (2.35d) can be relaxed:

$$0 \leq b_i \leq 1 \quad \forall i$$

leading to a relaxed mixed integer nonlinear programming problem. This relaxed problem can be tackled by any nonlinear programming solver (such as MINOS [11]). Numerical simulations show that the particular structure of problem

(2.35) imposes that relaxed binary variables b_i have a binary value at the optimum, i.e. $\hat{b}_i \in \{0, 1\}$.

2.5.4 Quadratic-Linear Criterion

QL technique is a state estimator that is similar to the previously considered QC method but involves linear terms rather than constant terms outside the tolerance region.

In this chapter, it is considered that the linear parts of the objective function of this estimator coincide with LAV objective function. Other works [23, 24] characterize these linear terms as tangents to the quadratic component so that the derivative of function $J(\boldsymbol{x})$ does not have any discontinuities. Note that the formulation proposed in this chapter presents two derivative discontinuities at points at $y_i(\boldsymbol{x}) = -T$ and $y_i(\boldsymbol{x}) = T$.

From the optimization perspective, the approach analyzed in this work has some benefits: (i) it can easily be formulated as a mathematical problem using a smaller number of binary variables and/or constraints than the quadratic-tangent technique, (ii) the resulting problem structure allows binary variables to be relaxed without altering the optimal solution, and (iii) numerical simulations suggest that the time required for CPU is significantly smaller than that required by the quadratic-tangent approach.

2.5.4.1 QL General Formulation

The QL general formulation is

$$\underset{\boldsymbol{x}}{\text{minimize}}\, J(\boldsymbol{x}) = \sum_{i=1}^{m} s_i(\boldsymbol{x}) \tag{2.36a}$$

subject to

$$\boldsymbol{f}(\boldsymbol{x}) = \boldsymbol{0} \tag{2.36b}$$

$$\boldsymbol{g}(\boldsymbol{x}) \leq \boldsymbol{0} \tag{2.36c}$$

$$s_i(\boldsymbol{x}) = \left\{ \begin{array}{ll} y_i^2(\boldsymbol{x}) & \text{if } |y_i(\boldsymbol{x})| \leq T \\ T \mid y_i(\boldsymbol{x}) \mid & \text{if } |y_i(\boldsymbol{x})| \geq T \end{array} \right\} \forall i \tag{2.36d}$$

Figure 2.5 depicts the objective function for QL estimator in the case of one single error. The violet and green curves correspond to a high and low weighting factor ω_i, respectively. Note that the behavior of function $J(\boldsymbol{x})$ for the QL estimator is the same as WLS if the measurement error $y_i(\boldsymbol{x})$ is within given bounds.

The objective function value of QL estimator for a given measurement has a quadratic shape (like WLS estimator) if the measurement error value is within a tolerance T; otherwise, this objective function component has an absolute value shape like LAV estimator (see Figure 2.5). Note that a tolerance T must be selected in advance.

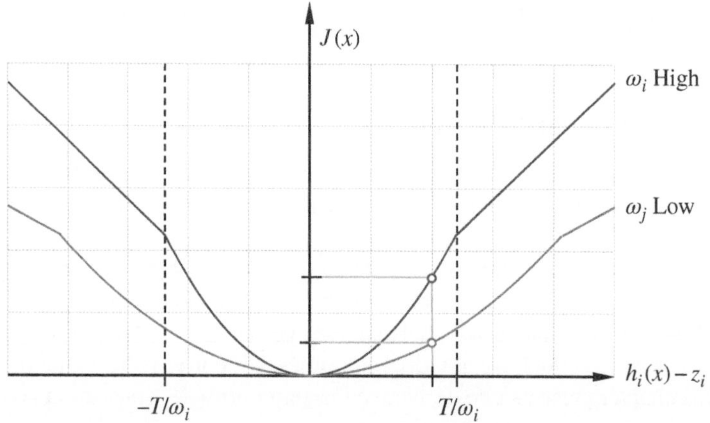

Figure 2.5 Objective function for the QL estimator as a function of the error (one measurement).

2.5.4.2 *QL Mathematical Programming Formulation*

The QL mathematical programming formulation is

$$\underset{\boldsymbol{x},\boldsymbol{s},\boldsymbol{b}}{\text{minimize}}\ J(\boldsymbol{x}) = \sum_{i=1}^{m} \left[(1-b_i)s_i^2 + b_i(Ts_i) \right] \tag{2.37a}$$

subject to

$$-s_i \le y_i(\boldsymbol{x}) \le s_i \quad \forall i \tag{2.37b}$$

$$\boldsymbol{f}(\boldsymbol{x}) = \boldsymbol{0} \tag{2.37c}$$

$$\boldsymbol{g}(\boldsymbol{x}) \le \boldsymbol{0} \tag{2.37d}$$

$$b_i = \{0, 1\} \quad \forall i \tag{2.37e}$$

If the general formulation (2.36) is to be recast as a mathematical programming problem, it is necessary to include (i) a binary variable vector $\boldsymbol{b}$, (ii) a positive variable vector $\boldsymbol{s}$, and (iii) two sets of constraints (2.37b). The resulting formulation (2.37) is a mixed integer nonlinear problem.

Again, the set of constraints (2.37e) of problem (2.37) can be relaxed, leading to a nonlinear problem, whose optimal solution generally meets constraints (2.37e).

2.5.5 Least Median of Squares

Least median of squares (LMS) is a robust estimator [2]. The objective function to be minimized is the squared measurement error whose value is the median of all squared measurement errors. The key idea underlying this technique is that the median of a set of values is a more robust estimate than the mean.

2.5.5.1 LMS General Formulation

The general formulation of the LMS estimator is

$$\underset{x}{\text{minimize}}\, J(\boldsymbol{x}) = y_{\nu}^{2}(\boldsymbol{x}) \tag{2.38a}$$

subject to

$$\boldsymbol{f}(\boldsymbol{x}) = \boldsymbol{0} \tag{2.38b}$$

$$\boldsymbol{g}(\boldsymbol{x}) \leq 0 \tag{2.38c}$$

$$y_{\nu}^{2}(\boldsymbol{x}) = \text{median}\left(y_{1}^{2}(\boldsymbol{x}), \ldots, y_{m}^{2}(\boldsymbol{x})\right) \tag{2.38d}$$

where the function median(x_1, x_2, …, x_n) computes the median value of set $\{x_1, x_2, \ldots, x_n\}$.

Reference [25] proposes a mathematical programming formulation for LMS estimator to be applied to a linear estimator. In this chapter, this formulation is applied to the state estimation problem, and the mathematical programming formulation of LMS estimator is presented below.

2.5.5.2 LMS Mathematical Programming Formulation

The LMS mathematical programming formulation is

$$\underset{\boldsymbol{x}, \boldsymbol{b}, T_{\text{LMS}}}{\text{minimize}}\, J(\boldsymbol{x}) = T_{\text{LMS}} \tag{2.39a}$$

subject to

$$-T_{\text{LMS}} - M\cdot(1 - b_i) \leq y_i(\boldsymbol{x}) \leq T_{\text{LMS}} + M\cdot(1 - b_i) \quad \forall i \tag{2.39b}$$

$$\boldsymbol{f}(\boldsymbol{x}) = \boldsymbol{0} \tag{2.39c}$$

$$\boldsymbol{g}(\boldsymbol{x}) \leq \boldsymbol{0} \tag{2.39d}$$

$$\sum_{i=1}^{m} b_i = \nu \tag{2.39e}$$

$$b_i = \{0, 1\} \quad \forall i \tag{2.39f}$$

where parameter M is a sufficiently large constant and parameter ν identifies the median and can be computed as [26]

$$\nu = \text{int}\left(\frac{m}{2} + \frac{n+1}{2}\right) \tag{2.40}$$

where n is the number of state variables and function int(x) denotes the integer part of x.

The rationale for formulation (2.39) can be graphically explained. Figure 2.6 depicts the set of measurement errors $y_i(\boldsymbol{x})$, which are sorted by value. The colored zone is delimited by the bounds $[-T_{\text{LMS}}, T_{\text{LMS}}]$ and includes the smaller measurement errors up to the position ν.

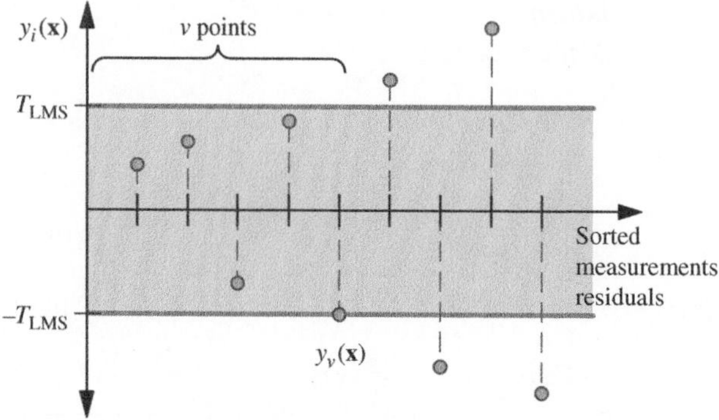

Figure 2.6 Graphical representation of the LMS estimator.

The objective function (2.39a) minimizes the variable T_{LMS}, i.e. the width of the colored zone. The vector constraints (2.39b) and (2.39e) enforce that the number of measurements contained in the zone $[-T_{\text{LMS}}, T_{\text{LMS}}]$ is equal to the parameter ν. The measurement residual $y_\nu(x)$ is thus at one edge of the interval $[-T_{\text{LMS}}, T_{\text{LMS}}]$.

In formulation (2.39), note that the minimization of the median of squared errors corresponds to the minimization of the median of absolute errors.

The proposed mathematical formulation requires the addition of the following optimization variables: (i) a binary variable vector $\boldsymbol{b}$ whose values identify those absolute errors $|y_i(\boldsymbol{x})|$, which are smaller than or equal to $|y_\nu(\boldsymbol{x})|$, and (ii) a variable T_{LMS} whose value is equal to $|y_\nu(\boldsymbol{x})|$. Three sets of constraints must also be included. Observe that the symbol T_{LMS} represents a variable to be optimized, not a predefined parameter.

2.5.6 Least Trimmed of Squares

An alternative to the LMS estimator is provided by the estimator that minimizes the sum of the smallest ordered squared errors up to the position ν, the so-called least trimmed of squares (LTS) estimator [3, 27].

2.5.6.1 *LTS General Formulation*

The general formulation of the LTS estimator is

$$\underset{\boldsymbol{x}}{\text{minimize}}\ J(\boldsymbol{x}) = \sum_{i=1}^{\nu} s_i^2(\boldsymbol{x}) \tag{2.41a}$$

subject to

$$\boldsymbol{f}(\boldsymbol{x}) = \boldsymbol{0} \tag{2.41b}$$

$$\boldsymbol{g}(\boldsymbol{x}) \leq 0 \tag{2.41c}$$

$$y_\nu^2(\boldsymbol{x}) = \text{median}\left(y_1^2(\boldsymbol{x}), \ldots, y_m^2(\boldsymbol{x})\right) \tag{2.41d}$$

$$s_i(\boldsymbol{x}) = \begin{Bmatrix} y_i(\boldsymbol{x}) & \text{if } y_i^2(\boldsymbol{x}) \le y_\nu^2(\boldsymbol{x}) \\ 0 & \text{if } y_i^2(\boldsymbol{x}) \ge y_\nu^2(\boldsymbol{x}) \end{Bmatrix} \forall i \tag{2.41e}$$

Note that the main difference between the LMS and LTS estimators is that the former considers only one squared measurement error in the objective function, whereas the latter takes into consideration about half of the squared measurement errors.

2.5.6.2 LTS Mathematical Programming Formulation

The mathematical programming formulation of the LTS method is

$$\underset{\boldsymbol{x},\boldsymbol{s},\boldsymbol{b}}{\text{minimize}}\, J(\boldsymbol{x}) = \sum_{i=1}^{m} s_i^2 \tag{2.42a}$$

subject to

$$-M\cdot(1-b_i) + s_i \le y_i(\boldsymbol{x}) \le M\cdot(1-b_i) + s_i \quad \forall i \tag{2.42b}$$

$$\boldsymbol{f}(\boldsymbol{x}) = \boldsymbol{0} \tag{2.42c}$$

$$\boldsymbol{g}(\boldsymbol{x}) \le \boldsymbol{0} \tag{2.42d}$$

$$\sum_{i=1}^{m} b_i = \nu \tag{2.42e}$$

$$b_i = \{0, 1\} \quad \forall i \tag{2.42f}$$

Again, parameter ν identifies the median and can be computed as [26]

$$\nu = \text{int}\left(\frac{m}{2} + \frac{n+1}{2}\right) \tag{2.43}$$

Problem (2.42) minimizes the sum of all squared variables s_i. Note that the number of optimization variables for the proposed LTS formulation is larger than that of any of the previous formulations.

2.5.7 Least Measurements Rejected

References [28, 29] propose a mathematical programming formulation for the least measurements rejected (LMR) estimator. The underlying idea is to find the largest set of measurements whose errors are within a given tolerance T, i.e. to minimize the number of measurement errors that are out of tolerance. Hereafter, this estimator is denominated as LMR.

2.5.7.1 LMR General Formulation

The general formulation of the LMR estimator is

$$\underset{\boldsymbol{x}}{\text{minimize}}\, J(\boldsymbol{x}) = \text{card}(\Omega_{\text{BM}}) \tag{2.44a}$$

subject to

$$f(\boldsymbol{x}) = \boldsymbol{0} \tag{2.44b}$$

$$g(\boldsymbol{x}) \leq \boldsymbol{0} \tag{2.44c}$$

where card(Ω) represents the cardinality of set Ω and Ω_{BM} is the set of those measurement errors that are out of tolerance, i.e. with $|y_i(\boldsymbol{x})| \geq T$.

2.5.7.2 LMR Mathematical Programming Formulation

The mathematical programming formulation for the LMR estimator proposed in [28, 29] is

$$\underset{\boldsymbol{x},\boldsymbol{b}}{\text{minimize}}\, J(\boldsymbol{x}) = \sum_{i=1}^{m} b_i \tag{2.45a}$$

subject to

$$-T - M \cdot b_i \leq y_i(\boldsymbol{x}) \leq T + M \cdot b_i \quad \forall i \tag{2.45b}$$

$$f(\boldsymbol{x}) = \boldsymbol{0} \tag{2.45c}$$

$$g(\boldsymbol{x}) \leq \boldsymbol{0} \tag{2.45d}$$

$$b_i = \{0, 1\} \quad \forall i \tag{2.45e}$$

In (2.45b), note that each binary variable b_i indicates whether or not the ith weighted measurement error ($y_i(\boldsymbol{x})$) is within the range $[-T, T]$. Specifically, the value $b_i = 0$ implies that $|y_i(\boldsymbol{x})| \leq T$. On the other hand, the value $b_i = 1$ implies that $|y_i(\boldsymbol{x})| > T$. Since the objective function (2.45a) minimizes the sum of all binary variables b_i, the LMR procedure searches the largest set of measurement errors within the range $[-T, T]$.

Once problem (2.45) has been solved and the set of out-of-tolerance measurement errors has been identified, a WLS estimation is performed to enhance the estimation quality, considering only those measurements whose associated binary variables have optimal values that are equal to zero [28].

2.5.8 Formulation Overview

Table 2.7 provides a general overview of the computational requirements needed to formulate the considered estimators as mathematical programming problems. This table contains the following information:

- The first column indicates the estimation techniques considered.
- The second and third columns indicate whether or not a large parameter M and/or a predefined tolerance T is required.
- The fourth and fifth columns show the number of continuous/binary optimization variables.

TABLE 2.7 Characterization of different state-estimation formulations.

	Parameter M	Tolerance T	Continuous variables	Binary variables	Additional constraints	Problem type
WLS	✗	✗	n	—	0	NLP
LAV	✗	✗	$n+m$	—	$2m$	NLP
QC	✗	✓	n	m	0	MINLP
QL	✗	✓	$n+m$	m	$2m$	MINLP
LMS	✓	✗	$n+1$	m	$2m+1$	MINLP
LTS	✓	✗	$n+m$	m	$2m+1$	MINLP
LMR	✓	✓	n	m	$2m$	MINLP

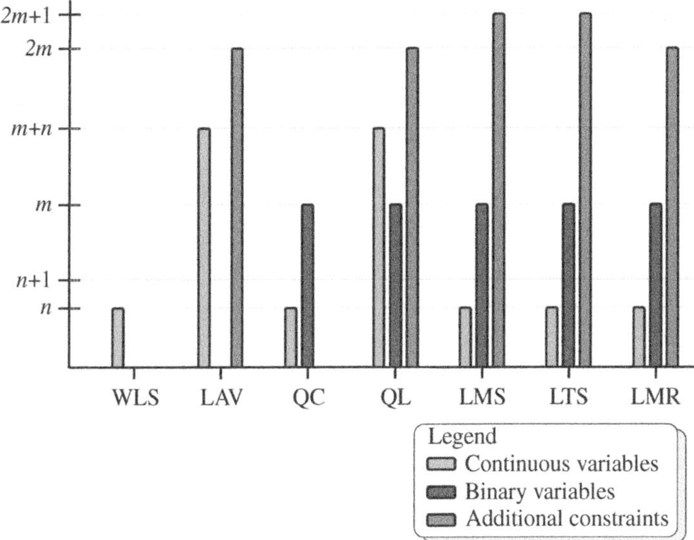

Figure 2.7 Problem size comparison for different estimators.

- The sixth column indicates the number of additional constraints required.
- Finally, the seventh column denotes the nature of the resulting mathematical programming problem.

The acronyms NLP and MINLP represent nonlinear problem and mixed integer nonlinear problem, respectively. The symbol ✓ indicates "needed," while the symbol ✗ indicates "not needed."

Figure 2.7 provides a graphical comparison between methods in terms of number of continuous/binary variables and quantity of additional constraints.

In Figure 2.7, observe that the LMS and LTS estimators require the highest number of both continuous/binary variables and additional constraints. The computational burden of these algorithms is thus expected to be heavier than that of the others.

2.5.9 Illustrative Example

For illustrative purposes, the state estimation procedures previously described are applied to a small 4-bus system. Network topology and measurement configuration is depicted in Figure 2.8.

The network is composed of two generating buses, two load buses, and four lines. The set of measurements includes four voltage measurements, one active/reactive power injection measurement, and four active/reactive power flow measurements (labeled as V_i, P_i/Q_i, and P_{ij}/Q_{ij}). If we consider that the measurement vector comprises 14 elements, then the redundancy ratio is $r = 14/(4 \times 2 - 1) = 2$. The network data (resistance, reactance, and total line charging susceptance) are provided in Table 2.8.

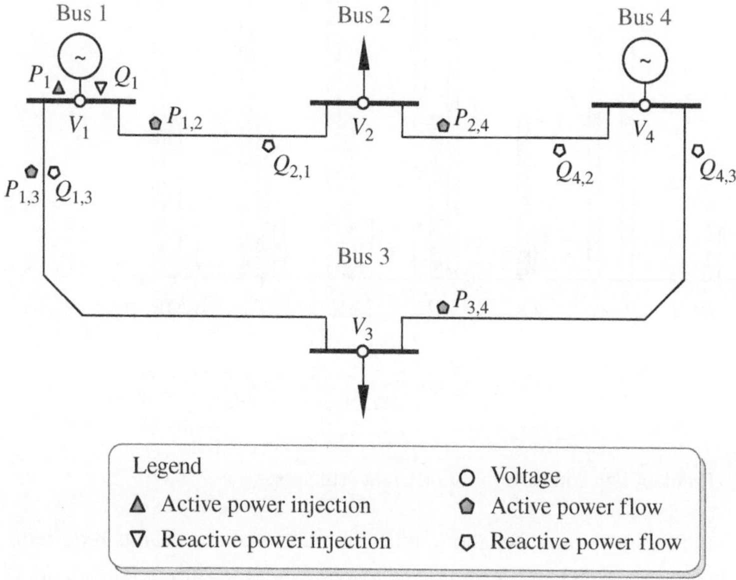

Figure 2.8 Example of alternative estimators: four-bus system.

TABLE 2.8 Example of alternative estimators: line characteristics.

Line	Resistance (p.u.)	Reactance (p.u.)	Susceptance (p.u.)
1–2	0.010 08	0.0504	0.1025
1–3	0.007 44	0.0372	0.0775
1–4	0.007 44	0.0372	0.0775
1–4	0.012 72	0.0636	0.1275

TABLE 2.9 Example of alternative estimators: operating point.

Bus no.	Voltage magnitude (p.u.)	Voltage angle (rad)
1	1.000	0.000
2	0.985	−0.008
3	0.973	−0.025
4	1.020	0.038

TABLE 2.10 Example of alternative estimators: measurements.

Measurement	Value (p.u.)	Measurement	Value (p.u.)
V_1	0.993	$P_{2,4}$	−1.350
P_1	0.963	$Q_{2,1}$	−0.293
$P_{1,2}$	0.204	V_3	0.974
$P_{1,3}$	0.750	$P_{3,4}$	−1.076
Q_1	0.762	V_4	1.025
$Q_{1,3}$	0.556	$Q_{4,2}$	0.673
V_2	0.963	$Q_{4,3}$	0.506

Table 2.9 shows the true state of the network, obtained from a converged power flow solution [15]. Bearing in mind that the measurement standard deviations are 0.01 p.u., the actual measurement values are provided in Table 2.10.

Since the proposed 4-bus system does not comprise any zero-injection buses, no equality constraints are considered. The tolerance T used for the QC, QL, and LMR estimators is set to 1, and parameter M, which is used for the LMS, LTS, and LMR procedures, is set to 100. From (2.40), the median ν is computed using the following equation:

$$\nu = \text{int}\left(\frac{m}{2} + \frac{n+1}{2}\right) = \text{int}\left(\frac{14}{2} + \frac{7+1}{2}\right) = 11 \tag{2.46}$$

where function int(x) denotes the integer part of x.

Table 2.11 provides a brief description of the computational characteristics of each optimization problem, detailing the number and type of optimization variables, and the number of additional constraints. Nonlinear problems (WLS and LAV estimators) are solved using MINOS 5.5 [11] under GAMS 23.5 [12, 30], whereas mixed integer nonlinear problems (QC, QL, LMS, LTS, and LMR procedures) are solved using SBB [31] under GAMS.

Figures 2.9–2.12 depict the weighted measurement residuals $y_i(\boldsymbol{x})$ for the WLS, LAV, QC, and QL estimators (blue data points). The objective function shapes corresponding to each estimator are represented with dotted lines.

In Figures 2.11–2.12, note that the largest absolute residual $y_i(\boldsymbol{x})$ is located in the non-quadratic zone of the objective function for the QC and QL estimators. This large residual corresponds to measurement V_2.

TABLE 2.11 Example of alternative estimators: characterization.

	Continuous variables	Binary variables	Additional constraints
WLS	7	—	0
LAV	21	—	28
QC	7	14	0
QL	21	14	28
LMS	8	14	29
LTS	21	14	29
LMR	7	14	28

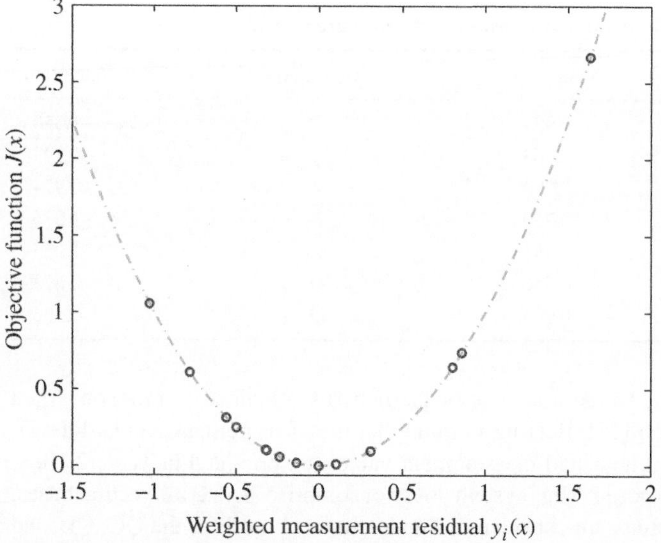

Figure 2.9 Example of alternative estimators: residuals of the WLS solution.

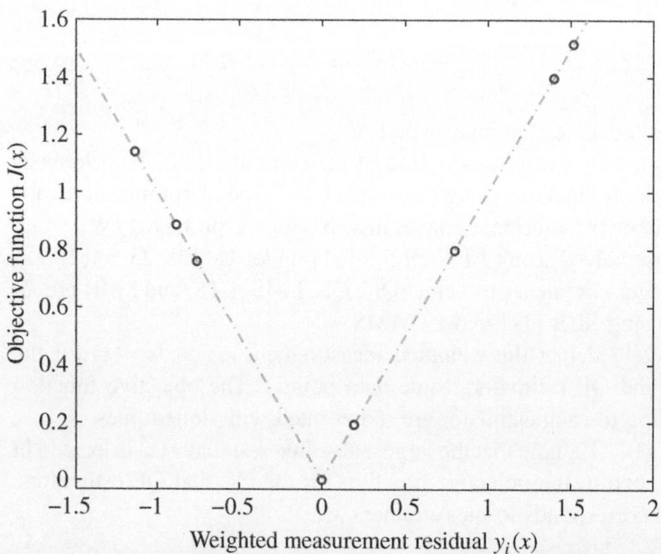

Figure 2.10 Example of alternative estimators: residuals of the LAV solution.

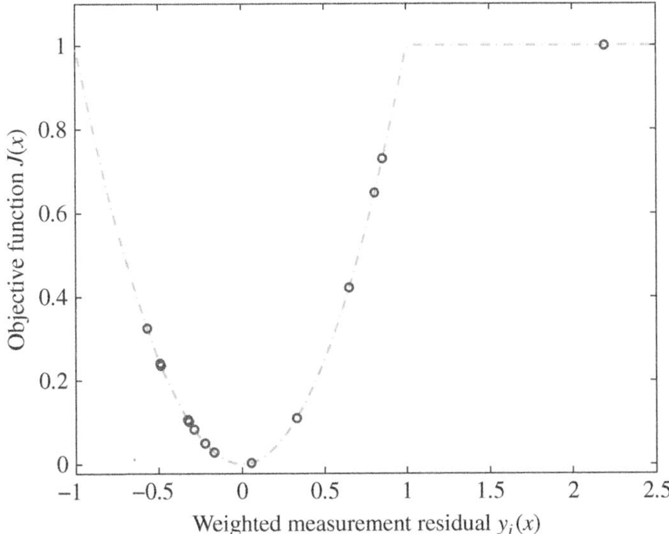

Figure 2.11 Example of alternative estimators: residuals of the QC solution.

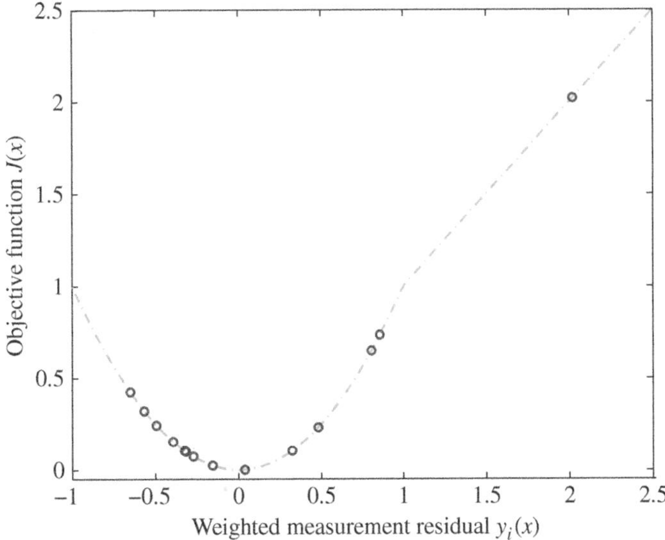

Figure 2.12 Example of alternative estimators: residuals of the QL solution.

With regard to the LMS, LTS, and LMR procedures, Table 2.12 provides the optimal value of each binary variable b_i corresponding to the ith measurement indicated in the first column of the same table.

In Table 2.12, note that the sum of the optimal binary variables for the LMS and LTS estimators is equal to 11, i.e. to the median. However, the LMR procedure only rejects one measurement. This residual corresponds to measurement V_2.

TABLE 2.12 Example of alternative estimators: optimal values for the binary variables.

	Estimators				Estimators		
	LMS	LTS	LMR		LMS	LTS	LMR
V_1	1	1	0	$P_{2,4}$	1	1	0
P_1	1	1	0	$Q_{2,1}$	1	1	0
$P_{1,2}$	1	1	0	V_3	1	1	0
$P_{1,3}$	1	1	0	$P_{3,4}$	1	1	0
Q_1	1	0	0	V_4	1	1	0
$Q_{1,3}$	1	0	0	$Q_{4,2}$	0	0	0
V_2	0	1	1	$Q_{4,3}$	0	1	0

TABLE 2.13 Example of alternative estimators: true and estimated state vectors.

		Estimators						
	x^{true}	WLS	LAV	QC	QL	LMS	LTS	LMR
V_1 (p.u.)	1.000	0.994	0.993	1.000	0.998	0.996	0.985	1.000
V_2 (p.u.)	0.985	0.979	0.978	0.985	0.983	0.981	0.970	0.985
V_3 (p.u.)	0.973	0.967	0.965	0.972	0.970	0.969	0.976	0.972
V_4 (p.u.)	1.020	1.014	1.013	1.020	1.018	1.030	1.023	1.020
θ_1 (rad)	0.000	0.000	0.000	0.000	0.000	0.000	0.000	0.000
θ_2 (rad)	−0.008	−0.008	−0.008	−0.008	−0.008	−0.008	−0.008	−0.008
θ_3 (rad)	−0.025	−0.025	−0.025	−0.025	−0.025	−0.025	−0.028	−0.025
θ_4 (rad)	0.038	0.038	0.038	0.038	0.038	0.035	0.034	0.038

Finally, Table 2.13 provides the estimated voltage magnitude and angle for the WLS, LAV, QC, QL, LMS, LTS, and LMR estimation procedures. For comparison purposes, the second column of this table shows the true state x^{true}.

In Table 2.13, note that the best estimates are provided by the QC and LMR estimators, while the performance of the LTS procedure is poor. Note that this conclusion is withdrawn considering only one measurement scenario, a particular metering configuration, and a small 4-bus system. In order to obtain statistically sound conclusions, the following section provides a detailed analysis of both the numerical and computational behaviors of the aforementioned algorithms, considering 100 measurement scenarios, different measurement configurations, and a large system.

2.5.10 Case Study

In this section, the performance of the mathematical programming estimation procedures described above is analyzed from both numerical and computational perspectives. These state estimation methods are applied to a realistic 118-bus system.

This system corresponds to the IEEE 118-bus network, which represents a significant area of the American Electric Power system.

For statistical consistency, each analysis is carried out by considering 100 measurement scenarios. Each measurement scenario is synthetically generated from the solution of a converged power flow by adding Gaussian-distributed random errors to the corresponding true values.

Each scenario involves:

i. A random active/reactive power consumption level.
ii. Random locations of voltage and active/reactive power meters (ensuring observability of the whole system).
iii. A random redundancy level.
iv. Gaussian-distributed random errors in all measurements (standard deviations of 0.01 and 0.02 p.u. for voltage and power measurements, respectively).

The computational analysis of this chapter has been performed using MINOS [11] and SBB [31] solvers under GAMS [12, 30] on a Linux-based server with four processors clocking at 2.9 GHz and 64 GB of RAM.

2.5.10.1 Estimation Assessment

This subsection analyzes the performance of the estimators described in this chapter: WLS, LAV, QC, QL, LMS, LTS, and LMR.

To compare the accuracy provided by each estimator with regard to the true values for each scenario ω, the metrics $\epsilon^V_{\text{abs},\omega}$ and $\epsilon^\theta_{\text{abs},\omega}$ are considered:

$$\epsilon^V_{\text{abs},\omega} = \frac{\sum_{i=1}^{N} \left|\hat{V}_{i,\omega} - V^{\text{true}}_{i,\omega}\right|}{N} \tag{2.47}$$

$$\epsilon^\theta_{\text{abs},\omega} = \frac{\sum_{i=2}^{N} \left|\hat{\theta}_{i,\omega} - \theta^{\text{true}}_{i,\omega}\right|}{N-1} \tag{2.48}$$

where $\hat{V}_{i,\omega}$ and $\hat{\theta}_{i,\omega}$ are the estimated voltage magnitude and angle for the ith bus in scenario ω, $V^{\text{true}}_{i,\omega}$ and $\theta^{\text{true}}_{i,\omega}$ correspond to the solution of the converged power flow in scenario ω, and N is the number of buses in the whole system. Note that $\epsilon^V_{\text{abs},\omega}$ ($\epsilon^\theta_{\text{abs},\omega}$) is the average absolute error of the voltage magnitude (voltage angle) estimates as compared with the true values for scenario ω. In (2.48), note that the reference angle bus is located at bus 1.

2.5.11 Results

The estimation problems previously mentioned are solved by using optimization software and by considering that parameters M and T are set to 100 and 2, respectively. The maximum number of iterations and the duality gap for MINLP problems is set to 1000 and 1%, respectively.

In order to remove mathematical unfeasibilities, the solutions to LMS and LTS mathematical problems are obtained by imposing that the set of measurements whose residuals are smaller than $|y_\nu(\boldsymbol{x})|$ provides whole system observability. This observability is achieved by forcing some binary variables to be equal 1 (i.e. $b_i = 1$). These fixed variables correspond to a set of voltage and active power flow measurements that make the system observable.

Both numerical accuracy and computation performance are analyzed and compared. Tables 2.14 and 2.15 and Figures 2.13–2.15 provide results regarding estimation accuracy and computational performance:

- Table 2.14 provides the mean and standard deviation of metrics $\epsilon^V_{\text{abs},\omega}$ and $\epsilon^\theta_{\text{abs},\omega}$ for each estimator considered.
- Table 2.15 provides the minimum, mean, maximum, and standard deviation for the CPU time required (measured in seconds) for each estimator.
- Figure 2.13 depicts the histogram of voltage magnitude absolute error $\epsilon^V_{\text{abs},\omega}$ for each estimator considered.
- Figure 2.14 likewise shows the histogram of voltage angle estimation accuracy (measured in terms of absolute error $\epsilon^\theta_{\text{abs},\omega}$) for each estimation technique.
- Finally, Figure 2.15 provides the histogram of the CPU time required (measured in seconds) for each estimation procedure.

TABLE 2.14 Case study: estimation accuracy results.

Method	$\text{mean}(\epsilon^V_{\text{abs}})$ (p.u.)	$\text{std}(\epsilon^V_{\text{abs}})$ (p.u.)	$\text{mean}(\epsilon^\theta_{\text{abs}})$ (rad)	$\text{std}(\epsilon^\theta_{\text{abs}})$ (rad)
WLS	0.0015	0.0012	0.0019	0.0015
LAV	0.0019	0.0015	0.0023	0.0018
QC	0.0017	0.0013	0.0020	0.0016
QL	0.0016	0.0012	0.0019	0.0015
LMS	0.0095	0.0083	0.0052	0.0043
LTS	0.0053	0.0052	0.0029	0.0024
LMR	0.0015	0.0012	0.0019	0.0015

TABLE 2.15 Case study: computational performance results.

Method	Minimum (s)	Mean (s)	Maximum (s)	Std. dev. (s)
WLS	0.84	0.96	1.37	0.09
LAV	0.37	0.56	0.76	0.07
QC	0.17	0.24	0.37	0.04
QL	0.84	1.04	1.54	0.14
LMS	3.56	5.05	8.39	0.85
LTS	1.00	1.25	1.87	0.16
LMR	0.81	5.01	24.39	6.02

In Tables 2.14 and 2.15 and Figures 2.13–2.15, we can observe the following:

1. In terms of estimation accuracy, the WLS, QC, QL, and LMR procedures are more precise than the other procedures in estimating both voltage magnitudes and angles. On the other hand, the accuracy provided by the LMS algorithm is worse than that provided by other estimators.

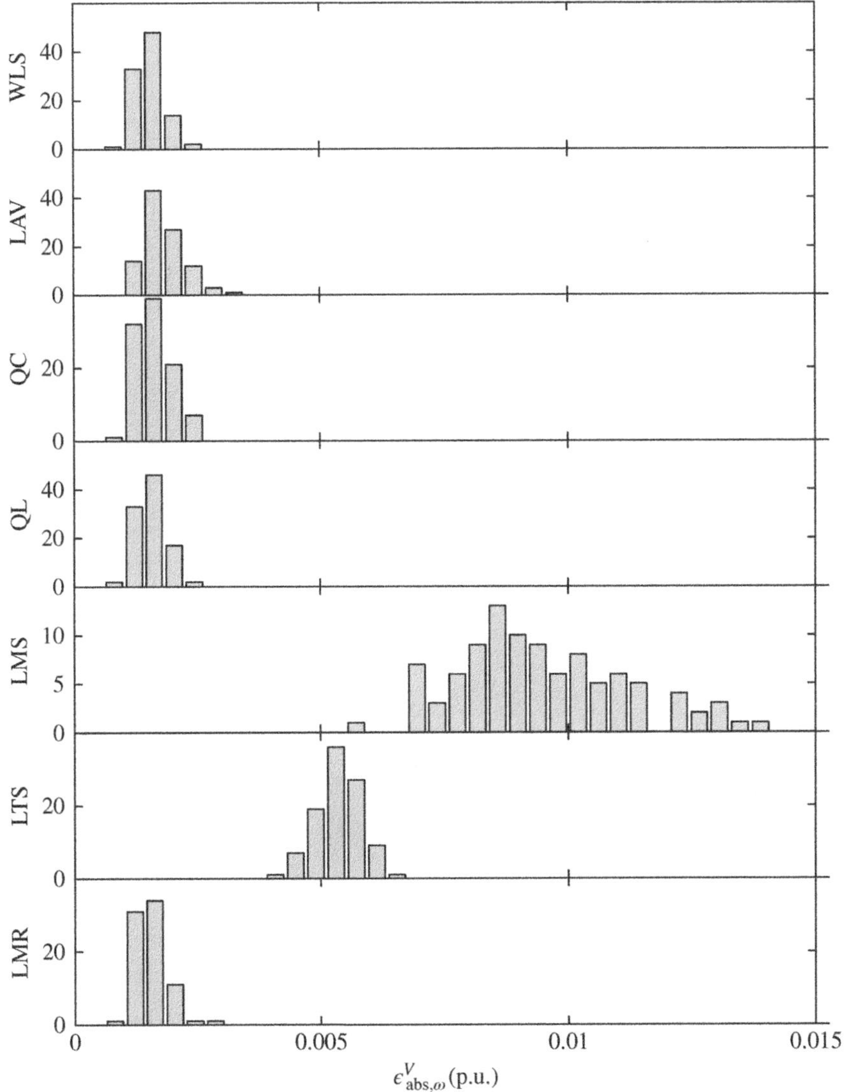

Figure 2.13 Histogram of voltage magnitude estimation accuracy for each method.

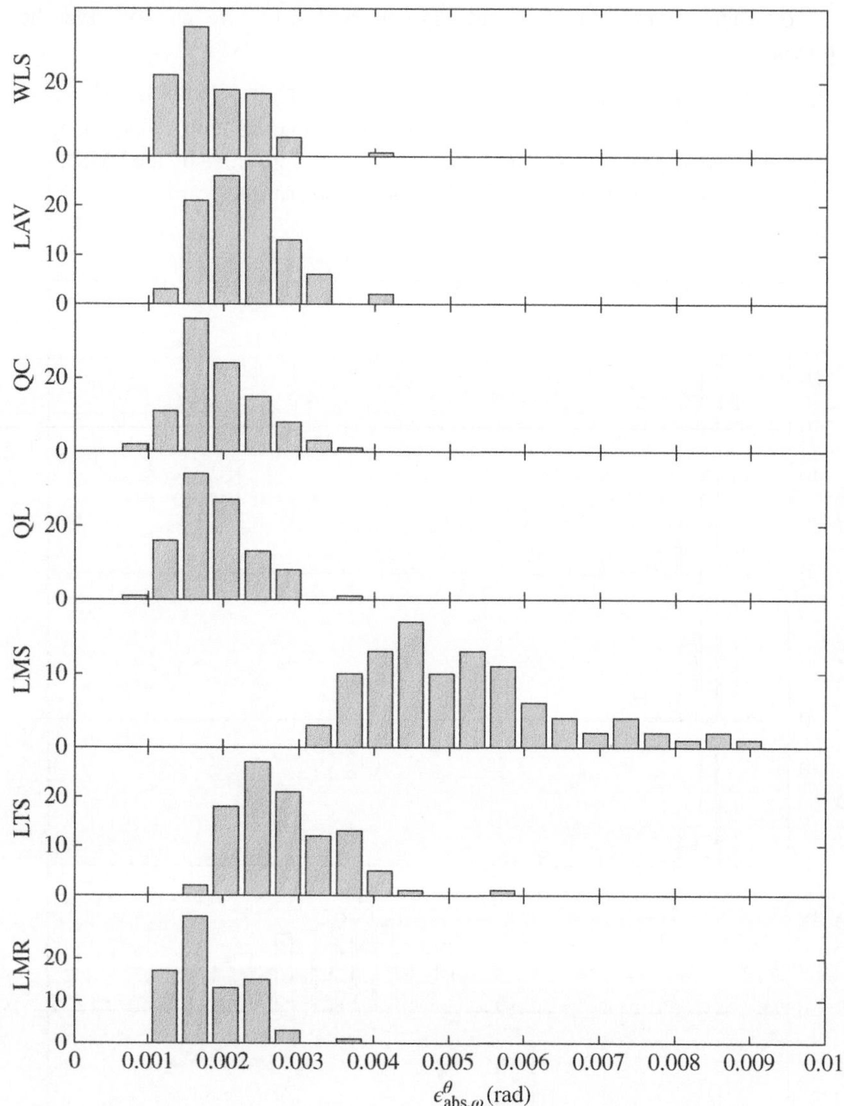

Figure 2.14 Histogram of voltage angle estimation accuracy for each method.

Note that the absolute errors are notably small for all the estimators, with the exception of the LMS and LTS approaches, whose absolute errors are five and three times higher than those of other procedures.

Note also that the accuracy variability of the LMS estimator is high for both the magnitude and the angle estimates.

2. From the computational point of view, the LAV and QC approaches are more efficient than the other procedures. Observe that these two techniques

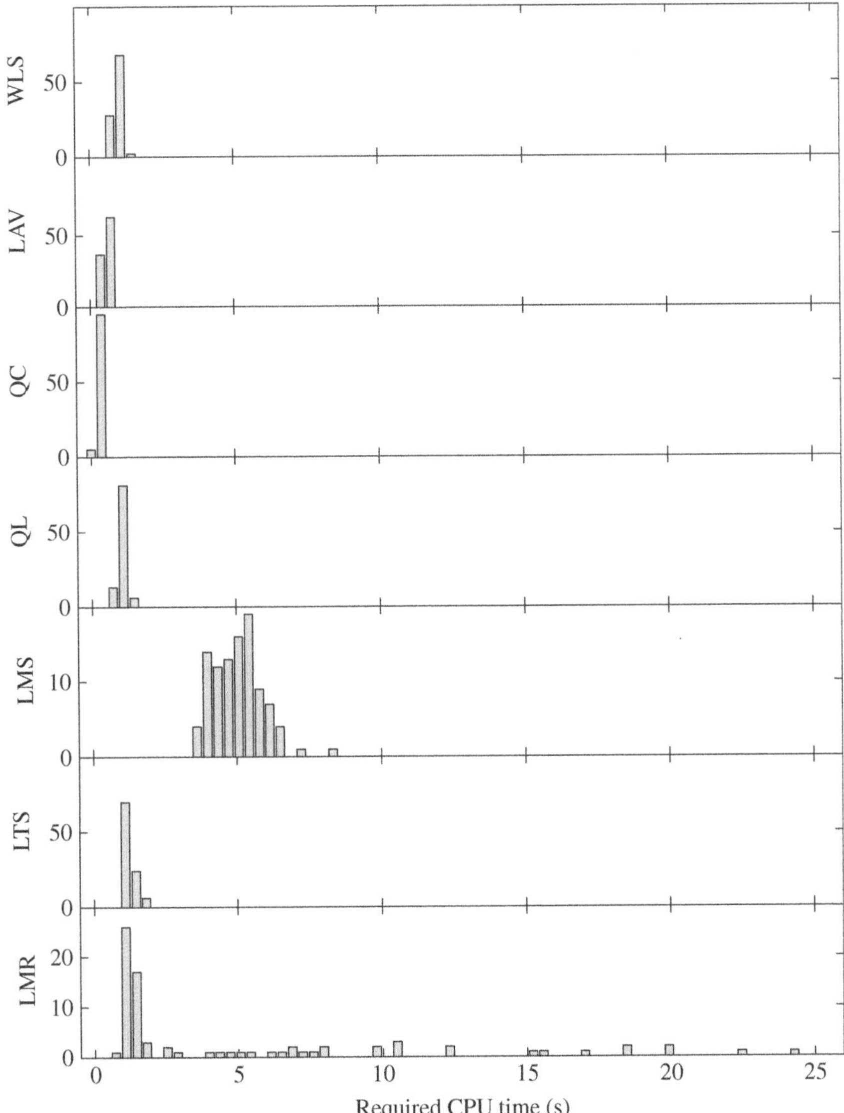

Figure 2.15 Histogram of the computation time for each estimator.

slightly outperform the conventional WLS method. However, the LMS and LTS methods require a larger amount of CPU time, owing to the non-convexities of their mathematical programming formulations and the non-differentiable objective functions.

In Figure 2.15, note that the variability of the computation time required for the LMR procedure is significantly high, but its average value coincides with that of the LMS technique.

2.5.11.1 Performance Analysis: Bad Data

All estimators presented in this chapter (with the exception of the WLS) are robust when dealing with bad data, i.e. a gross error does not significantly influence the estimated state.

If WLS estimator is used, then bad measurement detection and identification procedures must be employed after the estimation process in order to eliminate any bad data that may populate the measurement set. These detection and identification procedures are generally based on χ^2 test and on normalized residuals, respectively (see [18]).

According to the traditional bad measurement detection procedure, if the objective function value at the estimated state $J(\hat{x})$ is smaller than $\chi_\gamma^{1-\alpha}$ (where γ is the number of degrees of freedom and $1-\alpha$ is the confidence interval), the measurement set is assumed to be error-free. If not, a procedure is carried out to identify the erroneous measurements.

Therefore, if a bad measurement is corrupting the measurement set but its magnitude is not sufficiently large to satisfy $J(\hat{x}) > \chi_\gamma^{1-\alpha}$, the bad data detection process concludes, and the identification procedure is not performed. In this case, the measurement set is corrupted by a bad measurement that the WLS estimation procedure cannot detect.

In the study below, two bad measurements are present in each measurement scenario. The magnitude of the corresponding error is sufficiently small to satisfy the condition $J(\hat{x}) < \chi_\gamma^{1-\alpha}$, i.e. the bad data identification procedure does not detect any error, and the bad measurement is not therefore removed from the measurement set.

One hundred measurement scenarios have been considered, and Tables 2.16 and 2.17 provide the results concerning estimation accuracy and computational performance, respectively. Note that the format of these tables is similar to that of Tables 2.14 and 2.15.

The following observations can be made about Tables 2.16 and 2.17:

1. As expected, the WLS approach does not provide the most accurate results. The estimates computed using the QC and QL techniques are more precise than that obtained with the conventional WLS method.
2. The QC and LAV approaches are the most efficient ones from the computational perspective. The computational burden of the LMS technique is higher than that of any of the other procedures.

2.5.12 Conclusions

Considering recent advances in computational techniques, this work addresses the electric state estimation problem from a mathematical programming perspective.

In this chapter, the most common state estimators are formulated as optimization problems and implemented, proving to be computationally efficient and numerically accurate.

TABLE 2.16 Case study: estimation accuracy results with bad measurements.

Method	mean$\left(\epsilon_{\text{abs}}^{V}\right)$ (p.u.)	std$\left(\epsilon_{\text{abs}}^{V}\right)$ (p.u.)	mean$\left(\epsilon_{\text{abs}}^{\theta}\right)$ (rad)	std$\left(\epsilon_{\text{abs}}^{\theta}\right)$ (rad)
WLS	0.0019	0.0014	0.0017	0.0014
LAV	0.0019	0.0016	0.0019	0.0016
QC	0.0016	0.0012	0.0017	0.0014
QL	0.0016	0.0012	0.0017	0.0014
LMS	0.0103	0.0106	0.0053	0.0047
LTS	0.0050	0.0048	0.0025	0.0022
LMR	0.0018	0.0013	0.0017	0.0014

TABLE 2.17 Case study: computational performance results with bad measurements.

Method	Minimum (s)	Mean (s)	Maximum (s)	Std. dev. (s)
WLS	0.94	1.70	2.28	0.18
LAV	0.59	0.94	1.29	0.12
QC	0.22	0.31	0.45	0.05
QL	1.00	1.74	2.71	0.27
LMS	3.80	8.21	12.64	1.42
LTS	1.28	2.36	3.96	0.36
LMR	0.94	2.73	34.84	4.99

From the computational point of view, QC and LAV techniques perform faster than the conventional WLS estimator, saving up to 75% CPU time (compared with the WLS method directly solved as an optimization problem). On the other hand, mathematical programming formulation of some estimators (such as LMS and LTS approaches) involves non-convexities and a significant number of binary variables, resulting in higher computational burdens.

With regard to estimation accuracy, numerical simulations denote that the LMR and QL techniques provide an estimation accuracy level that is similar to that obtained using WLS method. On the other hand, if the measurement set is corrupted with errors that χ^2 test cannot detect, WLS estimation quality deteriorates, providing the worst estimation quality (if LMS and LTS approaches are not considered). It is observed that QL and QC techniques outperform the rest of estimators, followed by LMR procedure.

The computational analyses carried out in this chapter have led to the conclusion that alternative estimators are potential substitutes for traditional WLS method. The state of the art of current nonlinear optimization solvers and recent advances in computational equipments allow using robust estimators in real electric energy systems.

REFERENCES

1. Monticelli, A. (Feb. 2000). Electric power system state estimation. *Proceedings of the IEEE* 88 (2): 262–282.
2. Rousseeuw, P.J. (Dec. 1984). Least median of squares regression. *Journal of the American Statistical Association* 79 (388): 871–880.
3. Rousseeuw, P.J. and Leroy, A.M. (1987). *Robust Regression and Outlier Detection*. New York: Wiley.
4. Schweppe, F.C. (Jan. 1970). Power system static state estimation. Part III: implementation. *IEEE Transactions on Power Apparatus and Systems* 89 (1): 130–135.
5. Schweppe, F.C. and Rom, D. (Jan. 1970). Power system static state estimation. Part II: approximate model. *IEEE Transactions on Power Apparatus and Systems* 89 (1): 125–130.
6. Schweppe, F.C. and Wildes, J. (Jan. 1970). Power system static state estimation. Part I: exact model. *IEEE Transactions on Power Apparatus and Systems* 89 (1): 120–125.
7. Abur, A. and Gómez-Expósito, A. (2004). *Power System State Estimation: Theory and Implementations*. New York: Marcel Dekker.
8. Chatterjee, S. and Hadi, A.S. (2006). *Regression Analysis by Example*, 4e. New York: Wiley.
9. Conejo, A.J., de la Torre, S., and Cañas, M. (Feb. 2007). An optimization approach to multi-area state estimation. *IEEE Transactions on Power Systems* 22 (1): 213–221.
10. Drud, A. (2007). CONOPT. In: *GAMS—The Solver Manuals*. Washington, DC: GAMS Development Corporation.
11. Drud, A. (2008). *MINOS: A Solver for Large-Scale Nonlinear Optimization Problems*. Washington, DC: GAMS Development Corporation.
12. Brooke, A., Kendrick, D., Meeraus, A., and Raman, R. (1998). *GAMS: A User's Guide*. Washington, DC: GAMS Development Corporation.
13. Fourer, R., Gay, D.M., and Kernighan, B.W. (2002). *AMPL: A Modeling Language for Mathematical Programming*, 2e. New Mexico: Brooks/Cole Publishing Company.
14. Bisschop, J. and Roelofs, M. (2007). *AIMMS – The User's Guide*. Haarlem: Paragon Decision Technology B.V.
15. Caro, E., Conejo, A.J., and Mínguez, R. (2008). A mathematical programming approach to state estimation. In: *Optimization Advances in Electric Power Systems* (ed. E.D. Castronuovo), 1–26. New York: Nova Science Publishers Inc.
16. Holten, L., Gjelsvik, A., Aam, S. et al. (Nov. 1988). Comparison of different methods for state estimation. *IEEE Transactions on Power Systems* 3 (4): 1798–1806.
17. Larson, R., Tinney, W., Hadju, L., and Piercy, D. (Mar. 1970). State estimation in power systems. Part II: implementations and applications. *IEEE Transactions on Power Apparatus and Systems* 89 (3): 353–362.
18. Gómez-Expósito, A., Conejo, A.J., and Cañizares, C. (2008). *Electric Energy Systems: Analysis and Operation*. New York: CRC Press, Taylor & Francis Group.
19. Abur, A. (Aug. 1990). A bad data identification method for linear programming state estimation. *IEEE Transactions on Power Systems* 5 (3): 894–901.
20. Abur, A. and Çelik, M.K. (May 1993). Least absolute value state estimation with equality and in-equality constraints. *IEEE Transactions on Power Systems* 8 (2): 680–686.
21. Çelic, M.K. and Abur, A. (Feb. 1992). A robust WLAV state estimator using transformations. *IEEE Transactions on Power Systems* 7 (1): 106–113.

22. Jabr, R.A. and Pal, B.C. (Nov. 2003). Iteratively re-weighted least absolute value method for state estimation. *IEE Proceedings – Generation, Transmission and Distribution* 150 (4): 385–391.
23. Baldick, R., Clements, K.A., Pinjo-Dzigal, Z., and Davis, P.W. (Feb. 1997). Implementing non-quadratic objective functions for state estimation and bad data rejection. *IEEE Transactions on Power Systems* 12 (1): 376–382.
24. Zhuang, F. and Balasubramanian, R. (Jul. 1985). Bad data suppression in power system state estimation with a variable quadratic-constant criterion. *IEEE Transactions on Power Apparatus and Systems* 104 (4): 857–863.
25. Irving, M.R. (Aug. 2008). A tutorial introduction to robust estimators with mathematical programming solutions. Technical report, Brunel University.
26. Mili, L., Phaniraj, V., and Rousseeuw, P. (May 1991). Least median of squares estimation in power systems. *IEEE Transactions on Power Systems* 6 (2): 511–523.
27. Mili, L., Cheniae, M.G., and Rousseeuw, P.J. (May 1994). Robust state estimation of electric power systems. *IEEE Transactions on Circuits and Systems* 41 (5): 348–358.
28. Irving, M.R. (Aug. 2008). Robust state estimation using mixed integer programming. *IEEE Transactions on Power Systems* 23 (3): 1519–1520.
29. Irving, M.R. (Nov. 2009). Robust algorithm for generalized state estimation. *IEEE Transactions on Power Systems* 24 (4): 1886–1887.
30. The GAMS Development Corporation Website. (2011). http://www.gams.com (accessed September 2020).
31. Bussieck, M.R. and Drud, A.S. *SBB: A New Solver for Mixed Integer Nonlinear Programming*. Washington, DC: GAMS Development Corporation. https://www.gams.com/archives/presentations/present_sbb.pdf (accessed September 2020).

PART I

SYSTEM FAILURE MITIGATION

CHAPTER 3

SYSTEM STRESS AND CASCADING BLACKOUTS

Hyde M. Merrill[1] *and James W. Feltes*[2]

[1]*University of Utah and Merrill Energy, LLC, Salt Lake City, UT, USA*
[2]*Siemens Power Technologies International, Schenectady, NY, USA*

3.1 INTRODUCTION

The editor of this book was an astute engineer with in-depth understanding of state estimation. Why would he request a contribution from two engineers whose direct involvement, if any, with the topic was limited and occurred decades ago? The answer requires a brief story.

An iconic cascading blackout occurred in the northeastern United States and part of Canada in 1965. At that time Dr. Fred C. Schweppe, a brilliant and successful aerospace engineer, was looking for a change. He decided to get into electric power. He recognized that there was little communication between the aerospace and electric power technical communities and asked himself what aerospace tools he might bring with him. He took leave from his job at an aerospace lab, got a faculty appointment at MIT, took leave from that, and got a temporary job at American Electric Power, with the purpose of learning power. At AEP he fell in with an unusually astute and knowledgeable Cuban power engineer, Jorge F. Dopazo. Dopazo educated Schweppe in power, and Schweppe taught Dopazo the basics of state estimation, which Dopazo used to build one of the industry's first practical state estimators.[1]

[1] Schweppe and Dopazo were elected IEEE Fellows, in the same class, for their work on state estimation. One of the authors (HMM) did a dissertation on state estimation under Schweppe and after graduating worked for eight years under Dopazo and with Dr. A Mayer Sasson, another outstanding AEP engineer, on state estimation and other topics. No young engineer ever had better mentors!

Advances in Electric Power and Energy: Static State Estimation, First Edition.
Edited by Mohamed E. El-Hawary.

Published 2021 by John Wiley & Sons, Inc.

Informally, the state of a system is a minimal set of data that tells everything we need to know about the system. Schweppe defined the state of the bulk power system as the steady-state voltage magnitude and angle at each bus. Having estimated the state accurately from error-prone real-time measurements and knowing the topography of the system, the flows in every line and injections at every bus are readily calculated using Ohm's and Kirchhoff's laws, and dynamic analyses can also be done. Schweppe viewed his contribution to solving the cascading blackout problem as providing an accurate system state that others could rely on in developing and applying preventive and corrective actions.

In 1965 no one recognized how difficult the cascading blackout problem was. Schweppe, who was usually right about things, wrote an important 1978 paper on the future of the power system. He spent about 1/6 of his paper on blackouts and concluded, "[Cascading] blackouts will not exist in the year 2000.... There is a good chance that by the year 2000, the term [cascading blackout] will be considered to be a term out of the Dark Ages [1]." (He did not object to a good pun.)

This chapter describes an approach to measuring power system stress, the susceptibility of electric power systems to cascading blackouts. It has application in planning as well as in real time. For operations, it depends, as Schweppe expected, on data from a state estimator. But the additional computations go far beyond simply calculating flows and injections using Ohm's and Kirchhoff's laws. Perhaps the editor viewed stress as sufficiently different that it might be considered an additional element of the state of the power system. Perhaps this is why he invited this chapter. While this interpretation may be valid, we do not insist on it.

The context today, since 1965, large interconnected systems have experienced a new class of disruptions – wide-area cascading blackouts. These blackouts can cause billions of dollars in customer economic impacts as well as collateral damage. The industry continues to look for new methods of assessment and prevention, but in spite of great efforts, blackouts keep occurring.

This chapter draws on two bodies of knowledge – electric power engineering and network (graph) theory – to develop and apply a new failure network, an application of line outage distribution factors. This network represents how failures propagate from branch to branch for analyzing susceptibility to cascading blackouts. The properties of this network are studied in the context of the three elements of the power system and with metrics based on well-known data and tools.

3.2 CASCADING BLACKOUTS AND PREVIOUS WORK

3.2.1 Cascading Blackouts

A cascading blackout is an uncontrolled, unexpected chain of cause-and-effect events that interrupts bulk power service over a large area. Cascading blackouts are of concern because of the following:

- A blackout can cause billions of dollars of damage.
- It can take hours or days to restore the system.
- Vital health, police, defense, etc. functions are degraded.

3.2.2 Typical Events

Let us review briefly four examples of cascading blackouts. Typically, in a cascading blackout a branch is taken out of service by protection equipment. The flow this first branch was carrying is automatically and immediately redistributed to other branches. This sometimes increases their loading, depending on the directions of pre-contingency flows in the various branches. Even if they are not overloaded, increased flows increase the likelihood of their tripping, due to "hidden failures."[2] It is significant that the sequences of failures are not independent events. They are a chain of causes and effects. They are random because not every possible cause results in all possible effects, depending on whether hidden failures are present.

The 9 November 1965 Northeast blackout occurred when there were heavy flows from the Niagara Falls area to Toronto, Canada. Relays were set too low on five essentially parallel lines; the system operators did not know it. As a result, a line opened needlessly that caused first one other and then the last three lines to overload and open. The Ontario and New York systems formed islands. Ontario had too little generation, New York too much. They could not absorb the sudden imbalances and blacked out [2, 3].

The US–Canada blackout of 14 August 2003 occurred with high flows in Ohio and nearby areas. Three 345-kV lines failed between 15:05 and 15:46 EDT. Though not overloaded, they sagged into trees that were taller than they should have been. The economic effect in the United States was estimated at $4–$10 billion. Canada's GNP was measurably affected.

Precursor events to the 14 August 2003 blackout included procedural failures in two control centers and computer failures, affecting among other things a state estimator. Precursor outages weakened the system and increased stress. Notwithstanding these, the very thorough postmortem concluded that at "15:05 EDT ... the system was electrically secure ... Determining that the system was in a reliable operational state at 15:05 ... is extremely significant for determining the causes of the blackout" [3].

But in just 41 minutes, at 15:46, "the blackout might have been averted [but] it may already have been too late ... to make any difference" [3]. The final straw, loss of a fourth 345-kV line at 16:06, was a relay misinterpreting high current and low voltage as a short circuit and tripping its line.

[2] The term "hidden failures" apparently was coined to apply to failures in the protection system that are not manifested until the system is stressed, where they may be important contributors to cascading. The problem with the relays in the 1965 blackout is a classic example. Control systems, as well as practices and procedures, are subject to similar failures and have contributed to cascading failures. The term "hidden failures" is so felicitous that we will extend it to embrace these other failures as well.

"[D]etermining that the system was in a reliable operational state at 15:05" is extremely significant for showing that present methods of measuring susceptibility to cascading are inadequate. The postmortem seemed to recognize this, saying, "Although FirstEnergy's system was technically [i.e. by present $n - 1$ standards] in secure electrical condition before 15:05 EDT, it was still highly vulnerable" [3].

That is, the system satisfied the $n - 1$ criteria but was still highly vulnerable – what this chapter calls stressed. The metrics introduced below are intended to measure this stress.

When the 28 September 2003 outage in Italy began, Italy was importing about 6400 MW, mainly from France through Switzerland, in part to fill Italy's pumped storage plants. This was 300 MW more than the contracted amount. At 3:01 a.m. on Sunday morning, the Swiss 380-kV Mettlen–Lavorgo line sagged into a tree, flashed over, tripped, and could not be returned to service. "The appropriate countermeasure for the loss of the line was the shutting down of the pumps in the pump storage plants in Italy.... The pumping load in Italy amounted to about 3,500 MW. Shutting down the pumps in mutual support ... is operational practice [4]."

Instead of requesting this, the Swiss operator asked the Italians to reduce their imports by 300 MW to the contractual amount, which now was more than the interface could handle without the tripped line. At 3:25 a.m. the Sils–Soazza 380-kV line (which had picked up some of the load from the first outage) sagged into a tree and tripped. Other cascading outages followed immediately. Within 12 seconds, Italy was isolated from the rest of Europe, with about 6100 MW more demand than generation. The resulting dynamic power and frequency swings blacked out Italy about 2:30 minutes later [5]. The rest of the European system was much larger and withstood its own dynamic swings without serious effects.

The committee that reviewed the blackout found that "The system was complying with the N-1 rule at this time, with ETRANS [the Swiss coordination center] taking into account countermeasures available outside Switzerland" [4], specifically, shutting down the pumps in Italy.

Both the temperature and system loading in the southwestern part of the Western Interconnection (WI) of North America on 8 September 2011 were very high for what was supposed to be a "shoulder" month. An experienced technician at the North Gila substation of the Arizona Public Service Company was multitasking. He was interrupted after performing Task 6 of a complicated switching procedure but wrote the completion time on his schedule on the line for Task 8. When he returned to switching, he performed Task 9, inadvertently skipping Tasks 7 and 8. The resulting arcing and currents caused the Hassayampa–North Gila 500-kV line to trip. This caused the two Imperial Irrigation District (IID) Coachella 230/92-kV transformers to load to 130% of normal rating, and they both tripped 38 seconds later. Less than five minutes after the Hassayampa–North Gila line trip, as a consequence of that event and the Coachella trips, IID's Ramon 230/92-kV transformer overloaded and tripped. Other lines and devices also were tripping, and the cascading was well underway [6].

Cascading blackouts that might have happened but did not also are instructive. For instance, on 11 April 1965, five months before the famous 1965 blackout,

37 "Palm Sunday" tornados in Ohio, Michigan, and Indiana destroyed 27 transmission lines and two substations of the American Electric Power Company. Customers served from failed radial line lost power, but there was no cascading blackout [7]. It was Sunday of an off-peak month. The system was lightly loaded – unstressed by any definition. (It was not designed to survive 29 contingencies.)

3.2.3 Prior Work on Blackouts

Extensive work has been done on analyzing cascading blackouts, seeking ways to eliminate them or at least to reduce their frequency or extent and speed recovery. We will not pretend to be comprehensive but will mention highlights.

Many blackouts have been subject to postmortem analyses. One of the largest blackouts as of this writing was described in an illuminating three-volume report [3]. This report also has useful descriptions of a number of earlier events.

The US–Canadian power industry organized the NERC (now North American Electric Reliability Corp.) after the 1965 blackout to improve reliability, notably by producing criteria and collecting data. Its transmission planning criteria were based on previous criteria [2] and have evolved but have been reasonably stable for decades. Major changes in form and substance were made recently [8]. Preventing cascading blackouts always has been central. These efforts have undoubtedly helped, but the blackouts continue.

A useful state model was developed for treating power system security and reliability [9].

State estimation was introduced to provide accurate inputs to real-time procedures for increasing reliability [10]. This was arguably the single most significant and most radical post-1965 technical innovation. Within 10 years of Schweppe's first paper on the topic, a state estimator was commonly in the specs for new power system control centers. The authors know of no other power system innovation that has been accepted so rapidly, except possibly Tinney's work on sparsity.

Much labor has been invested in other efforts to analyze and solve the blackout problem. In particular, network theory has been applied to a variety of problems in power, including attempting to determine contributions of each generator to each load [11] and generating random networks for studying widespread failures [12].

Applications of network theory to cascading blackouts always seem to use models where buses are vertices and lines are edges [13–15]. They consider the probability of a random failure at one bus propagating to a neighboring bus. They are often based on abstract networks rather than real power systems. They do not seem to use Ohm's and Kirchhoff's laws to model the mechanisms for percolation or propagation of failures.

Two papers present work complementary to the approach described in this chapter [16, 17]. They identify and simulate problems that affect cascading: network loading, spinning reserve, controls, and "hidden failures," a term which as noted earlier generally refers to failures in the protection system, including relaying. The latter are not manifested until the system is stressed by a contingency. As do others, these papers importantly assume that the probability of cascading due to hidden failures (Pr) increases from Pr = 0 below a threshold loading (for instance, 100% of rating) to Pr = 1 above a second threshold (for instance, 140% of rating). However, cascading events have begun without overloading.

A team from many entities analyzed $n-k$ conditions on a three-area part of the 50 000-bus and 65 000-branch model of the Eastern Interconnection (EI) of North America. But analyzing all $n-k$ conditions is impossible. Many (~31 000) $n-2$ outage combinations were tested from a user-supplied list of contingencies, but this still represented a small fraction of the total potential $n-2$ combinations. Most $n-2$ events ended uneventfully after overloaded elements tripped. Some 38 led to voltage collapse or islanding, but none took more than three steps to do so [18]. In cascading blackouts, the consecutive outages are not independent; each is a consequence of its predecessors. The dangerous combinations are a small fraction of the total $n-2$ combinations.

What is wrong with the efforts described immediately above? Nothing, as far as they go. Importantly, they question the usual assumption that control and protective devices and manual system adjustments will work right and the loading at which a line trips. The first paper [16] prescribes better relaying, spinning reserve, and control. No surprise here, but perfection is not possible. The third [18] suggests planning or operating to $n-k$. But in real blackouts, $n-0$ or $n-1$ plus hidden failures is more common than $n-k$ alone. All three of the above papers use assumed probabilities of failures and cascading. But knowing true probabilities might not change their conclusions.

The leap that the efforts summarized above do not quite make is to say: "When the system is stressed, it is more susceptible to cascading due to a variety of failures (most of which we can't model). Therefore, let's focus on modeling and managing stress." We will return to this idea in the next section.

3.3 PROBLEM STATEMENT AND APPROACH

3.3.1 Diagnosis of the Cascading Blackout Problem [19]

The authors have studied many cascading blackouts and non-blackouts around the world. All share the characteristics illustrated above and summarized below.

First, the electric power system is a man–machine energy conversion system of three major elements:

1. Current-generating and current-carrying hardware
2. Control and protective devices and systems
3. Practices and procedures

Second, cascading blackouts always involve overloads or failures of current-carrying hardware. But these usually (perhaps always) occur because:

- Blackouts are triggered by failures of one of the other two elements of the power system.

- Cascading follows only if the system is stressed, usually by high interregional transfers.
- Cascading blackouts are phenomena of large interconnections.

These three conditions are based on reviewing many blackouts. They are not proven by math or physics. They are easily dismissed as old news. But their importance is not fully appreciated or reflected in planning and operations.

For instance, stress is not limited to peak demand states. The Sunday 28 - September 2003 blackout of Italy began at 3:00 a.m. – off-peak hour, day, and month – but with the system stressed by high imports from France for supplying pumped storage [4]. The metrics and studies described below model stress whether off- or on-peak.

Modeling stress is the key contribution of this chapter. The industry does not have a good definition of stress. Stress can occur off-peak. Measuring stress – with respect to cascading – is all that this chapter is about.

Back to the three key conditions above, NERC planning standards generally assume that control and protective devices and systems (including remedial action schemes [RAS]) and policies and procedures (including operator corrective actions) will work properly. Exceptions: Stuck breaker and relay failures are listed, but as "Multiple Contingency" conditions (Category P3 though and P7, formerly Category C) *that systems generally were not required to withstand without loss of load* [8]. Most network upgrades and real-time operator studies have been for $n - 1$ conditions (Category B). (The recent major restructuring of the NERC conditions is more complicated. For instance, some P2 $n - 1$ events allow load to be lost, and some P3–P7 multiple outage events do not.)

In contrast, a NERC study found that 73.5% of significant disturbances were aggravated by protection system "hidden failures"[3] [16, 17]. Operator actions (including "situation awareness") failed dramatically on 14 August 2003 and other blackouts.

We observe that it is not just one failure in control and protective devices and systems, and in practices and procedures, that triggers a cascading failure. It is truly remarkable how many such failures are identified in the reports of the blackouts of 14 August 2003 and 8 September 2011. Even the relatively simple 1965 Northeast and 2003 Italy blackouts each included at least three such failures. (We say "simple" because in each of these two blackouts, two major systems were linked

[3] Medicine provides a problem that is not totally unlike ours. A recent paper claims that medical error is the third leading cause of death in the United States [20]. It notes that "medical error" is not an option for "cause of death" on death certificates. One commenter said that the "data shows that the vast majority of errors are caused by failures in the system." Another observed that the death rate declined in California during a two-week stoppage when only emergency medical care was available. Yet another revealed that death is inevitable. (One trusts that cascading blackouts are not.) Medical errors are much more critical when the patient is very sick ("stressed"?). One commenter dismissed all this as "junk science"; others also were skeptical. (Is it reasonable to expect the medical system to function perfectly? How about the power system? Is it reasonable to strive to do better?)

principally by several essentially parallel lines. The topology of the US–Canada blackout of 14 August 2003 was much more complicated.)

It is important that these failures generally have been accidental, but sabotage and cyberwar could create the same effects as hidden or other accidental failures.

Half a century ago a brilliant electrical engineer said, "Power systems have grown enormously and have become interconnected over vast regions. And we have had two severe blackouts and are undoubtedly headed for more." He also observed, "The more complex a society, the more chance there is that it will get fouled up [21]."

Illustrating his point, the large eastern and western North American interconnections have had cascading blackouts; the smaller ERCOT system in Texas has not. To be more precise, the latter is not synchronously connected to the others. A representative of ERCOT told the authors in 2008 that it had never had a cascading blackout. It had had voltage collapse, which can be a precursor to cascading, in the Bryan/College Station area in April 2003 and October 2006 (Drew, R.H. (25 Jan. 2008). Electric Reliability Council of Texas, Inc., Personal communication). The system was in extremis during the Christmas seasons of 1983 and 1989 and on 17 April 2006 and applied its Emergency Electric Curtailment Program, including controlled rotating load shedding, but again without cascading [22, 23].

3.3.2 An Approach to Blackouts: Focus on Stress

We have stated the problem – and the problem statement points to a solution. History, reviewed above, shows that cascading blackouts occur only when three conditions are met. (i) The system is stressed, especially by high interregional power transfers; (ii) there is a failure in the second or third elements, and (iii) the system is large.

- There are many, many ways how control and protective devices and systems and practices and procedures can fail:
  - Too many to analyze, individually and in interaction with each other
  - Too many to identify and fix
  - Increasing reliance on increasingly complex controls and protection
  - It is unreasonable to expect to design and operate the system perfectly.
- But these failures only trigger blackouts when a large system is stressed.
- So let us work on stress:
  - Let us figure out how to measure stress (the topic of this chapter).
  - Let us figure out how to reduce stress (the topic of future work).

Unfortunately, there is no such thing as a strain gauge for measuring the stress of the power system. In fact, stress is a complex, multidimensional concept. It is a function of the system state, including the network topology and parameters and the operating state (demand, generation, and status of the network elements).

We define stress as the susceptibility of the power network to cascading failures. The strain gauge analogy has been introduced. Another analogy, more apt for a large complex system, is the risk of an epidemic propagating through a network. We will first define the network and then define metrics for its susceptibility to cascading.

These metrics are applicable to real-time operations (dispatch), short-term operations planning (where options include switching, unit commitment, and redispatch), and long-term planning (where system reinforcements are considered).

3.3.3 Cascading Failure Networks

Figure 3.1 is a general representation of a network. It applies to both conventional models of the power network and to a model of cascading failures but in radically different ways.

In the traditional Ybus network, analyzed using the **Ybus** matrix, the vertices are buses, and the edges are branches (lines and transformers). This network is fundamental to all major power system analysis, including "the big three": power flow, stability, and short circuit. It measures how current and power flow along the branches from bus to bus and generators to loads.

The Ybus network is clumsy in measuring how cascading failures propagate, like an epidemic, generally from branch to branch rather than from bus to bus. For this we need a different network with different properties. A network that is useful for measuring stress is defined by vertices that are branches (subject to failure) with edges that measure how a failure in one branch affects others.

The well-known line outage distribution factors, sometimes called DFAX or DFAXes, define such a network. It is important that this network, in particular the values of the line outage distribution factors, is based on Ohm's law and Kirchhoff's laws. Much work on hidden failures is based on abstract networks that ignore the basic network physics.

3.3.4 Stress Metrics

Ybus networks measure admittance and impedance. The failure network (hereafter "the DFAX network") measures stress with two fundamental metrics: vulnerability and criticality. Each of these metrics is expressed as rank and degree.

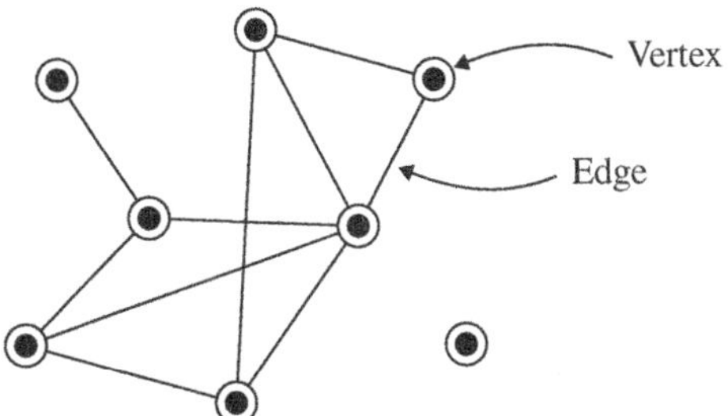

Figure 3.1 A generic network [24]. Source: Copyright ©2003 Society for Industrial and Applied Mathematics. Reprinted with permission. All rights reserved.

The terms “rank” and “degree” are from modern network (graph) theory. This theory is largely a twentieth century mathematical development, responding to two observations: Networks are everywhere, and they can be very large (and “big is different”). Since they are too big to understand, element by element, their properties must be measured by statistics.

In the next section we define rank and degree a bit differently than mathematicians do in order to better capture the specific physical realities of cascading in the power system. The term “metrics” is used instead of “statistics” to avoid giving the false impression that the results are to be viewed as probabilities.

Cascading blackouts are invariably due to low-probability events, so the law of large numbers does not apply. Furthermore, the probabilities are unknown. For these two reasons a probabilistic analysis would not be expected to reflect reality. Traditional transmission analysis recognizes this: the NERC criteria consider contingencies and the results thereof without attempting to assign probabilities to them. Some applications of graph theory to the cascading blackouts problem, as discussed earlier, are fundamentally probabilistic, where the probabilities are made up.

3.4 DFAXes, VULNERABILITY, AND CRITICALITY METRICS

DFAXes and the various metrics are illustrated using a nine-branch subsystem of the IEEE 118-bus test system. See Figure 3.2. The rest of the 118-bus system is below and to the right of this piece. The calculations we will present recognize the entire 118-bus system – specifically, pre- and post-contingency flows through the entire system.

3.4.1 DFAX Matrix

The traditional Ybus and DFAX networks are quite different.

For networks of reasonable size, **Ybus**, the basis of power flow and other network analysis programs, is sparse: the matrix only has nonzero admittance terms when two buses are connected by a branch. The number of nonzero terms is so small for systems of practical size that sparsity techniques are used to reduce dramatically the memory and the execution time for power flow computations. In addition, **Ybus** is symmetrical and square, $n \times n$, where n is the number of buses. See Table 3.1, where only the nonzero terms are shown.[4] Because the nine-branch

[4] For simplicity, we have included in the diagonal terms only the admittances for the branches connecting buses within the nine-branch subsystem. The rows and columns for the branches interconnecting the subsystem to the rest of the system will have a few other nonzero terms beyond the rows and column shown, and the diagonal terms for these branches will incorporate the admittances of these interconnecting branches.

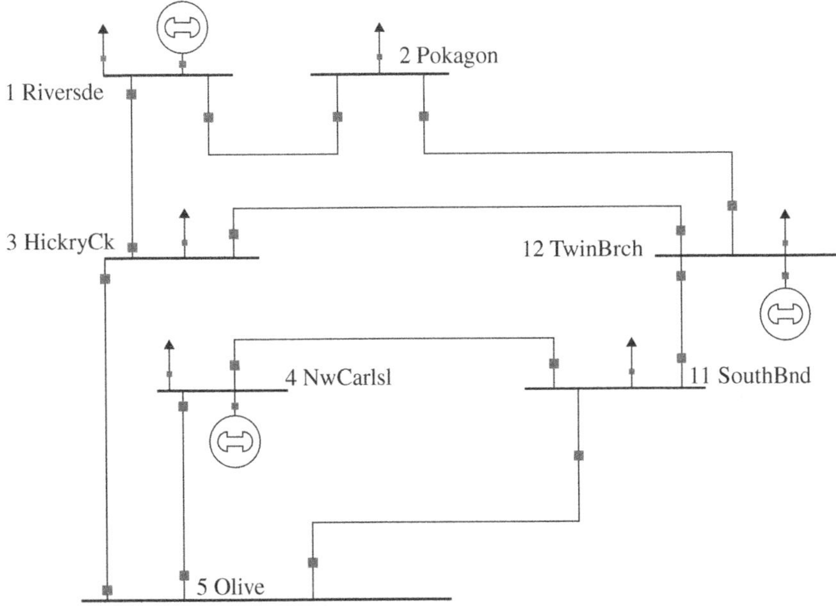

Figure 3.2 Nine-branch subsystem of 118-bus IEEE test system.

TABLE 3.1 Ybus structure for the nine-branch subsystem.

bus	1	2	3	4	5	11	12
1	Z-a-b	a	b				
2	a	Y-a-i					i
3	b		X-b-c-h		c		h
4				W-d-e	d	e	
5			c	d	V-c-d-f	f	
11				e	f	U-e-f-g	g
12		i	h			g	T-h-g-i

system is so small, its **Ybus** is much less sparse than for a practical system. Larger systems have a much higher percentage of zero terms.

Ybus is a Jacobian matrix, whose elements are d**I**/d**V**, where **I** and **V** are the current injections and voltages at each bus. (**Ybus**, **V**, and **I** are complex numbers.

A "dc power flow" approximation with the same structure relates real power injections and bus voltage angles. It is useful in many applications.) Power flow software uses **Ybus** to calculate **V** at each bus, taking as inputs the load and generation (power) injections at each bus. With **V** known, the current and power flows on each line can be calculated using Ohm's and Kirchhoff's laws.

The **DFAX** matrix is also a Jacobian matrix, **Δfm/Δfo**, where **fm** and **fo** are real power flows on measured and outaged branches. The matrix explicitly incorporates changes in the network, since the change in each outaged branch is caused by the outage of that branch. Since the post-contingency flow on the outaged branch is 0, **Δfo** is simply the vector of pre-contingency flows in each outaged branch, or **fo**($n-0$). For each monitored branch i and each outaged branch j, by definition

$$\text{DFAX}_{i,j} = \frac{fm_i(n-1) - fm_i(n-0)}{fo_j(n-0)} \tag{3.1}$$

The $n-1$ flow on each monitored line is calculated using (3.2):

$$fm_i(n-1) = fm_i(n-0) + \text{DFAX}_{i,j} \times fo_j(n-0) \tag{3.2}$$

These equations look more daunting than what they represent. The subscript i refers to a particular outaged branch, or row in Table 3.2. The subscript j points to a particular monitored branch, or column in Table 3.2. So for each possible outage (row) of Table 3.2, there is a DFAX which describes its effect on the flow on each monitored branch (column).

The **DFAX** matrix is full: an outage of any of the meshed branches affects the flows in all other meshed branches except in special cases. Radial branches are not monitored or outaged. The DFAXes are very small, essentially zero, for branches that are electrically far apart. This means that sparsity techniques can be used to reduce computation time and use of memory. **DFAX** is not symmetrical, nor is it necessarily square. This is because all monitored vertices are not necessarily outaged. The number of branches is greater than the number of buses. The **DFAX** matrix is computed by commercial power flow programs.

The astute reader will recognize that the DFAXes, like the DC power flow, assume that the transmission network is linear. It is not, but for many purposes, such as ours, the linear approximation is adequate. We will address this issue again

TABLE 3.2 DFAX matrix and pre-contingency flows for nine-branch subsystem.

			Monitored branches								
		From	Riversde 1	Riversde 1	Pokagon 2	HickryCk 3	HickryCk 3	NwCarlsl 4	NwCarlsl 4	Olive 5	SouthBnd 11
		To	Pokagon 2	HickryCk 3	TwinBrch 12	Olive 5	TwinBrch 12	Olive 5	SouthBnd 11	SouthBnd 11	TwinBrch 12
Outaged branches		n–0	–13.3	–40.7	–34.4	–71.3	–10.6	–107.5	66.5	80.0	35.7
From bus	To bus	flows	DFAXes								
Riversde 1	Pokagon 2	–13.3	–1.000	1.000	–1.000	0.534	0.467	–0.143	0.143	0.161	0.310
Riversde 1	HickryCk 3	–40.7	1.000	–1.000	1.000	–0.534	–0.467	0.143	–0.143	–0.161	–0.310
Pokagon 2	TwinBrch 12	–34.4	–1.000	1.000	–1.000	0.534	0.467	–0.143	0.143	0.161	0.310
HickryCk 3	Olive 5	–71.3	0.440	–0.440	0.440	–1.000	0.560	0.267	–0.267	–0.301	–0.582
HickryCk 3	TwinBrch 12	–10.6	0.407	–0.407	0.407	0.593	–1.000	–0.159	0.159	0.178	0.345
NwCarlsl 4	Olive 5	–107.5	–0.057	0.057	–0.057	0.129	–0.072	–1.000	1.000	–0.502	0.442
NwCarlsl 4	SouthBnd 11	66.5	0.057	–0.057	0.057	–0.129	0.072	1.000	–1.000	0.502	–0.442
Olive 5	SouthBnd 11	80.0	0.060	–0.060	0.060	–0.137	0.077	–0.472	0.472	–1.000	–0.468
SouthBnd 11	TwinBrch 12	35.7	0.113	–0.113	0.113	–0.257	0.144	0.404	–0.404	–0.455	–1.000

later. For a linear network, the network elements are passive, and the DFAXes are in the range −1.0 to +1.0 (expressed often as −100 to +100%, sometimes with the % symbol suppressed). Some network elements are active – e.g. phase-shifting transformers, FACTS devices, and DC lines. Some power flow programs take the nonlinear effects of these active elements into account when calculating the DFAXes. This is legitimate because we are modeling finite changes with Δs, not infinitesimal changes with derivatives. With active elements in the network, some DFAXes may be <-100 or $>+100$.

In Table 3.2, if branch 1-3 has an outage, then 100% of its real (MW) pre-contingency flows are transferred to monitored branches 1-2 and 2-12, making the post-contingency flows in those branches −54.0 and −75.1 MW, respectively. The signs of the DFAX indicate whether the transferred flows increase or decrease the flows in the monitored lines. The positive direction of flow in each branch is from the "from" bus to the "to" bus. Similarly, −53.4% of pre-contingency flows of branch 1-3 are added to the 3–5 line, changing its flows after the outage of line 1–3 to $-40.7 \times (-0.534) + (-71.3) = -49.6$ MW.

The reader will note that it is important to keep the signs straight! The following are worth thinking about for understanding the relationships between **Ybus** and **DFAX**.

What percent of the elements of Ybus for the nine-branch subsystem system is nonzero?

For a 13 000-bus system with the same ratio of branches to buses as the 9-branch system, the percent of nonzero elements of Ybus would be tiny. How tiny?

What programming implications does this have?

Should the row and column of Ybus for SouthBnd include terms for the line from SouthBnd to the unseen part of the system to the East?

The DFAXes for the effect of the outage of line 1–3 on the flows on lines 1–2 and 2–12 are equal to 1.0. This means that 100% of the pre-contingency flow on line 1–2 is distributed to *both* lines 1–2 and 2–12. How can this be? Is this a violation of Kirchhoff's current law?

What does this say about the cascading of failures?

Is this in any way similar to a forest fire or to the spread of an epidemic?

The sum of the DFAXes for an outage does not seem to be 1.0. Is this troublesome? Is this a violation of Kirchhoff's current law?

If the response of all monitored branches to a contingency is more or less than 1.0, have we created or lost some MWs somehow?

Can a non-Kirchhovian network be real and meaningful? Does it look like a perpetual motion machine?

Suppose one system's **DFAX** matrix has a higher percentage of large terms than another's. Does this indicate that the first system is more tightly coupled and therefore more at risk to cascading than the other, all else being equal?

What key information is missing from a **DFAX** matrix, for instance, the one in Table 3.2, which would give a stronger indication of the stress of the system, or of its risk of cascading failure?

3.4.2 Vulnerability and Criticality

The DFAX network by itself is an interesting but limited indicator of stress or risk of cascading. It is limited because it does not contain information about the loading of the network. A prerequisite for cascading is that the network be highly loaded or stressed.

Vulnerability and criticality are complementary measures of stress. Each of these is quantified in two metrics, rank and degree. These are computed for a loaded network – with actual flows from a state estimator for real-time purposes or with $n - 0$ (pre-contingency) flows from a base case for planning.

Vulnerability measures one monitored branch's exposure to the outage of any other branch. More precisely, it measures the maximum post-contingency loading for "$n - 1$" contingencies of all other branches, taken one at a time.

Criticality measures the effect of an outage of one branch on all other (monitored) branches. More precisely, it is the maximum $n - 1$ loading due to a particular contingency over all other (monitored) branches.

A mnemonic helps keep vulnerability and criticality straight. *V*ulnerability has to do with a *v*ictim. *C*riticality has to do with a *c*ulprit.

Vulnerability and criticality are measures of the stress of a system, specifically of its susceptibility to cascading. For instance, if a monitored branch has high vulnerability, then some contingency(ies) elsewhere will cause it to become highly loaded or overloaded. The higher its $n - 1$ loading, the more likely it is to be taken out of service after a contingency by correct or incorrect operation of control or protective devices or practices and procedures, the grid thus cascading $n - 1 - 1$ to $n - 2$ state.

Similarly, if a branch has high criticality, then its outage would cause one or more monitored branches to become highly loaded or overloaded. The higher the loading on these monitored branches, the more likely for one or more of them to be taken out of service by correct or incorrect operation of the control or protection systems or practices and procedures, the grid thus entering an $n - 2$ or $n - k$ state.

- How are vulnerability and criticality related?
- Are highly vulnerable branches likely to be highly critical and vice versa?
- Would increasing the rating of a vulnerable line reduce or increase its vulnerability and the risk of cascading if there is no redispatch of the generation?
- How about if there *is* a redispatch to take advantage of this increase in transmission capability?

- Would increasing the rating of a critical line reduce or increase criticality and the risk of cascading?
- Suppose you increase the transfer capability of a congested interface or path without changing the impedances of the relevant lines and without adding a new line, how would this affect DFAX, vulnerability, criticality, and the risk of cascading?

3.4.3 Rank and Degree

Vulnerability and criticality are measured in two ways, rank and degree. We will first illustrate these metrics with the example of Figure 3.2 and Table 3.3. We will then define them formally.

In Table 3.3, the columns measure the vulnerability of monitored branches. The rows measure the criticality of contingencies. The lower right portion of the table gives the post-contingency ($n-1$) flows for all monitored branches, for all outages, in per unit of the monitored branches' ratings. The rating of each branch is given in the nonhighlighted (white background) sixth row of this table.

Note the following:

- Since in the example every monitored line is outaged, the **DFAX** matrix and the matrix of Table 3.3 are square. But they do not have to be square.
- Neither the **DFAX** matrix nor the matrix of Table 3.3 is symmetrical.

In this example, monitored branch ratings are the same in both directions. If they are not, then one uses the rating that is consistent with the sign of the $n-1$ flow. The MVA ratings given in the 118-bus case were modified slightly to be more interesting for our example.

Table 3.3 presents the following four stress metrics:

- Vulnerability rank, RankV
- Vulnerability degree, DegreeV,
- Criticality rank, RankC
- Criticality degree, DegreeC

TABLE 3.3 $N-1$ **flows, rank and degree of vulnerability and criticality, for nine-branch subsystem.**

Degree threshold = 0.75				Vulnerability of monitored Branches								
			RankV	0.405	0.563	0.563	0.673	0.327	1.321	0.671	0.998	0.579
			DegreeV	0	0	0	0	0	7	0	3	0
				Riversde (1) Pokagon (2)	Riversde (1) HickryCk (3)	Pokagon (2) TwinBrch (12)	HickryCk (3) Olive (5)	HickryCk (3) TwinBrch (12)	NwCarlsl (4) Olive (5)	NwCarlsl (4) SouthBnd (11)	Olive (5) SouthBnd (11)	SouthBnd (11) TwinBrch (12)
Criticality of outaged branches			Ratings (MW)	133	133	133	133	155	110	155	134	133
RankC	DegreeC	From bus	To bus	|n-1 flows| (per unit of limits)								
0.960	1	Riversde 1	Pokagon 2	0.000	0.405	0.158	0.588	0.109	0.960	0.416	0.580	0.237
1.030	1	Riversde 1	HickryCk 3	0.405	0.000	0.563	0.372	0.054	1.030	0.465	0.645	0.362
0.933	1	Pokagon 2	TwinBrch 12	0.158	0.563	0.000	0.673	0.172	0.933	0.396	0.555	0.187
1.151	2	HickryCk 3	Olive 5	0.335	0.070	0.493	0.000	0.327	1.151	0.550	0.756	0.579
0.962	1	HickryCk 3	TwinBrch 12	0.132	0.273	0.290	0.582	0.000	0.962	0.417	0.582	0.240
0.998	1	NwCarlsl 4	Olive 5	0.054	0.351	0.212	0.639	0.018	0.000	0.264	0.998	0.089
0.845	1	NwCarlsl 4	SouthBnd 11	0.072	0.333	0.230	0.599	0.037	0.373	0.000	0.845	0.047
1.321	1	Olive 5	SouthBnd 11	0.064	0.341	0.222	0.617	0.029	1.321	0.671	0.000	0.014
0.846	1	SouthBnd 11	TwinBrch 12	0.070	0.335	0.228	0.604	0.035	0.846	0.335	0.475	0.000

The vulnerability metrics for each monitored branch are computed by considering each row.

- RankV is the highest $n-1$ flow for each monitored line, for example, 0.998 for monitored branch 5-11.
- DegreeV is the number of outages for which $n-1$ flows are greater than 0.75, the Degree threshold given in the upper left corner of this table, for example, 7 for monitored branch 4-5.

The criticality metrics for each outage are computed by considering each column.

- RankC is the highest $n-1$ flow for each outage, for example, 1.321 for outage 5-11.
- DegreeC is the number of monitored lines for which $n-1$ flows are greater than a threshold of 0.75.
- The DegreeV and DegreeC thresholds are not cast in stone. This is discussed later in this chapter.

Note the following:

- For both vulnerability and criticality, RankV and RankC are *the highest* $n-1$ branch loading.
- DegreeV and DegreeC are *the number* of $n-1$ loadings that exceed a threshold.
- Rank and degree are valid but different measures of stress or risk of cascading.
- Vulnerability points to one risk: branches becoming highly loaded because of contingencies.
- Criticality measures a different risk: outages that cause other branches to become more highly loaded.
- The four metrics together provide a picture of the system stress – which, not surprisingly, is not a simple thing to get one's arms around. In particular, the criticality and vulnerability of a particular branch are independent of each other. One recalls the old story of the six blind men describing an elephant. We hope that today's engineers can integrate the different observations better than the blind men did.
- We will see later that these metrics give a clear-cut diagnosis of a system's stress and can be used to mitigate it.

The example above is easy to understand. The reasoning is intuitively satisfying. Having seen the example, one is prepared to deal with the formal descriptions that follow.

In network analysis "rank" is often used as a verb or adjective having to do with the relative weight (which depends on the context) of a vertex. We define *vulnerability* rank (a noun) to be the maximum post-contingency flow (weight) on a given monitored branch, expressed in terms of its rating, considering all single outages (one at a time). See (3.3). In these equations $fm_{i,j}$ is the post-contingency ($n - 1$) flow on branch i after the outage of branch j:

$$\text{RankV}_i = \max\left(\frac{|fm_{i,j}(n-1)|}{\text{rating}_i}\right) \text{ over all outages } j \tag{3.3}$$

Mathematicians usually take "degree" to mean the number of vertices that are connected to a given vertex. We define *vulnerability* degree slightly differently as the number of branches that are connected to a given monitored branch (through the DFAXes) *whose outages cause post-contingency flows, expressed in terms of its rating, that exceed a threshold* – in other words, the number of outages that cause relatively high $n - 1$ flows (see (3.4)):

$$\text{DegreeV}_i = \text{count\_if}\left(\frac{|fm_{i,j}(n-1)|}{\text{rating}_i}\right) > \text{threshold, over all outages } j \tag{3.4}$$

We define an outaged vertex's *criticality* rank as the maximum post-outage flow over all monitored branches, expressed in terms of each monitored branch's rating (3.5). We define an outaged branch's *criticality* degree as the number of monitored branches whose post-outage flows, in terms of their ratings, exceed a threshold for this particular outage (3.6).

It appears that (3.3) and (3.5) are the same, but they are not. In (3.3), the maximization is over all contingencies. In (3.5), the maximization is over all monitored branches. See the example above. The same comment applies to the counting operation in (3.4) and (3.6). The difference in the equations is the interchange of i and j subscripts:

$$\text{RankC}_j = \max\left(\frac{|fm_{i,j}(n-1)|}{\text{rating}_i}\right) \text{ for all monitored branches } i \tag{3.5}$$

$$\text{DegreeC}_j = \text{count\_if}\left(\frac{|fm_{i,j}(n-1)|}{\text{rating}_i}\right) > \text{threshold, over all monitored branches } i \tag{3.6}$$

For both vulnerability and criticality, rank is a floating point (non-integer) number. Degree is an integer. The following points are worth considering by the most serious readers.

- In Table 3.3, please spot-check the values of one or two of each of the four metrics.
- Please verify the $n - 1$ flow on branch 4-5 for the outage of branch 1-2.
- Why might it be reasonable to be concerned about RankV using a threshold below a branch's rating, say at 75% as in the table?
- Do these metrics provide an overall probability of cascading?

- Do these metrics identify the probability of a particular cascading chain occurring?
- Why are probabilities not assigned to various events?
- How might the metrics help planners or operators identify and reduce the risk of cascading?
- How might Table 3.3 help identify branches that need to be protected from failing as first contingencies and branches that need to be protected from failing as second contingencies, the result of first contingencies?
- What might one do to protect against a first contingency?
- What might one do to prevent a first contingency leading to a second contingency?
- Is it proved that these metrics are the best way or even a good way to measure stress (risk of cascading blackouts)?
- Do they seem to be reasonable ways to measure stress?

3.5 VALIDITY OF METRICS

3.5.1 Proof of Validity of Metrics

It has not been proven that the metrics defined above are the best or even a good way to measure stress relative to possible cascading blackouts. It also has not been proven that measuring and reducing stress is a better approach than identifying and fixing hidden failures or similar deficiencies in control and protective devices and in practices and procedures. Such propositions may not be subject to mathematical proof, though they may be supported by first principles and experience. We will argue later that the metrics defined above appear to be useful and valid.

In particular, these metrics do not pretend to measure the probability of cascading to occur, or the probability of its initiating with failure of a particular outage. We will discuss later reasons for eschewing probabilistic models in this application.

3.5.2 Examples of Metrics

In many related and unrelated situations, metrics are used to diagnose risk and help define preventative or corrective action. Three well-known examples are described below:

- Engineers use strain gauges to detect and measure slight deforming of structural elements before they fail catastrophically. Strain gauges do not attempt to measure or forecast the nonlinear effects that occur when a beam splinters or deforms permanently. The objective is early warning to avoid these catastrophic events.

- The medical profession has identified four or five metrics ("vital signs") that provide a quick diagnosis of a person's health: temperature, blood pressure, pulse, respiration rate, and pain. (The last is a subject of debate. It does not have the same degree of acceptance as the first four.) Their use is justified by the observation that illness often causes deviant values of one or more of them. However, such deviations are neither necessary nor sufficient conditions for illness to be fully developed or nascent. And medical folk do not compute the probability of a patient being ill if he has, for example, a temperature of 99.8° F. But they might look more carefully at someone with such a temperature.
- The NERC criteria were published after the 1965 Northeast Blackout (there was no NERC before 1965). They are the evolving products of generations of committee contemplations. Their principal objective is to prevent cascading blackouts – which continue to occur notwithstanding adherence to the criteria, at least in the planning context. See earlier comments in this chapter with regard to the 1965 Northeast and 2003 US–Canada blackouts.

All three of the above sets of metrics or indicators share the following attributes, in engineering parlance:

- The physics of severe or catastrophic failures are understood.
- They involve complex nonlinear deviations from norms.
- The metrics are used to identify risk or vulnerability of incipient failures.
- These metrics measure, in one way or another, whether a system is stressed.
- They are used to help prescribe preventative or corrective actions.
- They are not foolproof (they are neither necessary nor sufficient indicators).
- They are justified by experience and knowledge of first principles.
- At least the second and third are subject to intense continuing debate and evolution.

3.5.3 What Makes Metrics Valid and Useful

One study concludes that the following are characteristics of valid and useful metrics or indicators in the context of measuring the health of a population [25]:

- "Definition. The indicator must be well defined, and the definition must be uniformly applied internationally" (in the cascading blackout context, it must be applicable to different systems).
- "Validity. The indicator must be valid (it must actually measure what it is supposed to measure), reliable (replicable and consistent between settings), and readily interpretable."
- "Feasibility. The gathering of the required information must be technologically feasible and affordable and must not overburden the system."

- "Utility. The indicator should provide information that is useful to decision-makers and can be acted upon at various levels (local, national, and international)."

These four criteria seem appropriate for defining metrics in the cascading blackouts context as well. But they left one off: watch for unintended consequences. US medicine's dalliance with pain as a metric brought us the opioid crisis. There are parallels in power.

The metrics are *precisely defined*, with just one arbitrary parameter whose value is not critical. As will be shown below, they give comparable and consistent results over systems of very different characteristics (Peru, parts of the Eastern and Western Interconnections of North America – the EI and WI) and for the entire WI (results are not presented here but are in preparation for publication later).

Calculating the metrics is *feasible*. The major calculations of the **DFAX** matrix and the $n - 0$ flows are performed using power flow and state estimation software that is available in most control centers and power flow software that transmission planners use constantly. The required network model data (topography and admittances, etc.), though difficult to develop and maintain, also is available to operators and planners and is used constantly. The most time-consuming operation is computing and storing the **DFAX** matrix. This takes a few moments on an ordinary laptop for a large system like the southwest portion of the WI and less than an hour for the entire WI. Computing and analyzing the stress metrics take considerably less computer time. A well-documented pre-commercial-grade program for doing the latter, importantly incorporating sparsity techniques, was produced in the Western Electric Coordinating Council (WECC) and University of Utah project described below.

The next sections show why the metrics of vulnerability and criticality, rank and degree, are valid and useful.

3.5.4 Why the Vulnerability and Criticality Metrics Are Valid and Useful

A *necessary condition* for cascading to spread beyond $n - 1$ to $n - 1 - k$ is for cascading to occur from $n - 1$ to $n - 1 - 1$. (Here k represents any sequence of positive integers denoting a causal sequence of failures.) That is, until one contingency ($n - 1$) causes at least one second contingency, ($n - 1 - 1$), cascading cannot continue.[5] This seemingly trivial observation is of practical importance because the number of $n - 1$ contingencies, while in the tens of thousands for our largest systems, is

[5] Some will argue that the 2003 US–Canada blackout was considerably more complex than this model and had multiple independent initiating contingencies, e.g. $n - 2$ or more. We agree that the hidden failures were plentiful, and that independent and related generation and transmission events stressed the system, that is, made it more vulnerable. But the cascading itself began when a 345-kV line, loaded to just 44% of its rating, tripped at 15:05 p.m. This $n - 1$ event led in a slow cascade to the loss of two other 345-kV lines. These caused fast cascading at 16:06 p.m., consistent with our model.

tractable. The vulnerability and criticality metrics analyze *all of them* to measure each monitored branch's exposure to cascading. In contrast, the number of possible $n-1-k$ contingencies is astronomical for large systems. But our metrics show, and Bhat et al. [18] show through dint of much labor, that only a tiny fraction will cascade.

A *sufficient condition* for cascading to occur from $n-1$ to $n-1-k$ for a particular first contingency is that RankC and DegreeC (with its threshold) be high enough. If they are, the post-contingency monitored flows will activate one or more monitored branches' protection or will cause the branch or branches to fail. Either way, this is an $n-1-k$ event. Higher RankC means more stress because for at least one monitored branch, the post-contingency flow is higher. Higher DegreeC means more stress because the first contingency results in more monitored branches with higher RankV.

By a similar argument, a *sufficient condition* for a particular monitored branch to cascade from $n-1$ to $n-1-k$ is that its RankV and DegreeV be high enough.

What constitutes "high enough" is unknown, due in part to hidden failures, but see the Peru study later in this chapter.

We will show later that the metrics provide not just a way of measuring the stress of individual branches but of measuring the stress of the system as a whole. This is from complementary perspectives, both vulnerability and criticality.

The state of the system – including existence of hidden failures in control and protection systems (e.g. relaying) and practices and procedures (e.g. tree trimming, situation awareness) – has led to cascading with RankV less than 1. This occurs when a branch normally is loaded well below its rating. Suppose it is normally loaded up to a maximum of, say, 75% of its rating. Then a hidden failure at 80% will not become evident until the branch is loaded to that level by a contingency. (Such events were among the direct causes of the 2003 US–Canada blackout.)

It is impossible to guarantee that hidden failures do not exist. But such failures are more likely to wreak their havoc as rank and degree increase.

It is not necessary to model explicitly well-known nonlinear and dynamic effects in order to measure stress. When nonlinear and dynamic effects become important, by definition the system is stressed. High $n-1$ steady-state loading, measured with linear models, brings a system to this state. For further contemplation, consider the following:

- Are the rank and degree metrics of vulnerability and criticality well defined?
- Can the metrics be applied to a variety of systems?
- Are the metrics valid (do they actually measure what they are supposed to measure)?
- Are they reliable (replicable and consistent between studies of different demand levels and dispatches)?
- Are the metrics readily interpretable?

- Are the metrics feasible? (Is the necessary data readily available?)
- Are the metrics useful? (Will they provide information that should be useful to decision-makers and that can be acted upon at various levels (local, regional, for planning and operations)?

3.6 STUDIES WITH METRICS

These ideas were developed and tested on studies of three very different systems. Not only are the systems different, but also the studies themselves are quite different. These studies demonstrate clearly that all of the questions above have affirmative answers. The first study, of part of the Eastern Interconnection (EI) of North America, was a proof-of-concept study. The second, part of a planning study of the entire national electric system of Peru, was quite extensive. It developed and tested the metrics described above. These two studies were documented in an unpublished white paper [19]. The third, a series of analyses of the Western Interconnection (WI) of North America, was a joint effort of the University of Utah and the WECC. Preliminary results have been published [26]; a more complete paper is in preparation. Accordingly, this chapter will include only brief references to this third study, and to follow-up research being done at the University of Utah.

3.6.1 Line Outage Distribution Factor Properties

DFAXes were described earlier. Some of their properties will be discussed below.

Large positive or negative values of DFAX make cascading more likely. Tighter coupling is more likely to make branch i overload and go out of service if branch j has an outage, all else being equal. The **DFAX** matrix and network define how easy or difficult it is for a failure to cascade through the system. For the failure network, the **DFAX** matrix is analogous to the **Ybus** for the power flow network. For example, with a DFAX of zero, the outage of branch j will not cause any change in the flows of branch i; a failure of branch j will not propagate to branch i. Tight coupling, with a high value of the DFAX between two branches, means that cascading between the two is likely.

Figure 3.3 contrasts the distribution functions of the absolute values of DFAX for two very different systems [19]. The first is a 1706-branch high-voltage (230 kV and up) portion of the EI of the United States and Canada. Although only this subset of the branches of the EI is outaged and monitored, the **DFAX** matrix reflects the changes in flows due to all of the transmission systems throughout the entire interconnection, just as the entire 118-bus IEEE network affects the DFAXes between branches of the nine-branch subsystem.

The second system, of 169 nonradial vertices, is from one of several plans considered for the future national interconnected system of Peru, developed in a 2012 planning study [27]. All of the meshed system is modeled, with voltages from 500 kV to well below 100 kV.

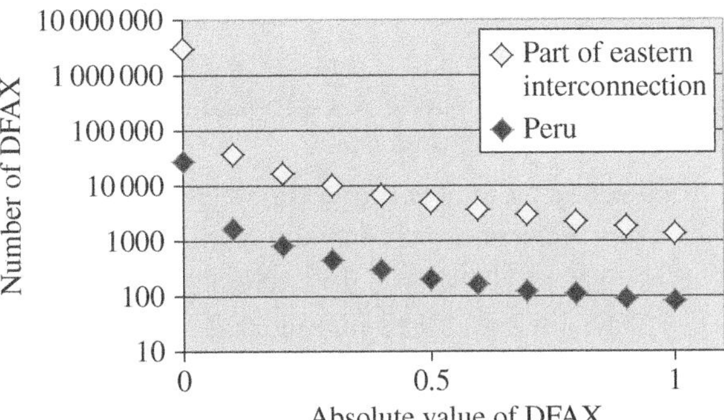

Figure 3.3 DFAX distribution functions for large and small systems [19]. Source: Hyde M. Merrill, reprinted with permission.

Each symbol in Figure 3.3 represents the number of DFAXes with absolute values equal to or greater than the value shown on the *x*-axis. The shapes of the distributions are similar and in fact are similar to the distributions of DFAX for part of the WI of North America [26]. This is remarkable because the three systems could hardly be more different.

For Peru and EI systems, the DFAX *density* functions are plotted using log–log axes in Figure 3.4. Each looks like a straight line for most of its range. This is how a "power law" appears, though strictly speaking a power law continues to infinity at both ends, which these curves do not. Nonetheless, such truncated power law distributions are often encountered in network studies [28]. One interesting characteristic they have is "fat tails," where the probability of DFAX (in this case) having large values is greater than for a normal (Gaussian) distribution.

Before continuing, it seems relevant to point out the important differences among these three systems and how they are modeled.

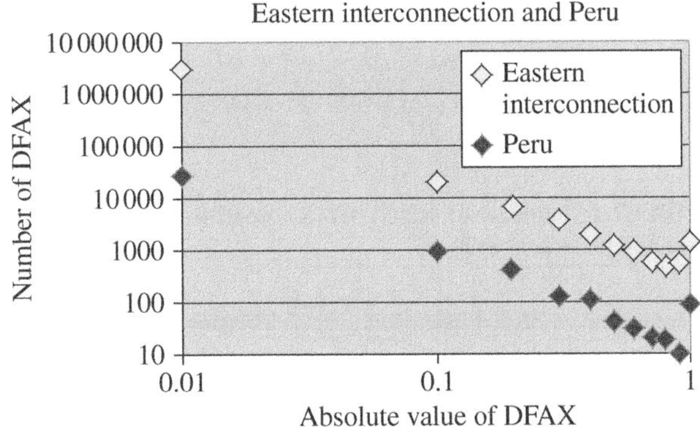

Figure 3.4 DFAX *density* functions resemble power law functions.

- The Peruvian system has an order of magnitude fewer meshed branches than the other two.
- The system studied is at a significant step in its evolution from a mostly radial system to a meshed one.
- Because of the Peru context, system planning and operating criteria may be very different from criteria in North America. (This is an observation, not a criticism. The criteria may appropriately be different.)
- The DFAX for the EI study consider outages and monitored branches only at 230 kV and above. The other two systems include all branches in the power flow cases, including those below 100 kV.
- The 8 September 2011 WI case is a reconstruction of the system state just before a major blackout on that date, with all the imperfections and realities of real-time operation.
- All of the other cases are long-range planning cases (EI and Peru) or short-range planning cases (WI).
- The DFAXes and metrics of Peru are for the entire system, while the North American DFAXes and metrics, though they reflect the paths and flows through the entire interconnections, monitor and outage only large subsystems of the entire interconnections.
- For the WI, seven base cases were studied over a five-year period, including five cases with different seasons and load levels for a single year.
- For Peru, 12 months of production and power flow simulation studies were used – all with the same network (the DFAX did not change) but reflecting different peak, shoulder, and minimum demand and dispatches for each month, so the metrics changed.

With not much in common beyond Ohm's and Kirchhoff's laws, it is truly remarkable how similar the results of analysis are from system to system. Here we will describe just one set of observations, having to do only with the DFAX. We will make more comparisons later, applying metrics.

Figure 3.3 and Table 3.4 seem to say that there is much more exposure to cascading in the larger IE system than in the Peru system. The statistics differ in an important way. The modeled portion of the EI, with 10 times the branches, has about 20 times the large DFAXes (absolute values ≥ 0.7). So the EI has 20 times

TABLE 3.4 Number of large DFAX compared with number of monitored branches.

	DFAX	Monitored Branches	DFAX ≥ 0.7	DFAX ≥ 0.7 per branch
Eastern Inter.	2 909 000	1706	2907	1.70
Peru	28 055	169	130	0.77

the number of tightly coupled branches in the failure sense, that is, that a large fraction of the pre-contingency loading of one branch will transfer to the other branch.

This may be largely due to the relative sizes of the two networks. But perhaps it is also because the EI network is more meshed than the Peru network. See the last column in the Table 3.4. It would seem worthwhile to analyze how the last two columns in the table differ from system to system, and why.

It may be argued that Table 3.4 also shows that the percent of DFAX with low values is also larger for the EI system. This is true but irrelevant: Low-value DFAXes do not contribute to cascading, and high-value DFAXes do.

Of course, DFAXes are not metrics. Stress depends critically on system loading, which the metrics capture but which the DFAXes do not.

Nonetheless the strong differences in the next-to-last column regarding network size are consistent with the statement by Vannevar Bush, cited earlier, that larger systems are more vulnerable to major failures. The differences also coincide with historical experience that cascading blackouts are large-system phenomena.

3.6.2 Eastern Interconnection Study [19]

Two ways to assess the validity of a metric are as follows:

1. To check whether the metrics give reasonable indications of stress.
2. To test whether the metric agrees with known attributes of a system.

For a portion of the EI of North America, Figure 3.5 is the distribution function of DegreeV as stress (power transferred from one subregion to another) changes, using a threshold of 75% of each observed branch's rating. The horizontal axis is DegreeV. The vertical axis is the number of monitored branches whose DegreeV equals or exceeds a particular value. For example, with no interregional transfers (diamond symbol), about 40 monitored branches have DegreeV ≥ 6.

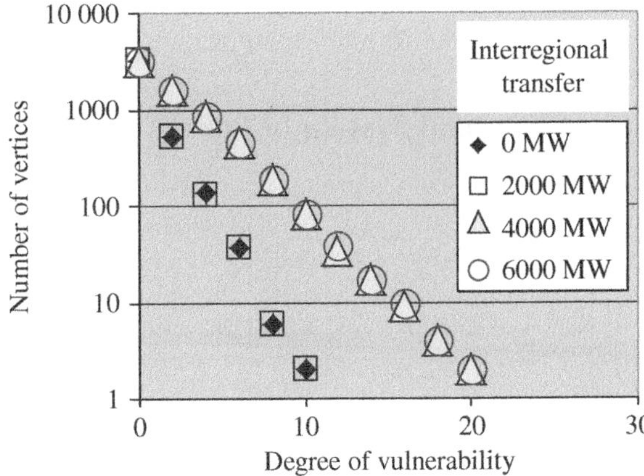

Figure 3.5 Degree of vulnerability is affected by interregional transfers.

When transfers increase from 0 to 2000 MW, there is no change in DegreeV. Between 2000 and 4000 MW, the DegreeV nearly doubles for any particular number of branches. This is especially significant for the relatively small number of branches with high DegreeV. For example, with 2000 MW of transfers, some 40 monitored branches have DegreeV ≥ 6. With 4000 MW of transfers, about 32 monitored branches have DegreeV ≥12. This is nearly the same number of branches but with double the DegreeV.

Looking at Figure 3.5 slightly differently makes the effect of increasing transfers even more dramatic. With transfers of 2000 MW, only two monitored branches have DegreeV ≥ 10. With transfers of 4000 MW, about 85 monitored branches have Degree ≥ 10. The difference is more than an order of magnitude.

History reveals that high stress, notably due to interregional transfers, is a necessary condition for cascading to occur. This analysis shows that as transfers increase, the DegreeV of any set number of monitored branches increases by a significant amount. Furthermore, the number of monitored branches with a particular DegreeV also increases by more than an order of magnitude for high values of DegreeV. These results are dramatic but intuitively reasonable.

In this example, the DegreeV metric gives reasonable indications of stress. The metric increases dramatically with transfers as shown by computation, and the stress or vulnerability to cascading increases with transfers, as history shows.

In addition, beyond 4000 MW, this metric does not change. Apparently when transfers reach 4000 MW, every branch affected by transfers is highly loaded, at the $n - 1$ or post-contingency state.

The tipping point between 2000 and 4000 MW is interesting but not surprising. It means that many branches become stressed at about the same level of transfers. In a well-designed system, one would expect branches to reach their ratings at about the same level of loading. A branch that reached its rating much sooner than others would have been fixed over the years, so that it alone would not unduly limit the transfers. A branch that reached its rating much later than others might be considered to be overdesigned.

Finally, the tipping point at around 3000 MW is consistent with the known transfer capability of that part of the system, based on conventional studies. That is, the metric agrees with known attributes of the system.

3.6.3 National System of Peru [19]

A thorough study of the grid of Peru used 12 monthly pre-contingency dispatches, for peak, minimum, and shoulder loads. Three system expansion plans were evaluated for a 10-year horizon: light, medium, and heavy expansions.[6]

[6] Options include both upgrading existing branches and adding new lines or transformers. The light expansion consists only of the simplest or most urgent options, targeted at low load growth. Medium and heavy expansions are for normal or high demand growth. One could regard planning as a two-party game where nature chooses the demand growth rate and the planners choose the plan. Since the planners must play first, it is possible that the plan may not be perfectly in sync with the demand growth. For simplicity, heavy expansion plans are not included in this discussion.

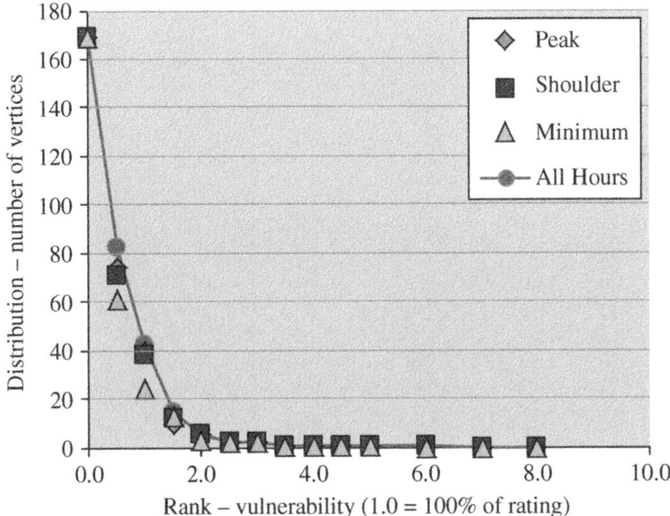

Figure 3.6 Rank of vulnerability for peak, shoulder, and minimum loads.

Figure 3.6 is the distribution function of rank of vulnerability (RankV) for one month for a future with high load growth and a light transmission expansion plan – a highly stressed extreme case. As one might expect, RankV is high but substantially less during minimum load hours on weekends and in the middle of the night.

It is interesting that RankV is about the same during shoulder and peak load hours. As has been observed for US systems, the economic dispatch often loads the system with economy flows during shoulder (nonpeak) hours. Furthermore, the load factor in Peru is fairly high because of mining and refining loads.

About 25% of the branches have RankV between 1.0 and 2.0 (between 100 and 200% of the branches' ratings). These are likely the principal concern.

Only a very few have very high RankV (over 2.0). These may be spurious. Many systems have lower-voltage networks essentially in parallel with high-voltage branches. An outage of such high-voltage branches could severely overload the underlying system and is the most likely way to get such a high RankV. This is sometimes prevented in some systems by special protection or remedial action schemes (RAS) that are not recognized in the DFAX or stress metrics. Often the parallel lower-voltage lines are operated "normally open." If so they should be represented as radial lines in the stress analysis, neither monitored nor outaged. Note, however, that the $n-2$, $n-3$, and $n-4$ failures in the 8 September 2011 Arizona–Southern California cascading blackout were of transformers into a complex but essentially parallel lower-voltage system. Without their loss the cascading might have fizzled out.

Figure 3.7 shows the criticality degree metrics (DegreeC) for the same case. The desired threshold for degree was 1.0 (100% of branch rating). To avoid

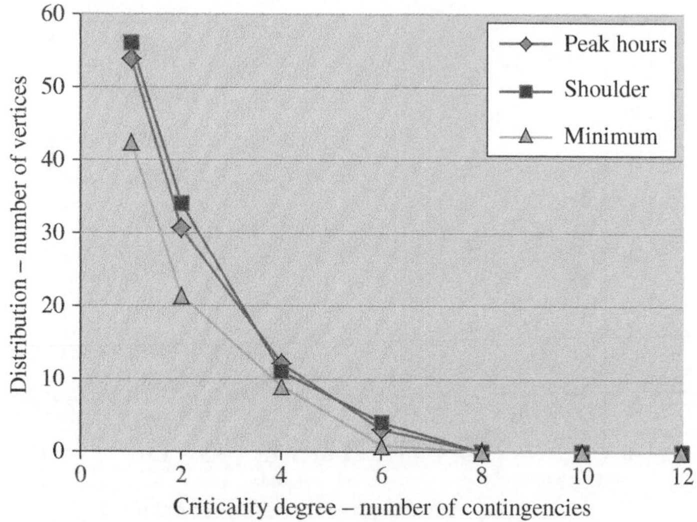

Figure 3.7 Degree of criticality for peak, shoulder, and minimum loads.

numerical issues with the production simulation program, which did not necessarily satisfy branch ratings precisely, the threshold was set at 1.02. In essence, DegreeR and DegreeV counted overloads. For this worst-case scenario (little system reinforcement, high demand growth), criticality is generally high but lower during minimum load hours.

There are about 10 branches whose outages would cause 4 or more parallel branches to overload – but they are not necessarily the same 10 for each load level. This has obvious implications for planning and operations. If these are not spurious, and if each overload resulted in the monitored branches going out of service, any one of those 10 contingencies would lead immediately to an $n-5$ state. Systems should not be exposed to $n-5$ conditions.

The figure shows that three branches, if outaged at peak load, would cause six or more monitored branches to overload. Four would do so at shoulder load and one at minimum load. No contingency has DegreeC ≥ 8.

It is interesting that though the degree of criticality is about the same for peak and shoulder hours, the vertices involved are not the same. The patterns of flows are different. Clearly one cannot study peak conditions alone.

Before discussing Table 3.5, we must digress briefly. Electric generators are dispatched or operated in real time so as to minimize the system-wide fuel cost. This is called "economic dispatch." If the economic dispatch would cause one or more vertices to overload, the system is said to be "congested." (Most power systems experience congestion from time to time.) A constraint is added to the mathematics, and the generation is redispatched in real time (at a higher fuel cost) to avoid overloading under $n-0$ conditions. Some systems go a step further with a "security-constrained" dispatch, which is $n-1$ compliant so that after a

TABLE 3.5 Load growth and system expansion, congestion, and stress [19].

		Scenarios					
Congestion		High	Medium		Low		Very Low
Case		1L	1M	2L	2M	4L	4M
Demand growth		High	High	Normal	Normal	Low	Low
Transmission expansion		Light	Medium	Light	Medium	Light	Medium
Rank		Number of monitored vertices					
Vulnerability	>1.02	42	50	33	42	12	11
	>1.5	15	15	9	13	3	3
Criticality	>1.02	62	114	48	105	16	16
	>1.5	11	15	5	12	3	2
		Maximum degree, number of contingencies					
Vulnerability		10	40	12	40	5	5
Criticality		6	10	6	10	4	4

Source: @ IEEE.

contingency, before operator action or redispatch, no branch is overloaded. System planners reinforce the network so as to keep congestion and redispatch at reasonably low levels in the future.

Table 3.5 compares several scenarios. Each of the three ovals in the table corresponds to a different load growth. Each oval covers two alternative expansion plans ("light" compared with "medium" expansions). Therefore each oval covers two different levels of congestion and stress. Six stress metrics are shown.

For the high and normal (expected) load growth futures (the first two ovals), the top line shows that the stronger network had less congestion but greater stress. Reinforcing the network, to reduce congestion, makes it more vulnerable to cascading.

Why is this so? As discussed above, the stronger network was reinforced to reduce congestion. The production simulation program recognized this; generation was dispatched with more power flowing across the system. The $n-0$ loading was higher. The stress increased, measured with $n-1$ metrics. All else essentially equal, greater transfers mean greater stress, and the stress metrics quantify this.

The technical conclusion is that reinforcements that allow more power to flow also may make it easier for failures to cascade through the system. There are two types of such reinforcements. The first is to increase the rating of a congested branch. For instance, in one occasion the rating of an EHV line in the PJM area of the EI was increased by lowering the ground with a bulldozer for a distance of 0.1 mile – about 160 m. The redispatch that this permitted would not change the DFAX but could raise the $n-0$ flows in branches in series and in parallel (since transfer capability is based on $n-1$ conditions) with this line, increasing both their vulnerability and their criticality.

A second type of reinforcement is to build a new branch. This increases the transfer capability in an area and changes DFAXes. Again, dispatchers may react by using the increased transfer capability, increasing the flows both in parallel and in series with the new branch.

For both types of expansion, there may be a conflict between the objectives of reducing congestion and reducing risk of cascading. The actions that reduce congestion (network expansion leading to redispatch to reduce fuel costs) may increase the risk of cascading (making it easier for failures to propagate and making the system less secure).

This conflict did not occur for the third oval. This is because the dispatches were about the same for the "light" and "medium" systems. There was little congestion for the former and therefore little redispatch for the latter. The "medium" system might be excessive for low load growth.

The middle block of the table, under the heading "Number of Monitored Vertices", presents statistics on RankV and RankC for two thresholds, 1.02× branch rating and 1.5× branch rating.

Another surprising observation: the table shows that the vulnerability and criticality metrics increase as load, generation, and transmission increase, e.g. case 2L vs. 1M and 4L vs. 2M. Mere growth seems to make the system more stressed, even when the network is reinforced to keep up.

3.6.4 Western Interconnection of North America [26]

A joint WECC/University of Utah study in 2016–2017 did an in-depth development of metrics for a very large system. Five base cases were analyzed for 2016, representing slightly different networks for each season (reflecting maintenance) for summer high and low loads, winter high and low loads, and spring high loads. A similar sixth case for 2012 was analyzed. Finally, metrics were computed for a seventh case representing the system state just before the 8 September 2011 Southern California–Arizona blackout.

Because of this blackout, the analyses focused on the southwestern part of the WI – "SW WI" – though all of the WI was modeled, and stress analyses of all of the WI have been done. Very interesting preliminary results were published early in 2017. Since then the WI study was completed, with significant additional findings that are being prepared for publication.

The reader is invited to read the 2017 paper, whose results will only be summarized briefly here. One important characteristic of SW WI is that it generally imports power from elsewhere in the WI. There are two general findings:

- The stress metrics for the southwestern part of the WI were intuitively reasonable. Stress was highest for the 2016 summer peak case and lowest for the 2016 winter and summer off-peak cases.
- The 2011 precascading case showed stress that was second only to the 2016 summer peak case. This is remarkable because the 2016 summer peak case was built in a way to be more extreme than realistic.

3.6.5 Tipping Points

We now return to the notion of tipping point that was found in the EI study.

The five WI operating cases for 2016 were produced by a well-developed, established annual process involving inputs and careful cooperation among the various power companies in the WI. The five cases have different load levels, presumably comparable from area to area and presumably reasonable and consistent generation dispatches. Because of this, the five 2016 cases allow an "apples to apples" comparison of the stresses on one network, as a function of changes in demand (from about 26 000 MW to about 61 000 MW) and dispatch. It is highly unusual to have access to data of such realism, comparability, and detail. Figure 3.8 supports the following four important conclusions:

- The number of vertices with RankV ≥ a threshold was proportional to the log of demand.
- The slope of this relationship went up dramatically at a certain level of demand, a tipping point.
- The slope of this relationship depends on the threshold chosen (75, 100, and 125% of branch rating).
- The demand at the tipping point was the same for all thresholds.

In other words, the stress of the system increases as demand increases. The rate of this increase goes up dramatically once demand reaches a certain level – clearly

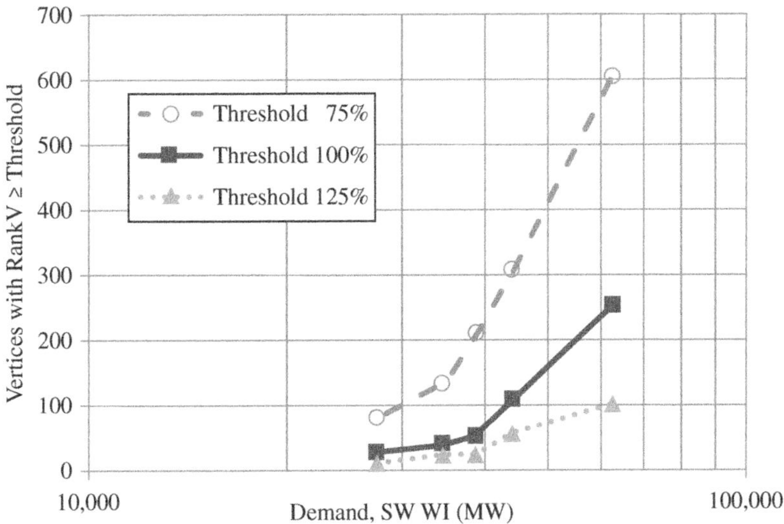

Figure 3.8 RankV and tipping points as a function of demand, SW WI.

an important data for operators and planners. And these conclusions are not affected by the arbitrary choice of the threshold.

Note, however, that if a threshold is set too low, false positives will obscure the DegreeV and DegreeC metrics. To avoid this problem, thresholds should be set above the normal operating values of branch loading. Choosing the threshold is no more difficult than adjusting a microscope to give a clear picture.

3.6.6 Pre-Blackout Stress

A tremendous amount of real-time power system data is recorded automatically and constantly. Mathematical models are used constantly to operate the system. With all of that, to reconstruct the state of the power system at a particular time is very time-consuming and very expensive. It requires collaboration of many organizations and people. The intermediate and final results may expose individuals and organizations to significant legal and economic risks and publication may expose the power system to risks from vandals and terrorists.

As a result, such data are confidential and are very closely held. It is unusual to be granted access to such data. What we present has been vetted and is printable.

As part of the WECC-U of Utah study, a stress analysis was performed using a reconstruction of the state of the system before the 8 September 2011 Arizona–Southern California cascading blackouts. Partial results of that study have been published [26]. Key observations from that study are as follows:

- The southwestern part of the WI (SW WI) in the reconstructed base case was highly stressed.
- The 500-kV line that was first to fail in the actual event was flagged as highly critical using the stress metrics.
- Branches diagnosed as vulnerable to failure of that line included the two Coachella 230/92-kV transformers that in the event were the first two branches to cascade. Each of them had DegreeV = 4, with threshold 100% of ratings, meaning that any of four different contingencies would cause them to overload.
- The area where $n-2$, $n-3$, and $n-4$ cascading occurred was identified as the most vulnerable and the second most critical area in SW WI.
- All of these do not mean that the stress metrics will predict the next cascading failure. The cascading does not have to begin with the most vulnerable or most critical branches, though in this case it apparently did.
- However these branches and this area should have received careful attention, which they apparently did not.
- For instance the NERC/FERC evaluation of the blackout observed that the 500-kV line that was first to fail had failed often in the recent past without cascading. The stress metrics revealed that the conditions were different on 8 September and that the line was critical that day. The operators apparently did not know this.

- These conclusions say nothing about the vulnerability and criticality of the various branches and areas today. Much has changed in that region since 2011.

3.7 SUMMARY

In a nutshell, we have shown that the metrics can measure the stress on a system, identify the most stressed subsystems, and flag the most vulnerable and critical branches. Specifically:

- **Vulnerability:** Vulnerability metrics are consistent with what one would expect. They are higher when the system is under obvious stress, that is, at peak load periods, and lower at minimum load periods, that is, their values are reasonable indicators of stress.
- **Criticality:** For a critical branch to overload is much more serious than for an overload in a branch that is merely vulnerable.
- **Rank:** Rank is affected by demand and dispatch. There is no "bright line" limit, notwithstanding current standards that branches are not to be loaded above their ratings. The actual load-carrying capability of a branch depends on ambient temperature, wind, tree trimming, etc. Cascading has occurred when branches were below their nominal rating (e.g. certain trips in 1995 and 2003). In other situations, overloaded branches have not tripped. But rank that is higher than usual is an indicator of stress.
- **Degree:** This "counting" metric depends on the values chosen for thresholds. With thresholds too low, every contingency has a high degree. Set too high, nothing registers. Since the thresholds can easily be "zoomed," this is not troublesome. Contingencies with high order are particularly interesting: if the thresholds coincide with the relaying, then their failures could lead immediately to $n - m$ conditions (for monitored branches that are essentially in parallel) or to higher-likelihood $n - 2$ conditions (for monitored branches that are essentially in series).
- **Tipping points**: Analyses for the WI and EI show that metrics increase dramatically past tipping points or points of inflection, that is, there comes an operational level where the system transitions to much more at risk.
- **Observations**: Branches that are highly vulnerable are not necessarily highly critical and vice versa. These two metrics represent two different kinds of stress. The metrics are based on $n - 1$ analysis (or $n - 2$, depending on interpretation). Cascading by definition is $n - 1–2–3$.... But the metrics do not pretend to simulate the cascading events, any more than a strain gauge simulates the sequence of breaking bonds in a beam. The metrics identify states that are more exposed to untoward contingencies.

3.8 APPLICATION OF STRESS METRICS

Stress metrics can be used in power system dispatching, operations planning, transmission planning, and generation planning. They can be used to monitor and anticipate the stress of the system and of individual branches.

They also may be useful in identifying ways to reduce system stress. For example:

- The effect of conventional real-time or planning decisions on vulnerability and criticality can be tested.
- Optimal power flow (OPF) software can be modified to reduce vulnerability or criticality by reducing or limiting flows on vulnerable or critical branches through dispatching.
- Dispatch of phase-shifting transformers, switching operations, etc. on stress may be possible.
- Reducing stress by *opening* branches may be useful. In discussing the Peru study, it was noted that a weaker network may be less stressed than a stronger one. This counter-intuitive observation is related to Braess' paradox, which observes that weakening a network by cutting an edge may increase its throughput.
- Other options may be developed to decrease stress, rather than congestion, which is an objective of planning and operations today.

3.9 CONCLUSIONS

- Cascading blackouts:
  - They are expensive in monetary and nonmonetary terms.
  - The central objective of the NERC planning criteria and of bulk transmission planning in industrialized countries is to prevent cascading blackouts.
  - Cascading blackouts are the power industry's oldest unsolved technical problem.
- The three fundamental elements of the power system are as follows:
  - Current-carrying hardware
  - Control and protective devices
  - Practices and procedures
- The three necessary requirements for a cascading blackout to occur are the following:
  - Failures in control and protective devices, or practices and procedures.
  - The power system is under stress.

- The power system must be large.
- The possibilities of failures in control and protective devices and in practices and procedures are legion.
  - With few exceptions, planning studies and operating procedures assume that they will not occur.
  - There are very many ways that the triggering elements can fail, and requisite models and remedial actions to detect, analyze, and fix them are not available.
  - Therefore the cascading blackout problem is attacked on a different front, by measuring and managing the stress.
- Future work should be done on operating the existing system in such a way as to reduce stress and inventing new options for reducing stress in the planning context.
- In a cascading failure network, the vertices are branches of the traditional power network, (lines and transformers) and the edges are line outage distribution factors, commonly called DFAX.
  - The DFAX network is too big to analyze elementally, and "big is different."
  - So as is the practice in other applications of modern network theory, metrics are chosen to measure what is of concern, in this case, stress.
  - Four stress metrics are defined: rank and order of vulnerability and criticality.
- These metrics, applied to representative base cases developed by planners, give intuitively reasonable results. Applied to a precascading outage reconstruction, they identified the line whose outage triggered the cascading as highly critical and the subsystem where the propagating occurred as highly vulnerable.
  - The metrics can be applied in real-time operations, in short-run operations planning, and in long term transmission planning.
- Traditional planning options are for maintaining or increasing the ability of the power system to transfer power.
  - Eliminating cascading calls for developing and applying, in planning and in real time, options for increasing the "resistance" of the system to cascading failures.

ACKNOWLEDGMENTS

We acknowledge with gratitude the contributions to the study of the system of Peru described in this chapter of our distinguished colleagues Jefferson Chávez, Freddy Portal, Edward Angelino, and Eduardo Antúnez de Mayolo of the Committee for the Economic Operation of the National Interconnected Electrical System (Comité de Operación Económica del Sistema Interconectado Nacional, or COES-SINAC), Lima, Peru. Their vision, ideas, and hard work in the research described and the far-thinking funding that was provided by COES-SINAC quite literally made this chapter possible.

The authors are also grateful to the Western Electricity Coordinating Council for supporting a study of the WI and providing the necessary data. The leadership and technical contributions of Donald Davies, Chief Senior Engineer at WECC, are also particularly acknowledged. We are grateful for the contributions of our colleagues Prof. Marc Bodson and graduate student Md Abid Hossein of the University of Utah.

Our colleagues may not agree with all of our conclusions, which do not necessarily represent the positions of any of our organizations, but this work could not have been done without their help.

REFERENCES

1. Schweppe, F.C. (1978). Power systems '2000': hierarchical control strategies. *IEEE Spectrum* July: 42–47.
2. Loehr, G.P. (May/Jun. 2017). The "good" blackout – the Northeast power failure of 9 November 1965. *IEEE Power and Energy Magazine* 15 (3): 84–96.
3. U.S.-Canada Power System Outage Task Force (2004). Final report on the August 14, 2003 blackout in the United States and Canada, Apr. 2004.
4. Investigation Committee on the 28 September 2003 blackout in Italy (2004). Final report on the 28 September 2003 blackout in Italy. Union for the Coordination of Transmission of Electricity, Apr. 2004.
5. Investigation Committee on the 28 September 2003 blackout in Italy (2003). Interim report on the 28 September 2003 blackout in Italy. Union for the Coordination of Transmission of Electricity, Brussels, 27 Oct. 2003.
6. Federal Energy Regulatory Commission and North American Electricity Reliability Corporation (2012). Arizona-Southern California Outages on September 8, 2011, Apr. 2012.
7. *American Electric Power Company Annual Report for 1965*. (1966). New York: American Electric Power Company.
8. Standard TPL-001-4 – Transmission system planning performance requirements. *Reliability Standards for the Bulk Electric Systems of North America*, NERC, Atlanta (17 Aug. 2016).
9. DyLiacco, T.R. (May 1967). The adaptive reliability control system. *IEEE Transactions on Power Apparatus and Systems* PAS-86: 517–531.
10. Schweppe, F.C. and Handschin, E. (1974). Static state estimation in electric power systems. *IEEE Proceedings* 62 (7): 972–982.
11. Chai, S.-K. and Sekar, A. (2001). Graph theory application to deregulated power system. Proceedings of the 33rd Southeastern Symposium on System Theory (Cat. No.01EX460), Athens, OH, pp. 117–121.
12. Wang, Z., Thomas, R.J., and Scaglione, A. *Generating Random Topology Power Grids*. Ithaca, NY: ECE Cornell University.

13. Sun, K. (2005). Complex networks theory: a new method of research in power grid. *IEEE/PES Transmission and Distribution Conference and Exposition*, Dalian, China.
14. Dobson, I., Wierzbicki, K.R., Kim, J., and Ren, H. (2007). Towards quantifying cascading blackout risk. *iREP Symposium – Bulk Power System Dynamics and Control*, Charleston SC.
15. Dobson, I. (2007). Where is the edge for cascading failure?: challenges and opportunities for quantifying blackout risk. *IEEE PES Gen Mtg*, Tampa, FL.
16. Mili, L., Qiu, Q., and Phadke, A.G. (2004). Risk assessment of catastrophic failures in electric power systems. *International Journal of Critical Infrastructures* 1 (1): 38–63.
17. Chen, J., Thorp, J.D., and Dobson, I. (2005). Cascadig dynamics and mitigation assessment in power system disturbances via a hidden failure model. *International Journal of Electrical Power & Energy Systems* 27 (4): 318–326.
18. Bhatt, N., Sarawgi, S., O'Keefe, R., et al. (2009). Assessing vulnerability to cascading outages. *Proceedings of IEEE/PES Power Systems Conference and Exposition* (Mar. 2009).
19. Merrill, H.M. and Feltes, J.W. (28 Sep. 2016). Cascading blackouts: stress, vulnerability, and criticality. www.merrillenergy.com (accessed 20 February 2020).
20. Makary, M.A. and Daniel, M. (3 May 2016). Medical error – the third leading cause of death in the US. *BMJ* 353: i2139.
21. Bush, V. (1970). *Pieces of the Action*. New York: Morrow.
22. ERCOT (Electric Reliability Council of Texas) (21–23 Dec 1989). *ERCOT Emergency Operation*. ERCOT.
23. ERCOT (2006). Investigation into April 17, 2006 rolling blackouts in the Electric Reliability Council of Texas region (preliminary report), Public Utility Commission of Texas, 24 April 2006, 68 pp. http://www.ercot.com/content/meetings/tac/keydocs/2006/0508/RollBlackouts_April_17_2006_04.pdf (accessed 20 February 2020).
24. Newman, M.E.J. (2003). The structure and function of complex networks. *SIAM Review* 45 (2): 167–256.
25. Larson, C. and Mercer, A. (9 Nov. 2004). Global health indicators: an overview. *Canadian Medical Association Journal* 171 (10): 1199–1200.
26. Hossain, M.A., Merrill, H.M., and Bodson, M. (2017). Evaluation of metrics of susceptibility to cascading blackouts. *Power & Energy Conference at Urbana*, Urbana (Feb. 2017).
27. COES SINAC (2012). Draft update of the 2013-2022 transmission plan (in Spanish). Committee for the Economic Operation of the National Electric System, Lima Peru (31 May 2012).
28. Watts, D.J. (2003). *Six Degrees: The Science of a Connected Age*, 104–107. New York: W. W. Norton.

CHAPTER 4

MODEL-BASED ANOMALY DETECTION FOR POWER SYSTEM STATE ESTIMATION

Aditya Ashok[1], *Manimaran Govindarasu*[2], *and Venkataramana Ajjarapu*[2]

[1]*Pacific Northwest National Laboratory, Richland, WA, USA*
[2]*Department of Electrical and Computer Engineering, Iowa State University, Ames, IA, USA*

4.1 INTRODUCTION

State estimation (SE) is one of the centerpieces of the energy management systems (EMS) in the control centers to obtain a real-time snapshot of the grid operating conditions and enables grid operators to perform various what-if analyses and market calculations to operate the grid securely and efficiently. SE being a fundamental and a critical network application in the EMS, there are strict operational standards defined by the North American Electric Reliability Corporation (NERC) to ensure that utilities and independent system operators (ISOs) have reliable and accurate telemetry data to compute a valid SE solution. In order to ensure that SE executes reliably and continuously, utilities rely on their supervisory control and data acquisition (SCADA) networks to collect remote telemetry data (status and analogs) from various substations throughout the bulk power system. Because of this dependence on the SCADA data, SE is vulnerable to cyberattacks that could impact data integrity and/or availability.

Traditionally, the formulation of power system SE has been set up to handle gross errors in measurement devices using standard bad data detection techniques leveraging the redundancy in the SCADA measurement configurations. However,

Advances in Electric Power and Energy: Static State Estimation, First Edition.
Edited by Mohamed E. El-Hawary.

Published 2021 by John Wiley & Sons, Inc.

recent literature has shown how SE is vulnerable to stealthy cyberattacks that go undetected by the bad data detection techniques being currently used [1]. Considering the criticality and importance of SE to power grid operations, over the last decade, a considerable amount of focus has been put on research efforts that addressed various aspects such as attack prevention, detection, and mitigation to increase the resilience of SE. Therefore, as part of this chapter, we first identify and describe relevant state-of-the-art research that addresses attack prevention and attack detection and mitigation to provide an overview. Then, we will elaborate in detail about a model-based anomaly detection approach that is complementary to existing bad data detection approaches to detect stealthy cyberattacks on SE. This discussion will include the methodology to detect attacks, a detailed workflow, and a performance evaluation through a case study on the standard IEEE 14-bus system. Finally, we summarize the overall contributions in this chapter.

4.2 CYBERATTACKS ON STATE ESTIMATION

As the smart grid continues to evolve leveraging advances in computation, automation, and communication capabilities, the degree of interconnectivity has led to an increased attack surface as sophisticated and persistent cyberattacks increasingly target several critical infrastructures such as the power grid. The Stuxnet worm [2] was an example of a highly sophisticated and stealthy cyberattack on industrial control system controllers that caused physical damage and avoided detection for a long time period. Similarly, carefully orchestrated, coordinated attacks, such as the recent cyberattacks on a distribution grid in Ukraine [3], could impact the reliability and security of the grid. There has been a significant body of research over the last decade that has looked at the vulnerability of SE to such stealthy attacks that bypass existing bad data detection mechanisms and could potentially cause reliability and market impacts to the grid. The following section will review some of the most pertinent literature that addresses cyberattacks on SE.

4.2.1 State of the Art: Literature Survey

As mentioned before, SE using the traditional weighted least squares (WLS) formulation inherently was designed to detect certain types of bad measurements through statistical analysis using the largest normalized residuals (LNR) test and the chi-squared test. However, there are scenarios where the analog measurements or the status measurements could be manipulated by a resourceful and intelligent adversary in a stealthy and intelligent manner to inject false data that cannot be detected by these tests. Before we introduce the attack models, we describe the basic equations that are used for SE below:

$$z = h(x) + e \tag{4.1}$$

$$z = Hx + e \tag{4.2}$$

$$\hat{x} = \left(H^T R^{-1} H\right)^{-1} H^T R^{-1} z \tag{4.3}$$

$$r = z - H\hat{x} \tag{4.4}$$

$$\| r = z - H\hat{x} \| \leq \tau \tag{4.5}$$

Equation (4.1) shows the relationship between the set of measurements (z) and the state variables (x), where $h(x)$ is a nonlinear vector function that expresses z in terms of x and e represents the measurement errors (assumed to be Gaussian with zero mean). Under DC power flow model assumptions, z and x are related by Eq. (4.2), where H represents the measurement Jacobian matrix. H is also referred to as the measurement configuration that shows the mapping between the measurements and the state variables. If there are n state variables in the system, H typically is of the order $m \times n$, where $m > n$. Typically, SCADA measurement configurations have redundant measurements to cover for data availability, accuracy, and bad data detection. Equation (4.3) expresses the state estimates in terms of z, H, and the measurement error covariance matrix R. The order of z and e are $m \times 1$, the order of H is $m \times n$, and the order of x and $\hat{x}$ is $n \times 1$. In WLS formulation, the state estimates are considered valid only if the vector of measurement residuals r is less than a threshold τ as shown in Eq. (4.4).

4.2.1.1 Cyberattacks on Analog Measurements

Liu et al. showed that there exists the possibility of a stealthy false data injection attack that manipulated targeted measurements to manipulate state estimates [1]. Equations (4.6)–(4.8) show the attack template for the stealthy false data injections, where the injection vector a does not create any change in measurement residuals, as shown in Eq. (4.8), if it satisfies Eq. (4.7). One of the key assumptions in this attack model is that the attacker has access to the measurement configuration H:

$$z_{\text{attack}} = z + a \tag{4.6}$$

$$a = Hc \tag{4.7}$$

$$\| z_{\text{attack}} - H\hat{x}_{\text{attack}} \| = \| z - H\hat{x} \| \leq \tau \tag{4.8}$$

Figure 4.1 presents an intuitive explanation of this attack using the basic equation for the linear case of power system SE. For a stealthy false data injection attack on a state variable x_2, the measurements that need to be manipulated consistently are determined by the nonzero entries in its corresponding column vector (H_{i2}), and this is tied to the measurement configuration. Therefore, for manipulating x_2, z_1, z_2, and z_m need to be injected correspondingly with false data as per Eq. (4.7).

There have been several papers that looked the problem of stealthy false data injection attacks on SE after the original work by Liu et al. [1]. These papers could be grouped under one or more of the following three categories, namely, vulnerability assessment, impact analysis, and attack detection and mitigation. A detailed survey of most of the work in these categories has been captured in [4]. Table 4.1

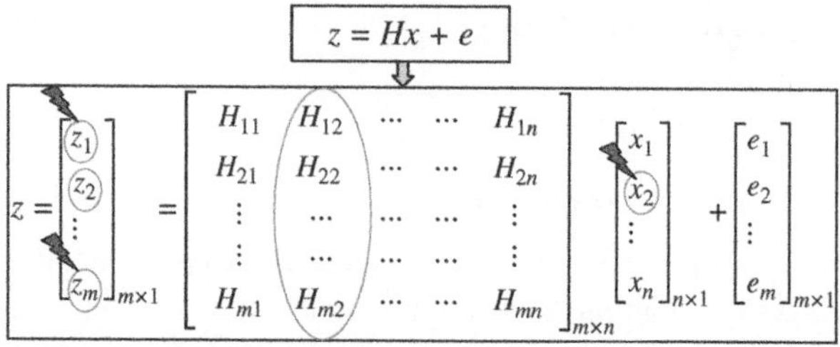

Figure 4.1 Intuitive understanding of stealthy false data injection attacks.

TABLE 4.1 Synthesis of related work.

Key contribution	Relevant literature
Vulnerability assessment	
1. Characterize attack model mathematically	[5–9]
2. Identify algorithm to enumerate attack vectors	[5, 9–11]
Impact analysis	
1. Quantify impact in operational metrics (state estimate errors, system operating limit violations)	[5, 8]
2. Quantify impact in market prices (locational marginal price fluctuations)	[6, 12, 13]
Attack detection and mitigation	
1. Online anomaly detection and/or bad data detection methods for attack detection	[6, 10, 14, 15]
2. Offline measurement placements to increase redundancy	[9, 16, 17]
3. Placement of infrastructure security mechanisms for improving security	[11, 16]

presents a condensed synthesis of the key contributions from the most relevant papers in literature that were published after the initial work done by Liu et al. [1].

4.2.1.2 *Cyberattacks on Topology Measurements*

One of the core assumptions that most of the papers make including the one by Liu et al. [1] is that the topology of the grid is accurate. The topology of the grid that is used for SE and other network applications is referred to as the bus/branch model, which is obtained after a translation from the node/breaker model that is populated by SCADA status measurements from the remote substations. There are several factors such as operator errors, maintenance, and cyberattacks that could result in an incorrect network topology in the EMS. While it is relatively harder to create topology errors that go undetected by the bad data detection methods, there are a few papers that have shown how stealthy topology-based attacks could be created to impact SE and cause significant impacts on grid operations and markets [8, 18].

As mentioned previously, the SCADA measurement configuration has redundant measurements in order to account for measurement loss due to network availability or reliability issues and also to improve the overall accuracy of the

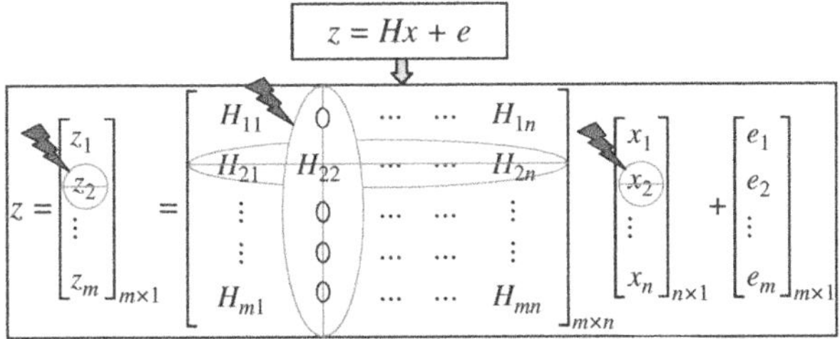

Figure 4.2 Intuitive understanding of stealthy topology attacks.

estimation. It is to be recalled here that the measurement Jacobian matrix H is a tall matrix of the order $m \times n$, where the total number of measurements m is always larger than the number of states to be estimated n. This ensures that even when certain measurements are unavailable, the redundant measurements can provide the equivalent information without impacting the "observability" of the system. However, there could be cases where the redundancy in H might be low either due to issues such as network availability or due to lack of adequate data coverage in areas external to a utility SCADA system. In those cases, this leads to the situation where some measurements in the measurement configuration become a "critical measurement." By definition, "critical measurements" are those whose removal affects the system observability. This is the aspect that is leveraged in a topology attack, where the attacker could use knowledge about H to identify branches in the system that are "critical measurements" to create undetectable topology changes.

Figure 4.2 intuitively shows how the removal of critical measurements leads to a stealthy topology attack. Mathematically, the unavailability of a critical measurement results in a zero column vector in the H matrix causing a reduction of its rank, thereby reducing the observability of the system through SE. In the example shown in Figure 4.2, if z_2 is a critical measurement, its removal causes the second column of H to become zeros, thereby resulting in removal of x_2 from SE process. Because of z_2 being a "critical measurement," this topology change would not be captured as part of the traditional bad data detection method, namely, the LNR test, as there would be no redundant row in H. Therefore, such a stealthy topology change on a transmission line branch could potentially impact the system operations and markets by estimating SE based on an incorrect topology. Additional details about how critical measurements can be obtained from the measurement configuration are described in [8].

4.3 ATTACK-RESILIENT STATE ESTIMATION

The previous section identified how SE could be impacted by stealthy false data injection attacks and stealthy topology attacks. In this section, we will look at ways that will increase the resiliency of SE against these types of attacks. Specifically, we will look at how existing literature addresses various aspects of attack prevention and attack detection to address this problem.

Fundamentally, the approaches to an attack-resilient SE process can be classified into two types: offline and online. Offline approaches mostly target attack prevention; however, they are limited by the assumptions about the attacker's resources. Therefore, relying only on offline attack prevention approaches would be inadequate to secure SE if the attacker resources exceed the assumptions made as part of the solution design. On the other hand, though online approaches do not make very strict assumptions about attacker resources, they are also constrained by their assumptions about the attack magnitudes, false positives, and false negatives as part of tuning the new bad data detection methods or the model-based anomaly detection. Therefore, a holistic approach should ideally involve a combination of attack-resilient planning approaches to improve attack prevention capabilities in conjunction with attack-resilient anomaly detection approaches to improve attack detection and mitigation resulting in a defense-in-depth architecture for an attack-resilient SE. The following sections will provide a brief overview of relevant methods for both attack prevention and detection/mitigation respectively.

4.3.1 Attack Prevention

Offline approaches that are part of existing literature as part of attack prevention are based on increasing the redundancy of the measurement configuration and protecting the redundancy by securing the critical measurements. Another aspect of attack prevention involves dynamically modifying the measurement configuration, weights, and the SE algorithms to increase attack difficulty. These attack prevention approaches are briefly discussed below.

4.3.1.1 Measurement Design

Conceptually, measurement design involves the selection of measurements, both net power flow injection measurements, as well as voltage and transmission line power flow measurements, across various buses in the power grid such that the overall costs are minimized subject to redundancy, SE accuracy, and bad data detection constraints for a range of scenarios including measurement loss and contingencies. The topic of SE measurement design has been well researched and has a wealth of relevant literature spread over several decades [19]. Although measurement design has been looked at with a traditional perspective for SE, recent literature that addresses measurement design involves the strategic placement of secure phasor measurement units (PMU) to satisfy redundancy and bad data detection requirements against stealthy false data injection and topology-based attacks [20–24]. The intuition behind how measurement placement prevents against false data injection attacks is the fact that the number of measurements that need to be compromised to create a stealthy attack is proportional to the degree of redundancy in the measurement Jacobian H matrix. More redundancy in the rows of H also implies that the measurement configuration does not contain critical measurements, thereby also increasing the difficulty of creating a stealthy topology attacks.

4.3.1.1.1 Protecting Critical Measurements While increasing the redundancy is a way to prevent stealthy attacks that involve the compromise of a small selection of measurements, another approach to prevent against these attacks involves the identification of measurements that need to be secured against a set of attack scenarios that involve the compromise of a set of measurements given certain assumptions about the capabilities of the attacker. Existing literature that addresses this aspect of attack prevention focuses on this aspect to guarantee that the securing of these measurements will eliminate all possible stealthy attacks corresponding to the attacker resource assumptions made [11, 16, 17].

4.3.1.1.2 MTD-Based State Estimation An interesting aspect of attack prevention has been explored recently in literature by leveraging and applying the idea of moving target defense (MTD) in [25, 26]. MTD involves the exploitation of diversity and randomness to create a dynamic perspective of the target systems, thereby increasing the difficulty of a successful attack. In this context, the aspects that could be exploited involve the measurement set that is used for SE, the formulation of SE that could be used, and the measurement weights that are employed during estimation. Intuitively, more measurements in the measurement set improve overall SE accuracy. However, the static nature of measurement configurations leads to stealthy false data injection attacks that go undetected. Therefore, attack difficulty could be increased by randomizing the measurements in the measurement set that are selected for SE. Such an approach could leverage parallel executions of SE for the same data with varying measurement configurations and finally compare the normalized residuals to check for anomalies [25]. Similarly, the same idea could be extended to have parallel executions with SE formulations other than the traditional WLS to compare and contrast results, thereby preventing stealthy attacks.

4.3.2 Attack Detection/Mitigation

Literature on online approaches focuses on developing new bad data detection methods and robust formulations of SE and also on developing anomaly detection that independently validates the output of SE by leveraging information that is external to traditional SCADA measurements. Therefore, they address attack detection and mitigation aspects and are briefly described below.

4.3.2.1 Bad Data Detection

One of the key factors in a false data injection attack is the inability of the traditional bad detection method, namely, the LNR test to detect them. To address this, researchers have developed novel bad data detection methods that perform better than LNR test to ensure that measurement manipulations as part of false data injection attacks are detected [6, 10, 13, 15, 27].

4.3.2.2 Model-Based Anomaly Detection

The novel bad data detection approaches developed for SE handle both scenarios where measurements are erroneous due to meter errors, as well as stealthy and malicious manipulations. This requires rigorous validation across a range of scenarios to be accepted and implemented in industry. Therefore, as part of our prior work, we have looked at developing a model-based anomaly detection approach that complements existing bad data detection approaches to detect stealthy false data injection attacks and could be readily implemented in real-world settings [14]. The basic intuition behind this approach is to leverage information that is independent of traditional SCADA measurements, such as short-term load forecasts, and available secure PMU data, to validate the output of SE to detect and recover from anomalies. This approach will be described in greater detail in Section 4.4.

4.3.2.3 Dynamic State Estimation

Existing literature has looked at other formulations of SE as potential solutions to address the problem of stealthy false data injection attacks. Traditionally, power system SE relies on using the static formulation of SE with no dynamics in the estimation process, whereas conventional formulations of SE used in control and estimation theory involve consideration of dynamics and therefore can track variations across time steps. This aspect is leveraged in [15], where the approach relies on tracking the deviation of measurements historically and with respect to the previous time step and by calculating a distance metric appropriately. Several other approaches have shown the bad data detection capabilities of SE formulations that are based on variations of traditional and extended Kalman filters [28–31]. Another approach involves decomposing the SE into subsystems with some degree of overlap between states and using a robust estimator in each subsystem to improve attack detection [32].

4.4 MODEL-BASED ANOMALY DETECTION

In this section, we describe an online model-based anomaly detection approach that utilizes information that is essentially independent of traditional SCADA measurements to detect potential false data injection attacks and validate the outputs of SE [14]. One of the key limitations behind offline approaches are that they place restrictive assumptions on knowing the attacker's resources for identifying measurements that are to be protected or to place new measurements. Though there are other online approaches such as bad data detection approaches, they still use the untrusted SCADA measurements to detect attacks and validate SE. In contrast, the model-based anomaly detection approach described below uses short-term load forecasts, generation schedule information from power markets, and available secure PMU data to coarsely validate SE outputs and enable detection of stealthy false data injection attacks.

4.4.1 Overall Methodology

Figure 4.3 shows an overview of the model-based anomaly detection approach for detecting stealthy attacks on SE measurements. The anomaly detection algorithm runs in parallel with SE at the same execution rate (every five minutes) and leverages the topology output from the topology processor in the EMS, which resolves the status measurements from SCADA into a bus/branch model. As shown, the algorithm relies on using short-term load forecast information that is available for real-time markets at five minute intervals as a key input. In addition, it also uses generator schedules that are available as part of real-time power markets as additional inputs that would be used along with the load forecasts at the individual bus level to predict the state variables for a given period of execution. The difference between the predicted state variables and the output of SE is computed and compared for every period of execution with thresholds that are obtained through statistical characterization of historical data.

Also, the algorithm will leverage available secure PMU data from selected locations as direct inputs to replace the predicted state variables to improve its detection accuracy for certain key state variables. If this difference is within the thresholds defined, then the validated SE outputs are fed to other network applications in the EMS such as contingency analysis, economic dispatch, optimal power flow, etc. If the difference exceeds thresholds, the algorithm flags it as an anomaly and raises EMS alarms so that SE could be rerun with the suspected bad data being eliminated or with pseudo-measurements. It is to be noted here that the proposed anomaly

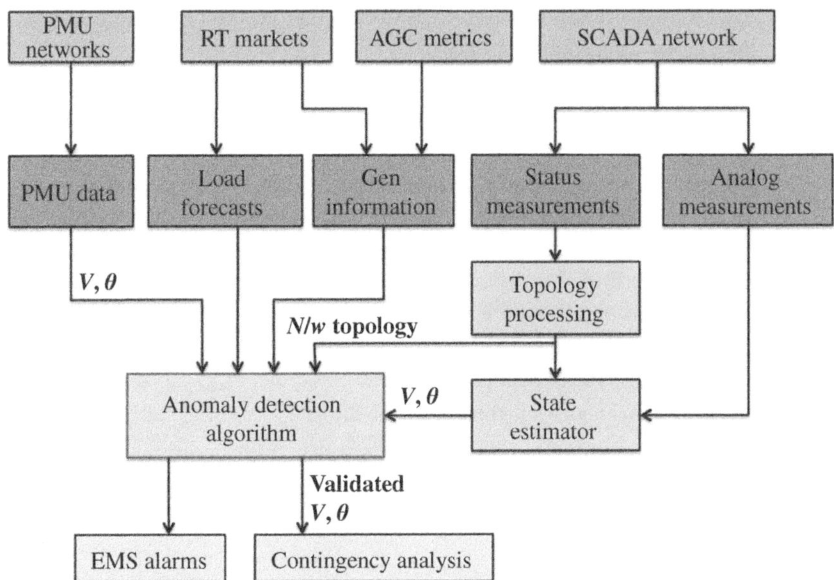

Figure 4.3 Overview of model-based anomaly detection approach. Source: Reproduced with permission from [14]. @ IEEE.

detection algorithm is complementary to existing bad data detection methods such as the LNR test. Its main purpose is to identify stealthy attacks that are above a threshold and is not meant to be a replacement for existing bad data detection in SE.

One of the key factors that affects the performance of the proposed algorithm is the accuracy of the short-term load forecasts. It is well known that typically load forecasts are an order of magnitude less accurate when compared to the accuracy of SCADA measurements [33, 34]. Therefore, the proposed anomaly detection algorithm can only detect attacks that are greater than certain thresholds and also needs to be tuned properly to ensure that it is not overly biased by false positives or false negatives. The availability of more accurate forecasts and/or secure PMU data for certain buses in the measurement configuration greatly improves the detection accuracy at those buses.

4.4.2 Detailed Workflow

Figure 4.4 describes the overall workflow that is performed as part of the anomaly detection algorithm in detail. The workflow represents the set of actions that are executed as part of every period of execution in parallel with the execution of SE. For purposes of simplification, we assume that the interval of SE execution is the same as that of short-term load forecasts. The various steps as part of the workflow are as follows:

Step 1: Update inputs to algorithm
Prior to each run, the updated network topology output is obtained from the topology processor in EMS. Then the current inputs for load forecasts and the generation schedules are obtained from the real-time markets.

Step 2: Perform economic dispatch
Based on the input data obtained in step 1, an economic dispatch computation is performed to obtain the generator dispatches for the schedules and the load forecasts. Traditional economic dispatch formulations use a linear optimization objective function to minimize overall cost of generation with the power balance equality constraints and generator limit inequality constraints. Equations (4.9)–(4.12) show the standard economic dispatch formulation, where c_i represents the generator cost curves for each of the m generators in the system. P_{gen_i} corresponds to the real power output of the ith generator, P_{loss} represents the total active power loss expressed as a function of generator outputs, $P_{\text{load\_total}}$ represents the total system load from the forecasts, and $P_{\text{gen}_i}^{\min}$ and $P_{\text{gen}_i}^{\max}$ represent each generator's minimum and maximum limits, respectively:

$$\min \sum_{i=1}^{m} c_i\left(P_{\text{gen}_i}\right) \tag{4.9}$$

$$\sum_{i=1}^{m} P_{\text{gen}_i} - P_{\text{loss}}\left(P_{\text{gen}}\right) - P_{\text{load\_total}} = 0 \tag{4.10}$$

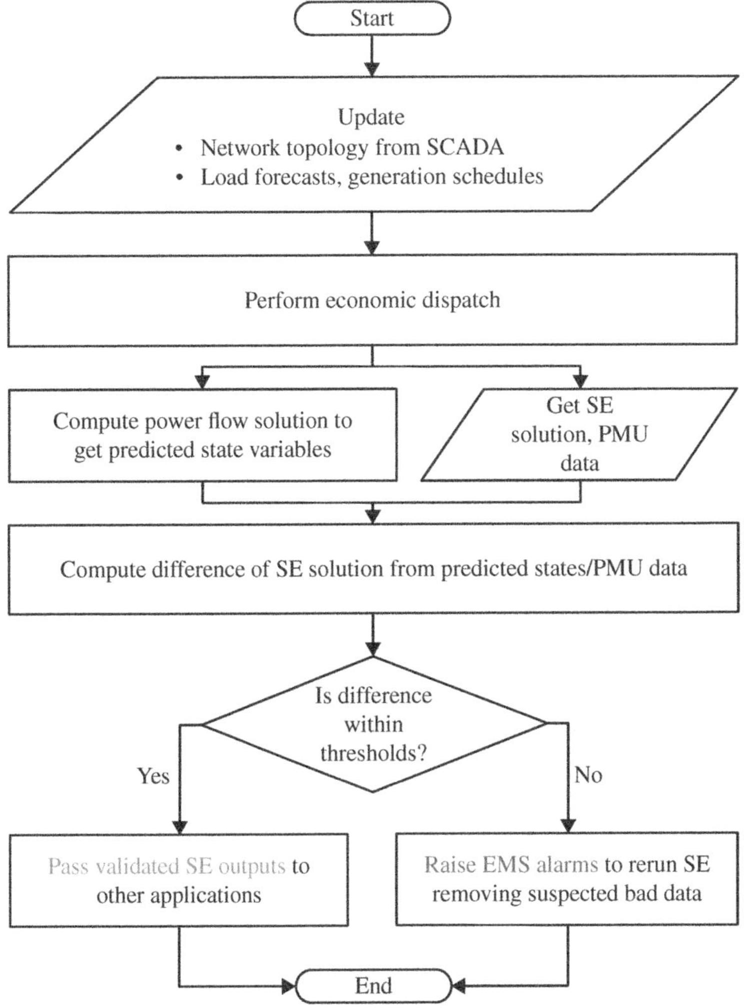

Figure 4.4 Detailed workflow of model-based anomaly detection.

$$P_{\mathrm{gen}_i}^{\min} \le P_{\mathrm{gen}_i} \le P_{\mathrm{gen}_i}^{\max}, \quad \forall i = 1, \ldots, m \tag{4.11}$$

$$P_{\mathrm{gen}_i} \ge 0, \quad \forall i = 1, \ldots, m \tag{4.12}$$

Step 3: Compute power flow solution to obtain predicted state variables
The output of the economic dispatch from step 2 provides the generation outputs and is used in combination with the individual load forecasts as inputs to compute a power flow solution to obtain the predicted state variables, namely, the voltage magnitudes and the phase angles. The power flow problem consists of $N-1$ unknown phase angles (1 bus is chosen

as the reference) and $N - N_G$ unknown voltage magnitudes (generator buses are PV buses). Therefore, there are a total of $2N - N_G - 1$ unknowns as part of the power flow problem, and an iterative method such as the Newton–Raphson is typically used to compute the solution. Equations (4.13)–(4.15) describe the power flow problem, where x represents the unknowns and J represents the system's Jacobian matrix. P_k and Q_k represent the real and reactive power injection at bus k, where P_{gen_k} and P_{load_k} represent the generation and load at that bus, respectively. Equation (4.16) shows an expanded version of Eq. (4.13) where the mismatches between the real and reactive power injections are solved iteratively:

$$f(\underline{x}) = 0, \quad \underline{x} = \begin{bmatrix} \underline{\theta} \\ \underline{V} \end{bmatrix} \tag{4.13}$$

$$\underline{x}^{(i+1)} = \underline{x}^{(i)} - J^{-1} \underline{f}\left(\underline{x}^{(i)}\right) \tag{4.14}$$

$$P_k = P_{\text{gen}_k} - P_{\text{load}_k}, \quad \forall k = 2, \ldots, N \tag{4.15}$$

$$Q_k = Q_{\text{gen}_k} - Q_{\text{load}_k}, \quad \forall k = N_{G+1}, \ldots, N \tag{4.16}$$

$$f(\underline{x}) = \begin{bmatrix} P_2(\underline{x}) - P_2 \\ \cdot \\ P_N(\underline{x}) - P_N \\ Q_{N_G+1}(\underline{x}) - Q_{N_G+1} \\ \cdot \\ Q_{2N-N_G-1}(\underline{x}) - Q_{2N-N_G-1} \end{bmatrix} = \begin{bmatrix} 0 \\ \cdot \\ 0 \\ 0 \\ \cdot \\ 0 \end{bmatrix} \tag{4.17}$$

Step 4: Compute difference of SE solution from predicted states/PMU data
The power flow outputs obtained from step 3 provide the predicted state variables. These are used along with the outputs of SE to compute the difference between the two (shown in Eq. (4.18)). If PMU data is available for certain buses, the corresponding voltage magnitudes and phase angles at that locations from the PMU data serve as replacements for the predicted states before this difference is computed. This enhances the accuracy of anomaly detection for those state variables. x_{dev} represents this difference between the predicted state variables or PMU data ($x_{\text{pred/pmu}}$) and SE outputs (x_{act}). It is to be noted here that the order of all these vectors is $n \times 1$, where n represents the total number of state variables:

$$x_{\text{dev}} = x_{\text{pred/pmu}} - x_{\text{act}} \tag{4.18}$$

Step 5: Compare difference with thresholds to determine anomalies
The final step in the workflow involves a comparison of the differences obtained in step 4 (x_{dev}) with the detection thresholds (τ_dev_i) as shown in Eq. (4.19). These detection thresholds are obtained through a

statistical characterization process that uses historical SE outputs and short-term load forecast information. Based on the statistical data characterization, we can obtain the mean (μ_i) and standard deviation (σ_{dev_i}) of the difference between the predicted states and SE outputs. From this information, appropriate detection thresholds ($\tau_i = k{*}\sigma_{\text{dev}_i}$) are obtained (the value of k is obtained empirically) for each state variable such that the false positives and false negatives are under specified thresholds:

$$|x_{\text{dev}_i}| > \tau\_\text{dev}_i \tag{4.19}$$

If x_{dev_i} is less than the thresholds (τ_{dev_i}), the validated SE outputs can be passed to other network applications. Else, the corresponding deviation is flagged as an anomaly. To mitigate the situation, an on-demand rerun of SE is initiated with additional pseudo-measurements added around the buses where the anomaly was detected to check for bad data again until the malicious measurements are removed.

4.4.3 Case Study and Performance Evaluation on IEEE 14-Bus System

In this section, we evaluate the performance of the proposed anomaly detection algorithm using the IEEE 14-bus system and provide some basic insights about the factors that affect it.

4.4.3.1 System Model

Figure 4.5 shows the one-line diagram of the IEEE 14-bus system along with a measurement configuration that would be used for the case study and performance evaluation. The measurement configuration consists of net power flow injection measurements, which are represented by the arrow marks, and individual power flow measurements on transmission lines, which are represented by little solid squares. For the purposes of this case study, we assume a DC power flow model, i.e. the voltage magnitudes are set at 1 p.u., and only real power measurements are considered. Therefore, only the phase angles are treated as unknowns. This implies that there are 13 unknown phase angles, considering one reference angle as part of the 14-bus system. It is to be noted here that this simplification is only to make the performance evaluation easy to comprehend and does not imply a limitation of the proposed anomaly detection algorithm as it can be readily applied also to the full AC model.

4.4.3.2 Experimental Setup and Parameters

As part of the parameters for the proposed anomaly detection algorithm, we leveraged the short-term load forecast information, namely, the mean (μ) and the standard deviation (σ) from [33]. We obtained the accuracy parameters for traditional SCADA measurements and PMU measurements, namely, the mean and standard deviation based on information in [34]. All these parameters are listed in Table 4.2.

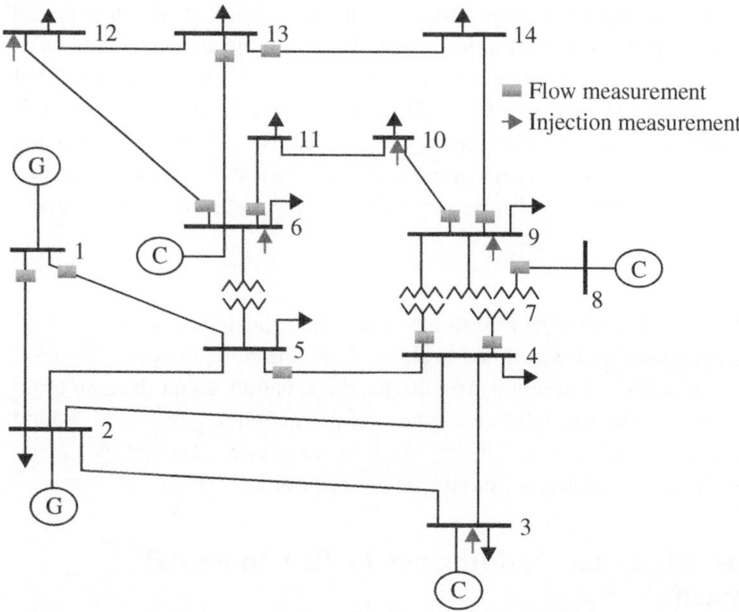

Figure 4.5 IEEE 14-bus system with a measurement configuration. Source: Reproduced with permission from [14]. @ IEEE.

TABLE 4.2 Experimental parameters for case study.

Parameter	Values
Load forecast accuracy	$\mu = 0.00427$
	$\sigma = 0.051194$
SE measurements	17
SE measurement accuracy	$\mu = 0.0$
	$\sigma = 0.001$
PMU measurement accuracy	$\mu = 0.0$
	$\sigma = 0.0001$

The experimental analysis was performed by analyzing the differences between the predicted state estimates and actual SE outputs over an entire day's load curve. Five minute load information from New England Independent System Operator (NE-ISO) data from [35] was normalized, converted into p.u., and used for our case study. We assumed that the load forecasts had a five minute resolution and also that the SE was run once every five minutes. Therefore, for the period of an entire day, there are 288 execution intervals. For each period, an economic dispatch would be computed using the real-time load information, and based on that the power flow would be calculated. Based on the accuracy parameters assumed for SE measurements, appropriate SCADA measurement data would be generated as per a measurement error model with Gaussian distribution with zero mean. In our use cases, we also assume that buses 2, 6, 8, and 9 (this provides complete observability with minimal PMUs) have PMU data available. The generation of PMU measurement data is similar to SCADA measurements but with their accuracy parameters respectively in place of SCADA measurement parameters.

With respect to the anomaly detection algorithm, the load forecasts are generated by adding the contribution from the load forecast error to the real-time load information. The short-term load forecast error follows a truncated normal distribution with mean and standard deviation as specified in Table 4.2. In order to obtain the thresholds for the detection of anomalies for each state variables, we performed statistical characterization using historical load profile data from NEISO for a week [35]. This characterization provides the mean μ_{dev_i} and standard deviations σ_{dev_i} for the difference between predicted state variables and SE outputs for the historical data without any attacks. Using this data, detection thresholds are chosen such that they appropriately to ensure that they meet target false positive rate (FPR) and false negative rate (FNR) specifications. Typically, the detection thresholds are between 1σ and 2σ for each state variable and are identified empirically using repeated simulations.

In order to simulate the stealthy false data injection attacks, we specifically inject false data in the measurements that correspond to 3 state variables out of the 13, namely, θ_2, θ_4, and θ_{13}. The actual measurements that would have these data injections depend on the attack model as defined in Eq. (4.7). In order to clearly distinguish between the phases with normal operation and the attacks, we perform false data injection attacks only during a small subset of the day (100 out of 288 total samples).

4.4.3.3 *Illustrative Example for an Execution Interval*

Let us look at the various steps that would be performed as part of the anomaly detection algorithm as defined in Section 4.4.2 using an illustrative example:

Step 1: Update inputs to algorithm
Let us assume that Figure 4.5 shows the current topology. The total system load for this execution interval is $P_{\text{load}} = 2.5326$ p. u. The load forecast for this interval is $P_{\text{load\_forecast}} = 2.6046$ p. u.

Step 2: Perform economic dispatch
After computing the solution for the economic dispatch, the generation outputs are $P_{g1} = 1.5$ p. u. and $P_{g2} = 1.1046$ p. u.

Step 3: Compute power flow solution to obtain predicted state variables
Based on the economic dispatch outputs, power flow is computed. Table 4.3 represents the output of the phase angles based on power flow solution.

Step 4: Compute difference of SE solution from predicted states/PMU data
Based on the power flow results, we can generate the SCADA measurements and obtain SE outputs. Similarly, based on load forecast outputs

TABLE 4.3 Power flow solution for the example.

State variable	Value (rad)
θ_2	−0.0440
θ_3	−0.1162
θ_4	−0.1267
θ_5	−0.1167
θ_6	−0.2743
θ_7	−0.2126
θ_8	−0.2126
θ_9	−0.2587
θ_{10}	−0.2853
θ_{11}	−0.3051
θ_{12}	−0.3245
θ_{13}	−0.2925
θ_{14}	−0.2764

TABLE 4.4 SE outputs and predicted states for the example.

State variable	SE outputs (rad)	Predicted states (rad)	Differences (rad)
θ_2	−0.0440	−0.0431	0.0091
θ_3	−0.1161	−0.1180	0.0019
θ_4	−0.1267	−0.1281	0.0115
θ_5	−0.1167	−0.1180	0.0014
θ_6	−0.2739	−0.2799	0.0060
θ_7	−0.2124	−0.2151	0.0027
θ_8	−0.2121	−0.2151	0.0029
θ_9	−0.2585	−0.2618	0.0033
θ_{10}	−0.2851	−0.2904	0.0053
θ_{11}	−0.3045	−0.3132	0.0087
θ_{12}	−0.3238	−0.3340	0.0102
θ_{13}	−0.2921	−0.2991	0.0171
θ_{14}	−0.2760	−0.2810	0.0049

and economic dispatch outputs, the predicted states are calculated. Table 4.4 shows the SE outputs, the predicted states, and the difference between them.

Step 5: Compare difference with thresholds to determine anomalies

Finally, Table 4.5 shows the comparison with the detection thresholds for all the state variables. We can see clearly that the difference between predicted and actual SE output for θ_2 (0.0091 rad) is greater than its detection threshold (0.0063 rad). From this we can conclude that there is a false data injection attack on the measurements corresponding to it.

TABLE 4.5 Comparison of differences with detection thresholds.

State variable	Differences (rad)	Detection thresholds (rad)
θ_2	**0.0091**	**0.0063**
θ_3	0.0019	0.0078
θ_4	**0.0115**	**0.0072**
θ_5	0.0014	0.0066
θ_6	0.0060	0.0128
θ_7	0.0027	0.0103
θ_8	0.0029	0.0103
θ_9	0.0033	0.0122
θ_{10}	0.0053	0.0132
θ_{11}	0.0087	0.0141
θ_{12}	0.0102	0.0162
θ_{13}	0.0171	0.0137
θ_{14}	0.0049	0.0127

4.4.3.4 Detecting Stealthy False Data Injection Attacks

Figure 4.6 shows plots of the differences between the predicted values based on the anomaly detection algorithm and the SE output over the course of a day (288 samples) for state variable θ_2. The top and bottom subplots show the differences during no attack and attack scenarios, respectively. The two straight lines indicate the detection thresholds for θ_2 that were obtained based on statistical characterization of historical data for a week. From the bottom subplot, we can see that the differences between predicted states and SE outputs fall outside the detection threshold enabling the detection of false data injection attacks clearly during the attack period.

4.4.3.5 False Positive Rate (FPR) and True Positive Rate (TPR) Analysis

Careful analysis and tuning of false positives and false negatives are essential aspects of the performance evaluation of every anomaly detection algorithm. By definition, false positives are cases where the anomaly detection algorithm detects an anomaly when there is none, and false negatives involve cases where the algorithm fails to detect an anomaly when there is one. In general, positives are cases where the algorithm detects an anomaly and involves a combination of true positives (TP) and false negatives (FN). Similarly, negatives are cases where the algorithm does not detect an anomaly, which involves a combination of true negatives (TN) and false positives (FP). With these definitions, we can define false positive rate as the ratio of false positives over the total negatives as shown in Eq. (4.20). False negative rate can be defined as the ratio of false negatives over total positives as shown in Eq. (4.21). A complementary measure of FNR is the true positive rate (TPR) or detection rate. TPR can be defined as the ratio of true positives over the total positives as shown in Eq. (4.22):

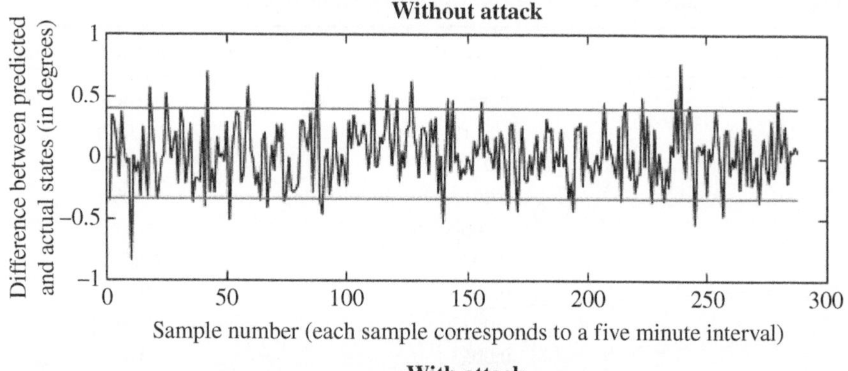

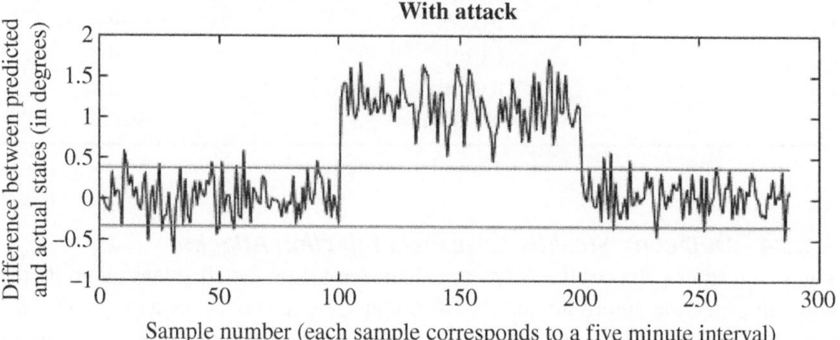

Figure 4.6 Plotting deviation between predicted states and SE outputs over a day for θ_2.

$$\text{FPR} = \text{FP}/(\text{TN} + \text{FP}) \tag{4.20}$$

$$\text{FNR} = \text{FN}/(\text{TP} + \text{FN}) \tag{4.21}$$

$$\text{TPR} = 1 - \text{FNR} = \text{TP}/(\text{TP} + \text{FN}) \tag{4.22}$$

In the context of the proposed anomaly detection algorithm, the accuracy of load forecasts and the minimum attack magnitude that needs to be detected strongly influences the FPR and TPR. Figure 4.7 shows how FPR is influenced by the detection thresholds for a single state variable (θ_2). The detection threshold has been varied from 1.0σ to 2.0σ to show how FPR reduces with an increase in threshold. Similarly, Figure 4.8 shows how the detection rate varies according to the magnitude of attack for a state variable (θ_2) for multiple detection thresholds. It can be seen that as the attack magnitude increases, the detection rate approaches 100%. Also, we can see that higher detection thresholds approach higher TPR much faster.

As we can see from Figures 4.7 and 4.8, a very high detection threshold results in a low FPR; this results in an increase in false negatives and reduces

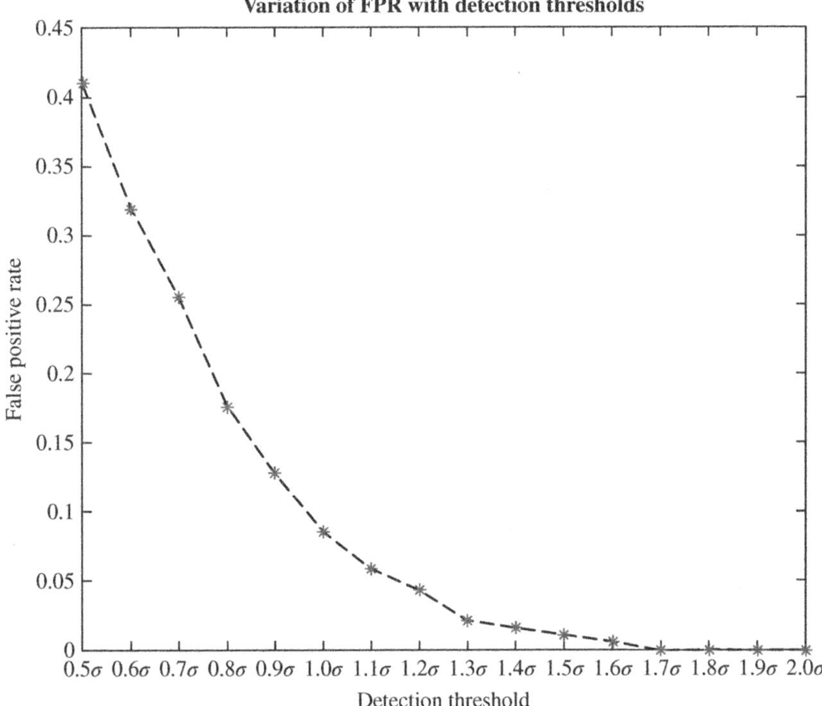

Figure 4.7 Variation of FPR with respect to detection thresholds for θ_2.

the detection rate or the TPR. Similarly, a very low detection threshold results in a high TPR at the expense of a very high FPR. Therefore, a careful balance of both FPR and TPR is essential for optimal performance of the anomaly detection algorithm. For practical deployments, it is essential to clearly identify the input design parameters such that they meet target FPR and TPR specification. For the proposed anomaly detection algorithm, the detection threshold and the minimum attack magnitude that can be detected for each state variable are the design parameters that need to be selected optimally. An empirical methodology to identify these parameters to meet specified FPR and TPR criteria is outlined in [14].

4.5 CONCLUSIONS

As part of this book chapter, we first provide a broad survey of relevant state-of-the-art literature that addresses the vulnerability of SE to stealthy false data injection attacks and topology-based attacks. Second, we briefly describe various offline attack prevention approaches to enhance the redundancy of SE against those

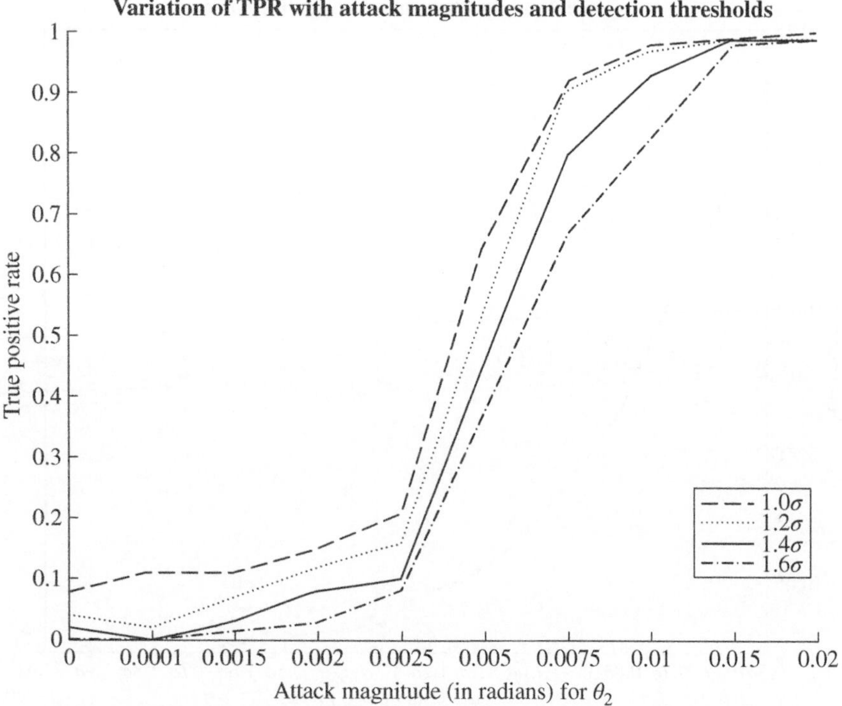

Figure 4.8 Variation of TPR with respect to attack magnitude for various detection thresholds for θ_2.

stealthy cyberattacks and online approaches for attack detection and mitigation that address anomaly detection, bad data detection, and other formulations of SE. Third, we elaborately describe a model-based anomaly detection approach that uses short-term load forecasts, generation schedules, and available secure PMU data to detect anomalies due to stealthy cyberattacks. Finally, we provide insights into the performance of the anomaly detection algorithm using a case study on the IEEE 14-bus system.

The security and resiliency of power system SE is of paramount importance as its output is used by several other network applications in the EMS such as real-time contingency analysis, power markets, etc. A holistic approach to attack-resilient SE should involve a combination of attack-resilient planning approaches to improve attack prevention capabilities in conjunction with attack-resilient anomaly detection approaches and robust SE formulations to improve attack detection and mitigation resulting in a defense-in-depth architecture. Emerging research topics that show promise for the future include PMU-based linear state estimator, integrated hybrid SE formulations with PMU data and SCADA, robust and dynamic SE formulations, and MTD-based approaches for SE that leverage redundancy and randomization of measurements.

REFERENCES

1. Liu, Y., Ning, P., and Reiter, M.K. (2009). False data injection attacks against state estimation in electric power grids. In: *Proceedings of the 16th ACM Conference on Computer and Communications Security*, ser. CCS '09, 21–32. New York, NY: ACM. http://doi.acm.org/10.1145/1653662.1653666.
2. Langner, R. (May–Jun. 2011). Stuxnet: dissecting a cyberwarfare weapon. *IEEE Security and Privacy* 9 (3): 49–51.
3. Industrial Control Systems Cyber Emergency Response Team (ICS-CERT). (Feb. 2016). Cyber-Attack Against Ukrainian Critical Infrastructure – Alert (IR-ALERT-H-16-056-01). https://ics-cert.us-cert.gov/alerts/IR-ALERT-H-16-056-01 (accessed 7 September 2020).
4. Deng, R., Xiao, G., Lu, R. et al. (Apr. 2017). False data injection on state estimation in power systems – attacks, impacts, and defense: a survey. *IEEE Transactions on Industrial Informatics* 13 (2): 411–423.
5. Liu, Y., Ning, P., and Reiter, M.K. (Jun. 2011). False data injection attacks against state estimation in electric power grids. *ACM Transactions on Information and System Security* 14 (1): 13:1–13:33. http://doi.acm.org/10.1145/1952982.1952995.
6. Kosut, O., Jia, L., Thomas, R., and Tong, L. (Dec. 2011). Malicious data attacks on the smart grid. *IEEE Transactions on Smart Grid* 2 (4): 645–658.
7. Hug, G. and Giampapa, J. (Sep. 2012). Vulnerability assessment of AC state estimation with respect to false data injection cyber-attacks. *IEEE Transactions on Smart Grid* 3 (3): 1362–1370.
8. Ashok, A. and Govindarasu, M. (2012). Cyber attacks on power system state estimation through topology errors. *2012 IEEE Power and Energy Society General Meeting* (July 2012). San Diego, CA: IEEE, pp. 1–8.
9. Giani, A., Bitar, E., Garcia, M. et al. (Sep. 2013). Smart grid data integrity attacks. *IEEE Transactions on Smart Grid* 4 (3): 1244–1253.
10. Teixeira, A., Amin, S., Sandberg, H., Johansson, K., and Sastry, S. (2010). Cyber security analysis of state estimators in electric power systems. *2010 49th IEEE Conference on Decision and Control (CDC)* (December 2010). Atlanta, GA: IEEE, pp. 5991–5998.
11. Dan, G. and Sandberg, H. (2010). Stealth attacks and protection schemes for state estimators in power systems. *2010 First IEEE International Conference on Smart Grid Communications (SmartGridComm)* (October 2010). Gaithersburg, MD: IEEE, pp. 214–219.
12. Xie, L., Mo, Y., and Sinopoli, B. (2010). False data injection attacks in electricity markets. *2010 First IEEE International Conference on Smart Grid Communications (SmartGridComm)* (October 2010). Gaithersburg, MD: IEEE, pp. 226–231.
13. Huang, Y., Li, H., Campbell, K., and Han, Z. (2011). Defending false data injection attack on smart grid network using adaptive cusum test. *2011 45th Annual Conference on Information Sciences and Systems (CISS)* (March 2011). Baltimore, MD: IEEE, pp. 1–6.
14. Ashok, A., Govindarasu, M., and Ajjarapu, V. (2016). Online detection of stealthy false data injection attacks in power system state estimation. *IEEE Transactions on Smart Grid* 9 (3): 1636–1646.
15. Chaojun, G., Jirutitijaroen, P., and Motani, M. (Sep. 2015). Detecting false data injection attacks in AC state estimation. *IEEE Transactions on Smart Grid* 6 (5): 2476–2483.

16. Bobba, R., Rogers, K., Wang, Q., Khurana, H., Nahrstedt, K., and Overbye, T. (2010). Detecting false data injection attacks on DC state estimation. *Preprints of the First Workshop on Secure Control Systems, CPSWEEK 2010*, Stockholm, Sweden.
17. Kim, T. and Poor, H. (Jun. 2011). Strategic protection against data injection attacks on power grids. *IEEE Transactions on Smart Grid* 2 (2): 326–333.
18. Kim, J. and Tong, L. (Jul. 2013). On topology attack of a smart grid: undetectable attacks and countermeasures. *IEEE Journal on Selected Areas in Communications* 31 (7): 1294–1305.
19. Ashok, A., Govindarasu, M., and Ajjarapu, V. (2016). Attack-resilient measurement design methodology for state estimation to increase robustness against cyber attacks. *2016 IEEE Power and Energy Society General Meeting (PESGM)* (July 2016). Boston, MA: IEEE, pp. 1–5.
20. Chen, J. and Abur, A. (Nov. 2006). Placement of PMUs to enable bad data detection in state estimation. *IEEE Transactions on Power Systems* 21 (4): 1608–1615.
21. Gou, B. (2008). Generalized integer linear programming formulation for optimal PMU placement. *IEEE Transactions on Power Systems* 23 (3): 1099–1104.
22. Chakrabarti, S. and Kyriakides, E. (2008). Optimal placement of phasor measurement units for power system observability. *IEEE Transactions on Power Systems* 23 (3): 1433–1440.
23. Abbasy, N. and Ismail, H. (2009). A unified approach for the optimal PMU location for power system state estimation. *IEEE Transactions on Power Systems* 24 (2): 806–813.
24. Emami, R. and Abur, A. (2010). Robust measurement design by placing synchronized phasor measurements on network branches. *IEEE Transactions on Power Systems* 25 (1): 38–43.
25. Yao, Y. and Li, Z. (2016). MTD-inspired state estimation based on random measurements selection. *2016 North American Power Symposium (NAPS)* (September 2016). Scottsdale, AZ: Association for Computing Machinery, pp. 1–6.
26. Rahman, M.A., Al-Shaer, E., and Bobba, R.B. (2014). Moving target defense for hardening the security of the power system state estimation. In: *Proceedings of the First ACM Workshop on Moving Target Defense*, ser. MTD '14, 59–68. New York, NY: ACM http://doi.acm.org/10.1145/2663474.2663482.
27. Rawat, D.B. and Bajracharya, C. (Oct. 2015). Detection of false data injection attacks in smart grid communication systems. *IEEE Signal Processing Letters* 22 (10): 1652–1656.
28. Zhao, J., Netto, M., and Mili, L. (2016). A robust iterated extended Kalman filter for power system dynamic state estimation. *IEEE Transactions on Power Systems* (Jul. 2017) 32 (4): 3205–3216.
29. Ghahremani, E. and Kamwa, I. (2011). Dynamic state estimation in power system by applying the extended Kalman filter with unknown inputs to phasor measurements. *IEEE Transactions on Power Systems* 26 (4): 2556–2566.
30. Wang, S., Gao, W., and Meliopoulos, A.P.S. (2012). An alternative method for power system dynamic state estimation based on unscented transform. *IEEE Transactions on Power Systems* 27 (2): 942–950.
31. Singh, A.K. and Pal, B.C. (Mar. 2014). Decentralized Dynamic State Estimation in Power Systems Using Unscented Transformation. *IEEE Transactions on Power Systems* 29 (2): 794-804. doi: 10.1109/TPWRS.2013.2281323.
32. Chakhchoukh, Y. and Ishii, H. (Nov. 2016). Enhancing robustness to cyber-attacks in power systems through multiple least trimmed squares state estimations. *IEEE Transactions on Power Systems* 31 (6): 4395–4405. doi: 10.1109/TPWRS.2015.2503736.

33. California ISO. (Oct. 2010). Integration of renewable resources: technical appendices for California ISO renewable integration studies. http://www.caiso.com/Documents/DraftTechnicalAppendices_RenewableIntegrationStudies-OperationalRequirementsandGenerationFleetCapability.pdf (accessed 7 September 2020).
34. KEMA. (Jan. 2006). Metrics for determining the impact of phasor measurements on power system state estimation-eastern interconnection phasor project. http://citeseerx.ist.psu.edu/viewdoc/download;jsessionid=E16B2AA774EE72DA716AA84B180ACED5?doi=10.1.1.134.1493&rep=rep1&type=pdf (accessed 7 September 2020).
35. New England ISO. (Apr. 2014). Energy, load, and demand reports: five-minute system demand. https://www.iso-ne.com/isoexpress/web/reports/load-and-demand/-/tree/dmnd-five-minute-sys (accessed 7 September 2020).

CHAPTER 5

PROTECTION, CONTROL, AND OPERATION OF MICROGRIDS

A. P. Sakis Meliopoulos[1], *Yu Liu*[2], *Sungyun Choi*[3], *and George J. Cokkinides*[1]

[1]*School of Electrical and Computer Engineering, Georgia Institute of Technology, Atlanta, GA, USA*

[2]*School of Information Science and Technology, ShanghaiTech University, Shanghai, China*

[3]*School of Electrical Engineering, Korea University, Seoul, Korea*

Acronyms

DDSE	Distributed dynamic state estimation
DER	Distributed energy resource
DERMS	Distributed energy resources management system
DG	Distributed generation
MU	Merging unit
IED	Intelligent electronic device
PMU	Phasor measurement unit
UMPCU	Universal monitoring protection and control unit

5.1 PRELUDE

Microgrids are defined as a collection of resources and loads that can be organized and controlled as an entity to appear as an intelligent set of resources and loads. Typically, the resources of a microgrid are generating resources that are interfaced

Advances in Electric Power and Energy: Static State Estimation, First Edition.
Edited by Mohamed E. El-Hawary.

Published 2021 by John Wiley & Sons, Inc.

with inverters. They can be controlled to achieve operating constraints. The protection, control, and operation of microgrids pose significant challenges. For example, protection of microgrids is especially challenging as the presence of inverters result in reduced fault current contribution from microgrids (comparable with load currents) and the failure of legacy protection schemes that rely on separation of fault currents and load currents. In addition protection schemes must be reliable and speedy under grid-connected operation as well as islanded operation. The control and operation of microgrids must be accomplished in a way that makes the microgrid a valuable asset for the power grid and the owners. At the same time the overall protection, control, and operation infrastructure must be of cost commensurate with the economic value of the microgrid, which is very small.

This chapter addresses the issues associated with abovementioned problems for microgrids. The chapter provides a review of the state of the art in microgrid protection, control, and operation. Following the state-of-the-art review, it provides new approaches for integrated protection, control, and operation and in particular a method based on dynamic state estimation.

During fault conditions it is necessary to clear the fault current as soon as possible. However, the microsources in a microgrid are usually interfaced through power electronic (PE) devices that cannot contribute a significant amount of fault current. Additionally, the common protection device that is used in microgrids is the overcurrent sensing device, which may not even respond to such low levels of current. To overcome this challenge, the automatic adaptive protection method that is based on pre-calculated or real-time calculated settings was proposed. Other way to address the low fault current is by increasing the fault current level using a dedicated device.

The protection issues encountered in a microgrid is dependent on whether the microgrid is grid connected or islanded and if the fault is an external fault to the microgrid or internal. Regardless of the connection type or fault location, the common struggle that protection devices in the microgrid encounter is the possible fault current sensitivity problem, where the fault level is too low to be detected. To combat the low current level, it is recommended to install higher rated PE, which can contribute more current under fault conditions.

The concept of adaptive protection for microgrids was discussed, which could be in a centralized or decentralized approach. Both the centralized and decentralized approaches have different communication architecture. In the centralized approach, the master controller polls the slave, such as an overcurrent relay, to obtain the data like equipment status and electrical values, and in turn periodically update the relay settings for all the slaves based on the current state of the microgrid. The sent relay setting update can be based on pre-calculated data during offline fault analysis of the microgrid or an online operating block.

The communication architecture used for the microgrid protection can be a centralized or decentralized architecture. In the centralized architecture, the central controller sends decision commands to the other IEDs in the system, which have different setting for different operating conditions. In the decentralized architecture, a central controller is not required; instead the IEDs make decisions on their own based on the information received from the other IEDs.

As was mentioned previously, the fault current level in the microgrids is limited to low levels because the microsources are interfaced with PE devices, which cannot supply a high current magnitude. Fault current sources (FCS) have been proposed, which are capacitive and have a slow charge, fast discharge characteristic. Under normal circumstances, the FCS device will remain idle in the power network, when there is a fault in the microgrid and the system voltage drops, which activates the FCS. The FCS attempts to restore the system voltage by injecting current into the network. The injected current leads to a high enough current magnitude for a fuse or circuit breakers to clear the fault.

Depending on the distributed energy resource (DER) that is installed in the microgrid, fault currents as high as 130–150% of the fault current without DER can be reached. In order to reduce these fault currents, fault current limiters are proposed, which could either be passive or active limiters. In the passive solution, the fault current is reduced by increasing the impedance of the current path at nominal and post fault conditions. This approach has the disadvantage of generating losses, and voltage drops during normal operating conditions. Under the active limiter solution, highly nonlinear devices that quickly increase their impedance during faults are used.

An essential part of microgrids is the aspect of microgrid control. The functionalities of the control can be categorized into the low level, which is composed of local control and protection such as primary voltage and frequency control, battery management, etc.; the medium level, which is composed of microgrid control such as load shedding and active/reactive power control; and the upper level control, which is composed of the upstream network interface such as upstream coordination.

A microgrid control architecture has been proposed, which is composed of the microgrid controller, which is responsible for the control and monitoring of the DERs; the microgrid center controller, which provides an interface between the microgrid and the microgrid operators; the distribution management system (DMS), which collaborates between the distribution system operator (DSO), energy services company (ESCO), and the microgrid operator; and the microgrid operators.

The operation of the proposed microgrid architecture can be in a centralized or decentralized way depending on the availability and affordability of resources such as personnel and microgrid equipment. A centralized control that is composed of a single owner constitutes a single task of minimization of energy costs, while decentralized control that is composed of multiple owners does not have such unitary goals as each owner may have different goal than the other owners. Furthermore, under market participation, the centralized control has all units working in collaboration, whereas in a decentralized control some units may be in competition with each other.

Whether the system is operated in a centralized or decentralized way, it is necessary to perform forecasts of the electricity demand, heat demand, generation, and external prices for the next few hours. Forecasting therefore enables the system to better optimize revenues for the production process.

The DSO needs to have an overview of the system network operation conditions and be able to define appropriate control strategies. To achieve this functionality, state estimation, which plays an important role in the network management, is utilized. At the transmission level, state estimation is used to reduce the uncertainty of the available redundant measurements, while at the distribution level where there are typically a lot less data, state estimation is used to make up for the lack of measured data. While performing state estimation in the distribution system, it is possible for a large number of pseudo-measurements to give rise to convergence problems, in which case robust SE algorithms that is based on orthogonal transformation can be used.

When state estimation is performed, the solution obtained is primarily dependent on the quality of data measurements that were received. In order to account of the receipt if erroneous data that will eventually affect the solution that is obtained, the use of fuzzy state estimation (FSE) is used, which is based on data that is characterized by uncertainty applying fuzzy sets theory. With this approach, if the microgrid communication medium is lost resulting in the loss of data, the qualitative data obtained from FSE can be used to replace the missing measurements.

5.2 INTRODUCTION

Environmental concerns and increasing electric demand have inevitably promoted the proliferation of renewable and distributed energy resources (DERs) in transmission and distribution systems. However, renewable DERs such as wind and solar power have uncertainty in their availability by nature, and therefore, the installation of dispatchable sources such as gas turbine generators or energy storage systems has become essential parts in microgrids to maximize the utilization of renewable DERs in a timely manner. Furthermore, when energy balance between power supply and demand is lost, DERs must be managed and coordinated with demand to help maintain system frequency and keep system stability from disturbances. This need gives rise to DER management system (DERMS). It needs to be pointed out that both the dispatchable sources and the demand-side management are achievable with the centralized approach by the DERMS. In other words, the centralized DERMS monitors, protects, and controls a microgrid according to its operational purposes by collecting useful information (e.g. present operating conditions, forecasted data, operational limits, prices, and model data).

The fundamental information of DERMS is the real-time operating conditions. In the bulk power system, the real-time operating conditions can be computed from SCADA data (many times supplemented with phasor measurement unit [PMU] data) and single-phase circuit models by state estimators. The state estimation has played a crucial role not only in providing minimum operational states that represent the entire system operation but also in filtering out measurement errors by means of measurement redundancy. Similarly, the DERMS of microgrids should use this state estimation technique, but in this case, detailed modeling of electric components, based on three-phase circuits, is essential to consider imbalanced operation and unsymmetric structure. Further, the state estimation in microgrids with high penetration of renewable DERs interfaced with PEs must extract state variables based on transient characteristics of microgrid, and thus, the dynamic state estimation is best suitable for this case.

Despite the necessity for the dynamic state estimation in microgrid operations, its implementation in the centralized DERMS is almost impossible even if the microgrid is a small island. The main problem of the centralized dynamic state estimation is due to the fact that each measurement must be sampled with high sampling rates (e.g. 1 kHz), thus significantly increasing the amount of data to be collected to the centralized DERMS. In the end, the state estimation process is slowed down or even stalled as a result of high congestion of communication and computational burden of processing units of the DERMS. It is within this context that the chapter proposes the distributed approach for obtaining the real-time operating conditions of microgrids and performing adaptive protection over renewable DERs on the basis of dynamic characteristics.

There have been numerous efforts for distribution system state estimation [1–3]. In [1], the authors provided a survey of literature related to distribution system state estimation techniques, addressing the importance of advanced metering infrastructure and computational intelligence methods. However, the authors did not consider the dynamic characteristics of renewable DERs. The characteristics of PE loads [2] were considered in distribution system state estimation, but full dynamics that include electrical transients were not taken into account. In [3], the three-phase, current-based distribution system state estimation was presented.

In addition to distribution system state estimation, the distributed state estimation approach has also been researched [4–8]. In [4], the authors presented a distributed state estimation algorithm based on synchronized phasor measurement, providing how to formulate the distributed approach, where to locate PMUs, and where to determine the slack bus. The overlapping zone-based distributed state estimation approach [5] was proposed for scalability and cost efficiency. The local state estimation was performed in parallel, and then, the resultant data were exchanged among zones through overlapped points until global convergence was achieved. However, this method is not proper for real-time operation and monitoring since it takes hundreds of seconds to obtain optimal solution. In [6], the authors classified multi-area state estimation after a detailed survey of literature. The classification was made based on area overlapping levels, computing architecture, measurement synchronization, solution methodology, and so on. Furthermore, a testbed for evaluating distributed state estimation and other distributed applications [7] was proposed, considering a large volume of data exchange. A fully distributed state estimation algorithm was proposed based on the assumption that all local state estimations did not need to be observable at the same time [8].

Meantime, recent trends of state estimation are moving toward on its employment in smart grids or microgrids [9, 10]. In [9], the authors presented the methodology to complement the low number of measurements in distribution systems by extracting pseudo-measurements from data that have different refreshing rates. Moreover, a multilevel state estimation for smart grids was described comprehensively in [10]. The authors presented the generalized approach that integrates existing state estimation algorithms at different levels of systems in order to monitor very large-scale interconnected power systems.

Although distribution system state estimation and distributed state estimation have been researched so far [1–10], there has been little effort to use the dynamic state estimation to capture full dynamics of grids for adaptive protection. Moreover, as grid becomes more active and unpredictable than before, the development of new intelligent protection methodology is of prime importance. In this sense, the chapter addresses the utilization of the distributed dynamic state estimation (DDSE) for the purpose of adaptive grid protection and real-time monitoring and operations.

The chapter, first, describes the implementation of the DDSE on the renewable DER, followed by the description of how to apply the DDSE in two points of view: (i) adaptive setting-less protection and (ii) real-time operation of microgrids achieved by the centralized DERMS. Then, the chapter presents several simulation experiments to verify the effectiveness of the proposed approach (i.e. the DDSE-based operation and protection scheme) by comparing it with the conventional centralized approach with respect to data accuracy and computational speed.

5.3 STATE OF THE ART IN MICROGRID PROTECTION AND CONTROL

5.3.1 Present Protection Methods and Limitations

Microgrid protection challenges include (i) bidirectionality of fault current due to DGs, (ii) reduction in fault current capacity due to converter-interfaced generations (CIGs), (iii) variable fault current level due to output changing of DGs, (iv) variable current level due to dynamic topology (grid-connected model or islanding model, sectionalizing of microgrids, etc.), and (v) looped feeders.

Next, the performances of legacy methods as well as several new methods in literatures are introduced as examples. These methods include legacy distance protection, legacy line differential protection (using alpha plane method), adaptive protection scheme with microgrid central protection unit, and differential energy-based protection. Details are further introduced as follows:

5.3.2 Performance of Legacy Protection Functions Applied to mGrids

Conceptually, microgrids can be protected with legacy protection functions.

Method 1: Legacy Distance Protection

An example microgrid system is shown in Figure 5.1. Legacy distance protection simply calculates the "distance" between the relay and the fault. The relay will trip the line if the calculated "distance" falls within the length of the circuit under protection. Besides, the settings of the distance relay need to guarantee selectivity (e.g. fault F_2 will instantaneously trip by relay III and IV, not relay I and II) and reliability (e.g. if relay III does not isolate fault F_2, relay I needs to operate as a backup protection).

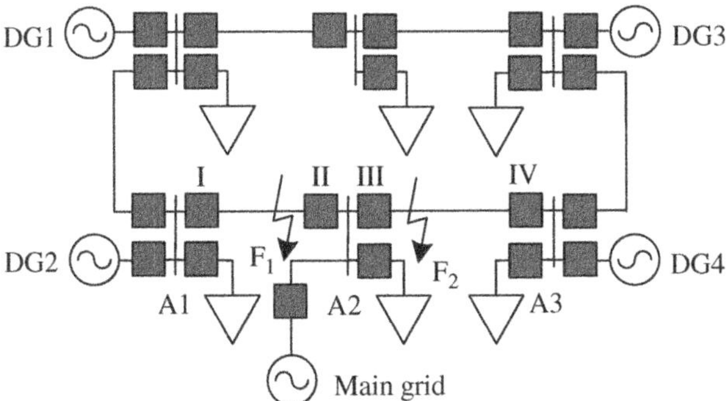

Figure 5.1 Example microgrid circuit, legacy distance protection.

Typical characteristics of a distance relay are shown in Figure 5.2. It includes three tripping zones, typical settings of which are 80, 125, and 260% of the positive sequence impedance of the circuit under protection. Also, faults with calculated impedance falling into zone 1 will be isolated instantaneously, while the faults with calculated impedance falling into zone 2 and zone 3 are isolated with shorter and longer delays, respectively.

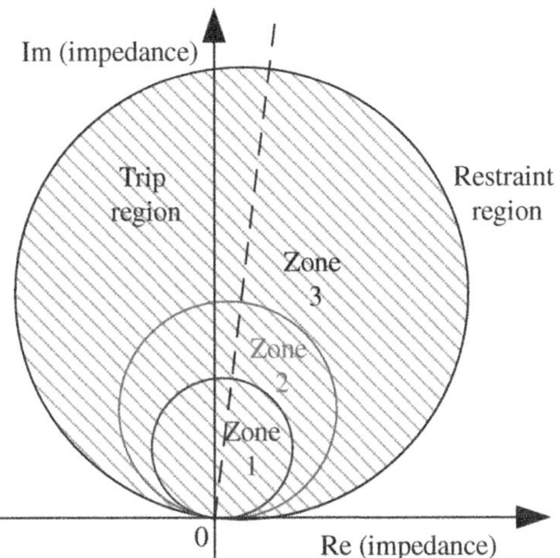

Figure 5.2 Distance relay characteristics.

However, distance relay may not operate correctly in microgrid circuits. That is mainly due to the fact that microgrid circuits are short, i.e. the trip region in Figure 5.2 is tiny. Therefore, when an internal fault happens, a small error brought by the asymmetry of the circuit or the fault impedance may cause the relay to operate with delay, sometime even refuse to operate.

Consider protecting circuit A1–A2 as an example. The length of the circuit is 330 ft. The positive (negative) sequence impedance of the circuit is $0.084\,10 + j0.013\,47\ \Omega$; the zero sequence impedance of the circuit is $0.094\,09 + j0.265\,28\Omega$. The settings of the relay I are zone 1 ($0.0681 \angle 9.10°\ \Omega$, 0.01 seconds delay), zone 2 ($0.1065 \angle 9.10°\ \Omega$, 0.15 seconds delay), and zone 3 ($0.2214 \angle 9.10°\ \Omega$, 0.5 seconds delay). The compensation factor is $k = 2.959 \angle 78.63°$.

An internal-bolted phase A to ground fault (0.01 Ω fault impedance) happened at 0.5 seconds, at the midpoint of circuit A1–A2. The performance of the distance protection is shown in Figure 5.3. From the figure, we can observe that the calculated impedance falls outside of tripping region during the fault, which means that the protection method fails to detect this internal fault.

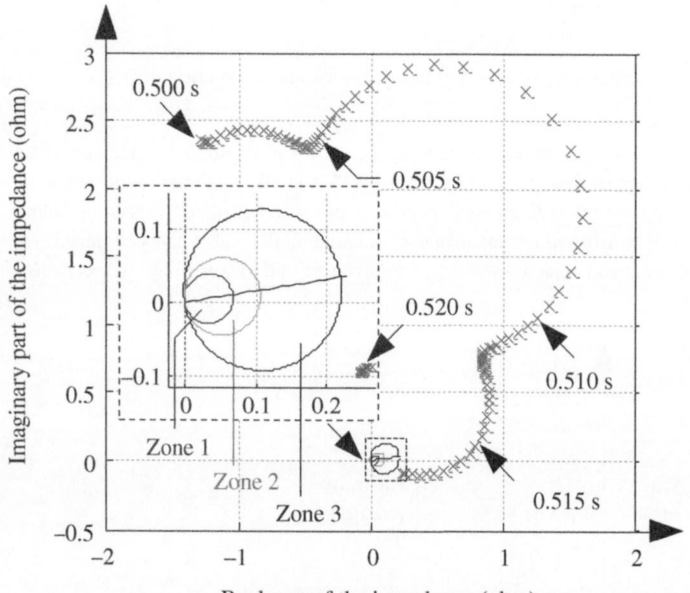

Figure 5.3 Performance of the distance protection, relay I, with a phase A to ground-bolted internal fault.

Method 2: Legacy Line Differential Protection
Consider the microgrid system in Figure 5.4 as an example. Legacy line differential protection is to examine the Kirchhoff current law of the circuit under protection, whether the sum of the currents at both terminals of the circuit is zero. There is an internal fault if the sum is not zero. To consider measurements error, CT saturation, etc., specific settings need to be considered to ensure that the relay will only operate corresponding to internal faults.

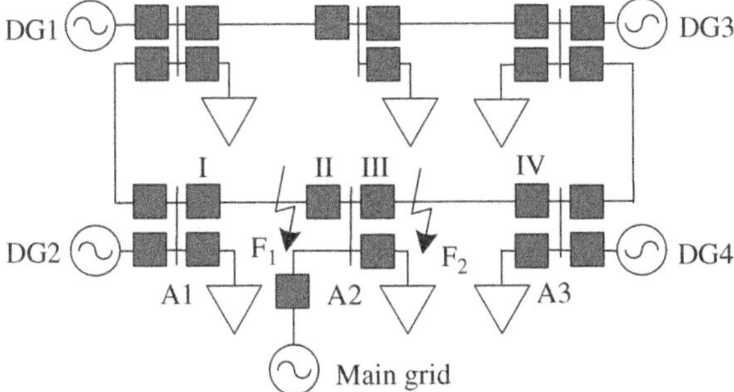

Figure 5.4 Example microgrid circuit, legacy line differential protection.

Typical characteristics of a line differential relay using the alpha plane method [11] are shown in Figure 5.4. The current phasor ratios between two terminals are depicted inside the complex plane. The restraint region is a sector with settings such as inner radius, outer radius, and total angular extent. During normal operating conditions, the ratio will stay near (−1,0). The relay will enable the trip logic of the circuit if the calculated phasor ratio enters the trip region and at the same time at least one of the following thresholds is exceeded: positive sequence current, negative sequence threshold, and zero sequence threshold. Also, a user-defined delay is introduced to ensure reliable operation of the relay. The circuit will be tripped if the trip logic is enabled for the user-defined delay.

Line differential relay performs well in microgrid circuit protections. However, it may have limited sensitivity during high impedance faults (Figure 5.5).

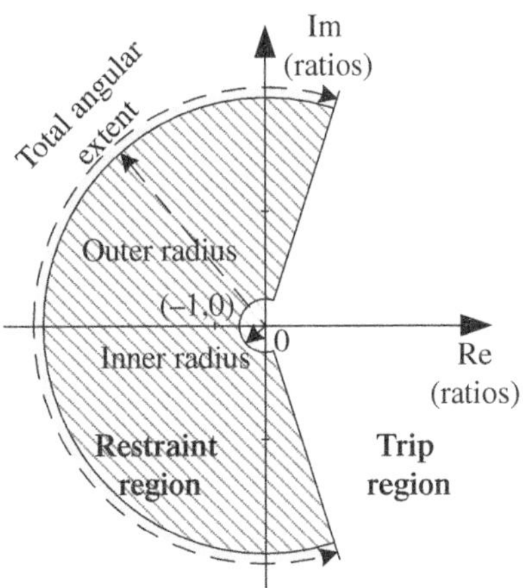

Figure 5.5 Line differential relay characteristics (alpha plane method).

Consider protecting circuit A1–A2 as an example. The settings of the relay are inner radius 1/6; outer radius 6; total angular extent 195°; phase current threshold, 1.2 times the rated current (1.2 × 120 A = 144 A); negative sequence current threshold, 0.1 times the rated current (0.1 × 120 A = 12 A); and zero sequence current threshold, 0.1 times the rated current (0.1 × 120 A = 12 A). The delay is selected as 1 cycle (16.67 ms for a 60 Hz system). Next, a bolted internal fault case and a high impedance internal fault case are considered as examples.

Bolted Internal Fault Case

An internal-bolted phase A to ground fault (0.01 Ω fault impedance) happened at 0.5 seconds, at the midpoint of circuit A1–A2. The performance of the line differential protection is shown in Figure 5.6. Here only the current phasor ratio of phase A is depicted (ratio of phase B and phase C stay near (−1,0) and will not affect trip decision). From the figure, we can observe that the phasor ratio enters the trip region at 0.507 seconds with thresholds exceeded and stays inside the trip region over one cycle. Therefore, the relay will correctly trip the fault at 0.507 seconds + 1 cycle = 0.524 seconds.

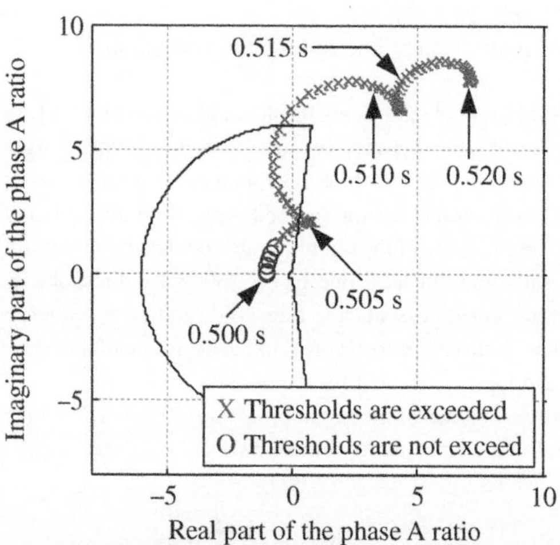

Figure 5.6 Performance of the line differential protection, relay I, with a phase A to ground-bolted internal fault.

High Impedance Internal Fault Case

Consider a phase A to ground high impedance internal fault with 50 Ω fault impedance occurs at the midpoint of circuit A1–A2 and at time 0.5 seconds. The performance of the relay is shown in Figure 5.7. We can see that the ratio hardly moves during the fault, with no thresholds exceeded. Therefore, line differential protection will fail to detect this internal fault.

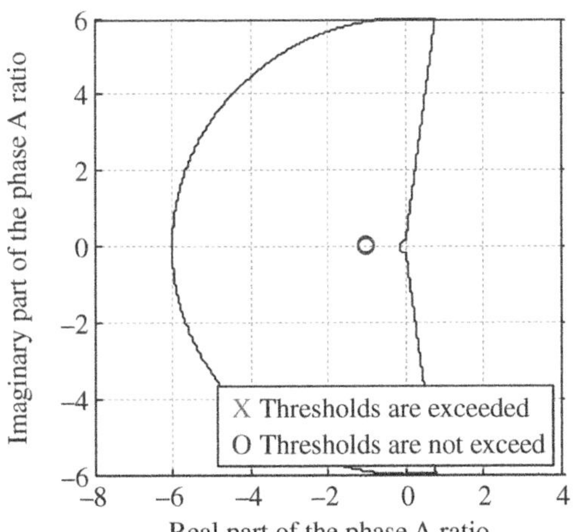

Figure 5.7 Performance of the line differential protection, relay I, with a phase A to ground high impedance internal fault.

Method 3: Adaptive Protection Scheme with Microgrid Central Protection Unit (MCPU)

This method is an extension of legacy time overcurrent protection scheme [12]. Legacy time overcurrent scheme will encounter many issues dealing with microgrid circuits, due to the variable fault current and different coordination due to different topology of the microgrid (grid-connected mode or islanded mode, sectionalizing of microgrids, etc.). This method solves the above limitations of legacy time overcurrent scheme by exchanging DG/topology information inside the whole microgrid circuits and adjusting the relay settings/coordination adaptively. This method is one of an "adaptive protection schemes."

However, this method does not consider variable fault current level due to output changing of DGs, and it may not perform well with short length circuits.

Details of this scheme are provided as follows:

Step 1: Calculate the Settings of Any Relay

$$I_{\text{relay}} = (I_{\text{faultGrid}} \cdot \text{OperatingMode}) + \sum_{i=1}^{m} (k_i \cdot I_{\text{faultDG}i} \cdot \text{Status}_{\text{DG}i}) \tag{5.1}$$

where $I_{\text{faultGrid}}$ is the fault current contribution from the grid, OperatingMode is 0 if the microgrid is in island mode and 1 if the microgrid is connected to the main grid, $I_{\text{faultDG}i}$ is the fault current contribution from the ith DG, $\text{Status}_{\text{DG}i}$ is 0 if the ith DG is connected to the grid and 1 if the ith DG is disconnected to the grid, and k_i is a coefficient corresponding to the relative location between the fault and the ith DG.

$I_{\text{faultGrid}}$ is calculated by the Thevenin equivalent at the location of the fault:

$$I_{\text{faultGrid}}(d) = \frac{V_{\text{th}}}{Z_{\text{th}}(d)} \tag{5.2}$$

where $Z_{th}(d)$ is the total Thevenin equivalent impedance as a function of the distance between the location of the fault and the PCC.

Also, $I_{\text{faultDG}i}$ is selected according to the maximum fault current limitation, which is 1.5 times the rated current of the ith DG.

The coefficient k_i is calculated as a function of the distance between the fault and the ith DG. Note that here it is assumed that the output voltage of DG during the fault is regulated by the stability and safety code of the grid. The value of the output voltage is V_{dm}. The coefficient k_i is defined as

$$k_{ri} = \frac{I_{dmi}}{I_{\text{faultDG}i}} \tag{5.3}$$

and I_{dmi} is calculated as

$$I_{dm} = \frac{V_{dm}}{Z_d \cdot x} \tag{5.4}$$

where x is the distance between the location of the fault and the ith DG.

From the above setting calculation, the settings of relays could be updated according to the topology of the microgrid (according to whether the microgrid is islanded, according to the network distance between the fault and the DGs, according to the network distance between the fault and the main grid).

Also, for all relays, two settings need to be calculated, one is the setting for "forward current" and another is the setting for "reverse current," to consider the effects of bidirectional fault current. Each setting is calculated according to Eq. (5.1).

Step 2: Coordination Consideration of Relays

Here the coordination of relays is arranged to make sure that for any fault, we have one relay for instantaneous tripping and one relay for backup. For each relay, we find its downstream relay according to the topology of the microgrid. These coordination data are stored prior to the fault and updated when there exists any topology change. Afterward, the backup relay of a specific relay is selected according to the upstream relay of that specific relay.

Similarly, for all relays, two groups of coordination data need to be stored, one is for "forward pair" and another is for "reverse pair," to consider the effects of bidirectional fault current.

For example, for a microgrid topology shown in Figure 5.8, the example branch data and the coordination data are shown in Tables 5.1 and 5.2, respectively. Here Ri means the relay at breaker CBi. Also, the delay time (i.e. 200 ms) is selected between the maximum time that could still make the system stable and the minimum time to guarantee selectivity of these CBs during external faults.

The meanings of the two tables are explained as follows. For example, branch 1 includes R2, R3, and the individual relays of components DG1, DG2, and Load1. For relay R2, forward direction, the downstream relay is R3; for relay R2, reserve direction, the downstream relay is R1. Therefore, for the fault current with forward direction (from the main grid to the microgrid), the backup relay for R3 is R2. For the fault current with reverse direction (from the microgrid to the main grid), the backup relay for R1 is R2.

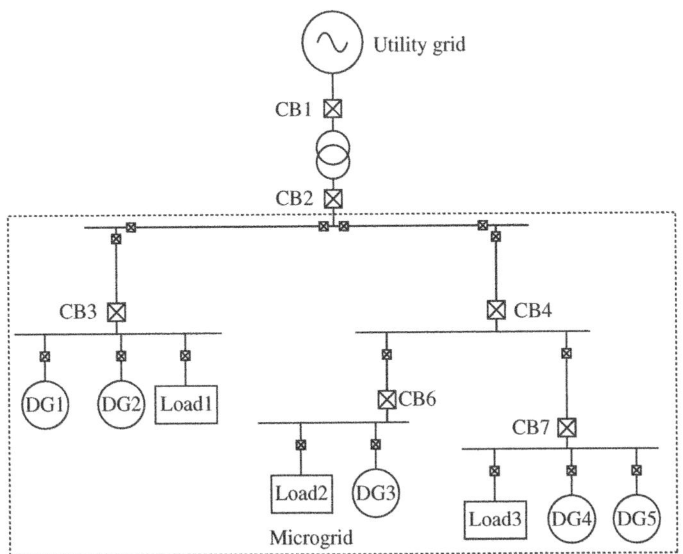

Figure 5.8 Example topology of a microgrid, adaptive protection scheme.

TABLE 5.1 Example branch data for the topology in Figure 5.1 (*A* > *B* means *B* is the downstream relay of *A*).

First branch	Second branch	Third branch
R2 > R3 > I.R.*	R2 > R4 > R6 > I.R.	R2 > R4 > R7 > I.R.

*I.R., Individual relays of components (small boxes in figures).

TABLE 5.2 Example coordination data for the topology in Figure 5.1.

First branch	R2	R3	I.R.*	
Forward pair	R3	I.R.	N/A	
Reverse pair	R1	R2	R3	
Forward current	$IrelayF_{R2}$	$IrelayF_{R3}$	$IrelayF_{I.R.}$	
Reverse current	$IrelayR_{R2}$	$IrelayR_{R3}$	$IrelayR_{I.R.}$	
Second branch	R2	R4	R6	I.R.
Forward pair	R4	R6	I.R.	N/A
Reverse pair	R1	R2	R4	R6
Forward current	$IrelayF_{R2}$	$IrelayF_{R4}$	$IrelayF_{R6}$	$IrelayF_{I.R.}$
Reverse current	$IrelayR_{R2}$	$IrelayR_{R4}$	$IrelayR_{R6}$	$IrelayR_{I.R.}$
Third branch	R2	R4	R7	I.R.
Forward pair	R4	R7	I.R.	N/A
Reverse pair	R1	R2	R4	R7
Forward current	$IrelayF_{R2}$	$IrelayF_{R4}$	$IrelayF_{R7}$	$IrelayF_{I.R.}$
Reverse current	$IrelayR_{R2}$	$IrelayR_{R4}$	$IrelayR_{R7}$	$IrelayR_{I.R.}$

*I.R., Individual relays of components (small boxes in figures).

Method 4: Differential Energy-Based Protection

This method calculates the difference of the spectral energy of the fault currents at both terminals of the circuit under protection [13]. The spectral energy is calculated with the *S*-transform. The trip signal is issued if the difference of the spectral energy exceeds a user-defined threshold (setting). This method also utilizes the idea of line differential protection: the sum of currents should be zero during normal operating conditions and nonzero during faults. The advantage of this method is that the spectral energy of the current is compared instead of the current phasors, so it only requires less accurate time synchronization technique.

Nevertheless, the disadvantage of the method is that it needs specific settings for each relay, and the settings can only be selected according to operator experiences. This leads to limited capability in detecting high impedance faults.

Details of this scheme are provided as follows:

Step 1: Calculate the Spectral Energy of the Interested Current

Use *S*-transform to calculate the spectral energy *E* of the current:

$$E = \{\mathrm{abs}(S(j,n))\}^2 \tag{5.5}$$

$$S(j,n) = \sum_{m=0}^{N-1} X(m+n) \cdot e^{\frac{-2\pi^2 m^2 \alpha^2}{n^2}} \cdot e^{(2\pi mj)i} \tag{5.6}$$

$$X(n) = \frac{1}{N}\sum_{m=0}^{N-1} x(k) \cdot e^{(-2\pi nk)i} \tag{5.7}$$

where α defines the resolution of the *S*-transform (in this chapter, a value of 0.7 gave the best result during analysis), $x(k)$ is the current waveform of one phase, and N is the number of samples per cycle.

Step 2: Calculate the Differential Spectral Energy of Currents Between Two Terminals

Take bus "S" and bus "T" as an example. The differential spectral energy can be calculated as

$$E_{\mathrm{B_S}} = \{\mathrm{abs}(S_{\mathrm{B_S}}(j,n))\}^2 \tag{5.8}$$

$$E_{\mathrm{B_T}} = \{\mathrm{abs}(S_{\mathrm{B_T}}(j,n))\}^2 \tag{5.9}$$

$$\text{Differential energy} = E_{\mathrm{B_S}} - E_{\mathrm{B_T}} \tag{5.10}$$

Step 3: Compare the Differential Energy to the User-Defined Threshold

Note that here the threshold is different for the grid-connected mode and the islanded mode of the microgrid. Detail settings and the performance of the relay are demonstrated with the following example.

An example microgrid is presented in Figure 5.9. Two cases are simulated: one fault in grid-connected mode and one fault in islanded mode. The setting for grid-connected mode is "1," and the setting for islanded mode is "–0.3."

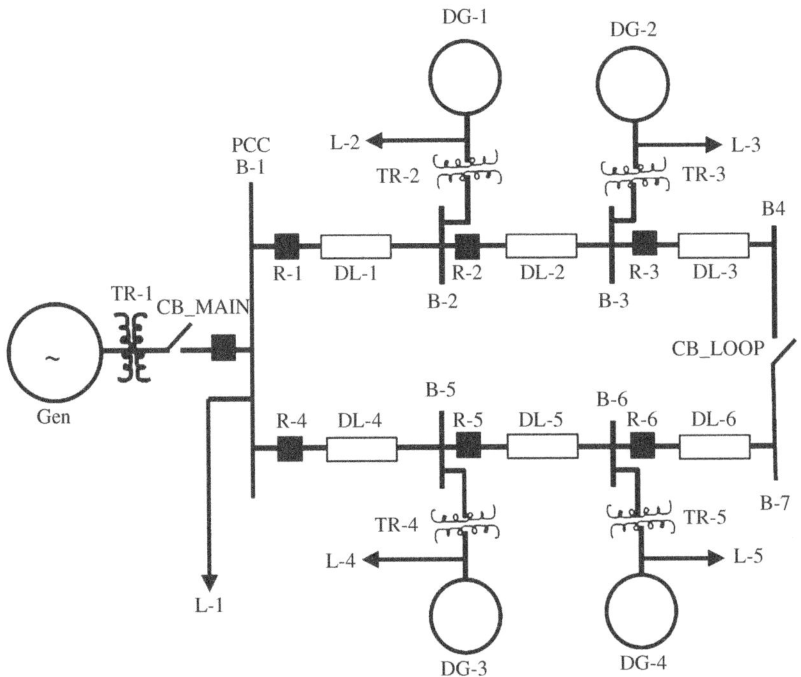

Figure 5.9 An example microgrid system, differential energy-based protection.

Case 1: Grid-Connected Mode

A phase A to ground fault occurs on circuit DL-1, between buses B-1 and B-2. The phase A currents measured at B-1 and B-2 are shown in Figures 5.10 and 5.11. The spectral energy of the fault current at B-1 and B-2 is shown in Figures 5.12 and 5.13.

The differential spectral energy is provided in Figure 5.14. The differential energy is low prior to the fault and exceeds the setting "1" during the fault. The relay will trip this fault 3.5 cycles after the fault occurs.

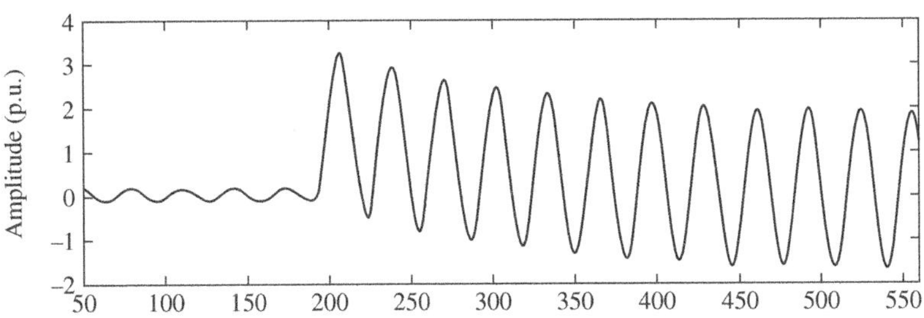

Figure 5.10 Phase A current measurement at bus B-1 of the line DL-1, grid-connected mode.

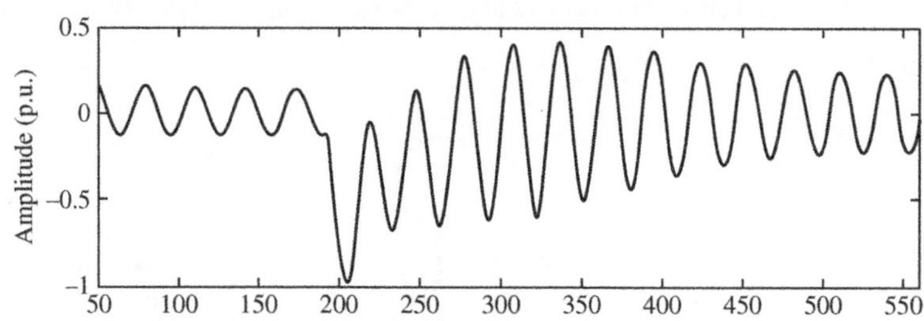

Figure 5.11 Phase A current measurement at bus B-2 of the line DL-1, grid-connected mode.

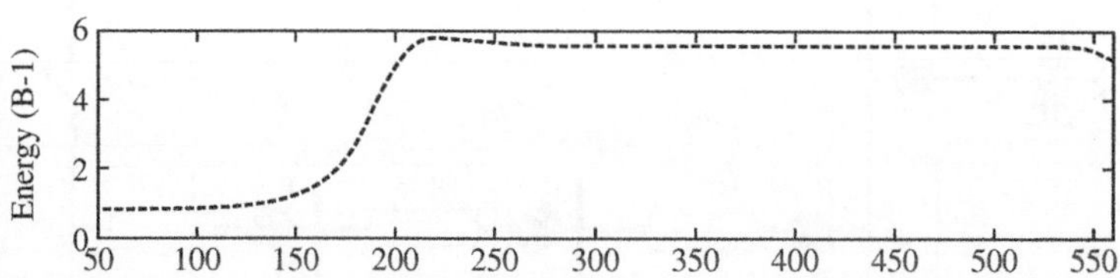

Figure 5.12 Phase A current spectral energy at bus B-1 of the line DL-1, grid-connected mode.

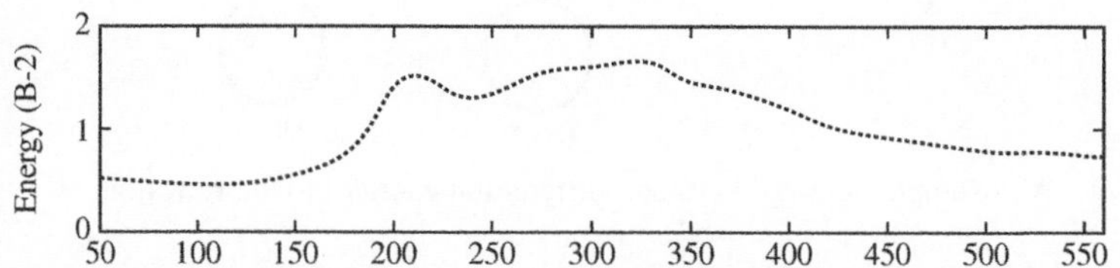

Figure 5.13 Phase A current spectral energy at bus B-2 of the line DL-1, grid-connected mode.

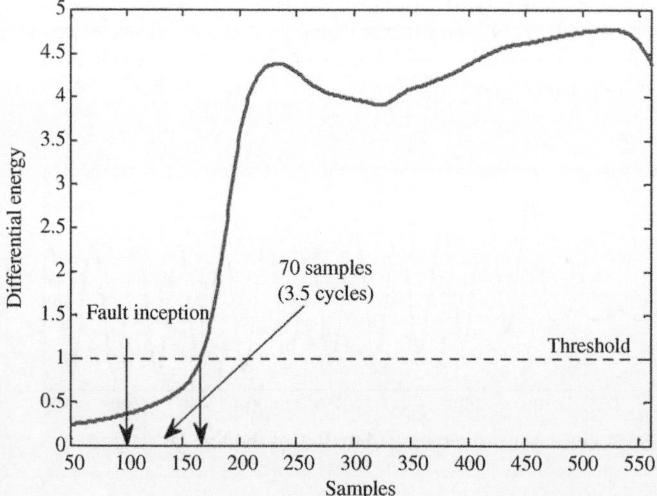

Figure 5.14 Phase A differential energy of the line DL-1, grid-connected mode.

Case 2: Islanded Mode

A phase A to ground fault occurs inside DL-1, between bus B-1 and bus B-2. The differential energy can be calculated similar as before. Figure 5.15 shows the differential energy. The differential energy is low (absolute value) prior to the fault and is large (absolute value) during the fault. The relay will trip this fault 3.6 cycles after the fault occurs.

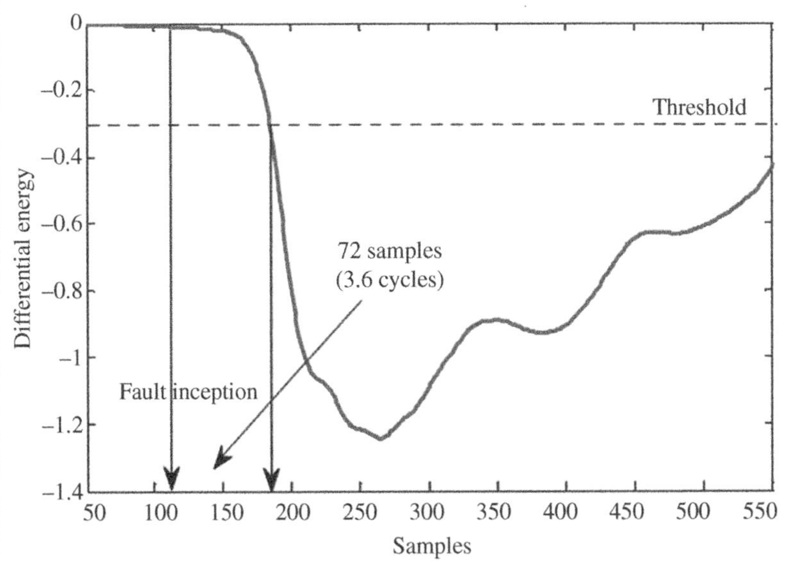

Figure 5.15 Phase A differential energy of the line DL-1, islanded mode.

5.3.3 Present Control Methods and Limitations

Control of microgrids includes the primary control and the secondary control of DERs. Primary control focuses on controlling DGs, storage devices, or controllable loads locally, while secondary control focuses on collaborations of devices inside the whole microgrid (providing set points to primary control). An example control architecture is shown in Figure 5.16, with the microsource controllers (MCs) acting as the primary control and the microgrid central controller (MGCC) acting as the secondary control.

Specifically for the control of CIGs, the main task is to allow active and reactive power sharing among all the generations and to regulate voltages. Since the communication lines are expensive and vulnerable, local controllers are preferable. The idea comes from the droop control of synchronous generators.

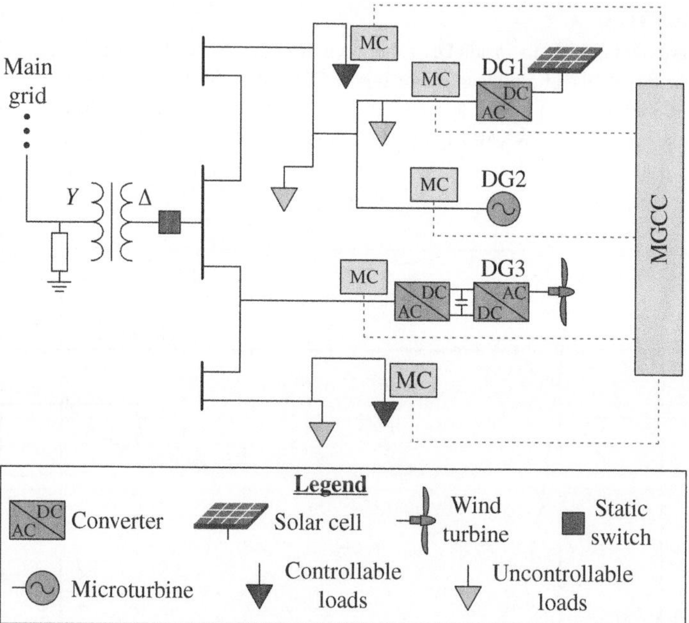

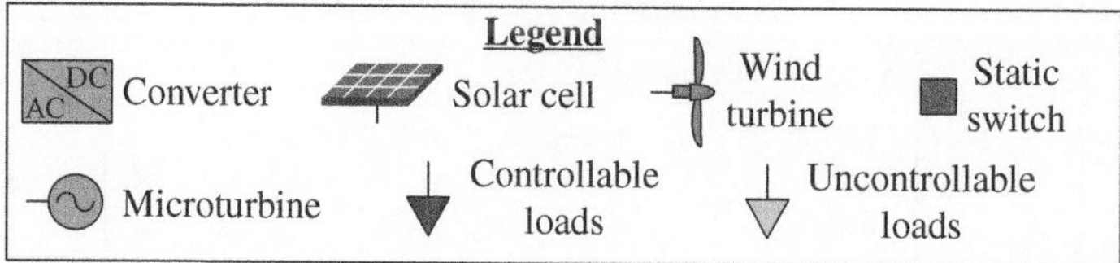

Figure 5.16 Control architecture of an example microgrid.

Case 1: High-Voltage Network

Take the system in Figure 5.17 as an example. Here we have the assumption that, for the circuit connecting the two sources, the reactance is much larger than the resistance (this is true for high-voltage network) so that the resistance of the circuit is neglected. The active and reactive power generated by source 1 are

$$P = \frac{U_1 U_2}{X} \sin \delta$$

$$Q = \frac{U_1^2}{X} - \frac{U_1 U_2}{X} \cos \delta$$

where δ is the phase angle difference between $\widetilde{U}_1$ and $\widetilde{U}_2$.

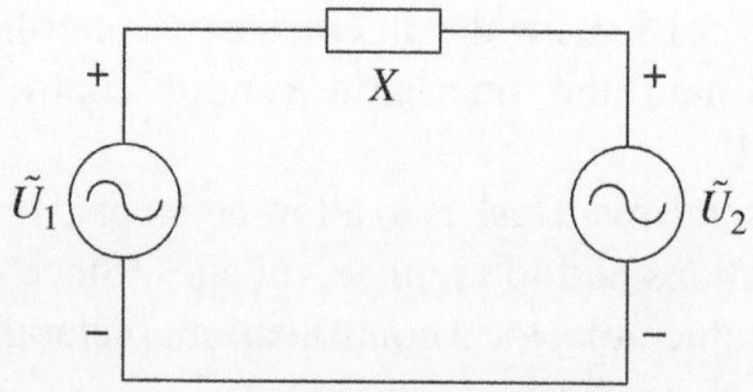

Figure 5.17 Two voltage sources and the inductive circuit connecting them.

Since δ is a small number, the above two equations can be further simplified as

$$P = \frac{U_1 U_2}{X}\delta$$

$$Q = \frac{U_1^2}{X} - \frac{U_1 U_2}{X}$$

From the fact that frequency is the time derivative of phase angle δ, we have the following decoupled relationship: frequency corresponding to active power and voltage corresponding to reactive power. The control method is to use active power/frequency droop and reactive power/voltage droop for the control of inverters (primary control) [14], as shown in Figure 5.18. This droop is known as the "conventional droop." Secondary control provides parameter settings for each component, such as f_0, u_0, droop slopes, etc. The primary control methods can be further categorized into the following two methods:

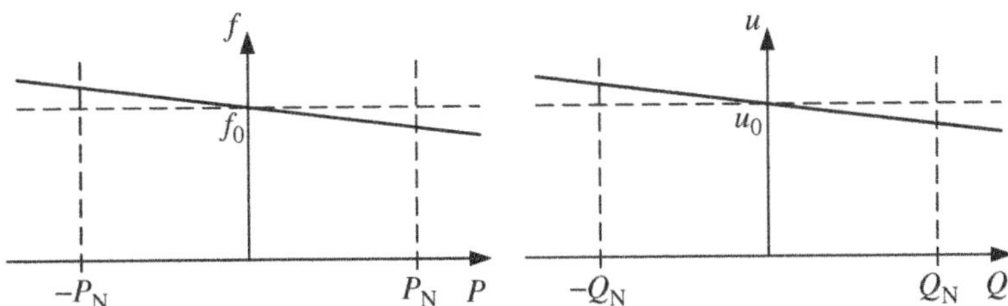

Figure 5.18 Active power/frequency droop and reactive power/voltage droop.

Method 1

For the active power/frequency droop, measuring frequency and defining active power output, as shown in Figure 5.19:

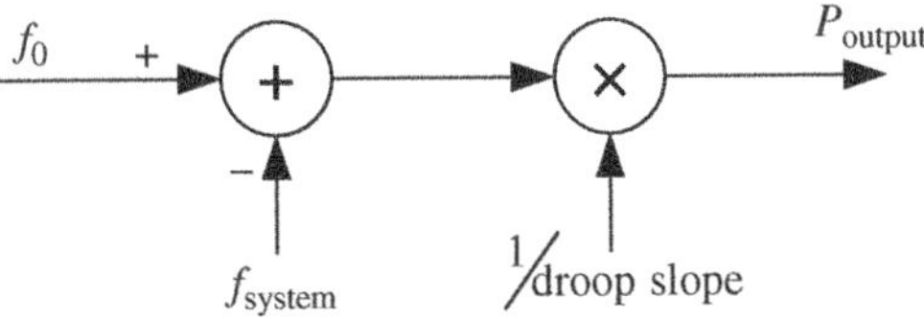

Figure 5.19 Active power/frequency droop diagram, method 1.

For the reactive power/voltage droop, measuring voltage and defining reactive power output, as shown in Figure 5.20:

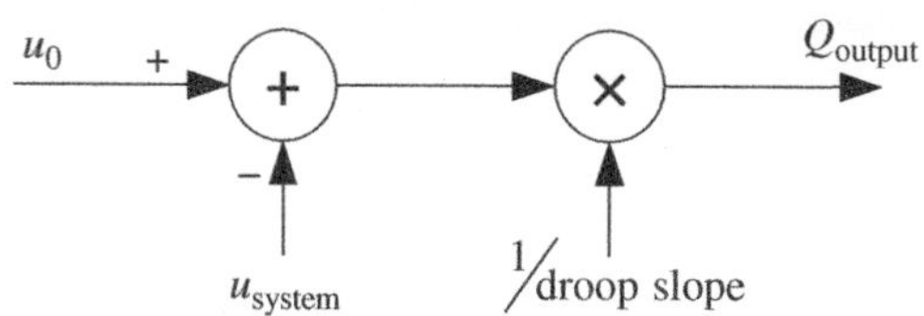

Figure 5.20 Reactive power/voltage droop diagram, method 1.

Method 2

In practice, obtaining accurate instantaneous real power measurements can be much easier than accurate instantaneous frequency measurements. Therefore, the droop control is designed as follows:

For the active power/frequency droop, measuring active power and defining frequency output, as shown in Figure 5.21.

For the reactive power/voltage droop, measuring reactive power and defining voltage output, as shown in Figure 5.21:

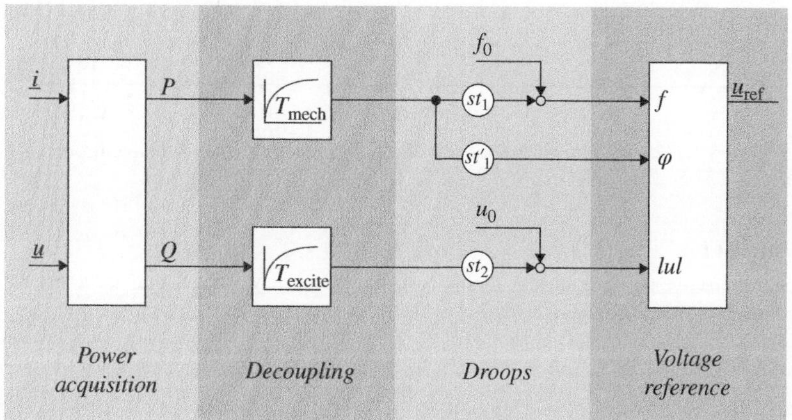

Figure 5.21 Droop control diagram, method 2 [15].

Note that for the secondary control and for both methods, the droop lines are shifted vertically or horizontally according to different scenarios. For active power/frequency droop, if the control selects the active power output to be a fixed reference P_{ref}, the droop line is shifted vertically; if the control selects the frequency to be a fixed reference f_{ref}, the droop line is shifted horizontally, as shown in Figure 5.22. Similarly, for reactive power/voltage droop, if the control selects the reactive power output to be a fixed reference Q_{ref}, the droop line is shifted vertically; if the control selects the voltage to be a fixed reference V_{ref}, the droop line is shifted horizontally, as shown in Figure 5.23.

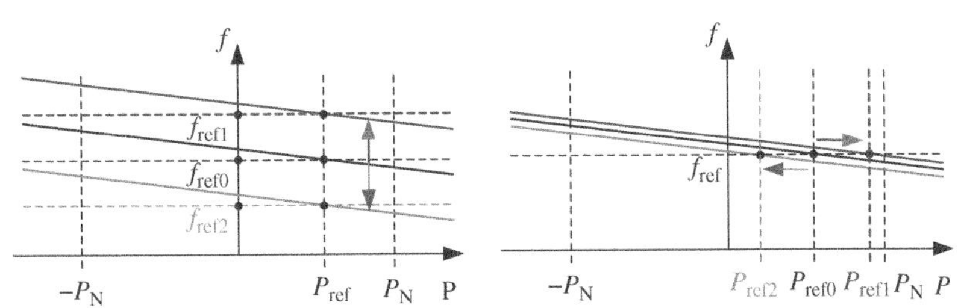

Figure 5.22 Secondary control, active power/frequency droop.

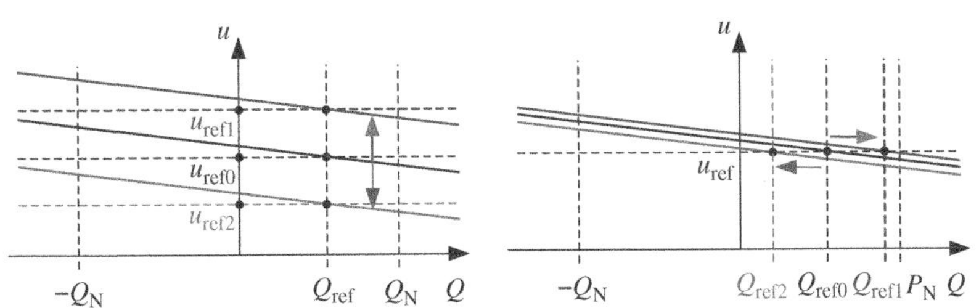

Figure 5.23 Secondary control, reactive power/voltage droop.

Case 2: Low-Voltage Network

Take the system in Figure 5.24 as an example. Here we have the assumption that, for the circuit connecting the two sources, the resistance is much larger than the reactance (this is true for low-voltage network) so that the reactance of the circuit is neglected. The active and reactive power generated by source 1 are

$$P = \frac{U_1^2}{R} - \frac{U_1 U_2}{R} \cos\delta$$

$$Q = \frac{U_1 U_2}{R} \sin\delta$$

where δ is the phase angle difference between $\tilde{U}_1$ and $\tilde{U}_2$.

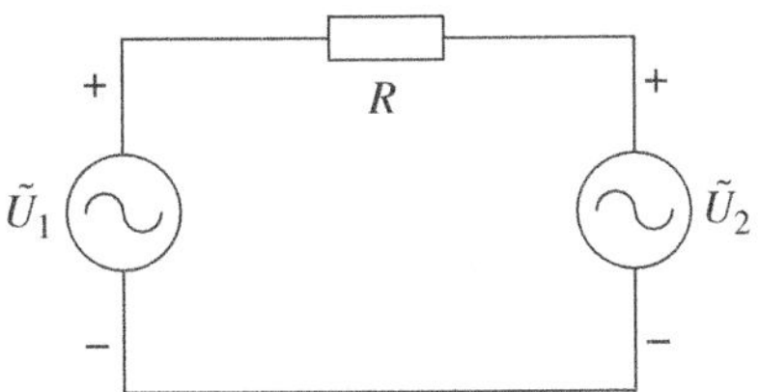

Figure 5.24 Two voltage sources and the resistive circuit connecting them.

Since δ is a small number, the above two equations can be further simplified as

$$P = \frac{U_1^2}{R} - \frac{U_1 U_2}{R}$$

$$Q = \frac{U_1 U_2}{R}\delta$$

Similar as before, we can conclude that the frequency corresponds to reactive power and the voltage corresponds to active power. The control method is to use reactive power/frequency droop and active power/voltage droop for the control of inverters (primary control) [16]. This droop is known as the "opposite droop" in contrast to the previously mentioned "conventional droop." Secondary control is also similar as in high-voltage network, to provide parameter settings for each component, such as f_0, u_0, droop slopes, etc.

However, if the "opposite droop" is applied to control the voltage, we cannot at the same time provide active power dispatch, since in this case the voltage and the active power are strongly coupled. Therefore, the standard is to still use conventional droops, which not only enable active power dispatch but also are compatible with high-voltage networks and rotating generators. However, voltage deviations are expected and depend on the structure of the grid.

To find proper droop parameters [16], consider the circuit in Figure 5.25.

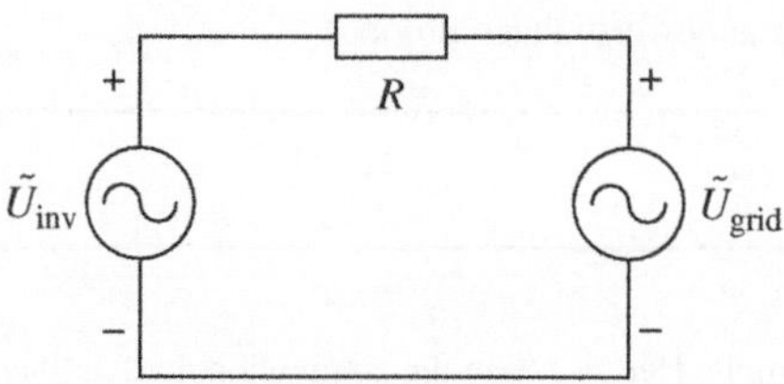

Figure 5.25 Two voltage sources (the inverter and the grid) and the resistive circuit.

The output power of the inverter is

$$P_{\text{inv}} = \frac{U_{\text{inv}}^2}{R} - \frac{U_{\text{inv}} U_{\text{grid}}}{R}\cos\delta$$

where δ is the phase angle difference between $\tilde{U}_1$ and $\tilde{U}_2$.

Since δ is a small value, $\cos\delta \approx 1$. Therefore,

$$P_{\text{inv}} = \frac{U_{\text{inv}}^2}{R} - \frac{U_{\text{inv}} U_{\text{grid}}}{R}$$

$$\Rightarrow$$

$$U_{\text{inv}}^2 - U_{\text{inv}} U_{\text{grid}} + P_{\text{inv}} R_{\text{line}} = 0$$

Two solutions of the above equation are

$$U_{\text{inv},1} = \frac{U_{\text{grid}}}{2} + \sqrt{\frac{U_{\text{grid}}^2}{4} + P_{\text{inv}} R}$$

$$U_{\text{inv},2} = \frac{U_{\text{grid}}}{2} - \sqrt{\frac{U^2_{\text{grid}}}{4} + P_{\text{inv}}R}$$

We can observe that $U_{\text{inv},\,1}$ is a solution that is slightly larger than U_{grid}, where $U_{\text{inv},\,2}$ is a negative solution. Considering the fact that $U_{\text{inv},\,2}$ is the magnitude of the voltage at the inverter, this solution is not valid.

Therefore,

$$U_{\text{inv}} = \frac{U_{\text{grid}}}{2} + \sqrt{\frac{U^2_{\text{grid}}}{4} + P_{\text{inv}}R}$$

Recall the expression of inverter real power output:

$$P_{\text{inv}} = \frac{U_{\text{inv}} - U_{\text{grid}}}{R} \cdot U_{\text{inv}}$$

From the reactive power/voltage droop control:

$$U_{\text{inv}} - U_{\text{grid}} = Q_{\text{inv}} \cdot q_{\text{droop}}$$

Recall the expression of inverter reactive power output:

$$Q_{\text{inv}} = \frac{U_{\text{inv}} U_{\text{grid}}}{R} \cdot \delta_{\text{inv,grid}}$$

and

$$\delta_{\text{inv,grid}} = \int \Delta f \, \mathrm{d}t$$

where Δf is the difference between the measured frequency at the inverter and the reference frequency.

From the real power/frequency droop control:

$$\Delta f = (P_{\text{set}} - P_{\text{inv}})p_{\text{droop}}$$

Therefore,

$$P_{\text{inv}} = \frac{\frac{U_{\text{inv}} U_{\text{grid}}}{R} \cdot \left[\int (P_{\text{set}} - P_{\text{inv}})p_{\text{droop}} \cdot \mathrm{d}t\right] \cdot q_{\text{droop}}}{R} \cdot U_{\text{inv}} = \frac{U^2_{\text{inv}} U_{\text{grid}} p_{\text{droop}} q_{\text{droop}}}{R^2}\left[\int (P_{\text{set}} - P_{\text{inv}}) \cdot \mathrm{d}t\right] = C\int (P_{\text{set}} - P_{\text{inv}}) \cdot \mathrm{d}t$$

where

$$C = \frac{U^2_{\text{inv}} U_{\text{grid}} p_{\text{droop}} q_{\text{droop}}}{R^2}$$

The solution of the above equation is

$$P_{\text{inv}} = P_{\text{set}}\left(1 - \mathrm{e}^{-Ct}\right)$$

Note that the solution is stable if $C > 0$, which means $p_{\text{droop}} q_{\text{droop}} > 0$. However, if we select $p_{\text{droop}} > 0$ and $q_{\text{droop}} > 0$, the control scheme will not be compatible with the interconnected grid or synchronous generators. Therefore, we select the droop parameter the same as conventional droops ($p_{\text{droop}} < 0$ and $q_{\text{droop}} < 0$).

Limitations

Even if the controller of the CIGs can mimic frequency responses and inertia, this approach may not be as good as expected because synchronizing torques are contributed by high transient currents during disturbances. For traditional power systems, synchronous machines can provide transient currents in the order of 500–1000% of load currents. On the contrary, the converters have to limit the transient currents to no more than approximately 200% of load currents and further decrease this value as time evolves [17], which lasts for an even shorter period of time than synchronous machines. Consequently, the CIGs' imitation of synchronous machines might not be quite effective.

Therefore, during system transients, the frequency and phase angle may oscillate quickly, and in this case the operational constraints of the inverters may be exceeded to the point of damaging the inverters or causing shutdown of inverters.

5.4 EMERGING TECHNOLOGIES

This section proposes the use of dynamic state estimation in two perspectives: adaptive protection and real-time operation by the DERMS.

Method 5: Dynamic State Estimation-Based Protection (EBP)

This method is to use dynamic state estimation algorithm to test the consistency between the measurements and the high-fidelity dynamic model of the circuit under protection [18]. The dynamic model is built according to physical structure of the circuit and is represented via a set of algebraic and differential equations. Also, the measurements are instantaneous measurements. Therefore, this method can capture electrical transients more accurately during faults since it does not use any approximations of symmetrical system model or phasors. The method can reliably detect high impedance faults.

Details of the method are described as follows:

Step 1: Build High-Fidelity Dynamic Model of the Circuit Under Protection

The model of the circuit is shown in Figure 5.26:

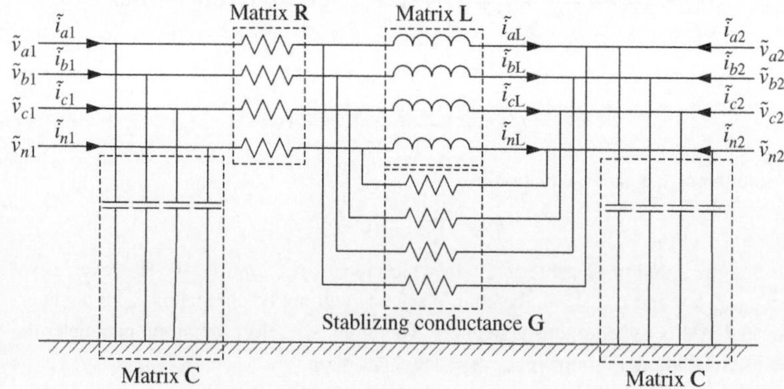

Figure 5.26 π-Equivalent microgrid circuit model.

The following differential equations in matrix form apply:

$$[\mathbf{i}_1 \ \ \mathbf{i}_2 \ \ \mathbf{0}]^T = \mathbf{A}\cdot[\mathbf{v}_1 \ \ \mathbf{v}_2 \ \ \mathbf{i}_L]^T + \mathbf{B}\cdot\frac{\mathrm{d}}{\mathrm{d}t}[\mathbf{v}_1 \ \ \mathbf{v}_2 \ \ \mathbf{i}_L]^T \tag{5.11}$$

where

$$\mathbf{i}_j = [i_{aj} \ \ i_{bj} \ \ i_{cj} \ \ i_{nj}]^T \ \mathbf{v}_j = [v_{aj} \ \ v_{bj} \ \ v_{cj} \ \ v_{nj}]^T \ (j = 1, 2),$$
$$\mathbf{i}_L = [i_{aL} \ \ i_{bL} \ \ i_{cL} \ \ i_{nL}]^T,$$

$$\mathbf{A} = \begin{bmatrix} \mathbf{0} & \mathbf{0} & \mathbf{I}_{4\times 4} \\ \mathbf{0} & \mathbf{0} & -\mathbf{I}_{4\times 4} \\ -\mathbf{I}_{4\times 4} & \mathbf{I}_{4\times 4} & \mathbf{R} \end{bmatrix} \mathbf{B} = \begin{bmatrix} \mathbf{C} & \mathbf{0} & \mathbf{G}\cdot\mathbf{L} \\ \mathbf{0} & \mathbf{C} & -\mathbf{G}\cdot\mathbf{L} \\ \mathbf{0} & \mathbf{0} & \mathbf{R}\cdot\mathbf{G}\cdot\mathbf{L}+\mathbf{L} \end{bmatrix},$$

$\mathbf{I}_{4\times 4}$ is the 4×4 identity matrix.

Afterwards, Eq. (5.11) is integrated using quadratic integration method to generate

$$[\mathbf{I}(t) \ \ \mathbf{I}(t-\Delta t)]^T = \mathbf{Y}\cdot[\mathbf{X}(t) \ \ \mathbf{X}(t-\Delta t)]^T - \mathbf{M}\cdot\mathbf{I}(t-2\Delta t) - \mathbf{N}\cdot\mathbf{X}(t-2\Delta t) \tag{5.12}$$

where

$\mathbf{I}(t) = [\mathbf{i}_1(t) \ \ \mathbf{i}_2(t) \ \ \mathbf{0}]^T \ \mathbf{X}(t) = [\mathbf{v}_1(t) \ \ \mathbf{v}_2(t) \ \ \mathbf{i}_L(t)]^T$, Δt is the sampling interval;

$$\mathbf{Y} = \begin{bmatrix} \mathbf{A}+2\mathbf{B}/\Delta t & -4\mathbf{B}/\Delta t \\ \mathbf{B}/(4\Delta t) & \mathbf{A}+\mathbf{B}/\Delta t \end{bmatrix} \mathbf{N} = \begin{bmatrix} -2\mathbf{B}/\Delta t+\mathbf{A} \\ 5\mathbf{B}/(4\Delta t)-\mathbf{A}/2 \end{bmatrix} \mathbf{M} = \begin{bmatrix} -\mathbf{I}_{12\times 12} \\ 0.5\mathbf{I}_{12\times 12} \end{bmatrix},$$

$\mathbf{I}_{12\times 12}$ is the 12×12 identity matrix.

Considering available measurements of a circuit under protection (three-phase voltages and currents at both terminals of the line), the equation becomes

$$\breve{\mathbf{z}}(t) = \mathbf{H}\cdot\mathbf{x}(t) - \mathbf{B}_{\mathrm{eq}}(t) \tag{5.13}$$

where

$$\breve{\mathbf{z}}(t) = [\breve{\mathbf{v}}_1(t) \ \ \breve{\mathbf{v}}_2(t) \ \ \breve{\mathbf{v}}_1(t-\Delta t) \ \ \breve{\mathbf{v}}_2(t-\Delta t) \ \ \mathbf{I}(t) \ \ \mathbf{I}(t-\Delta t)]^T$$

$$\mathbf{x}(t) = \begin{bmatrix} \mathbf{X}(t) \\ \mathbf{X}(t-\Delta t) \end{bmatrix}^T, \mathbf{H} = \begin{bmatrix} \mathbf{T} & \mathbf{0} & \mathbf{0} & \mathbf{0} & \mathbf{0} & \mathbf{0} \\ \mathbf{0} & \mathbf{T} & \mathbf{0} & \mathbf{0} & \mathbf{0} & \mathbf{0} \\ \mathbf{0} & \mathbf{0} & \mathbf{0} & \mathbf{T} & \mathbf{0} & \mathbf{0} \\ \mathbf{0} & \mathbf{0} & \mathbf{0} & \mathbf{0} & \mathbf{T} & \mathbf{0} \\ & & \mathbf{Y} & & & \end{bmatrix} \mathbf{B}_{\mathrm{eq}}(t) = \begin{bmatrix} \mathbf{0} \\ \mathbf{0} \\ \mathbf{0} \\ \mathbf{0} \\ \mathbf{b}_{\mathrm{eq}}(t) \end{bmatrix},$$

$$\mathbf{T} = \begin{bmatrix} 1 & 0 & 0 & -1 \\ 0 & 1 & 0 & -1 \\ 0 & 0 & 1 & -1 \\ 0 & 0 & 0 & 1 \end{bmatrix}$$

Step 2: Calculate Confidence Level

The best estimated state $\hat{\mathbf{x}}(t)$ is

$$\hat{\mathbf{x}}(t) = \left(\mathbf{H}^T\mathbf{W}\mathbf{H}\right)^{-1}\mathbf{H}^T\mathbf{W}\left(\mathbf{z}(t) + \mathbf{B}_{\mathrm{eq}}(t)\right) \tag{5.14}$$

where $\mathbf{W}$ is the weight matrix of the measurements $\mathbf{W} = \mathrm{diag}\left\{1/\sigma_1^2, 1/\sigma_2^2, \ldots, 1/\sigma_n^2\right\}$, and σ_i is the error standard deviation of measurement i.

The confidence level is calculated as

$$p(t) = P\left(\chi^2 \geq \zeta(t)\right) = 1 - P(\zeta(t), v) \tag{5.15}$$

$$\zeta(t) = \hat{\mathbf{s}}(t)^T\hat{\mathbf{s}}(t) \tag{5.16}$$

$$\hat{\mathbf{s}}(t) = \sqrt{\mathbf{W}} \cdot \hat{\mathbf{r}}(t) \tag{5.17}$$

$$\hat{\mathbf{r}}(t) = \mathbf{H} \cdot \hat{\mathbf{x}}(t) - \mathbf{B}_{\mathrm{eq}}(t) - \breve{\mathbf{z}}(t) \tag{5.18}$$

where $\hat{\mathbf{r}}(t)$ is the residual; $\hat{\mathbf{s}}(t)$ is the normalized residual; $P(\zeta(t), v)$ is the probability of χ^2 distribution of $\chi^2 \leq \zeta(t)$, with v degree of freedom; and v is the difference between the number of measurements and states.

A high confidence level suggests that the circuit under protection is healthy. A low or an oscillating confidence level suggests that there exists an internal fault inside the circuit. Normally a user-defined delay is selected to trip the circuit after the confidence level drops.

An example microgrid system is shown in Figure 5.27. Here the user-defined delay is selected as 1 cycle.

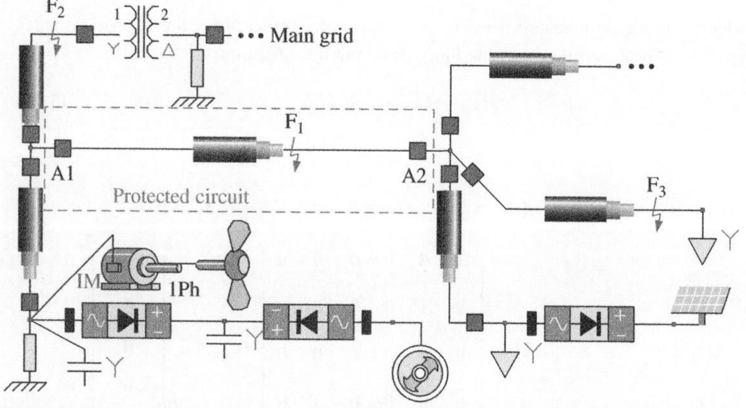

Figure 5.27 Example microgrid system, dynamic state estimation-based protection.

Case 1

An internal fault with 0.01 Ω impedance is simulated at 1.4 seconds. The fault location is the midpoint of circuit A1–A2.

The performance of this DSE-based method is shown in Figure 5.28. The method can correctly detect this internal fault. During this internal fault, the confidence level drops from 100 to 0% at 1.4 seconds, which clearly indicates an abnormality in the system.

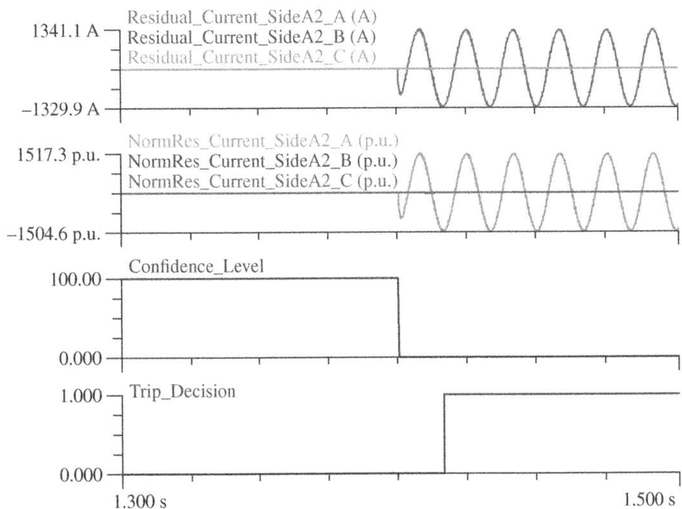

Figure 5.28 Dynamic state estimation-based protection results: low impedance internal fault.

Case 2

An internal fault with 50 Ω impedance is simulated at 1.4 seconds. The fault location is the midpoint of circuit A1–A2.

The performance of this DSE-based method is shown in Figure 5.29. The method can correctly detect this internal fault. During this internal fault, the confidence level drops from 100 to 0% and start oscillating at 1.4 seconds, which clearly indicates an abnormality in the system.

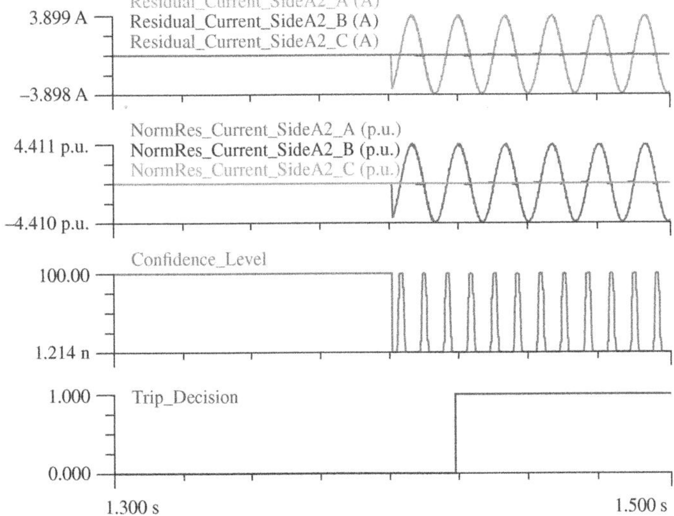

Figure 5.29 Dynamic state estimation-based protection results: high impedance internal fault.

5.4.1 Adaptive Setting-Less Protection

With modernized numerical relays, there has been an attempt to utilize dynamic state estimation for component protection, which is called setting-less protection [19–22]. This approach applies the dynamic state estimation and its statistical evaluation for the identification of internal faults. As illustrated in Figure 5.30, the setting-less protection keeps monitoring a device under protection regarding operating quantities (e.g. voltages, currents, power flows, or frequency) as well as present status (e.g. breaker status or operational modes), simultaneously computing the best estimates of dynamic operating conditions based on dynamic component models of a DER. Then, based on the well-known chi-square test, the goodness of fit of models to measurements can be quantified as the confidence level, which represents the health index of the DER; high confidence levels indicate that the DER operates normally, but low confidence levels indicate the existence of any internal faults.

For dynamic state estimation, the setting-less protection is based on dynamic states and models of power system components. Hence, state estimation is implemented based on a time-domain formulation, taking into account full dynamics

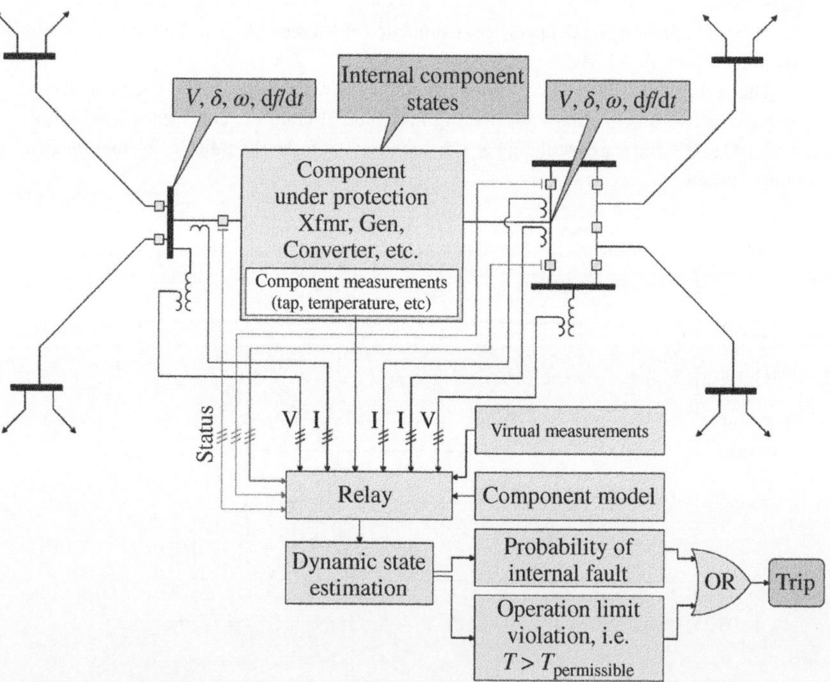

Figure 5.30 Adaptive setting-less protection scheme.

with electrical and mechanical transients. The component dynamic model is described by a set of algebraic and differential equations as follows:

$$\frac{dx}{dt} = f(i,x), \quad 0 = g(i,x), \quad x = [v \quad y]^T \tag{5.19}$$

where x is a vector of state variables, i is a vector of currents, v is a vector of voltages, and y is a vector of internal states. Note that all variables are time variant. This component dynamic model is used to formulate measurement models based on current measurements as follows:

$$z = c + \sum_i a_i x_i + \sum_j \sum_k b_{jk} x_j x_k + \eta \tag{5.20}$$

where z is a current measurement; c is a constant term; a_i are linear coefficient terms; b_{jk} are nonlinear coefficient terms, which are quadratic terms; x_i, x_j, and x_k are state variables; i, j, and k are indices of state variables; and η is a measurement error. As expressed in (5.20), nonlinear terms are expressed as quadratic terms. In addition to current measurement models, virtual measurement models can be formulated from (5.19) as follows:

$$0 = c + \sum_i a_i x_i + \sum_j \sum_k b_{jk} x_j x_k + \eta \tag{5.21}$$

From the practical point of view, the present-day numerical relays are typically based on high-speed microprocessors, so it is available to install the dynamic state estimation and proposed protection logics at currently used relays. Moreover, the modern microprocessor has computational capability to compute the dynamic state estimation with faster rates than 1 ms in case of being implemented on a single device. If the dynamic state estimation is implemented every 1 ms roughly, it can evaluate transients up to 500 Hz according to Nyquist–Shannon sampling theorem. Such frequency is enough to capture mechanical transients as well as electromagnetic ones given a full dynamic model of a component under protection.

It should be noted that the proposed protection has two major advantages over traditional protection: it does not require any additional settings other than initial models, and it can provide adaptive protection in recently restructuring grids such as microgrids or smart grids. In fact, the traditional protection requires initial settings like zone impedances, time dials, and thresholds as well as additional settings to provide adaptive protection under various operational circumstances. For instance, transformer protection requires proper algorithms or settings to prevent nuisance tripping from inrush currents or overexcited ones, and the distribution system needs additional protection scheme to detect high impedance fault. All these traditional adaptive methods, however, are based on a certain setting, which cannot explain all abnormal phenomena, eventually leading to desensitization of relays in special cases. In contrast, the proposed setting-less protection is based on model-based dynamic state estimation with initially provided component model data and real-time measurements. Therefore, as long as abnormal conditions are modeled in the component model, no additional settings for the abnormal conditions are required. For example, if a transformer model contains the nonlinearity of cores, then the setting-less protection can adaptively detect inrush currents, which are abnormal but unfaulty, and would not trip [22]. It should be noted that the only requirement for the setting-less protection is the nonlinear transformer model, which is initially defined.

As the second advantage over traditional protection, the setting-less protection can be applied on microgrids with increasing penetration of DERs. Since DERs are generally interfaced with PE devices, their introduction to microgrids yields various transients as well as unusual characteristics such as small fault currents by power converters. As a result, traditional protective algorithms or coordination might not operate properly. In extreme cases such as cold start of microgrids, most of protection relays should be desensitized until the grid is recovered to normal. In this period, power devices may be damaged in fault conditions that are not isolated. As a countermeasure of protecting DERs or devices even in such conditions, the setting-less component protection can be employed autonomously on each individual component.

In the previous works, [19] and [22], the setting-less protection was proposed and tested about its feasibility. The basic algorithm was verified with numerical simulations on several power components such as transmission lines, distribution lines, transformers, capacitor banks, and so on. The test results prove that the setting-less protection not only protects devices from various fault conditions, which include high impedance faults, but also maintains security by not tripping for conditions that are not faulty but abnormal (e.g. inrush currents).

5.4.2 Real-Time Operation by the DERMS

In the proposed protection concept (i.e. DDSE), the dynamic state estimation is implemented individually on a single DER by a universal monitoring protection and control unit (UMPCU), which is capable of monitoring the DER with a high sampling rate, protecting it from malfunctioning or from external faults, and controlling in accordance with operational purposes; this UMPCU can be implemented simply by modifying the present IEDs such as relays, digital fault recorders, or smart meters [23]. Figure 5.31 illustrates the installation of a UMPCU and its functions. The UMPCU performs the dynamic state estimation on a DER based on real-time measurement data, which is sampled with high sampling rates, and the predefined dynamic component model of the DER. Then, the estimated states, represented in time-based waveform, are converted to phasor forms with GPS-synchronized time tags, and then transmitted to the centralized DERMS, where various applications for microgrid operation (e.g. economic dispatch, optimal power flow, contingency analysis, or dynamic stability analysis) are installed.

The DERMS, first, executes the static state estimation based on the collected phasor data, once filtered by the DDSE, as well as unfiltered measurement data coming from conventional meters, relays, and digital fault recorders; note that

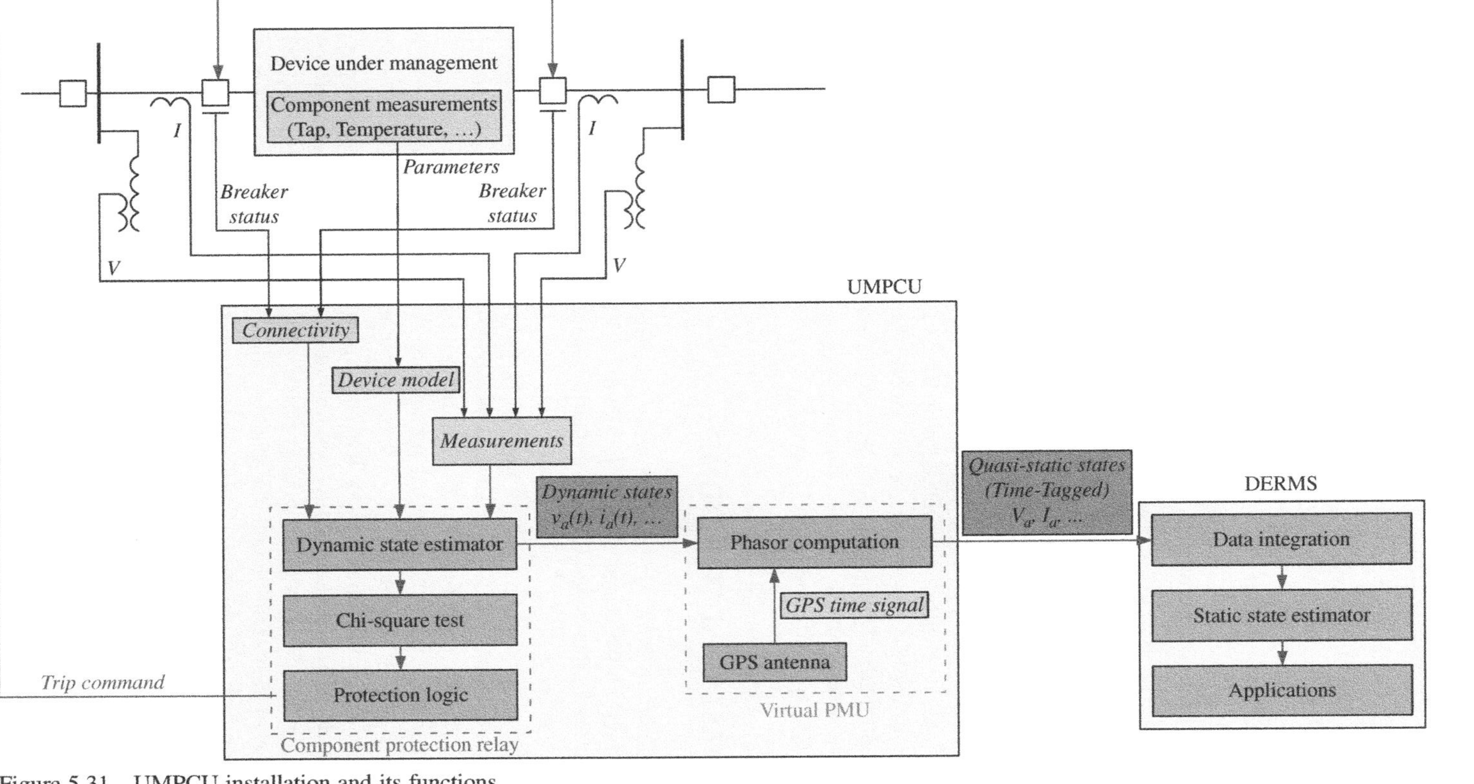

Figure 5.31 UMPCU installation and its functions.

the static state estimation is implemented on a centralized basis in the DERMS. The outcomes are real-time operating conditions of the microgrid, which are then used for the next stage applications.

It needs to be noted that phasor data estimated by UMPCUs are highly accurate since errors in raw measurement data are eliminated based on measurement redundancy and detailed dynamic models. Therefore, when these estimated data are used for static state estimation in the DERMS, the processing time can be drastically reduced, and further, more accurate real-time operating conditions can be obtained than when the DDSE is not applied before data collection. In addition, state variables that are estimated by the DDSE are minimum information that can describe the operation and status of a DER under management, thereby reducing communication congestion.

As mentioned earlier, the setting-less protection method has already been validated for its feasibility and effectiveness to adaptively protect various components in the previous works [19–22]. Hence, the focus of this chapter is on the applications of results of the DDSE. With the DDSE results and GPS time information, the UMPCU computes phasor values, which are the same as typical PMU data. Importantly, these phasor data are useful for the DERMS to obtain real-time operating states of more active distribution feeders (e.g. microgrids) than before with higher accuracy and faster speed.

5.5 TEST CASE FOR DDSE

In an attempt to show the facilitation of microgrid operations by the DDSE, this chapter presents a test case as shown in Figure 5.32. The test case represents a distribution feeder that contains two wind turbine generators. As described in the figure, the test case reflects real distribution feeders, which are unsymmetric and have few measuring sensors for voltage magnitude and active/reactive power flows. It is also assumed that a PMU is installed at the point interconnecting with the infinite bus for the purpose of time synchronization with the main grid.

In Figure 5.32, the nominal parameters of the infinite bus are 24.9 kV, 12 MVA, and 60 Hz, and 2 wind turbine generators have identical hardware of 660-kVA, 480-V induction motors. It needs to point out that there is no power converter such as back-to-back converters or doubly fed induction generators. The feeder lines are designed to have unsymmetric structure among different phasors as well as to include all mutual impedances. All details are tabulated in Tables 5.3 and 5.4. Finally, active/reactive power injections are listed in Table 5.5, indicating imbalanced operation of the test case. All power injections that are not presented in the table are zero.

For detailed simulation considering unsymmetric structure and imbalanced operation, time-based simulation should be implemented. This section utilizes PSCAD/EMTDC to simulate the test case with the time step of 50 μs, generating real-time measurement data every 16.7 ms (i.e. 60 Hz); this sampling frequency involves the assumption that high-speed communication infrastructure is installed,

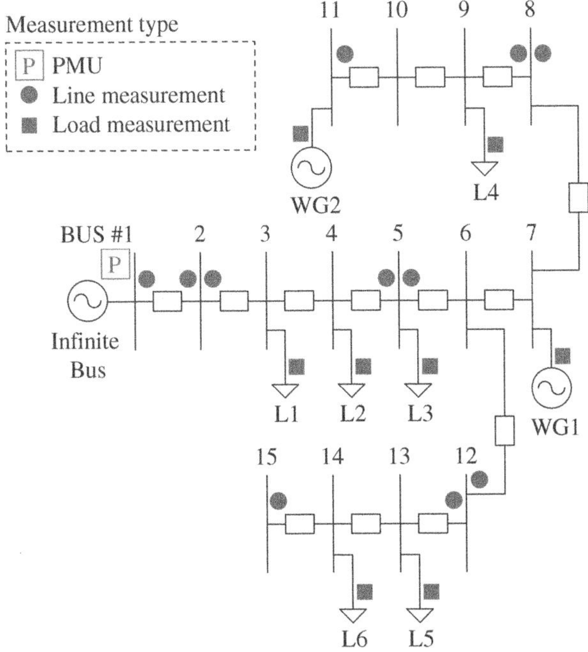

Figure 5.32 Test case for a distribution feeder. The acronyms WG and L represent a wind turbine generator and a load, respectively.

TABLE 5.3 Parameters for line impedances per unit length.

Parameter	Value (Ω)	Parameter	Value (Ω)
Series resistance [*A*]	1.3368	Mutual resistance [*A*–*B*]	0.2102
Series resistance [*B*]	1.3238	Mutual resistance [*B*–*C*]	0.2066
Series resistance [*C*]	1.3294	Mutual resistance [*C*–*A*]	0.2130
Series inductance [*A*]	1.3343	Mutual inductance [*A*–*B*]	0.5779
Series inductance [*B*]	1.3569	Mutual inductance [*B*–*C*]	0.4591
Series inductance [*C*]	1.3471	Mutual inductance [*C*–*A*]	0.5015
Shunt capacitance [*A*]	5.335	Mutual capacitance [*A*–*B*]	−1.5313
Shunt capacitance [*B*]	5.0979	Mutual capacitance [*B*–*C*]	−0.9943
Shunt capacitance [*C*]	4.888	Mutual capacitance [*C*–*A*]	−0.6212

facilitating data communication between the DERMS and sensors (or UMPCUs). Based on measurement data obtained from the simulation, the state estimation is implemented with MATLAB, and there was no effort in code optimization since it is out of scope of this chapter.

TABLE 5.4 Line length of distribution lines.

From bus	To bus	Length (p.u.)	From bus	To bus	Length (p.u.)
1	2	0.4886	8	9	0.058 71
2	3	0.0819	9	10	0.4834
3	4	0.2457	10	11	1.4503
4	5	6.1042	6	12	0.001 89
5	6	7.1023	12	13	1.4342
6	7	5.6301	13	14	0.3466
7	8	0.001 89	14	15	0.8456

TABLE 5.5 Power injection by loads.

Load number	Active power in kW (*A*/*B*/*C*)	Reactive power in kVar (*A*/*B*/*C*)
1	16.7/20/16.7	9.3/10/9.3
2	10/10/8.3	5/5/4.7
3	0/20.45/0	0/9.282/0
4	0/0/60.04	0/0/28.07
5	20/15/9.9	10/8/5.2
6	14/10/8.3	7/5/4.7

5.5.1 Formulation of Measurement Models

For the implementation of static state estimation in the DERMS, distribution lines and electric loads are mathematically modeled as follows:

$$\begin{bmatrix} \widetilde{I} \\ 0 \end{bmatrix} = \begin{bmatrix} \widetilde{Y}_{11} & \widetilde{Y}_{12} \\ \widetilde{Y}_{21} & \widetilde{Y}_{22} \end{bmatrix} \begin{bmatrix} \widetilde{V} \\ y \end{bmatrix} \tag{5.22}$$

where $\widetilde{I}$ is a vector of current phasors; $\widetilde{V}$ is a vector of external state variables, which are voltage phasors; $\widetilde{Y}_{11}$, $\widetilde{Y}_{12}$, $\widetilde{Y}_{21}$, and $\widetilde{Y}_{22}$ are matrices of linear coefficients; and y is a vector of internal state variables. The state variables are represented by Cartesian coordinates; for example, V_r and V_i indicate the real and imaginary part of a voltage phasor, respectively.

It should be noted that the wind turbine generator model is based primarily on mechanical and electromagnetic dynamics, so the model is suitable not for static state estimation, but rather for dynamic state estimation. In this manner, only measurement data at the terminal of the wind turbine generators are provided into the static state estimator in the DERMS.

Meanwhile, from line measurements at bus number 1, 2, 5, 8, 11, 12, and 15, measurement models of voltage magnitude squared and active/reactive power can be derived as follows:

$$z_{\mathrm{Vs}} = V_r\,V_r + V_i\,V_i + \eta \tag{5.23}$$

$$z_{\mathrm{P}} = V_r\,I_r + V_i\,I_i + \eta \tag{5.24}$$

$$z_{\mathrm{Q}} = V_i\,I_r - V_r\,I_i + \eta \tag{5.25}$$

where z_{Vs}, z_{P}, and z_{Q} are measured values with respect to voltage magnitude squared, active power, and reactive power, respectively; and η is a measurement error. Note that the sign of power and current measurements is positive when the flow is toward the corresponding device. In (5.24) and (5.25), current values can be substituted by (5.22), yielding the following equations:

$$z_{\mathrm{P}} = V_r\sum_k (Y_{kr}V_{kr} - Y_{ki}V_{ki}) + V_i\sum_k (Y_{ki}V_{kr} + Y_{kr}V_{ki}) + \eta \tag{5.26}$$

$$z_{\mathrm{Q}} = V_i\sum_k (Y_{kr}V_{kr} - Y_{ki}V_{ki}) - V_r\sum_k (Y_{ki}V_{kr} + Y_{kr}V_{ki}) + \eta \tag{5.27}$$

where the subscripts r and i represent the real and imaginary part, respectively, and k are indices of state variables. It is necessary to point out that the line measurements are nonlinear since they have quadratic terms.

In the meantime, the load measurements at bus number 3, 4, 5, 9, 13, and 14 represent power injections into loads. In case of bus number 15, a zero current injection is formulated as one of pseudo-measurements. The measurement models for the load measurements are formulated by multiplying bus voltages and sums of adjacent currents as follows:

$$z_{\mathrm{P}} = V_r\left(-\sum_j I_{jr}\right) + V_i\left(-\sum_j I_{ji}\right) + \eta \tag{5.28}$$

$$z_{\mathrm{Q}} = V_i\left(-\sum_j I_{jr}\right) - V_r\left(-\sum_j I_{ji}\right) + \eta \tag{5.29}$$

where j indicates the indices of adjacent buses.

At bus number 2, 6, 8, and 10, by Kirchhoff's current law, the following virtual measurements are formulated:

$$0 = V_r\left(-\sum_j I_{jr}\right) + V_i\left(-\sum_j I_{ji}\right) + \eta \tag{5.30}$$

$$0 = V_i\left(-\sum_j I_{jr}\right) - V_r\left(-\sum_j I_{ji}\right) + \eta \tag{5.31}$$

Finally, if the DDSE is employed on the wind turbine generators at bus number 7 or 11, additional phasor data are used to formulate measurement models as follows:

$$z_r = V_r + \eta = \left|\tilde{V}\right| \cos\theta + \eta \tag{5.32}$$

$$z_i = V_i + \eta = \left|\tilde{V}\right| \sin\theta + \eta \tag{5.33}$$

where θ indicates the phase angle of the phasor data.

5.5.2 Static State Estimation and Performance Evaluation

Once all measurement models are formulated, the well-known weighted least squares method is performed for state estimation. The objective function can be expressed as follows:

$$\min J = \eta^T W \eta = [h(x) - z]^T W [h(x) - z] \tag{5.34}$$

where W is a weight matrix whose diagonal entries are inverse of squared standard deviation of measurement errors and $h(x)$ is the function of state variables, x. The standard deviation values are set according to their nominal values and listed in Table 5.6.

Since there are nonlinear measurement models, the following iterative method is used to obtain best estimates of states:

$$x^{i+1} = x^i - \left(H^T W H\right)^{-1} H^T W \left(h\left(x^i\right) - z\right) \tag{5.35}$$

where H is the Jacobian matrix of $h(x)$ and i means the ith iteration.

After completing state estimation, the goodness of fit of measurements to models can be quantified by the chi-square test, which is based on the degree of freedom, ν, and the chi-square critical value, ζ. These variables are expressed as follows:

$$\nu = m - n \tag{5.36}$$

TABLE 5.6 Standard deviation of measurements.

Measurement type	Unit	Nominal value	Standard deviation
Voltage magnitude squared	$(\text{kV})^2$	206 L-G	2.0667
Line active power	MW	20	0.2
Line reactive power	MVar	10	0.1
Load active power injection	MW	0.02	0.0002
Load reactive power injection	MVar	0.02	0.0002
WG active power injection	MW	0.22	0.0022
WG reactive power injection	MVar	0.11	0.0011
Virtual	Per unit	1	0.001
PMU voltage	kV	14.4 L-G	0.0144

$$\zeta = \sum_{i}^{m} \left(\frac{h_i(\hat{x}) - z_i}{\sigma_i} \right)^2 \tag{5.37}$$

where m and n are the number of measurements and states, respectively, $\hat{x}$ is a vector of the best estimate of states, and σ_i is the standard deviation of the ith measurement.

The expected errors of states are given from the diagonal entries of the information matrix, expressed as follows:

$$I = \left(H^T WH\right)^{-1} \tag{5.38}$$

where I is the information matrix.

5.5.3 Test Scenarios

For reality, torques given to wind turbine generators are arbitrarily changed every second, which determines power generation capacity, affecting power flows in the test feeder. The test scenarios are as follows: in test case 1, the DDSE is not applied, but in test case 2, the proposed DDSE is done.

5.6 TEST RESULTS

5.6.1 Test Case 1

For test case 1, the confidence level maintains 100%, which indicates that measurement data are consistent with the models during testing time, and therefore, it can be concluded that the estimation results are credible. Figure 5.33 shows the expected errors of computed state variables at 0.516 seconds, which slightly matches the actual absolute errors in Figure 5.34. The results indicate that the actual errors of states are smaller than those expected statistically. Finally,

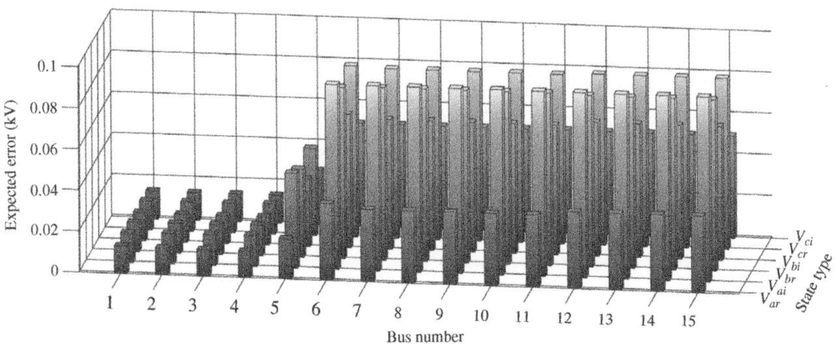

Figure 5.33 Expected errors of states (test case 1).

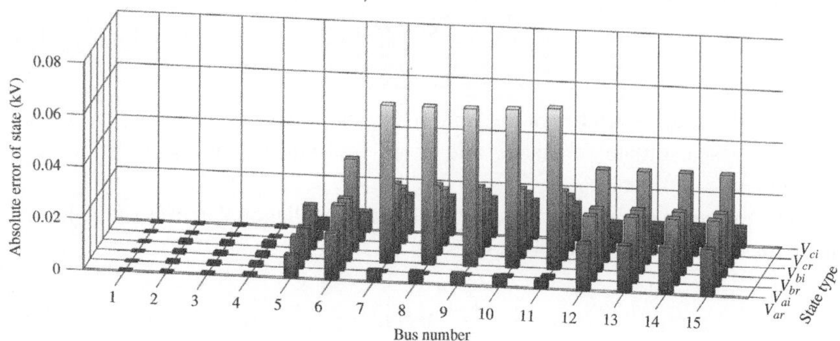

Figure 5.34 Actual absolute errors of states (test case 1).

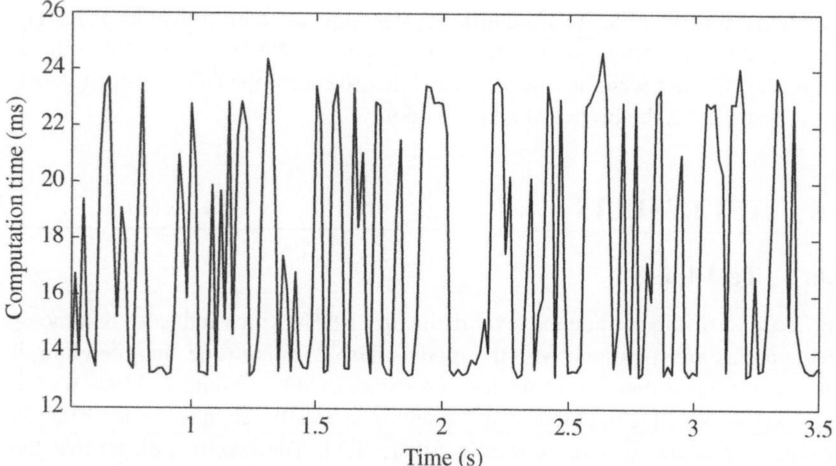

Figure 5.35 Computation time (test case 1).

Figure 5.35 illustrates the computation time for the state estimation and the chi-square test during testing time. It should be noted that initial conditions are reset to one every time instant because of the assumption that the microgrids or future distribution grids might experience frequent topology changes or various operational modes of power devices like converters. In summary, the average computation time is 17.409 ms, and the average absolute error of states at 0.516 seconds is 14.38 V.

5.6.2 Test Case 2

Similar to the test case 1, the confidence level keeps 100% during simulation time, and actual absolute errors of computed states are, averagely, smaller than the

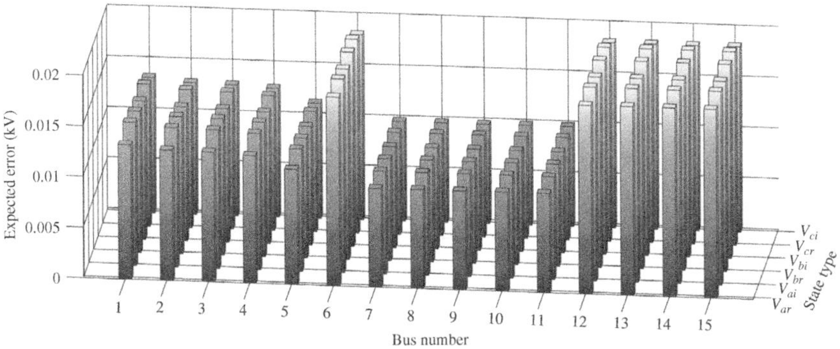

Figure 5.36 Expected errors of states (test case 2).

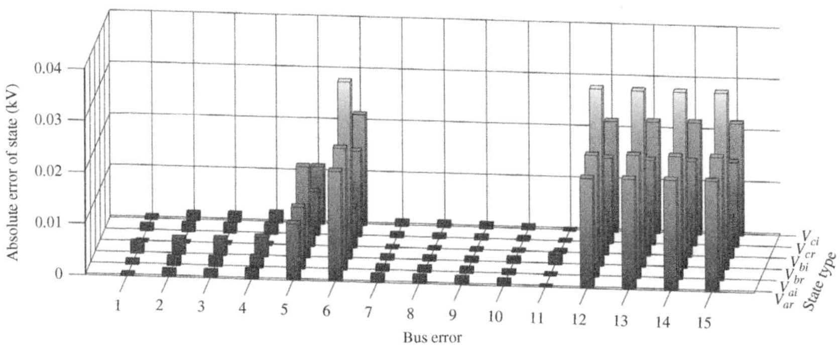

Figure 5.37 Actual absolute errors of states (test case 2).

expected errors of states as depicted in Figures 5.36 and 5.37. However, compared with the errors in test case 1, the expected errors as well as actual absolute errors drastically decrease with the DDSE. The average actual absolute error is 8.2605 V, which is nearly half of that of test case 1. Meanwhile, Figure 5.38 illustrates the computation time, and its average value is 7.22 ms, indicating that the time is less than half of that of test case 1. All in all, it can be concluded that the DDSE-based state estimation is faster and more accurate than the state estimation without the DDSE.

5.7 TEST CASE FOR ADAPTIVE SETTING-LESS PROTECTION

This section describes the application of adaptive setting-less protection on the induction machine, which is a general type of wind turbine generators in the distribution system, with experimental results presented.

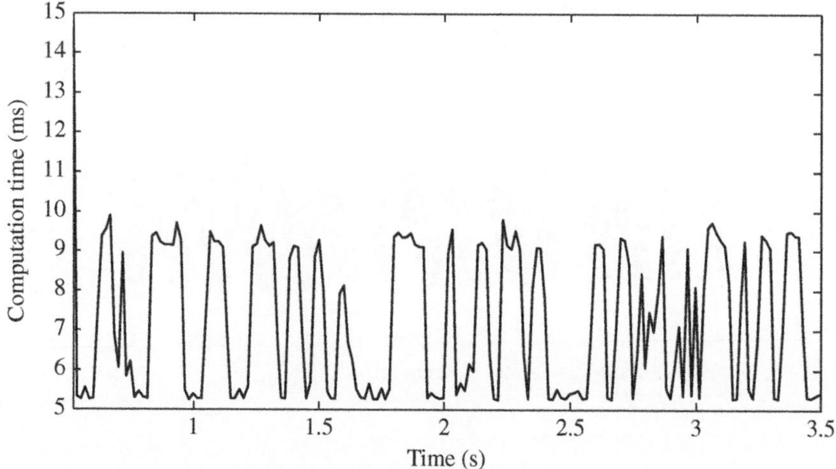

Figure 5.38 Computation time (test case 2).

5.7.1 Induction Machine Component Model

The detailed circuit model of the induction machine is illustrated in Figure 5.39 and can be written as follows:

$$\omega_b v_{ds} = \omega_b R_s i_{ds} + L_{ls} p\, i_{ds} + L_{md} p\, i_{md} - \omega_b \omega \psi_{qs} \tag{5.39}$$

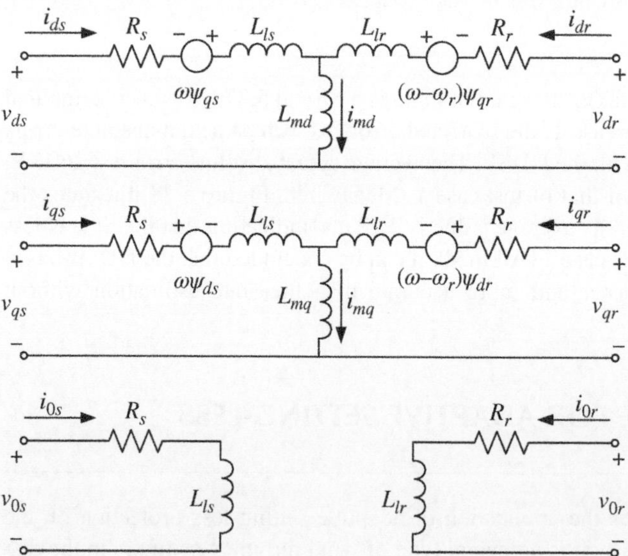

Figure 5.39 Equivalent circuit for a three-phase induction machine.

$$\omega_b v_{qs} = \omega_b R_s i_{qs} + L_{ls} p\, i_{qs} + L_{mq} p\, i_{mq} + \omega_b \omega \psi_{ds} \tag{5.40}$$

$$\omega_b v_{0s} = \omega_b R_s i_{0s} + L_{ls} p\, i_{0s} \tag{5.41}$$

$$\omega_b v_{dr} = \omega_b R_r i_{dr} + L_{lr} p\, i_{dr} + L_{md} p\, i_{md} - \omega_b (\omega - \omega_r) \psi_{qr} \tag{5.42}$$

$$\omega_b v_{qr} = \omega_b R_r i_{qr} + L_{lr} p\, i_{qr} + L_{mq} p\, i_{mq} + \omega_b (\omega - \omega_r) \psi_{dr} \tag{5.43}$$

$$\omega_b v_{0r} = \omega_b R_r i_{0r} + L_{lr} p\, i_{0r} \tag{5.44}$$

$$i_{ds} = i_{md} - i_{dr} \tag{5.45}$$

$$i_{qs} = i_{mq} - i_{qr} \tag{5.46}$$

$$\psi_{ds} = L_{md} i_{md} + L_{ls} i_{ds} \tag{5.47}$$

$$\psi_{qs} = L_{mq} i_{mq} + L_{ls} i_{qs} \tag{5.48}$$

$$\psi_{dr} = L_{md} i_{md} + L_{lr} i_{dr} \tag{5.49}$$

$$\psi_{qr} = L_{mq} i_{mq} + L_{lr} i_{qr} \tag{5.50}$$

$$2H\, p\omega_r = T_m - T_e - D\omega_r \tag{5.51}$$

$$T_e = \psi_{qr} i_{dr} - \psi_{dr} i_{qr} \tag{5.52}$$

where the subscripts d, q, and 0 denote the d-axis, q-axis, and zero sequence, respectively; the subscripts s and r denote variables and parameters related to the circuit of stator and rotor, respectively; v, i, and ψ are voltage, current, and magnetic flux linkage, respectively; i_{md} and i_{mq} are magnetizing currents for d-axis and q-axis, respectively; ω and ω_r are angular speed of the reference frame for stator and rotor circuits, respectively; T_m and T_e are mechanical and electric torque, respectively; ω_b is the base quantity of angular speed; L_{md} and L_{mq} are magnetizing inductance for d-axis and q-axis, respectively; R is resistance; L_l is leakage inductance; H is inertia constant; D is damping coefficient; and p indicates differential operator d/dt.

The induction machine tested is a three-phase, 60 Hz, two-pole wind turbine generator, which is located at bus number 7 as WG1 in Figure 5.28. The parameters are, in detail, listed in Table 5.7.

Equations (5.39)–(5.52), which represent the wind turbine generator, are integrated using the quadratic integration method [24, 25] over the three consecutive measured samples for $t - h$, t_m, and t, where t is current time, h is an integration time step, and t_m is intermediate time (i.e. the half point between t and $t - h$). As a result, a set of linear and quadratic algebraic equations is obtained as follows:

$$I(t) = LX(t) - b(t-h) + F(t) \tag{5.53}$$

TABLE 5.7 Parameters for the wind turbine generator.

Description	Value
Rated power	660 kVA
Rated voltage (line-to-line)	480 V
Stator/rotor turn ratio	1
Inertia constant, H	1 s
Stator resistance, R_s	0.0053 per unit
Rotor resistance, R_r	0.007 per unit
Magnetizing inductance, L_{md} or L_{mq}	4 per unit
Stator leakage inductance, L_{ls}	0.106 per unit
Rotor leakage inductance, L_{lr}	0.12 per unit

$$I(t) = \left[i(t)^T, 0, \ldots, 0, i(t_m)^T, 0, \ldots, 0\right]^T \tag{5.54}$$

$$i(t) = \left[i_{ds}(t), i_{qs}(t), i_{0s}(t), i_{dr}(t), i_{qr}(t), i_{0r}(t)\right]^T \tag{5.55}$$

$$X(t) = \left[v(t)^T, y(t)^T, v(t_m)^T, y(t_m)^T\right]^T \tag{5.56}$$

$$v(t) = \left[v_{ds}(t), v_{qs}(t), v_{0s}(t), v_{dr}(t), v_{qr}(t), v_{0r}(t)\right]^T \tag{5.57}$$

$$y(t) = \left[i_{md}(t), i_{mq}(t), \omega(t), \omega_r(t), T_m(t), T_e(t)\right]^T \tag{5.58}$$

$$b(t-h) = M\begin{bmatrix} i(t-h) \\ 0 \end{bmatrix} + N\begin{bmatrix} v(t-h) \\ y(t-h) \end{bmatrix} \tag{5.59}$$

$$F(t) = \left[f(t)^T, f(t_m)^T\right]^T \tag{5.60}$$

$$f(t) = \left[X(t)^T Q_1 X(t), \ldots, X(t)^T Q_{n/2} X(t)\right]^T \tag{5.61}$$

where L, M, and N are the coefficient matrices; $X(t)$ consists of state variables; and n is the number of state variables.

5.7.2 State Estimation Formulation

For testing adaptive setting-less protection on the wind turbine generator, the measurement set includes three-phase stator voltages and currents, three-phase rotor voltages and currents, mechanical and electric torques, angular velocity of the reference frame and the rotor, and virtual measurements from (5.53). In this case, the unit of all measurements is per unit, and the standard deviations for voltage,

TABLE 5.8 Test scenarios.

Case number	Description
3	Stator terminal fault (phase *A* to ground)
4	Stator terminal fault (three phase to ground)
5	Stator terminal fault (phase *A* to phase *B*)
6	High impedance fault (phase *A* to ground)
7	External fault at bus 13 (phase *A* to ground)

current, state (e.g. torque or angular velocity), and virtual measurements are 0.001, 0.01, 0.001, and 0.01 per unit, respectively. The sampling period for measurements is 300 μs.

5.7.3 Test Scenarios

A total of five scenarios are tested as listed in Table 5.8. In cases 3–6, various internal faults occurring at stator terminals of the induction machine are simulated. Especially, in case 6, the fault impedance is relatively high, which is 0.286 per unit. In case 7, external fault occurs at bus 13 (see Figure 5.28). In all test cases, the fault occurred at two seconds.

5.7.4 Test Results and Observation

The test results of each test scenario are presented in Figure 5.40. The adaptive setting-less protection method can determine the occurrence of internal faults according to the confidence level, which indicates the goodness of fit of measurements to component models. Hence, the confidence level becomes zero when internal faults occur in test cases 3–6. However, in case 7, the fault occurred outside the wind turbine generator, thereby maintaining a confidence level of 100% even during faults. As a result, the proposed protection scheme does not trip a breaker, and this decision is correct since the component protection zone only includes the induction machine.

In case of the high impedance fault (i.e. case 6), the fault current is approximately 0.707 kA rms, which is less than the rated current, 0.794 kA rms (see Figure 5.41). Hence, the traditional overcurrent relay might not operate in this fault condition, but the proposed protection method can detect it.

In the meantime, the conventional differential relay, which monitors terminal currents and neutral currents, can differentiate internal faults from external faults for test cases 3, 5, 6, and 7, but it cannot identify the three-phase internal faults in case 4. As shown in Figure 5.42, the sum of three-phase terminal currents is almost

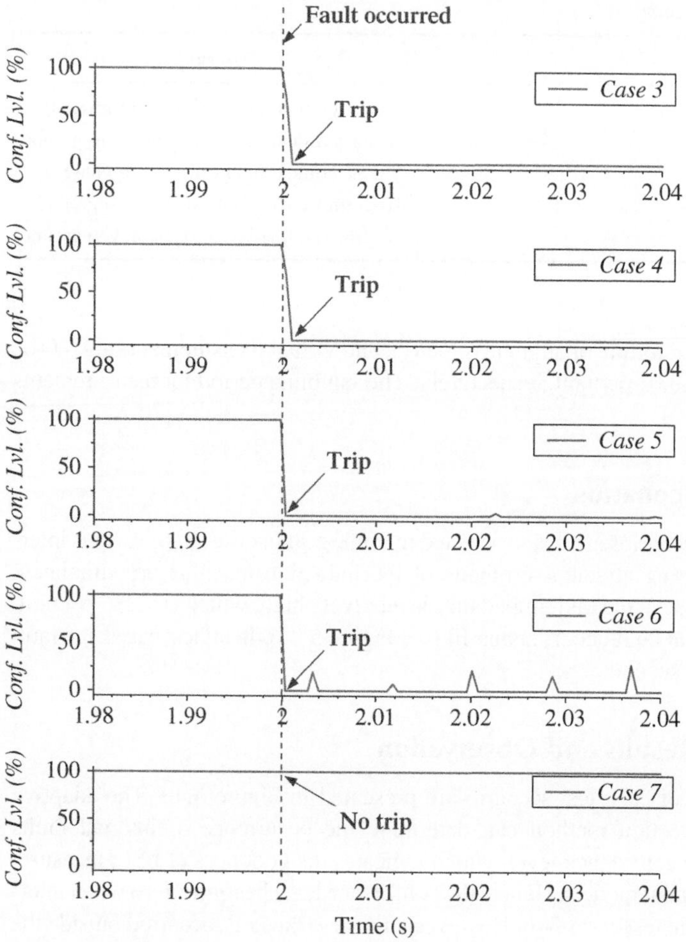

Figure 5.40 Confidence levels for test cases 3–7.

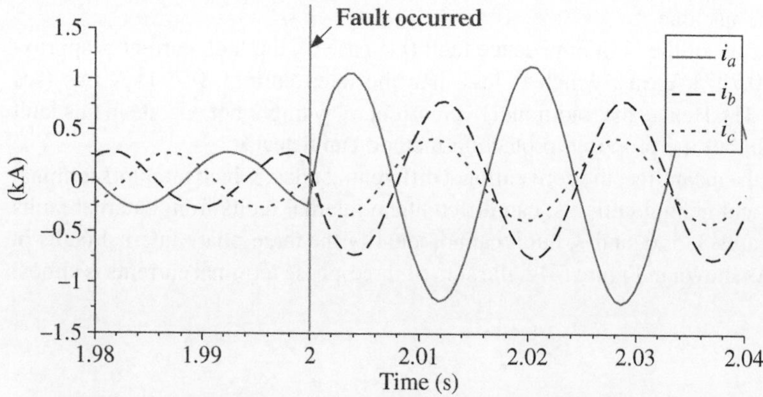

Figure 5.41 Terminal currents for test case 6 (high impedance fault).

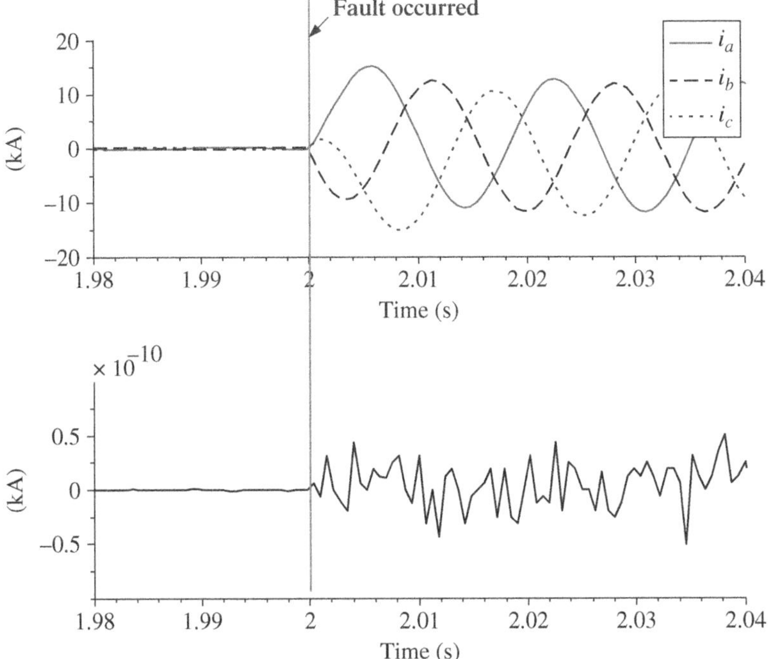

Figure 5.42 Top: terminal currents for test case 4 (three-phase-to-ground fault), bottom: sum of the terminal currents.

zero during the fault, and the neutral current of the induction machine is also zero. Accordingly, the differential relay cannot detect the three-phase internal fault.

5.8 CONCLUSIONS

The chapter proposes the effective real-time operation and protection by means of the DDSE approach. Any component in microgrids can be protected by the setting-less protection method, capable of tracking full dynamic characteristics of a device under protection. This method can provide adaptive protection in microgrids, where unpredictable fault conditions or abnormal states are emerging. It is important to point out that the setting-less protection is fundamentally based on the physical characteristics, thus requiring no additional settings for grid conditions.

In addition to contribution of the DDSE to adaptive protection, it also facilitates the real-time operation by reducing the state estimation computation time as well as by enhancing the accuracy of estimation results. In this sense, the DDSE can be of great importance to the real-time operation and management of microgrids in which the penetration of renewable DERs has recently increased.

REFERENCES

1. Hayes, B. and Prodanovic, M. (Oct. 2014). State estimation techniques for electric power distribution systems. *2014 European Modelling Symposium*, pp. 303–308, Pisa, IEEE. https://ieeexplore.ieee.org/abstract/document/7154016.
2. Yang, X., Wei, Z., Sun, G. et al. (Jul. 2014). Distribution system state estimation considering the characteristics of power electronic loads. *Proceedings of the IEEE PES General Meeting*, IEEE. https://ieeexplore.ieee.org/document/6939060.
3. Lu, C.N., Teng, J.H., and Liu, W.H.E. (Feb. 1995). Distribution system state estimation. *IEEE Trans. Power Syst.* 10 (1): 229–240.
4. Jiang, W., Vittal, V., and Heydt, G.T. (May 2007). A distributed state estimator utilizing synchronized phasor measurements. *IEEE Trans. Power Syst.* 22 (2): 563–571.
5. Nusrai, N., Lopatka, P., Irving, M.R. et al. (Jul. 2015). An overlapping zone-based state estimation method for distribution systems. *IEEE Trans. Smart Grid* 6 (4): 2126–2133.
6. Gomez-Exposito, A., de la Jaen, A.V., Gomez-Quiles, C. et al. (Apr. 2011). A taxanomy of multi-area state estimation methods. *Electr. Power Syst. Res.* 81 (4): 1060–1069.
7. Jin, S., Chen, Y., Rice, M. et al. (Jul. 2012). A testbed for deploying distributed state estimation in power grids. *Proceedings of the IEEE PES General Meeting*, IEEE. https://ieeexplore.ieee.org/document/6345456.
8. Xie, L., Choi, D., Kar, S., and Poor, H.V. (Sep. 2012). Fully distributed state estimation for wide-area monitoring systems. *IEEE Trans. Smart Grid* 3 (3): 1154–1169.
9. Gomez-Exposito, A., Gomez-Quiles, C., and Dzafic, I. (Jan. 2015). State estimation in two time scales for smart distribution systems. *IEEE Trans. Smart Grid* 6 (1): 421–430.
10. Gomez-Exposito, A., Abur, A., de la Jaen, A.V., and Gomez-Quiles, C. (Jun. 2011). A multilevel state estimation paradigm for smart grids. *Proc. IEEE* 99 (6): 952–976.
11. *SEL-387L Relay Instruction Manual.* (2011). Pullman, WA: Schweitzer Engineering Laboratories, Inc.
12. Ustun, T.S., Ozansoy, C., and Ustun, A. (May 2013). Fault current coefficient and time delay assignment for microgrid protection system with central protection unit. *IEEE Trans. Power Syst.* 28 (2): 598–606.
13. Samantaray, S.R., Joos, G., and Kamwa, I. (2012). Differential energy based microgrid protection against fault conditions. In: *IEEE PES Innovative Smart Grid Technologies (ISGT)*, 1–7. Washington, DC: IEEE.
14. Lasseter, B. (2004). Microgrids distributed power generation. *IEEE Power Engineering Society Winter Meeting Conference Proceedings*, Columbus, OH (2001), vol. 1, pp. 146–149. IEEE. https://ieeexplore.ieee.org/abstract/document/917020.
15. Engler, A. (17 Feb. 2004). Device for equal-rated parallel operation of single-or three-phase voltage sources. U.S. Patent No. 6,693,809.
16. Engler, A. (2005). Applicability of droops in low voltage grids. *Int. J. Distrib. Energy Resour.* 1 (1): 3–15.
17. Baran, M.E. and Mahajan, N.R. (Jan. 2007). Overcurrent protection on voltage-source-converter-based multiterminal DC distribution systems. *IEEE Trans. Power Del.* 22 (1): 406–412.
18. Liu, Y., Meliopoulos, S., Fan, R. et al. (2015). Dynamic state estimation based protection of microgrid circuits. *IEEE Power and Energy Society (PES) General Meeting*, Denver, CO (26–30 July 2015). IEEE. https://ieeexplore.ieee.org/abstract/document/7286513/.

19. Meliopoulos, A.P.S., Cokkinides, G.J., Tan, Z. et al. (Jan. 2013). Setting-less protection: feasibility study. *2013 46th Hawaii International Conference on System Sciences*, Wailea, HI, pp. 2345–2353 (18 March 2013). IEEE. https://ieeexplore.ieee.org/document/6480127.
20. Meliopoulos, A.P.S., Cokkinides, G.J., Huang, R. et al. (Jan. 2014). Grid modernization: seamless integration of protection, optimization and control. *2014 47th Hawaii International Conference on System Sciences*, Wailea, HI, pp. 2463–2474 (6–9 January 2014). IEEE. https://ieeexplore.ieee.org/document/6758908.
21. Meliopoulos, A.P.S., Polymeneas, E., Tan, Z. et al. (Dec. 2013). Advanced distribution management system. *IEEE Trans. Smart Grid* 4 (4): 2109–2117.
22. Choi, S., Lee, Y., Cokkinides, G.J. et al. (Jun. 2011). Dynamically adaptive transformer protection using dynamic state estimation. *Proceedings of the Protection, Automation & Control (PAC) World Conference*, Dublin, Ireland (27–30 June 2011).
23. Choi, S. and Meliopoulos, A.P.S. (Feb. 2017). Effective real-time operation and protection scheme of microgrids using distributed dynamic state estimation. *IEEE Trans. Power Del.* 32 (1): 504–514.
24. Meliopoulos, A.P., Cokkinides, G.J., and Stefopoulos, G.K. (Dec. 2005). Improved numerical integration method for power/power electronic systems based on three-point collocation. *Proceedings of the 44th IEEE Conference on Decision and Control, and European Control Conference* (15 December 2005), pp. 6780–6787. IEEE. https://ieeexplore.ieee.org/document/1583252.
25. Meliopoulos, A.P.S., Cokkinides, G.J., and Stefopoulos, G. (Jun. 2005). Quadratic integration method. *Proceedings of the International Power System Transients Conference*, Montreal.

PART II

ROBUST STATE ESTIMATION

CHAPTER 6

PSSE REDUX: CONVEX RELAXATION, DECENTRALIZED, ROBUST, AND DYNAMIC SOLVERS

Vassilis Kekatos[1], Gang Wang[2], Hao Zhu[3], and Georgios B. Giannakis[2]

[1]*The Bradley Department of Electrical and Computer Engineering, Virginia Tech, Blacksburg, VA, USA*

[2]*Digital Technology Center and the Department of Electrical and Computer Engineering, University of Minnesota, Minneapolis, MN, USA*

[3]*Department of Electrical and Computer Engineering, The University of Texas at Austin, Austin, TX, USA*

6.1 INTRODUCTION

With the advent of digital computers, power systems engineers in the 1960s tried computing the voltages at critical buses based on readings from current and potential transformers. Local personnel manually collected these readings and forwarded them by phone to a control center. Nevertheless, due to timing, modeling, and instrumentation inaccuracies, the power flow equations were always infeasible. In a seminal contribution [1], the statistical foundations were laid for a multitude of grid monitoring tasks, including topology detection, static state estimation, exact and linearized models, bad data analysis, centralized and decentralized implementations, and dynamic state tracking. Since then, different chapters, books, and review articles have nicely outlined the progress in the area; see, for example, [2–4]. The revolutionary monitoring capabilities enabled by synchrophasor units have been put forth in [5].

This chapter aspires to glean some of the recent advances in power system state estimation (PSSE), though our collection is not exhaustive by any means. The

Advances in Electric Power and Energy: Static State Estimation, First Edition.
Edited by Mohamed E. El-Hawary.

Published 2021 by John Wiley & Sons, Inc.

Cramér–Rao bound, a lower bound on the (co)variance of any unbiased estimator, is first derived for the PSSE setup. After reviewing the classical Gauss–Newton iterations, contemporary PSSE solvers leveraging relaxations to convex programs and successive convex approximations are explored. A disciplined paradigm for distributed and decentralized schemes is subsequently exemplified under linear(ized) and exact grid models. Novel bad data processing models and fresh perspectives linking critical measurements to cyberattacks on the state estimator are presented. Finally, spurred by advances in online convex optimization (OCO), model-free and model-based state trackers are reviewed.

Notation: Lower- (upper-) case boldface letters denote column vectors (matrices), and calligraphic letters stand for sets. Vectors $\mathbf{0}$, $\mathbf{1}$, and $\mathbf{e}_n$ denote the all-zero, all-one, and the nth canonical vectors of suitable dimensions, respectively. The conjugate of a complex-valued object (scalar, vector, or matrix) x is denoted by x^*; $\Re\{x\}$ and $\Im\{x\}$ are its real and imaginary parts, and $j := \sqrt{-1}$. Superscripts $\mathcal{T}$ and $\mathcal{H}$ stand for transpose and conjugate transpose, respectively, while Tr($\mathbf{X}$) is the trace of matrix $\mathbf{X}$. A diagonal matrix having vector $\mathbf{x}$ on its main diagonal is denoted by dg($\mathbf{x}$), whereas the vector of diagonal entries of $\mathbf{X}$ is dg($\mathbf{X}$). The range space of $\mathbf{X}$ is denoted by range($\mathbf{X}$), and its null space (kernel) by null($\mathbf{X}$). The notation $\mathcal{N}(\boldsymbol{\mu}, \boldsymbol{\Sigma})$ represents the Gaussian distribution with mean $\boldsymbol{\mu}$ and covariance matrix $\boldsymbol{\Sigma}$.

6.2 POWER GRID MODELING

This section introduces notation and briefly reviews the power flow equations; for detailed exposition, see, e.g. [2, 6] and references therein. A power system can be represented by the graph $\mathcal{G} = (\mathcal{B}, \mathcal{L})$, where the node set $\mathcal{B}$ comprises its N_b buses and the edge set $\mathcal{L}$ its N_l transmission lines. Given the focus on alternating current (AC) power systems, steady-state voltages and currents are represented by their single-phase equivalent phasors per unit.

A transmission line $(n, k) \in \mathcal{L}$ running across buses $n, k \in \mathcal{B}$ is modeled by its total series admittance $y_{nk} = g_{nk} + jb_{nk}$ and total shunt susceptance jb^s_{nk}. If $\mathcal{V}_n$ is the complex voltage at bus n, the current $\mathcal{I}_{nk}$ flowing from bus n to bus n over line (m, n) is

$$\mathcal{I}_{nk} = \left(y_{nk} + jb^\mathrm{s}_{nk}/2\right)\mathcal{V}_n - y_{nk}\mathcal{V}_n \tag{6.1}$$

The current $\mathcal{I}_{nm}$ coming from the other end of the line can be expressed symmetrically. That is not the case if the two buses are connected via a transformer with complex ratio ρ_{nk} followed by a line, where

$$\mathcal{I}_{nk} = \frac{y_{nk} + jb^\mathrm{s}_{nk}/2}{|\rho_{nk}|^2}\mathcal{V}_n - \frac{y_{nk}}{\rho^*_{nk}}\mathcal{V}_n \tag{6.2a}$$

$$\mathcal{I}_{nm} = \left(y_{nk} + jb^\mathrm{s}_{nk}/2\right)\mathcal{V}_k - \frac{y_{nk}}{\rho_{nk}}\mathcal{V}_n \tag{6.2b}$$

Kirchhoff's current law dictates that the current injected into bus n is $\mathcal{I}_n = \sum_{n \in \mathcal{B}_n} \mathcal{I}_{nk}$, where $\mathcal{B}_n$ denotes the set of buses directly connected to bus n. If vector $\mathbf{i} \in \mathbb{C}^{N_b}$ collects all nodal currents and $\mathbf{v} \in \mathbb{C}^{N_b}$ all nodal voltages, the two vectors are linearly related through the bus admittance matrix $\mathbf{Y} = \mathbf{G} + j\mathbf{B}$ as

$$\mathbf{i} = \mathbf{Y}\mathbf{v} \tag{6.3}$$

Similar to (6.3), line currents can be stacked in the $2N_l$-dimensional vector $\mathbf{i}_f$ and expressed as a linear function of nodal voltages

$$\mathbf{i}_f = \mathbf{Y}_f\mathbf{v} \tag{6.4}$$

for some properly defined $2N_l \times N_b$ complex matrix $\mathbf{Y}_f$ (cf. (6.1)–(6.2)).

The complex power injected into bus n will be denoted by $\mathcal{S}_n := P_n + jQ_n$. Since by definition $\mathcal{S}_n = \mathcal{V}_n\mathcal{I}_n^*$, the vector of complex power injections $\mathbf{s} = \mathbf{p} + j\mathbf{q}$ can be expressed as

$$\mathbf{s} = \mathrm{dg}(\mathbf{v})\mathbf{i}^* = \mathrm{dg}(\mathbf{v})\mathbf{Y}^*\mathbf{v}^* \tag{6.5}$$

The power flowing from bus n to bus n over line (m, n) is $\mathcal{S}_{nk} = \mathcal{V}_n\mathcal{I}_{nk}^*$.

If voltages are expressed in polar form $\mathcal{V}_n = V_n e^{j\theta_n}$, the power flow equations in (6.5) per real and imaginary entry can be written as

$$P_n = \sum_{n=1}^{N_b} V_n V_n [G_{nk} \cos(\theta_n - \theta_n) + B_{nk} \sin(\theta_n - \theta_n)] \tag{6.6a}$$

$$Q_n = \sum_{n=1}^{N_b} V_n V_n [G_{nk} \sin(\theta_n - \theta_n) - B_{nk} \cos(\theta_n - \theta_n)] \tag{6.6b}$$

Since power injections are invariant if voltages are shifted by a common angle, the voltage phase is arbitrarily set to zero at a particular bus called the reference bus.

Alternatively to (6.6), if voltages are expressed in rectangular coordinates $\mathcal{V}_n = V_{r,m} + jV_{i,m}$, power injections are quadratically related to voltages

$$P_n = V_{r,m} \sum_{n=1}^{N_b} (V_{r,n}G_{nk} - V_{i,n}B_{nk}) + V_{i,m} \sum_{n=1}^{N_b} (V_{i,n}G_{nk} + V_{r,n}B_{nk}) \tag{6.7a}$$

$$Q_n = V_{i,m} \sum_{n=1}^{N_b} (V_{r,n}G_{nk} - V_{i,n}B_{nk}) - V_{r,m} \sum_{k=1}^{N_b} (V_{i,n}G_{nk} + V_{r,n}B_{nk}) \tag{6.7b}$$

To compactly express (6.7), observe that $\mathcal{S}_n^* = \mathcal{V}_n^*\mathcal{I}_n = (\mathbf{v}^{\mathcal{H}}\mathbf{e}_n)(\mathbf{e}_n^{\mathcal{T}}\mathbf{i}) = \mathbf{v}^{\mathcal{H}}\mathbf{e}_n\mathbf{e}_n^{\mathcal{T}}\mathbf{Y}\mathbf{v}$ from which it readily follows that

$$P_n = \mathbf{v}^{\mathcal{H}}\mathbf{H}_{P_n}\mathbf{v} \tag{6.8a}$$

$$Q_n = \mathbf{v}^{\mathcal{H}}\mathbf{H}_{Q_n}\mathbf{v} \tag{6.8b}$$

where the involved matrices are defined as

$$\mathbf{H}_{P_n} := \frac{1}{2}\left(\mathbf{e}_n\mathbf{e}_n^{\mathcal{T}}\mathbf{Y} + \mathbf{Y}^{\mathcal{H}}\mathbf{e}_n\mathbf{e}_n^{\mathcal{T}}\right) \tag{6.9a}$$

$$\mathbf{H}_{Q_n} := \frac{1}{2j}\left(\mathbf{e}_n\mathbf{e}_n^{\mathcal{T}}\mathbf{Y} - \mathbf{Y}^{\mathcal{H}}\mathbf{e}_n\mathbf{e}_n^{\mathcal{T}}\right) \tag{6.9b}$$

Similar expressions hold for the squared voltage magnitude at bus n:

$$V_n^2 = \mathbf{v}^{\mathcal{H}}\mathbf{H}_{V_n}\mathbf{v}, \text{ where } \mathbf{H}_{V_n} := \mathbf{e}_n\mathbf{e}_n^{\mathcal{T}} \tag{6.10}$$

Realizing that a line current can also be provided as $\mathcal{I}_{nk} = \mathbf{e}_{nk}^{\mathcal{T}}\mathbf{i}_{\text{f}}$, the power flow on line (n, k) seen from bus n is expressed as $\mathcal{S}_{nk}^* = \mathcal{V}_n^*\mathcal{I}_{nk} = \left(\mathbf{v}^{\mathcal{H}}\mathbf{e}_n\right)\left(\mathbf{e}_{nk}^{\mathcal{T}}\mathbf{i}_{\text{f}}\right) = \mathbf{v}^{\mathcal{H}}\mathbf{e}_n\mathbf{e}_{nk}^{\mathcal{T}}\mathbf{Y}_{\text{f}}\mathbf{v}$, from which it follows that

$$P_{nk} = \mathbf{v}^{\mathcal{H}}\mathbf{H}_{P_{nk}}\mathbf{v} \tag{6.11a}$$

$$Q_{nk} = \mathbf{v}^{\mathcal{H}}\mathbf{H}_{Q_{nk}}\mathbf{v} \tag{6.11b}$$

where $\mathbf{H}_{P_{nk}}$ and $\mathbf{H}_{Q_{nk}}$ are defined by substituting $\mathbf{e}_n^{\mathcal{T}}\mathbf{Y}$ and $\mathbf{Y}^{\mathcal{H}}\mathbf{e}_n$ by $\mathbf{e}_{nk}^{\mathcal{T}}\mathbf{Y}_{\text{f}}$ and $\mathbf{Y}_{\text{f}}^{\mathcal{H}}\mathbf{e}_{nk}$ in (6.9) accordingly.

Equations (6.8), (6.10), and (6.11) explain how power injections, flows, and squared voltage magnitudes are quadratic functions of voltage phasors as described by $\mathbf{v}^{\mathcal{H}}\mathbf{H}_m\mathbf{v}$ for certain complex $N_{\text{b}} \times N_{\text{b}}$ matrices $\mathbf{H}_m$. Regardless if $\mathbf{Y}$ and/or $\mathbf{Y}_{\text{f}}$ are symmetric or Hermitian, $\mathbf{H}_m$ are Hermitian by definition. This means that $\mathbf{H}_m = \mathbf{H}_m^{\mathcal{H}}$, or equivalently, $\Re\{\mathbf{H}_m\}^{\mathcal{T}} = \Re\{\mathbf{H}_m\}$, and $\Im\{\mathbf{H}_m\}^{\mathcal{T}} = -\Im\{\mathbf{H}_m\}$. It can be easily verified that the quadratic functions can be expressed in terms of real-valued quantities as

$$\mathbf{v}^{\mathcal{H}}\mathbf{H}_m\mathbf{v} = \overline{\mathbf{v}}^{\mathcal{T}}\overline{\mathbf{H}}_m\overline{\mathbf{v}} \tag{6.12}$$

for the expanded real-valued voltage vector $\overline{\mathbf{v}} := \left[\Re\{\mathbf{v}\}^{\mathcal{T}}\ \Im\{\mathbf{v}\}^{\mathcal{T}}\right]^{\mathcal{T}}$ and the real-valued counterpart of $\mathbf{H}_m$, namely,

$$\overline{\mathbf{H}}_m := \begin{bmatrix} \Re\{\mathbf{H}_m\} & -\Im\{\mathbf{H}_m\} \\ \Im\{\mathbf{H}_m\} & \Re\{\mathbf{H}_m\} \end{bmatrix} \tag{6.13}$$

6.3 PROBLEM STATEMENT

It was seen in Section 6.2 that given grid parameters collected in $\mathbf{Y}$ and $\mathbf{Y}_{\text{f}}$, all power system quantities can be expressed in terms of the voltage vector $\mathbf{v}$, which justifies its term as the system state. Meters installed across the grid measure electric quantities and forward their readings via remote terminal units to a control center for grid monitoring. Due to lack of synchronization, conventional meters cannot utilize the angle information of phasorial quantities. For this reason, legacy measurements involve phaseless power injections and flows along with voltage and current magnitudes at specific buses. The advent of the Global Positioning System (GPS) facilitated a precise timing signal across large geographical areas, thus enabling the revolutionary technology of synchrophasors or phasor measurement units (PMUs) [5]. Recovering bus voltages given network parameters and the available measurements constitutes the critical task of power system state estimation. This section formally states the problem, provides the Cramér–Rao bound on the variance of any unbiased estimator, and reviews the Gauss–Newton iterations. Solvers based on semidefinite relaxation (SDR) and successive convex approximations are subsequently explicated, and the section is wrapped up with issues germane to PMUs.

6.3.1 Weighted Least Squares Formulation

Consider M real-valued measurements $\{z_m\}_{m=1}^{M}$ related to the complex power system state $\mathbf{v}$ through the model

$$z_m = h_m(\mathbf{v}) + \epsilon_m \tag{6.14}$$

where $h_m(\mathbf{v}) : \mathbb{C}^{N_b} \rightarrow \mathbb{R}$ is a (non)linear function of $\mathbf{v}$ and ϵ_m captures the measurement noise and modeling inaccuracies. Collecting measurements and noise terms in vectors $\mathbf{z}$ and $\boldsymbol{\epsilon}$ accordingly, the vector form of (6.14) reads

$$\mathbf{z} = \mathbf{h}(\mathbf{v}) + \boldsymbol{\epsilon} \tag{6.15}$$

for the mapping $\mathbf{h} : \mathbb{C}^{N_b} \rightarrow \mathbb{R}^M$. Model (6.15) is instantiated for different types of measurements next.

Traditionally, the system state $\mathbf{v}$ is expressed in polar coordinates, namely, nodal voltage magnitudes and angles. Then $\mathbf{h}(\mathbf{v})$ maps the $2N_b$-dimensional state vector to SCADA measurements through the nonlinear equations (6.6). Expressing the states in polar form has been employed primarily due to two reasons. Firstly, the Jacobian matrix of $\mathbf{h}(\mathbf{v})$ is amenable to approximations. Secondly, voltage magnitude measurements are directly related to states. Nevertheless, due to recent computational reformulations, most of our exposition models voltages in the rectangular form. Then, as detailed in (6.12), the mth SCADA measurement z_m involves the quadratic function of the state $h_m(\mathbf{v}) = \mathbf{v}^{\mathcal{H}}\mathbf{H}_m\mathbf{v}$ for a Hermitian matrix $\mathbf{H}_m$.

Expressing voltages in rectangular coordinates is computationally advantageous when it comes to synchrophasors too. As evidenced by (6.3)–(6.4), PMU measurements feature *linear mappings* $h_m(\mathbf{v})$. If PMU measurements are expressed in rectangular coordinates, the model in (6.14) simplifies to

$$\mathbf{z} = \mathbf{H}\mathbf{v} + \boldsymbol{\epsilon} \tag{6.16}$$

for an $M \times N_b$ complex matrix $\mathbf{H}$ and complex-valued noise $\boldsymbol{\epsilon}$. Following the notation of (6.12)–(6.13), the linear measurement model of (6.16) can be expressed in terms of real-valued quantities as

$$\overline{\mathbf{z}} = \overline{(\mathbf{H}^*)}\overline{\mathbf{v}} + \overline{\boldsymbol{\epsilon}} \tag{6.17}$$

The random noise vector $\boldsymbol{\epsilon}$ in (6.15) is usually assumed independent of $\mathbf{h}(\mathbf{v})$, zero mean and *circularly symmetric*, that is, $\mathbb{E}\left[\boldsymbol{\epsilon}\boldsymbol{\epsilon}^{\mathcal{H}}\right] = \boldsymbol{\Sigma}_\epsilon$ and $\mathbb{E}\left[\boldsymbol{\epsilon}\boldsymbol{\epsilon}^{\mathcal{T}}\right] = \mathbf{0}$. The last assumption holds if, for example, the real and imaginary components of $\boldsymbol{\epsilon}$ are independent and have identical covariance matrices. This is true for a PMU measurement, where the actual state lies at the center of a *spherically shaped* noise cloud on the complex plane.

Moreover, the entries of $\boldsymbol{\epsilon}$ are oftentimes assumed uncorrelated, yielding a diagonal covariance $\boldsymbol{\Sigma}_\epsilon = \mathrm{dg}\left(\left\{\sigma_m^2\right\}\right)$ with σ_m^2 being the variance of the mth entry ϵ_m. However, that may not always be the case. For example, active and reactive powers at the same grid location are derived as products between the readings of a current transformer and a potential transformer. Further, noise terms may be correlated between the real and imaginary parts of the same phasor in a PMU.

Adopting the weighted least squares (WLS) criterion, power system state estimation can be formulated as

$$\underset{\mathbf{v}\in\mathbb{C}^{N_b}}{\text{minimize}} \; \| \boldsymbol{\Sigma}_\epsilon^{-1/2}(\mathbf{z} - \mathbf{h}(\mathbf{v})) \|_2^2 \tag{6.18}$$

where $\boldsymbol{\Sigma}_\epsilon^{-1/2}$ is the matrix square root of the inverse noise covariance matrix. If the noise is independent across measurements, then (6.18) simplifies to

$$\underset{\mathbf{v}}{\text{minimize}} \sum_{m=1}^{M} \frac{(z_m - h_m(\mathbf{v}))^2}{\sigma_m^2} \tag{6.19}$$

Either way, the PSSE task boils down to a (non)linear least squares (LS) fit. When the mapping $\mathbf{h}(\mathbf{v})$ is linear or when the entries of $\boldsymbol{\epsilon}$ are uncorrelated, the measurement model in (6.15) can be prewhitened. For example, the linear measurement model $\mathbf{z} = \mathbf{H}\mathbf{v} + \boldsymbol{\epsilon}$ can be equivalently transformed to

$$\boldsymbol{\Sigma}_\epsilon^{-1/2}\mathbf{z} = \left(\boldsymbol{\Sigma}_\epsilon^{-1/2}\mathbf{H}\right)\mathbf{v} + \boldsymbol{\Sigma}_\epsilon^{-1/2}\boldsymbol{\epsilon} \tag{6.20}$$

so that the associated noise $\boldsymbol{\Sigma}_\epsilon^{-1/2}\boldsymbol{\epsilon}$ is now uncorrelated. To ease the presentation, the noise covariance will be henceforth assumed $\boldsymbol{\Sigma}_\epsilon = \mathbf{I}_M$, yielding

$$\hat{\mathbf{v}} := \arg\min_{\mathbf{v}} \sum_{m=1}^{M} (z_m - h_m(\mathbf{v}))^2 \tag{6.21}$$

For Gaussian measurement noise $\boldsymbol{\epsilon} \sim \mathcal{N}(\mathbf{0}, \mathbf{I}_M)$, the minimizer of (6.21) coincides with the maximum likelihood estimate (MLE) of $\mathbf{v}$ [7].

6.3.2 Cramér–Rao Lower Bound Analysis

According to standard results in estimation theory [7], the variance of any *unbiased* estimator is lower bounded by the Cramér–Rao lower bound (CRLB). Appreciating its importance as a performance benchmark across different estimators, the ensuing result shown in the Appendix derives the CRLB for any unbiased power system state estimator based on the so-termed *Wirtinger's calculus* for complex analysis [8].

Proposition 6.1 Consider estimating the unknown state vector $\mathbf{v} \in \mathbb{C}^{N_b}$ from the noisy SCADA data $\{z_m\}_{m=1}^{M}$ of (6.15), where the Gaussian measurement error ϵ_m is independent across meters with mean zero and variance σ_m^2. The covariance matrix of any unbiased estimator $\hat{\mathbf{v}}$ satisfies

$$\mathrm{Cov}(\hat{\mathbf{v}}) \succeq \left[\mathbf{F}^{\dagger}(\mathbf{v}, \mathbf{v}^*)\right]_{1:N_b,1:N_b} \tag{6.22}$$

where the Fisher information matrix (FIM) is given as

$$\mathbf{F}(\mathbf{v}, \mathbf{v}^*) = \begin{bmatrix} \sum_{m=1}^{M} \frac{1}{\sigma_m^2} (\mathbf{H}_m \mathbf{v})(\mathbf{H}_m \mathbf{v})^{\mathcal{H}} & \sum_{m=1}^{M} \frac{1}{\sigma_m^2} (\mathbf{H}_m \mathbf{v})(\mathbf{H}_m^* \mathbf{v}^*)^{\mathcal{H}} \\ \sum_{m=1}^{M} \frac{1}{\sigma_m^2} (\mathbf{H}_m^* \mathbf{v}^*)(\mathbf{H}_m \mathbf{v})^{\mathcal{H}} & \sum_{m=1}^{M} \frac{1}{\sigma_m^2} (\mathbf{H}_m^* \mathbf{v}^*)(\mathbf{H}_m^* \mathbf{v}^*)^{\mathcal{H}} \end{bmatrix} \tag{6.23}$$

In addition, matrix $\mathbf{F}(\mathbf{v}, \mathbf{v}^*)$ has at least rank-one deficiency even when all possible SCADA measurements are available.

Although rank-deficient, the pseudo-inverse of $\mathbf{F}(\mathbf{v}, \mathbf{v}^*)$ qualifies as a valid lower bound on the mean-square error (MSE) of any unbiased estimator [9]. Rank deficiency of the FIM originates from the inherent voltage angle ambiguity: SCADA measurements remain invariant if nodal voltages are shifted globally by a unimodular phase constant. Fixing the angle of the reference bus waives this issue. It is also worth stressing that the CRLB in Proposition 6.1 is oftentimes attainable and benchmarks the optimal estimator performance [9]. Having derived the CRLB for the PSSE task, our next subsection deals with PSSE solvers.

6.3.3 Gauss–Newton Iterations

Consider for specificity model (6.16), though the real-valued model in (6.17) or the model involving polar coordinates could be employed as well. When the noise covariance matrix $\mathbf{\Sigma}_\epsilon = \mathbf{I}_M$, the PSSE task in (6.18) reduces to the nonlinear LS problem

$$\underset{\mathbf{v} \in \mathbb{C}^{N_b}}{\text{minimize}} \ \| \mathbf{z} - \mathbf{h}(\mathbf{v}) \|_2^2 \tag{6.24}$$

for which the Gauss–Newton iterations are known to offer the "workhorse" solution [2, chapter 2, 10, chapter 1]. According to the Gauss–Newton method, the function $\mathbf{h}(\mathbf{v})$ is linearized at a given point $\mathbf{v}^i \in \mathbb{C}^{N_b}$ using Taylor's expansion as

$$\tilde{\mathbf{h}}(\mathbf{v}, \mathbf{v}^i) := \mathbf{h}(\mathbf{v}^i) + \mathbf{J}^i(\mathbf{v} - \mathbf{v}^i)$$

where $\mathbf{J}^i := \nabla\mathbf{h}(\mathbf{v}^i)$ is the $M \times N_b$ Jacobian matrix of $\mathbf{h}$ evaluated at $\mathbf{v}^i$, whose (m, n)th entry is given by the Wirtinger derivative $\partial h_m/\partial \mathcal{V}_n$; see, e.g. [8] for Wirtinger's calculus. The Gauss–Newton method subsequently approximates the nonlinear LS fit in (6.24) with a linear one of $\tilde{\mathbf{h}}$ and relies on its minimizer to obtain the next iterate as

$$\begin{aligned}\mathbf{v}^{i+1} &\in \arg\min_{\mathbf{v}} \ \| \mathbf{z} - \tilde{\mathbf{h}}(\mathbf{v}, \mathbf{v}^i)\|^2 \\ &= \arg\min_{\mathbf{v}} \ \| \mathbf{z} - \mathbf{h}(\mathbf{v}^i)\|^2 - 2(\mathbf{v}-\mathbf{v}^i)^{\mathcal{H}}(\mathbf{J}^i)^{\mathcal{H}}(\mathbf{z}-\mathbf{h}(\mathbf{v}^i)) \\ &+ (\mathbf{v}-\mathbf{v}^i)^{\mathcal{H}}(\mathbf{J}^i)^{\mathcal{H}}\mathbf{J}^i(\mathbf{v}-\mathbf{v}^i)\end{aligned} \tag{6.25}$$

When matrix $(\mathbf{J}^i)^{\mathcal{H}}\mathbf{J}^i$ is invertible, $\mathbf{v}^{i+1}$ can be found in closed form as

$$\mathbf{v}^{i+1} = \mathbf{v}^i + \left[(\mathbf{J}^i)^{\mathcal{H}}\mathbf{J}^i\right]^{-1}(\mathbf{J}^i)^{\mathcal{H}}(\mathbf{z}-\mathbf{h}(\mathbf{v}^i)) \tag{6.26}$$

The state estimate is iteratively updated using (6.26) until some stopping criterion is satisfied.

If, on the other hand, the WLS cost (6.18) is minimized, the Gauss–Newton iterations can be similarly obtained by treating $\boldsymbol{\Sigma}_\epsilon^{-1/2}\mathbf{z}$ as $\mathbf{z}$ and $\boldsymbol{\Sigma}_\epsilon^{-1/2}\mathbf{h}(\mathbf{v})$ as $\mathbf{h}(\mathbf{v})$ in (6.24), yielding

$$\mathbf{v}^{i+1} = \mathbf{v}^i + \left[(\mathbf{J}^i)^{\mathcal{H}}\boldsymbol{\Sigma}_\epsilon^{-1}\mathbf{J}^i\right]^{-1}(\mathbf{J}^i)^{\mathcal{H}}\boldsymbol{\Sigma}_\epsilon^{-1}(\mathbf{z}-\mathbf{h}(\mathbf{v}^i)) \tag{6.27}$$

It is well known that the pure Gauss–Newton iterations in (6.26) or (6.27) may not guarantee convergence, which in fact largely depends on the starting point $\mathbf{v}^0$ [10, chapter 1.5]. A common way to improve convergence and ensure descent of the cost in (6.18) consists of including a backtracking line search in (6.27) to end up with

$$\mathbf{v}^{i+1} = \mathbf{v}^i + \mu^i\left[(\mathbf{J}^i)^{\mathcal{H}}\boldsymbol{\Sigma}_\epsilon^{-1}\mathbf{J}^i\right]^{-1}(\mathbf{J}^i)^{\mathcal{H}}\boldsymbol{\Sigma}_\epsilon^{-1}(\mathbf{z}-\mathbf{h}(\mathbf{v}^i)) \tag{6.28}$$

where the step size $\mu^i > 0$ is found through the backtracking line search rule [10, chapter 1.2]. Due to its intimate relationship with ordinary gradient descent alternatives for nonconvex optimization however, this Gauss–Newton iterative procedure can be trapped by local solutions [10, chapter 1.5]. In a nutshell, the grand challenge remains to develop PSSE solvers capable of attaining or approximating the global optimum at manageable computational complexity. A few recent proposals in this direction are presented next.

6.3.4 Semidefinite Relaxation

A method to tackle the nonlinear measurement model that can convert the PSSE problem of (6.21) to a convex semidefinite program (SDP) has been introduced in [11, 12]. Consider first expressing each measurement in $\mathbf{z}$ linearly in terms of the outer-product matrix $\mathbf{V} := \mathbf{v}\mathbf{v}^{\mathcal{H}}$. In this way, the quadratic models in (6.8), (6.10), and (6.11) can be transformed to linear ones in terms of the matrix variable $\mathbf{V}$. Thus, each noisy measurement in (6.14) can be written as $z_m = \mathbf{v}^{\mathcal{H}}\mathbf{H}_m\mathbf{v} + \epsilon_m = \mathrm{Tr}(\mathbf{H}_m\mathbf{V}) + \epsilon_m$. Rewriting the PSSE task in (6.21) accordingly in terms of $\mathbf{V}$ reduces to

$$\hat{\mathbf{V}}_1 := \arg \min_{\mathbf{V}\in\mathbb{C}^{N_b\times N_b}} \sum_{m=1}^{M} [z_m - \mathrm{Tr}(\mathbf{H}_m\mathbf{V})]^2 \tag{6.29a}$$

$$\text{s.to } \mathbf{V} \succeq \mathbf{0}, \text{ and } \mathrm{rank}(\mathbf{V}) = 1 \tag{6.29b}$$

where the positive semidefinite (PSD) and the rank-1 constraints jointly ensure that for any $\mathbf{V}$ obeying (6.29b), there always exists a vector $\mathbf{v}$ such that $\mathbf{V} = \mathbf{v}\mathbf{v}^{\mathcal{H}}$.

Although z_m and $\mathbf{V}$ are linearly related as in (6.29), nonconvexity is still present in two aspects: (i) the cost function in (6.29a) has degree 4 in the entries of $\mathbf{V}$, and (ii) the rank constraint in (6.29b) is nonconvex. Aiming for an SDP reformulation of (6.29), Schur's complement lemma (see, e.g. [13, appendix 5.5]) can be leveraged to tightly bound each summand in (6.29a) using an auxiliary variable $\chi_m > 0$. Collecting all χ_m's in $\chi \in \mathbb{R}^m$, the problem in (6.29) can be expressed as

$$\{\hat{\mathbf{V}}_2, \hat{\chi}_2\} := \arg \min_{\mathbf{V},\ \chi} \mathbf{1}^T\chi \tag{6.30a}$$

$$\text{s.to } \mathbf{V} \succeq \mathbf{0}, \text{and } \mathrm{rank}(\mathbf{V}) = 1 \tag{6.30b}$$

$$\begin{bmatrix} \chi_m & z_m - \mathrm{Tr}(\mathbf{H}_m\mathbf{V}) \\ z_m - \mathrm{Tr}(\mathbf{H}_m\mathbf{V}) & 1 \end{bmatrix} \succeq \mathbf{0}, \forall m \tag{6.30c}$$

The equivalence among all three SE problems (6.21), (6.29), and (6.30) has been shown in [11], where their optimal solutions satisfy

$$\hat{\mathbf{V}}_1 = \hat{\mathbf{V}}_2 = \hat{\mathbf{v}}\hat{\mathbf{v}}^{\mathcal{H}}, \text{ and } \hat{\chi}_{2,m} = \left[z_m - \mathrm{Tr}\left(\mathbf{H}_m\hat{\mathbf{V}}_2\right)\right]^2, \forall m \tag{6.31}$$

The only source of nonconvexity in the equivalent SE problem of (6.30) comes from the rank-1 constraint. Motivated by the technique of SDR (see, e.g. the seminal work of [14]), one can obtain the following convex SDP upon dropping the rank constraint:

$$\{\hat{\mathbf{V}}, \hat{\chi}\} := \arg \min_{\mathbf{V},\chi} \mathbf{1}^T\chi \tag{6.32a}$$

$$\text{s.to } \mathbf{V} \succeq \mathbf{0}, \text{ and } (6.30\text{c}) \tag{6.32b}$$

For the SDR-PSSE formulation in (6.32), a few assumptions have been made in [12] to establish its global optimality in a specific setup:

(as1) *The graph* $\mathcal{G} = (\mathcal{B}, \mathcal{L})$ *has a tree topology.*

(as2) *Every bus is equipped with a voltage magnitude meter.*

(as3) *All measurements in* $\mathbf{z}$ *are noise-free, that is* $\boldsymbol{\epsilon} = \mathbf{0}$.

Proposition 6.2 Under (as1)–(as3), solving the relaxed problem (6.32) attains the global optimum of the original PSSE problem (6.30) or (6.21); that is, $\text{rank}(\hat{\mathbf{V}}) = 1$.

Assumptions (as1)–(as3) may offer a close approximation of the realistic PSSE scenario, thanks to characteristics of transmission systems such as sparse connectivity, almost flat voltage profile, and high metering accuracy. Although they do not hold precisely in realistic transmission systems, near-optimality of the relaxed problem (6.32) has been numerically supported by extensive tests [12]. A more crucial issue is to recover a feasible SE solution from the relaxed problem (6.32), as $\hat{\mathbf{V}}$ is very likely to have rank greater than 1. This is possible either by finding the best rank-1 approximation to $\hat{\mathbf{V}}$ via eigenvalue decomposition or via randomization [15].

SDR endows SE with a convex SDP formulation for which efficient schemes are available to obtain the global optimum using, for example, the interior-point solver. The computational complexity for eigendecomposition is in the order of matrix multiplication, thus negligible compared to solving the SDP; see [15] and references therein. However, the polynomial complexity of solving the SDP could be a burden for real-time power system monitoring, which motivates well the distributed implementation of Section 6.4.2.

6.3.5 Penalized Semidefinite Relaxation

Building on an alternative formulation of the *power flow problem*, a penalized version of the aforementioned SDP-based state estimator has been devised in [16–18]. Commencing with the power flow task, it can be interpreted as a particular instance of PSSE, where:

- Measurements (henceforth termed specifications) are noiseless.
- Excluding the reference bus, buses are partitioned into the subset $\mathcal{B}_{\text{PV}}$ for which active injections and voltage magnitudes are specified, and the subset $\mathcal{B}_{\text{PQ}}$, for which active and reactive injections are specified.

The power flow task can be posed as the feasibility problem; that is,

$$\begin{aligned} &\text{find } \mathbf{v} \in \mathbb{C}^{N_\text{b}} \\ \text{s.to } & P_n = \mathbf{v}^{\mathcal{H}}\mathbf{H}_{P_n}\mathbf{v}, \quad \forall n \in \mathcal{B}_{\text{PV}} \cup \mathcal{B}_{\text{PQ}} \\ & Q_n = \mathbf{v}^{\mathcal{H}}\mathbf{H}_{Q_n}\mathbf{v}, \quad \forall n \in \mathcal{B}_{\text{PQ}} \\ & V_n^2 = \mathbf{v}^{\mathcal{H}}\mathbf{H}_{V_n}\mathbf{v}, \quad \forall n \in \mathcal{B}_{\text{PV}}, \text{and } V_{\text{ref}}^2 = V_0 \end{aligned} \tag{6.33}$$

Using the SDP reformulation presented earlier, the power flow task can be equivalently expressed as

$$\begin{aligned} &\text{find } \mathbf{V} \in \mathbb{C}^{N_\text{b} \times N_\text{b}} \\ &\text{s.to } z_m = \text{Tr}(\mathbf{H}_m\mathbf{V}), \quad m = 1, \ldots, 2N_\text{b} - 1 \\ &\mathbf{V} \succeq \mathbf{0}, \text{and rank}(\mathbf{V}) = 1 \end{aligned} \tag{6.34}$$

where the specifications (constraints) of (6.33) have been generically captured by the pairs $\{(z_m; \mathbf{H}_m)\}_{m=1}^{2N_b-1}$.

Although the optimization in (6.34) is nonconvex, a convex relaxation can be obtained by dropping the rank constraint. To promote rank-one solutions, the feasibility problem is further reduced to [17]

$$\hat{\mathbf{V}} := \arg \min_{\mathbf{V} \succeq \mathbf{0}} \ \mathrm{Tr}(\mathbf{H}_0 \mathbf{V}) \tag{6.35}$$

$$\text{s.to } z_m = \mathrm{Tr}(\mathbf{H}_m \mathbf{V}), m = 1, \ldots, 2N_b - 1 \tag{6.36}$$

If the Hermitian matrix $\mathbf{H}_0$ is selected such that $\mathbf{H}_0 \succeq \mathbf{0}$, $\mathrm{rank}(\mathbf{H}_0) = N_b - 1$, and $\mathbf{H}_0 \mathbf{1} = \mathbf{0}$, then any state $\mathbf{v}$ close to the flat voltage profile $\mathbf{1} + j\mathbf{0}$ can be recovered from the rank-one minimizer $\hat{\mathbf{V}}$ of (6.35); see [17, theorem 2]. The stated conditions exclude the case $\mathbf{H}_0 = \mathbf{I}_{N_b}$ that would have led to the nuclear-norm heuristic commonly used in low-rank matrix completion.

Spurred by this observation, the PSSE task can be posed as [16]

$$\hat{\mathbf{V}}_\mu := \arg \min_{\mathbf{V} \succeq \mathbf{0}} \mathrm{Tr}(\mathbf{H}_0 \mathbf{V}) + \mu \sum_{m=1}^{M} f_m(z_m - \mathrm{Tr}(\mathbf{H}_m \mathbf{V})) \tag{6.37}$$

for some regularization parameter $\mu \geq 0$, where M now can be larger than $2N_b - 1$. The second term in the cost of (6.37) is a data-fitting term ensuring that the recovered state is consistent with the collected measurements based on selected criteria. Two cases for f_m of special interest are the LS fit $f_m^{\mathrm{LS}}(\epsilon) := \epsilon^2$ and the least absolute value (LAV) one $f_m^{\mathrm{LAV}}(\epsilon) := |\epsilon|$, $\forall m = 1, \ldots, M$. On the other hand, the first term in (6.37) can be understood as a *regularizer* to promote rank-one solutions $\hat{\mathbf{V}}_\mu$; see more details in [16].

In the noiseless setup, where all measurements comply with the model $z_m = \mathbf{v}_0^{\mathcal{H}} \mathbf{H}_m \mathbf{v}_0$, the minimization in (6.37) has been shown to possess a rank-one minimizer $\hat{\mathbf{V}}_\mu = \hat{\mathbf{v}}_\mu \hat{\mathbf{v}}_\mu^{\mathcal{H}}$ for all $\mu \geq 0$ under both f_m^{LS} and f_m^{LAV}; see details in [16]. Interestingly, the solution $\hat{\mathbf{v}}_\mu$ obtained under the LS fit does not coincide with $\mathbf{v}_0$ for any $\mu \geq 0$, whereas the LAV solution $\hat{\mathbf{v}}_\mu$ provides the actual state $\mathbf{v}_0$ for a sufficiently large μ. Error bounds between $\hat{\mathbf{v}}_\mu$ and $\mathbf{v}_0 \mathbf{v}_0^{\mathcal{H}}$ under the regularized LAV solution for noisy measurements are established in [18].

6.3.6 Feasible Point Pursuit

The feasible point pursuit (FPP) method studied in [19] offers another computationally manageable solver for approximating the globally optimal PSSE. As a special case of the convex–concave procedure [20], FPP is an iterative algorithm for handling general nonconvex quadratically constrained quadratic programs (QCQPs) [21]. It approximates the feasible solutions of a nonconvex QCQP by means of a sequence of convexified QCQPs obtained with successive convex inner restrictions of the original nonconvex feasibility set [21].

The first step in applying FPP to PSSE is a reformulation of (6.18) into a standard QCQP [22, 23]:

$$\underset{\mathbf{v}\in\mathbb{C}^{N_b},\ \chi\in\mathbb{R}^M}{\text{minimize}}\ \chi^{\mathcal{T}}\boldsymbol{\Sigma}_\epsilon^{-1}\chi \tag{6.38a}$$

$$\text{s.to}\ \ \mathbf{v}^{\mathcal{H}}\mathbf{H}_m\mathbf{v} \le z_m + \chi_m, \qquad 1 \le m \le M \tag{6.38b}$$

$$\mathbf{v}^{\mathcal{H}}(-\mathbf{H}_m)\mathbf{v} \le -z_m + \chi_m, \quad 1 \le m \le M \tag{6.38c}$$

where vector $\chi \in \mathbb{R}^M$ collects the auxiliary variables $\{\chi_m \ge 0\}_{m=1}^M$. For power flow and power injection measurements, the corresponding Hermitian measurement matrices $\{\mathbf{H}_m\}$ are indefinite in general; thus, both constraints (6.38b) and (6.38c) are nonconvex. On the contrary, squared voltage magnitude measurements relate to positive semidefinite matrices $\{\mathbf{H}_m\}$, so only the related constraint (6.38c) is nonconvex. Either way, problem (6.38) is NP-hard and hence computationally intractable [24].

Using eigendecomposition, every measurement matrix $\mathbf{H}_m$ can be expressed as the sum of a positive and a negative semidefinite matrix as $\mathbf{H}_m = \mathbf{H}_m^+ + \mathbf{H}_m^-$, so the constraints in (6.38) are rewritten as

$$\mathbf{v}^{\mathcal{H}}\mathbf{H}_m^+\mathbf{v} + \mathbf{v}^{\mathcal{H}}\mathbf{H}_m^-\mathbf{v} \le z_m + \chi_m \tag{6.39a}$$

$$\mathbf{v}^{\mathcal{H}}\mathbf{H}_m^+\mathbf{v} + \mathbf{v}^{\mathcal{H}}\mathbf{H}_m^-\mathbf{v} \ge z_m - \chi_m \tag{6.39b}$$

for $m = 1, \ldots, M$. Observe now that since $\mathbf{v}^{\mathcal{H}}\mathbf{H}_m^-\mathbf{v}$ is a concave function of $\mathbf{v}$, it is upper bounded by its first-order (linear) approximation at any point $\mathbf{v}^i$; that is,

$$\mathbf{v}^{\mathcal{H}}\mathbf{H}_m^-\mathbf{v} \le 2\Re\left\{(\mathbf{v}^i)^{\mathcal{H}}\mathbf{H}_m^-\mathbf{v}\right\} - (\mathbf{v}^i)^{\mathcal{H}}\mathbf{H}_m^-\mathbf{v}^i$$

The concave function $\mathbf{v}^{\mathcal{H}}(-\mathbf{H}_m^+)\mathbf{v}$ can be upper bounded similarly.

Based on this observation, the FPP technique replaces the concave functions in the constraints of (6.39) with their linear upper bounds. The point of linearization at every iteration is chosen to be the previous state estimate. Specifically, initializing with $\mathbf{v}^0$, the FPP produces the iterates

$$\begin{aligned}\{\mathbf{v}^{i+1},\chi^{i+1}\} &:= \arg\min_{\mathbf{v},\chi\ge\mathbf{0}} \chi^{\mathcal{T}}\boldsymbol{\Sigma}_\epsilon^{-1}\chi \\ \text{s.to}\ \ &\mathbf{v}^{\mathcal{H}}\mathbf{H}_m^+\mathbf{v} + 2\Re\left\{(\mathbf{v}^i)^{\mathcal{H}}\mathbf{H}_m^-\mathbf{v}\right\} \le z_m + (\mathbf{v}^i)^{\mathcal{H}}\mathbf{H}_m^-\mathbf{v}^i + \chi_m, \ \ \forall m \\ &\mathbf{v}^{\mathcal{H}}\mathbf{H}_m^-\mathbf{v} + 2\Re\left\{(\mathbf{v}^i)^{\mathcal{H}}\mathbf{H}_m^+\mathbf{v}\right\} \ge z_m + (\mathbf{v}^i)^{\mathcal{H}}\mathbf{H}_m^+\mathbf{v}^i - \chi_m, \ \ \forall m\end{aligned} \tag{6.40}$$

At every iteration, the FPP technique solves the now convex QCQP in (6.40). The procedure has been shown to globally converge to a stationary point of the WLS formulation (6.18) of the PSSE task [22].

Figure 6.1 compares Gauss–Newton iterations, the SDR-based solver, and the FPP-based solver on the IEEE 14- and 30-bus systems [25]. The actual nodal voltage magnitudes and angles were generated uniformly at random over [0.9,1.1]

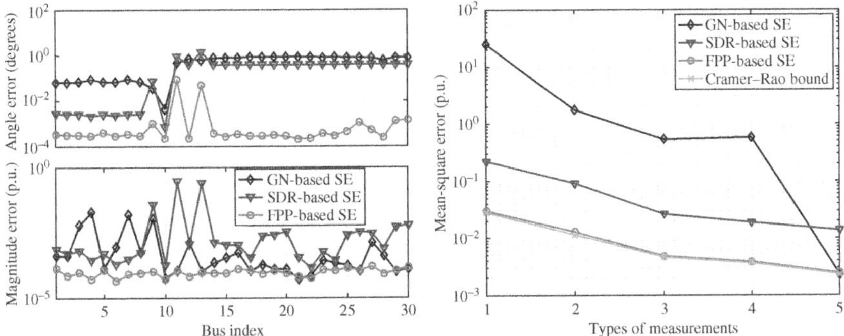

Figure 6.1 Left: Voltage magnitude and angle estimation errors per bus for the IEEE 30-bus system. Right: MSEs and CRLB versus types of measurements used for the IEEE 14-bus system using (i) Gauss–Newton iterations, (ii) the SDR-based PSSE, and (iii) the FPP-based PSSE.

and $[-0.4\pi, 0.4\pi]$, respectively. Independent zero-mean Gaussian noise with standard deviation 0.05 for power and 0.02 for voltage measurements was assumed, and all reported results were averaged over 100 independent Monte Carlo realizations. The measurements for the IEEE 30-bus system include all nodal voltage magnitudes and the active power flows at both sending and receiving ends. The left panel of Figure 6.1 depicts that the magnitude and angle estimation errors attained by the FPP solver are consistently below its competing alternatives [22].

The second experiment examines the MSE performance of the three approaches relative to the CRLB of (6.22) for the IEEE 14-bus test system. Initially, all voltage magnitudes as well as all sending- and receiving-end active power flows were measured, which corresponds to the base case 1 in the x-axis of the right panel of Figure 6.1. To show the MSE performance relative for an increasing number of measurements, additional types of measurements were included in a deterministic manner. All types of SCADA measurements were ordered as $\{V_n^2, P_{kn}, P_{nk}, Q_{kn}, Q_{nk}, P_n, Q_n\}$. Each x-axis value in the right panel of Figure 6.1 implies that the number of ordered types of measurements was used in the experiment to obtain the corresponding MSEs.

6.3.7 Synchrophasors

To incorporate synchrophasors into the PSSE formulation, let $\zeta_n = \mathbf{\Phi}_n \mathbf{v} + \boldsymbol{\varepsilon}_n$ collect the noisy PMU data at bus n (cf. (6.16)). The related measurement matrix is $\mathbf{\Phi}_n$, and the measurement noise $\boldsymbol{\varepsilon}_n$ is assumed to be complex zero-mean Gaussian, independent from the noise $\boldsymbol{\epsilon}$ in legacy meters and across buses. Following the normalization convention in (6.21), the noise vector $\boldsymbol{\varepsilon}_n$ is assumed prewhitened, such that all PMU measurements exhibit the same accuracy. The PSSE task now amounts to estimating $\mathbf{v}$ given both $\mathbf{z}$ and $\{\zeta_n\}_{n\in\mathcal{P}}$, where $\mathcal{P} \subseteq \mathcal{B}$ denotes the subset of the

PMU-instrumented buses. Hence, the MLE cost in (6.21) needs to be augmented by the log-likelihood induced by PMU data as

$$\hat{\mathbf{v}} := \arg\min_{\mathbf{v}} \sum_{m=1}^{M} (z_m - h_m(\mathbf{v}))^2 + \sum_{n \in \mathcal{P}} \|\boldsymbol{\zeta}_n - \boldsymbol{\Phi}_n \mathbf{v}\|_2^2 \tag{6.41}$$

The SDR methodology is again well motivated to convexify the augmented PSSE problem (6.41) into

$$\underset{\mathbf{V},\mathbf{v},\boldsymbol{\chi}}{\text{minimize}} \ \mathbf{1}^T\boldsymbol{\chi} + \sum_{n \in \mathcal{P}} \left[\text{Tr}\left(\boldsymbol{\Phi}_n^{\mathcal{H}}\boldsymbol{\Phi}_n\mathbf{V}\right) - 2\Re\{\boldsymbol{\zeta}_n^{\mathcal{H}}\boldsymbol{\Phi}_n\mathbf{v}\}\right] \tag{6.42a}$$

$$\text{s.to} \begin{bmatrix} \mathbf{V} & \mathbf{v} \\ \mathbf{v}^{\mathcal{H}} & 1 \end{bmatrix} \succeq \mathbf{0}, \ \text{and (6.30c)} \tag{6.42b}$$

By Schur's complement, the left SDP constraint in (6.42b) can be expressed equivalently as $\mathbf{V} \succeq \mathbf{v}\mathbf{v}^{\mathcal{H}}$. If the latter constraint is enforced with equality, the matrix $\mathbf{V}$ becomes rank-one. Imposing a rank-one constraint in (6.42) renders it equivalent to the augmented PSSE task of (6.41). The SDP here also offers the advantages of (6.32), in terms of the near-optimality and the distributed implementation deferred to Section 6.4.2. To recover a feasible solution, one can again use the best rank-1 approximation or adopt the randomization technique as elucidated in [12].

Alternatively, the two types of measurements can be jointly utilized upon interpreting the SCADA-based estimate as a prior for PMU-based estimation [5, 26]. Specifically, if $\hat{\mathbf{v}}_s$ is the SCADA-based estimate, the prior probability density function of the actual state can be postulated to be a circularly symmetric complex Gaussian with mean $\hat{\mathbf{v}}_s$ and covariance $\hat{\boldsymbol{\Sigma}}_s$.

Given PMU data and the SCADA-based prior, the state can be estimated following a maximum *a posteriori* probability (MAP) approach as

$$\hat{\mathbf{v}} := \arg\min_{\mathbf{v}} \ (\mathbf{v} - \hat{\mathbf{v}}_s)^{\mathcal{H}}\hat{\boldsymbol{\Sigma}}_s^{-1}(\mathbf{v} - \hat{\mathbf{v}}_s) + \sum_{n \in \mathcal{P}} \|\boldsymbol{\zeta}_n - \boldsymbol{\Phi}_n \mathbf{v}\|_2^2 \tag{6.43}$$

where the first summand is the negative logarithm of the prior distribution and the second one is the negative log-likelihood from the PMU data. In essence, the approach in (6.43) treats the SCADA-based estimate as pseudo-measurements relying on the model $\hat{\mathbf{v}}_s = \mathbf{v} + \boldsymbol{\epsilon}$ with circularly symmetric zero-mean noise having $\mathbb{E}\left[\boldsymbol{\epsilon}\boldsymbol{\epsilon}^{\mathcal{H}}\right] = \hat{\boldsymbol{\Sigma}}_s$.

6.4 DISTRIBUTED SOLVERS

Upcoming power system requirements call for decentralized solvers. Measurements are now collected at much finer spatiotemporal scales and the number of states increases exponentially as monitoring schemes extend to low-voltage distribution grids [27]. Tightly interconnected power systems call for the close coordination of regional control centers [28], while operators and utilities perform their computational operations *on the cloud.*

This section reviews advances in distributed PSSE solvers. As the name suggests, distributed PSSE solutions spread the computational load across different processors or control centers to speed up time, implement memory-intensive tasks, and/or guarantee privacy. A network of processors may be coordinated by one or more supervising control centers in a hierarchical fashion, or completely autonomously, by exchanging information between processors. To clarify terminology, the latter architecture will be henceforth identified as decentralized.

Distributed solvers with a hierarchical structure have been proposed since the statistical formulation of PSSE [1, part III]. Different versions of this original scheme were later developed in [28–32]. Decentralized schemes include block Jacobi iterations [33, 34], an approximate algorithm building on the related optimality conditions [35], or matrix-splitting techniques for facilitating matrix inversion across areas running Gauss–Newton iterations [36]. Most of the aforementioned approaches presume local identifiability (i.e. each area is identifiable even when shared measurements are excluded) or their convergence is not guaranteed. Assuming a ring topology, every second agent updates its state iteratively through the auxiliary problem principle in [37]. Local observability is waived in the consensus-type solver of [38], where each control center maintains a copy of the entire high-dimensional state vector resulting in slow convergence. For a relatively recent review on distributed PSSE solves, see also [39].

6.4.1 Distributed Linear Estimator

Consider an interconnected system partitioned in K areas supervised by separate control centers. Without loss of generality, an area may be thought of as an independent system operator region, a balancing authority, a power distribution center, or a substation [5]. Area k collects M_k measurements obeying the linear model

$$\mathbf{z}_k = \mathbf{H}_k \mathbf{v}_k + \boldsymbol{\epsilon}_k \tag{6.44}$$

where vector $\mathbf{v}_k \in \mathbb{C}^{N_k}$ collects the system states related to $\mathbf{z}_k$ through the complex matrix $\mathbf{H}_k$. The random noise vector $\boldsymbol{\epsilon}_k$ is zero mean with identity covariance upon prewhitening, if measurements are uncorrelated across areas. The model in (6.44) is exact for PMU measurements, but it may also correspond to a single Gauss–Newton iteration as explained in Section 6.3.3.

Performing PSSE locally at area k amounts to solving

$$\underset{\mathbf{v}_k \in \mathcal{X}_k}{\text{minimize}}\ f_k(\mathbf{v}_k) \tag{6.45}$$

where the convex set $\mathcal{X}_k$ captures possible prior information, such as zero-injection buses or short circuits [2, 3]. If $f_k(\mathbf{v}_k) = \| \mathbf{z}_k - \mathbf{H}_k \mathbf{v}_k \|_2^2/2$, the minimizer of (6.45) is the least squares estimate (LSE) of $\mathbf{v}_k$, which is also the MLE of $\mathbf{v}_k$ for Gaussian $\boldsymbol{\epsilon}_k$.

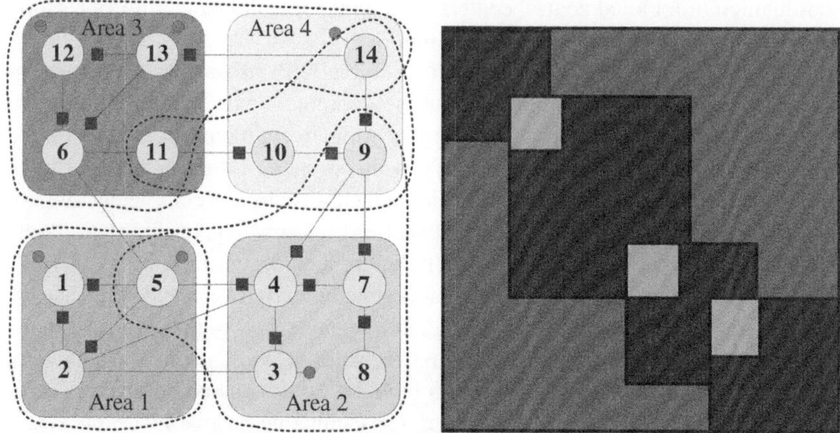

Figure 6.2 Left: The IEEE 14-bus system partitioned into four areas [25, 31]. Dotted lassos show the buses belonging to area state vectors $\mathbf{v}_k$'s. PMU bus voltage (line current) measurements depicted by circles (squares). Right: The matrix structure for the left system with the distributed SDR solver: square denotes the overall $\mathbf{V}$, while dark and light ones correspond to the four area submatrices $\{\mathbf{V}_k\}$ and their overlaps. Source: Reproduced with permission of IEEE.

As illustrated in Figure 6.2, the per-area state vectors $\{\mathbf{v}_k\}_{k=1}^K$ overlap partially. Although area 4 supervises buses {9,10,14}, it also collects the current reading on lines (10, 11). Thus, its state vector $\mathbf{v}_4$ extends to bus {11} that is nominally supervised by area 3. To set up notation, define $\mathcal{S}_{kl}$ as the shared states for a pair of neighboring areas (k, l). Let also $\mathbf{v}_k[l]$ ($\mathbf{v}_l[k]$) denote the sub-vector of $\mathbf{v}_k$ ($\mathbf{v}_l$) consisting of their overlapping variables ordered as they appear in $\mathbf{v}$. For example, $\mathbf{v}_3[4] = \mathbf{v}_4[3]$ contain the bus voltages of {11}. Solving the K problems in (6.45) separately is apparently suboptimal since the estimates of shared states will disagree, tie-line measurements have to be ignored, and boundary states may thus become unobservable.

Coupling the per-area PSSE tasks can be posed as

$$\begin{aligned} \underset{\{\mathbf{v}_k \in \mathcal{X}_k\}}{\text{minimize}} \quad & \sum_{k=1}^{K} f_k(\mathbf{v}_k) \\ \text{s.to} \quad & \mathbf{v}_k[l] = \mathbf{v}_l[k], \quad \forall l \in \mathcal{B}_k, \forall k \end{aligned} \tag{6.46}$$

where $\mathcal{N}_k$ is the set of areas sharing states with area k. The equality constraints of (6.46) guarantee consensus over the shared variables. Although (6.46) is amenable to decentralized implementations (cf. [37]), areas need a coordination protocol for their updates. To enable a truly decentralized solution, we follow the seminal approach of [40, 41]. An auxiliary variable $\mathbf{v}_{kl}$ is introduced per pair of connected areas (k, l); the symbols $\mathbf{v}_{kl}$ and $\mathbf{v}_{kl}$ are used interchangeably. The optimization in (6.46) can then be written as

$$\underset{\{\mathbf{v}_k \in \mathcal{X}_k\}, \{\mathbf{v}_{kl}\}}{\text{minimize}} \sum_{k=1}^{K} f_k(\mathbf{v}_k) \tag{6.47}$$
$$\text{s.to } \mathbf{v}_k[l] = \mathbf{v}_{kl}, \quad \forall l \in \mathcal{B}_k, k = 1, \ldots, K$$

Problem (6.47) can be solved using the alternating direction method of multipliers (ADMM) [42, 43]. In its general form, ADMM tackles convex optimization problems of the form

$$\underset{\mathbf{x} \in \mathcal{X}, \mathbf{z} \in \mathcal{Z}}{\text{minimize}} \; f(\mathbf{x}) + g(\mathbf{z}) \tag{6.48a}$$

$$\text{s.to } \mathbf{Ax} + \mathbf{Bz} = \mathbf{c} \tag{6.48b}$$

for given matrices and vectors ($\mathbf{A}$, $\mathbf{B}$, $\mathbf{c}$) of proper dimensions. Upon assigning a Lagrange multiplier $\boldsymbol{\lambda}$ for the coupling constraint in (6.48a), the ($\mathbf{x}$, $\mathbf{z}$) minimizing (6.48a) are found through the next iterations for some $\mu > 0$:

$$\mathbf{x}^{i+1} := \arg\min_{\mathbf{x} \in \mathcal{X}} f(\mathbf{x}) + \frac{\mu}{2} \| \mathbf{Ax} + \mathbf{Bz}^i - \mathbf{c} + \boldsymbol{\lambda}^i \|_2^2 \tag{6.49a}$$

$$\mathbf{z}^{i+1} := \arg\min_{\mathbf{z} \in \mathcal{Z}} g(\mathbf{z}) + \frac{\mu}{2} \| \mathbf{Ax}^{i+1} + \mathbf{Bz} - \mathbf{c} + \boldsymbol{\lambda}^i \|_2^2 \tag{6.49b}$$

$$\boldsymbol{\lambda}^{i+1} := \boldsymbol{\lambda}^i + \mathbf{Ax}^{i+1} + \mathbf{Bz}^{i+1} - \mathbf{c} \tag{6.49c}$$

Toward applying the ADMM iterations to (6.46), identify variables $\{\mathbf{v}_k\}$ as $\mathbf{x}$ in (6.48a) and $\{\mathbf{v}_{kl}\}$ as $\mathbf{x}$ with $g(\mathbf{z}) = 0$. Moreover, introduce Lagrange multipliers $\boldsymbol{\lambda}_{k,l}$ for each constraint in (6.47). Observe that $\boldsymbol{\lambda}_{k,l}$ and $\boldsymbol{\lambda}_{k,l}$ correspond to the distinct constraints $\mathbf{v}_k[l] = \mathbf{v}_{kl}$ and $\mathbf{v}_l[k] = \mathbf{v}_{kl}$, respectively. According to (6.49a), the per-area state vectors $\{\mathbf{v}_k\}$ can be updated separately as

$$\mathbf{v}_k^{i+1} := \arg\min_{\mathbf{v}_k \in \mathcal{X}_k} f_k(\mathbf{v}_k) + \frac{\mu}{2} \sum_{l \in \mathcal{B}_k} \| \mathbf{v}_k[l] - \mathbf{v}_{kl}^i + \boldsymbol{\lambda}_{k,l}^i \|_2^2 \tag{6.50}$$

From (6.49b) and assuming every state is shared by at most two areas, the auxiliary variables $\mathbf{v}_{kl}$ can be readily found in closed form given by

$$\mathbf{v}_{kl}^{i+1} = \frac{1}{2} \left(\mathbf{v}_k^{i+1}[l] + \mathbf{v}_l^{i+1}[k] + \boldsymbol{\lambda}_{k,l}^i + \boldsymbol{\lambda}_{l,k}^i \right) \tag{6.51}$$

while the two related multipliers are updated as

$$\boldsymbol{\lambda}_{k,l}^{i+1} := \boldsymbol{\lambda}_{k,l}^i + \left(\mathbf{v}_k^{i+1}[l] - \mathbf{v}_{kl}^{i+1} \right) \tag{6.52a}$$

$$\boldsymbol{\lambda}_{l,k}^{i+1} := \boldsymbol{\lambda}_{l,k}^i + \left(\mathbf{v}_l^{i+1}[k] - \mathbf{v}_{kl}^{i+1} \right) \tag{6.52b}$$

Adding (6.52) by parts and combining it with (6.51) yield $\boldsymbol{\lambda}_{k,l}^{i+1} = -\boldsymbol{\lambda}_{l,k}^{i+1}$ at all iterations i if the multipliers are initialized at zero. Hence, the auxiliary variable $\mathbf{v}_{kl}$ ends up being the average of the shared states; that is,

$$\mathbf{v}_{kl}^{i+1} = \frac{1}{2}\left(\mathbf{v}_k^{i+1}[l] + \mathbf{v}_l^{i+1}[k]\right) \tag{6.53}$$

To summarize, at every iteration i,

i. Each control area solves (6.50). If $f_k(\mathbf{v}_k)$ is the LS fit and for unconstrained problems, the per-area states are updated as the LSEs using legacy software. The second summand in (6.50) can be interpreted as pseudo-measurements on the shared states forcing them to consent across areas.

ii. Neighboring areas exchange their updated shared states. This step involves minimal communication, and no grid models need to be shared. Every area updates its copies of the auxiliary variables $\mathbf{v}_{kl}$ using (6.53).

iii. Every area updates the Lagrange multipliers λ_{kl} based on the deviation of the local from the auxiliary variable as in (6.52).

For convex pairs $\{ f_k(\mathbf{v}_k), \mathcal{X}_k\}_{k=1}^K$, the aforementioned iterates reach the optimal cost in (6.46) under mild conditions. If the overall power system is observable, the ADMM iterates converge to the unique LSE. The approach has been extended in [44] for joint PSSE and breaker status verification.

The decentralized algorithms were tested on a 4200-bus power grid synthetically built from the IEEE 14- and 300-bus systems. Each of the 300 buses of the latter was assumed to be a different area and was replaced by a copy of the IEEE 14-bus grid. Additionally, every branch of the IEEE 300-bus grid was an inter-area line whose terminal buses are randomly selected from the two incident to this line areas. Two performance metrics were adopted: the per-area error to the centralized solution of (6.46), denoted by $e_{k,c}^t := \| \mathbf{v}_k^{(c)} - \mathbf{v}_k^t \|_2 / N_k$, and the per-area error to the true underlying state defined as $e_{k,o}^t := \| \mathbf{v}_k - \mathbf{v}_k^t \|_2 / N_k$. Figure 6.3 shows the corresponding error curves averaged over 300 areas. The decentralized LSE approached the underlying state at an accuracy of 10^{-3} in approximately 10 iterations or 6.2 ms on an Intel Duo Core at 2.2 GHz (4GB RAM) computer using MATLAB, while the centralized LSE finished in 93.4 ms.

6.4.2 Distributed SDR-Based Estimator

Although the SDR-PSSE approach incurs polynomial complexity when implemented as a convex SDP, its worst-case complexity is still $\mathcal{O}\left(M^4\sqrt{N_\text{b}}\, \log\left(1/\epsilon\right)\right)$ for a given solution accuracy $\epsilon > 0$ [15]. For typical power networks, the number of measurements M is on the order of the number of buses N_b, and thus the worst-case complexity becomes $\mathcal{O}\left(N_\text{b}^{4.5} \log\left(1/\epsilon\right)\right)$. This complexity could be prohibitive for large-scale power systems, which motivates accelerating the SDR-PSSE method using distributed parallel implementations.

Following the area partition in Figure 6.2, the mth measurement per area k can be written as

$$z_{k,m} = h_{k,m}(\mathbf{v}_k) + \epsilon_{k,m} = \text{Tr}(\mathbf{H}_{k,m}\mathbf{V}_k) + \epsilon_{k,m}, \forall k, m$$

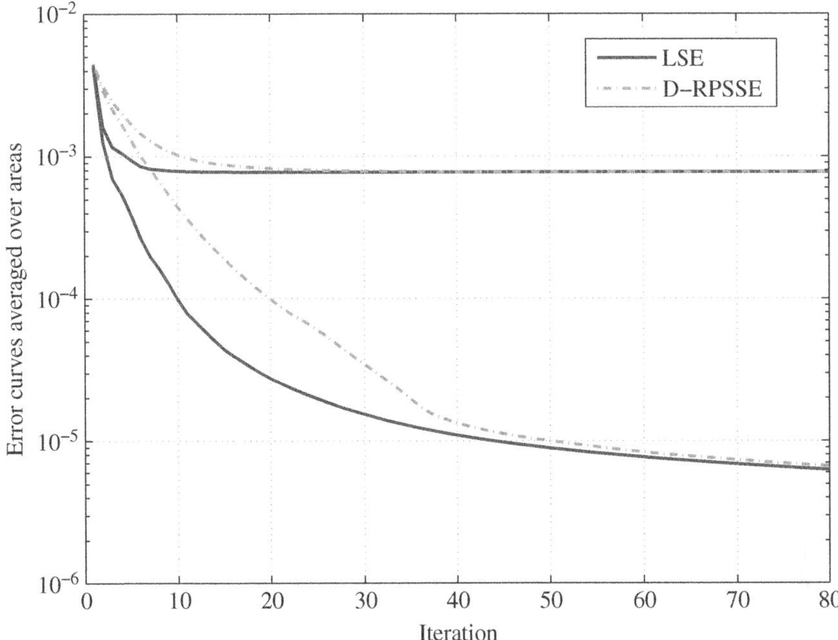

Figure 6.3 Average error curves $\sum_{k=1}^{300} e_{k,c}^t/300$ (bottom) and $\sum_{k=1}^{300} e_{k,o}^t/300$ (top) for the LSE and its robust counterpart D-RPSSE (see Section 6.5.1) on a 4200-bus grid.

where $\mathbf{V}_k$ denotes a submatrix of $\mathbf{V}$ formed by extracting the rows and columns corresponding to buses in area k, and likewise for each $\mathbf{H}_{k,m}$. Due to the overlap among the subsets of buses, the outer product $\mathbf{V}_k$ of area k overlaps also with $\mathbf{V}_l$ for each neighboring area $l \in \mathcal{N}_k$, as shown in Figure 6.2.

By reducing the measurements at area k to submatrix $\mathbf{V}_k$, one can define the PSSE error cost $f_k(\mathbf{V}_k) := \sum_{m=1}^{M_k} [z_{k,m} - \text{Tr}(\mathbf{H}_{k,m}\mathbf{V}_k)]^2$ per area k, which only involves the local matrix $\mathbf{V}_k$. Hence, the centralized PSSE problem in (6.32) becomes equivalent to

$$\hat{\mathbf{V}} = \arg\min_{\mathbf{V}\succeq\mathbf{0}} \sum_{k=1}^{K} f_k(\mathbf{V}_k) \tag{6.54}$$

This equivalent formulation effectively expresses the overall PSSE cost as the superposition of each local cost f_k. Nonetheless, even with such a decomposable cost, the main challenge to implement (6.54) in a distributed manner lies in the PSD constraint that couples the overlapping local matrices $\{\mathbf{V}_k\}$ (cf. Figure 6.2). If all submatrices $\{\mathbf{V}_k\}$ were non-overlapping, the cost would be decomposable as in (6.54), and the PSD of $\mathbf{V}$ would boil down to a PSD constraint per area k, as in

$$\hat{\mathbf{V}} = \arg \min_{\{\mathbf{V}_k \succeq \mathbf{0}\}} \sum_{k=1}^{K} f_k(\mathbf{V}_k) \tag{6.55}$$

Similar to PSSE for linearized measurements in (6.46), the formulation in (6.55) can be decomposed into subproblems, thanks to the separable PSD constraints. It is not always equivalent to the centralized (6.54) though, because the PSD property of all submatrices does not necessarily lead to a PSD overall matrix. Nonetheless, the decomposable problem (6.55) is still a valid SDR-PSSE reformulation, since with the additional per-area constraints rank($\mathbf{V}_k$) = 1, it is actually equivalent to (6.30). While it is totally legitimate to use (6.55) as the relaxed SDP formulation for (6.30), the two relaxed problems are actually equivalent under mild conditions.

The fresh idea here is to explore valid network topologies to facilitate such PSD constraint decomposition. To this end, it will be instrumental to leverage results on completing partial Hermitian matrices to obtain PSD ones [45]. Upon obtaining the underlying graph formed by the specified entries in the partial Hermitian matrices, these results rely on the so-termed graph *chordal* property to establish the equivalence between the positive semidefiniteness of the overall matrix and that of all submatrices corresponding to the graph's maximal cliques. Interestingly, this technique was recently used for developing distributed SDP-based optimal power flow (OPF) solvers in [46–48].

Construct first a new graph $\mathcal{B}'$ over $\mathcal{B}$, with all its edges corresponding to the entries in $\{\mathbf{V}_k\}$. The graph $\mathcal{G}'$ amounts to having all buses within each subset $\mathcal{N}_k$ to form a clique. Furthermore, the following are assumed:

(as4) *The graph with all the control areas as nodes and their edges defined by the neighborhood subset* $\{\mathcal{N}_k\}_{k=1}^K$ *forms a tree.*

(as5) *Each control area has at least one bus that does not overlap with any neighboring area.*

Proposition 6.3 Under (as4)–(as5), the two relaxed problems (6.54) and (6.55) are equivalent.

Proposition 6.3 can be proved by following the arguments in [12] to show that the entire PSD matrix $\mathbf{V}$ can be "completed" using only the PSD submatrices $\mathbf{V}_k$. The key point is that in most power networks even those not obeying **(as4)** and **(as5)**, (6.55) can achieve the same accuracy as the centralized one. At the same time, decomposing the PSD constraint in (6.55) is of paramount importance for developing distributed solvers. One can adopt the consensus reformulation to design the distributed solver for (6.55) as in (6.46) of Section 6.4.1. Accordingly, the ADMM iterations can be employed to solve (6.55) through iterative information exchanges among neighboring areas, and this is the basis of the distributed SDR-PSSE method.

This distributed SDR-PSSE method was tested on the IEEE 118-bus system using the three-area partition in [49]. All three areas measure their local bus voltage magnitudes, as well as real and reactive power flow levels at all lines. The overlaps among the three areas form a tree communication graph used to construct the

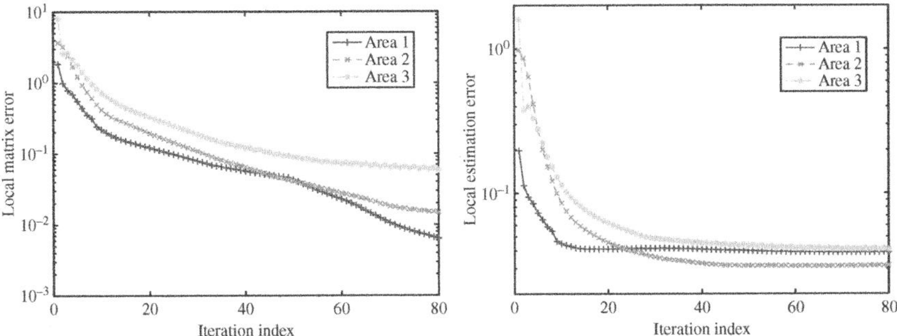

Figure 6.4 (Left) Per area state matrix error and (right) state vector estimation error versus the number of ADMM iterations for the distributed SDR-PSSE solver using the IEEE 118-bus system.

area-coupling constraints. To demonstrate convergence of the ADMM iterations to the centralized SE solution $\hat{\mathbf{V}}$ of (6.55), the local matrix Frobenius error norm $\|\hat{\mathbf{V}}^i_{(k)} - \hat{\mathbf{V}}_{(k)}\|_F$ is plotted versus the iteration index i in the left panel of Figure 6.4 for every control area k. Clearly, all local iterates converge to (approximately with a linear rate) their counterparts in the centralized solution. As the task of interest is to estimate the voltages, the local estimation error for the state vector $\|\hat{\mathbf{v}}^i_{(k)} - \mathbf{v}_{(k)}\|_2$ is also depicted in the right panel of Figure 6.4, where $\hat{\mathbf{v}}_{(k)}$ is the estimate of bus voltages at area k obtained from the iterate $\hat{\mathbf{V}}^i_{(k)}$ using the eigendecomposition method. Interestingly, the estimation error costs converge within the estimation accuracy of around 10^{-2} after about 20 iterations (less than 10 iterations for area 1), even though the local matrix has not yet converged. In addition, these error costs decrease much faster in the first couple of iterations. This demonstrates that even with only a limited number of iterations, the PSSE accuracy can be greatly boosted in practice, which in turn makes inter-area communication overhead more affordable.

6.5 ROBUST ESTIMATORS AND CYBERATTACKS

Bad data, also known as *outliers* in the statistics parlance, can challenge PSSE due to communication delays, instrument mis-calibration, and/or line parameter uncertainty. In today's cyber-enabled power systems, smart meter and synchrophasor data could be also purposefully manipulated to mislead system operators. This section reviews conventional and contemporary approaches to coping with outliers.

6.5.1 Bad Data Detection and Identification

Bad data processing in PSSE relies mainly on the linear measurement model $\mathbf{z} = \mathbf{H}\mathbf{v} + \boldsymbol{\epsilon}$, where $\mathbf{H} \in \mathbb{R}^{M \times N}$. Recall that this model is exact for PMU measurements [cf. (6.16)–(6.17)], but approximate per Gauss–Newton iteration or under the linearized grid model. In addition, the aforementioned model assumes real-valued states and measurements, slightly abusing the symbols introduced in (6.17). This is to keep the notation uncluttered and cover both cases of exact and inexact grid models. Albeit the nominal measurement noise vector is henceforth assumed zero mean with identity covariance, results extend to colored noise as per (6.20).

To capture bad data, the measurement model is now augmented as

$$\mathbf{z} = \mathbf{H}\mathbf{v} + \mathbf{o} + \boldsymbol{\epsilon} \tag{6.56}$$

where $\mathbf{o}$ is an unknown vector whose mth entry o_m is deterministically nonzero only if z_m is a bad datum [50–52]. Therefore, vector $\mathbf{o}$ is *sparse*, i.e. many of its entries are zero. Under the outlier-cognizant model in (6.56), the unconstrained LSE, $\hat{\mathbf{v}}_{\text{LSE}} = \left(\mathbf{H}^{\mathcal{T}}\mathbf{H}\right)^{-1}\mathbf{H}^{\mathcal{T}}\mathbf{z}$, yields the residual error

$$\mathbf{r} := \mathbf{z} - \mathbf{H}\hat{\mathbf{v}}_{\text{LSE}} = \mathbf{P}\mathbf{z} = \mathbf{P}(\mathbf{o} + \boldsymbol{\epsilon}) \tag{6.57}$$

with $\mathbf{P} := \mathbf{I}_M - \mathbf{H}\left(\mathbf{H}^{\mathcal{T}}\mathbf{H}\right)^{-1}\mathbf{H}^{\mathcal{T}}$ being the so-called projection matrix onto the orthogonal subspace of range($\mathbf{H}$). The last equality in (6.57) stems from the fact that $\mathbf{P}\mathbf{H} = \mathbf{0}$. As a projection matrix, $\mathbf{P}$ is idempotent, that is, $\mathbf{P} = \mathbf{P}^2$; Hermitian PSD with $(M - N)$ eigenvalues equal to one and N zero eigenvalues, while its diagonal entries satisfy $\mathbf{P}_{m,m} \in [0, 1]$ for $m = 1, \ldots, M$; see, e.g. [53].

For $\boldsymbol{\epsilon} \sim \mathcal{N}(\mathbf{0}, \mathbf{I}_M)$, it apparently holds that $\mathbf{P}\boldsymbol{\epsilon} \sim \mathcal{N}(\mathbf{0}, \mathbf{P})$. The mean-squared residual error is (see also [52] for its Bayesian counterpart)

$$\mathbb{E}\left[\|\mathbf{r}\|_2^2\right] = \mathbb{E}\left[\|\mathbf{P}\boldsymbol{\epsilon}\|_2^2\right] + \|\mathbf{P}\mathbf{o}\|_2^2 = (M - N) + \|\mathbf{P}\mathbf{o}\|_2^2 \tag{6.58}$$

In the absence of bad data, or if $\mathbf{o} \in$ range($\mathbf{H}$), the squared residual error follows a χ^2 distribution with mean $(M - N)$. The χ^2-test compares $\|\mathbf{r}\|_2^2$ against a threshold to detect the presence of bad data [2, 3].

Finding both $\mathbf{v}$ and $\mathbf{o}$ from measurements in (6.56) may seem impossible, given that the number of unknowns exceeds the number of equations. Leveraging the sparsity of $\mathbf{o}$ though, interesting results can be obtained [51]. If τ_0 bad data are expected, one would ideally wish to solve

$$\{\hat{\mathbf{v}}, \hat{\mathbf{o}}\} \in \arg\min_{\mathbf{v},\mathbf{o}} \left\{\frac{1}{2}\|\mathbf{z} - \mathbf{H}\mathbf{v} - \mathbf{o}\|_2^2 : \|\mathbf{o}\|_0 \leq \tau_0\right\} \tag{6.59}$$

But the ℓ_0-(pseudo)norm $\|\mathbf{o}\|_0$ counting the number of nonzero entries of $\mathbf{o}$ renders (6.59) NP-hard in general; see also Definition 6.2 in Section 6.5.2.

For the special case of $\tau_0 = 1$, problem (6.59) can be efficiently handled. Consider the scenario where the only nonzero entry of $\hat{\mathbf{o}}$ is the mth one, and denote the related $\hat{\mathbf{v}}$ minimizer by $\hat{\mathbf{v}}_{(m)}$. Apparently, the mth entry of the $\hat{\mathbf{o}}$ minimizer is $\hat{o}_m := z_m - \mathbf{h}_m^{\mathcal{T}}\hat{\mathbf{v}}_{(m)}$. This choice nulls the mth residual i.e., $z_m - \mathbf{h}_m^{\mathcal{T}}\hat{\mathbf{v}}_{(m)} - \hat{o}_m = 0$. With the mth residual zeroed, the cost in (6.59) becomes $\|\mathbf{r}_{(m)}\|_2^2 := \|\mathbf{z}_{(m)} - \mathbf{H}_{(m)}\hat{\mathbf{v}}_{(m)}\|_2^2$, where $\mathbf{z}_{(m)}$ is obtained from $\mathbf{z}$ upon dropping its mth entry and $\mathbf{H}_{(m)}$ by removing the mth row of $\mathbf{H}$. The problem in (6.59) is then equivalent to

$$\underset{m}{\text{minimize}}\ \frac{1}{2}\ \|\mathbf{r}_{(m)}\|_2^2 \tag{6.60}$$

Problem (6.60) can be solved by exhaustively finding all M LSEs excluding one measurement at a time. Fortunately, a classical result from the adaptive filtering literature relates the error $\|\mathbf{r}_{(m)}\|_2^2$ to the error attained using all outlier-free measurements $\|\mathbf{r}\|_2^2 := \|\mathbf{Pz}\|_2^2$; see, e.g. [54, chapter 9]

$$\|\mathbf{r}\|_2^2 = \|\mathbf{r}_{(m)}\|_2^2 + r_m\hat{o}_m \tag{6.61}$$

The same result links the *a posteriori* error r_m to the a priori error $\hat{o}_m$ as $r_m = \mathbf{P}_{m,m}\hat{o}_m$. Through these links, solving (6.60) is equivalent to

$$r_{\max} := \underset{m}{\text{maximize}}\ \frac{|\ r_m\ |}{\sqrt{\mathbf{P}_{m,m}}} \tag{6.62}$$

In words, a single bad datum can be identified by properly normalizing the entries of the original residual vector $\mathbf{r} = \mathbf{Pz}$.

Interestingly, the task in (6.62) coincides with the largest normalized residual (LNR) test that compares $r_{\max}$ to a prescribed threshold to identify a single bad datum [2, section 5.7]. The threshold is derived after recognizing that in the absence of bad data, $r_m / \sqrt{\mathbf{P}_{m,m}}$ is standard normal for all m.

The LNR test does not generalize for multiple bad data and problem (6.59) becomes computationally intractable for larger τ_0's. Heuristically, if a measurement is deemed as outlying, PSSE is repeated after discarding this bad datum, the LNR test is reapplied, and the process iterates till no corrupted data are identified. Alternatively, the *least median squares* and the *least trimmed squares* estimators have provable breakdown points and superior efficiency under Gaussian data; see, e.g. [55] and references therein. Nevertheless, their complexity scales unfavorably with the network size.

Leveraging compressed sensing [56], a practical robust estimator can be found if the ℓ_0-pseudonorm is surrogated by the convex ℓ_1-norm as [49, 51]

$$\underset{\mathbf{v},\mathbf{o}}{\text{minimize}} \left\{ \frac{1}{2}\|\mathbf{z} - \mathbf{Hv} - \mathbf{o}\|_2^2 : \|\mathbf{o}\|_1 \leq \tau_1 \right\} \tag{6.63}$$

for a selected constant $\tau_1 > 0$, or in its Lagrangian form

$$\{\hat{\mathbf{v}}, \hat{\mathbf{o}}\} \in \arg\min_{\mathbf{v},\mathbf{o}}\ \frac{1}{2}\|\mathbf{z} - \mathbf{Hv} - \mathbf{o}\|_2^2 + \lambda\ \|\mathbf{o}\|_1 \tag{6.64}$$

for some trade-off parameter $\lambda > 0$. The estimates of (6.64) offer joint state estimation and bad data identification. Even when some measurements are deemed as corrupted, their effect has been already suppressed. The optimization task in (6.64) can be handled by off-the-shelf software or solvers customized to the compressed sensing setup. When $\lambda \to \infty$, the minimizer $\hat{\mathbf{o}}$ becomes zero, and thus $\hat{\mathbf{v}}$ reduces to the LSE. On the contrary, by letting $\lambda \to 0^+$, the solution $\hat{\mathbf{v}}$ coincides with the LAV estimator [53, 57]; presented earlier in (6.37), namely,

$$\hat{\mathbf{v}}_{\text{LAV}} := \arg\min_{\mathbf{v}} \ \|\mathbf{z} - \mathbf{H}\mathbf{v}\|_1 \tag{6.65}$$

For finite $\lambda > 0$, the $\hat{\mathbf{v}}$ minimizer of (6.64) is equivalent to Huber's M-estimator; see [51] and references therein. Based on this connection and for Gaussian ϵ, parameter λ can be set to 1.34, which makes the estimator 95% asymptotically efficient for outlier-free measurements [58, p. 26]. Huber's estimate can be alternatively expressed as the $\mathbf{v}$-minimizer of [59], minimize$_{\mathbf{v},\omega}$ $\frac{1}{2}\ \|\omega\|_2^2 + \lambda\ \|\mathbf{v}\ - \mathbf{H}\mathbf{v} - \omega\|_1$. The bad data identification performance of this minimization has been analyzed in [60].

Table 6.1 compares several bad data analysis methods on the IEEE 14-bus grid of Figure 6.2 under the next four scenarios: (S0) no bad data, (S1) bad data on line (4, 7), (S2) bad data on line current (4, 7) and bus voltage 5, and (S3) bad data on bus voltage 5 and line currents (4, 7) and (10, 11). In all scenarios, bad data are simulated by multiplying the real and imaginary parts of the actual measurement by 1.2. The performance metric here is the ℓ_2-norm between the true state and the PSSE, which is averaged over 1000 Monte Carlo runs. Four algorithms were tested: (i) an ideal but practically infeasible genie-aided LSE (GA-LSE), which ignores the corrupted measurements; (ii) the regular LSE; (iii) the LNR test-based (LNRT) estimator with the test threshold set to 3.0 [2]; and (iv) Huber's estimator of (6.64) with $\lambda = 1.34$. For (S0)–(S1), the estimators perform comparably. The few corrupted measurements in (S2)–(S3) can deteriorate LSE's performance, while Huber's estimator performs slightly better than LNRT. Computationally, Huber's estimator was run within 1.3 ms, while the LNRT required 1.5 ms. The computing times were also measured for the IEEE 118-bus grid without corrupted data. Interestingly, the average time on the IEEE 118-bus grid without corrupted data is 3.2 ms and 81 ms, respectively.

Toward a robust decentralized state estimator, the ADMM-based framework of Section 6.4.1 can be engaged here too. If the measurement model for the kth area is $\mathbf{z}_k = \mathbf{H}_k\mathbf{v}_k + \mathbf{o}_k + \epsilon_k$, the centralized problem boils down to

$$\underset{\{\mathbf{v}_k \in \mathcal{X}_k,\, \mathbf{o}_k\}}{\text{minimize}} \sum_{k=1}^{K} \frac{1}{2} \|\mathbf{z}_k - \mathbf{H}_k\mathbf{v}_k - \mathbf{o}_k\|_2^2 + \lambda\ \|\mathbf{o}_k\|_1 \tag{6.66}$$

To allow for decentralized implementation, the optimization in (6.66) can be reformulated as

TABLE 6.1 Mean-square estimation error in the presence of bad data.

Method	GA-LSE	LSE	LNRT	Huber's
(S0)	0.0278	0.0278	0.0286	0.0281
(S1)	0.0313	0.0318	0.0331	0.0322
(S2)	0.0336	0.1431	0.0404	0.0390
(S3)	0.0367	0.1434	0.0407	0.0390

$$\text{minimize} \sum_{k=1}^{K} \frac{1}{2} \|\mathbf{z}_k - \mathbf{H}_k \mathbf{v}_k - \mathbf{o}_k\|_2^2 + \lambda \, \|\boldsymbol{\omega}_k\|_1 \tag{6.67a}$$

$$\text{over}\{\mathbf{v}_k \in \mathcal{X}_k, \mathbf{o}_k, \boldsymbol{\omega}_k\}, \{\mathbf{v}_{kl}\} \tag{6.67b}$$

$$\text{s.to } \mathbf{v}_k[l] = \mathbf{v}_{kl}, \text{for all } l \in \mathcal{B}_k, k = 1, \ldots, K \tag{6.67c}$$

$$\mathbf{o}_k = \boldsymbol{\omega}_k, \text{for all } k = 1, \ldots, K \tag{6.67d}$$

As in Section 6.4.1, the constraints in (6.67c) and the auxiliary variables $\{\mathbf{v}_{kl}\}$ enforce consensus of shared states. On the other hand, the variables $\{\mathbf{o}_k\}$ are duplicated as $\{\boldsymbol{\omega}_k\}$ in (6.67d). Then, variables $\{\mathbf{v}_k, \mathbf{o}_k\}$ are put together in the x-update of ADMM in (6.49a), whereas $\{\mathbf{v}_{kl}, \boldsymbol{\omega}_k\}$ fall into the z-update in (6.49b). In this fashion, costs are separable over variable groups, and the minimization involving the ℓ_1-norm enjoys a closed-form solution expressed in terms of the soft thresholding operator [49].

6.5.2 Observability and Cyberattacks

In the cyber–physical smart grid context, bad data are not simply unintentional errors, but can also take the form of malicious data injections [61]. Amid these challenges, the intertwined issues of critical measurements and stealth cyberattacks on PSSE are discussed next.

It has been tacitly assumed so far that the power system is *observable*. A power system is observable if distinct states $\mathbf{v} \neq \mathbf{v}'$ are mapped to distinct measurements $\mathbf{h}(\mathbf{v}) \neq \mathbf{h}(\mathbf{v}')$ under a noiseless setup. Equivalently, if the so-called measurement distance function is defined as [62]

$$D(\mathbf{h}) := \underset{\mathbf{v} \neq \mathbf{v}'}{\text{minimize}} \; \| \, \mathbf{h}(\mathbf{v}) - \mathbf{h}(\mathbf{v}') \|_0 \tag{6.68}$$

the power system is observable if and only if $D(\mathbf{h}) \geq 1$. Given the network topology and the mapping $\mathbf{h}(\mathbf{v})$, the well-studied topic of observability analysis aims at determining whether the system state is uniquely identifiable, at least locally in a neighborhood of the current estimate [2, chapter 4]. If not, mapping *observable islands*, meaning maximally connected sub-grids with observable internal flows, is important as well.

Observability analysis relies on the decoupled linearized grid model and is accomplished through topological or numerical tests [63, 64]. Apparently, under the linear or linearized model $\mathbf{h}(\mathbf{v}) = \mathbf{H}\mathbf{v}$, the state $\mathbf{v}$ is uniquely identifiable if and only if $\mathbf{H}$ is full column rank. Phase shift ambiguities can be waived by fixing the angle at a reference bus.

In the presence of bad data and/or cyberattacks, observability analysis may not suffice. Consider the noiseless measurement model $\mathbf{z} = \mathbf{h}(\mathbf{v}) + \mathbf{o}$, where the nonzero entries of vector $\mathbf{o}$ correspond to bad data or compromised meters, and let us proceed with the following definitions.

Definition 6.1 (Observable attack [62])

The attack vector $\mathbf{o}$ is deemed as observable if for every state $\mathbf{v}$ there is no $\mathbf{v}' \neq \mathbf{v}$, such that $\mathbf{h}(\mathbf{v}) + \mathbf{o} = \mathbf{h}(\mathbf{v}')$.

Definition 6.2 (Identifiable attack [62])

The attack vector $\mathbf{o}$ is identifiable if for every $\mathbf{v}$ there is no $(\mathbf{v}', \mathbf{o}')$ with $\mathbf{v}' \neq \mathbf{v}$ and $\|\mathbf{o}'\|_0 \leq \|\mathbf{o}\|_0$, such that $\mathbf{h}(\mathbf{v}) + \mathbf{o} = \mathbf{h}(\mathbf{v}') + \mathbf{o}'$.

If the outlier vector $\mathbf{o}$ is observable, the operator can tell that the collected measurements do not correspond to a system state and can hence decide that an attack has been launched. Nevertheless, the attacked meters can be pinpointed only under the stronger conditions of Definition 6.2.

The resilience of the measurement mapping $\mathbf{h}(\mathbf{v})$ against attacks can be characterized through $D(\mathbf{h})$ in (6.68): the maximum number of counterfeited meters for an attack to be observable is $K_o = D(\mathbf{h}) - 1$, and to be identifiable, it is $K_i = \left\lfloor \frac{D(\mathbf{h})-1}{2} \right\rfloor$; see [60, 62]. Here, the floor function $\lfloor x \rfloor$ returns the greatest integer less than or equal to x.

Consider the linear mapping $\mathbf{h}(\mathbf{v}) = \mathbf{H}\mathbf{v}$. Measurement m is termed *critical* if once removed from the measurement set, it renders the power system non-identifiable. In other words, although $\mathbf{H}$ is full column rank, its submatrix $\mathbf{H}_{(m)}$ is not. It trivially follows that $D(\mathbf{h}) = 1$, and the system operator can be arbitrarily misled even if only measurement m is attacked. Due to the typically sparse structure of $\mathbf{H}$, critical measurements or multiple simultaneously corrupted data do exist [2]. It was pointed out in [65] that if an attack $\mathbf{o}$ can be constructed to lie in the range($\mathbf{H}$), it comprises a "stealth attack." Although finding $D(\mathbf{h})$ is not trivial in general, a polynomial-time algorithm leveraging a graph-theoretic approach is devised in [52].

6.6 POWER SYSTEM STATE TRACKING

The PSSE methods reviewed so far ignore system dynamics and do not exploit historical information. Dynamic PSSE is well motivated thanks to its improved robustness, observability, and predictive ability when additional temporal information is available [66]. Recently proposed model-free and model-based state tracking schemes are outlined next.

6.6.1 Model-Free State Tracking via Online Learning

In complex future power systems, one may not choose to explicitly commit to a model for the underlying system dynamics. The framework of OCO, particularly popular in machine learning, can account for unmodeled dynamics and is thus briefly presented next [67].

The OCO model considers a multistage game between a player and an adversary. In the PSSE context, the utility or the system operator assumes the role of the player, while the loads and renewable generations can be viewed as the adversary.

At time t, the player first selects an action $\mathbf{V}_t$ from a given action set $\mathcal{V}$, and the adversary subsequently reveals a convex loss function $f_t : \mathcal{V} \rightarrow \mathbb{R}$. In this round, the player suffers a loss $f_t(\mathbf{V}_t)$. The ultimate goal for the player is to minimize the *regret* $R_f(T)$ over T rounds:

$$R_f(T) := \sum_{t=1}^{T} f_t(\mathbf{V}_t) - \underset{\mathbf{V} \in \mathcal{V}}{\text{minimize}} \sum_{t=1}^{T} f_t(\mathbf{V}) \tag{6.69}$$

The regret is basically the accumulated cost incurred by the player relative to that by a single fixed action $\mathbf{V}^0 := \arg\min_{\mathbf{V}\in\mathcal{V}} \sum_{t=1}^{T} f_t(\mathbf{V})$. This fixed action is selected with the advantage of knowing the loss functions $\{f_t\}_{t=1}^{T}$ in hindsight. Under appropriate conditions, judiciously designed online optimization algorithms can achieve sublinear regret; that is, $R_f(T)/T \rightarrow 0$ as $T \rightarrow +\infty$.

Building on the SDR-PSSE formulation of Section 6.3.4, the ensuing method considers streaming data for real-time PSSE. The data referring to and collected over the control period t are $\{(z_{m_t}; \mathbf{H}_{m_t})\}_{m_t=1}^{M_t}$ with $t = 1, \ldots, T$. The number and type of measurements can change over time, while the matrix corresponding to measurement m may change over time as indicated by $\{\mathbf{H}_{m_t}\}_{m_t=1}^{M_t}$ due to topology reconfigurations. The online PSSE task can be now formulated as

$$\underset{\mathbf{V} \succeq \mathbf{0}}{\text{minimize}} \sum_{t=1}^{T} f_t(\mathbf{V}) \tag{6.70}$$

where $f_t(\mathbf{V}) := \sum_{m_t=1}^{M_t} [z_{m_t} - \text{Tr}(\mathbf{H}_{m_t}\mathbf{V})]^2$. Online PSSE aims at improving the static estimates by capitalizing on previous measurements as well as tracking slow time-varying variations in generation and demand.

Minimizing the cost in (6.70) may be computationally cumbersome for real-time implementation. An efficient alternative based on online gradient descent amounts to iteratively minimizing a regularized first-order approximation of the instantaneous cost instead [68]

$$\mathbf{V}_{t+1} := \arg\min_{\mathbf{V} \succeq \mathbf{0}} \text{Tr}\left(\mathbf{V}^{\mathcal{H}} \nabla f_t(\mathbf{V}_t)\right) + \frac{1}{2\mu_t} \|\mathbf{V} - \mathbf{V}_t\|_F^2 \tag{6.71}$$

for $t = 1, \ldots,$ and suitably selected step sizes $\mu_t > 0$. Interestingly, the optimization in (6.71) admits a closed-form solution given by

$$\mathbf{V}_{t+1} = \text{Proj}_{\mathbb{S}^+} [\mathbf{V}_t - \mu_t \nabla f_t(\mathbf{V}_t)] \tag{6.72}$$

with $\text{Proj}_{\mathbb{S}^+}$ denoting the projection onto the positive semidefinite cone, which can be performed using eigendecomposition followed by setting negative eigenvalues to zero. It is worth mentioning that the online PSSE in (6.72) enjoys sublinear regret [68]. Upon finding $\mathbf{V}_t$, a state estimate $\mathbf{v}_t$ can be obtained by eigendecomposition or randomization as in Section 6.3.4. With an additional nuclear-norm regularization term promoting low-rank solutions in (6.71), online ADMM alternatives were devised in [69]. Interestingly, online learning tools has recently been advocated for numerous real-time energy management tasks in [70–72].

6.6.2 Model-Based State Tracking

Although the previous model-free solver can recover slow time-varying states, model-based approaches facilitate tracking of fast time-varying system states. A typical state space model for power system dynamics is [73]

$$\mathbf{v}_{t+1} = \mathbf{F}_t\mathbf{v}_t + \mathbf{g}_t + \boldsymbol{\omega}_t \tag{6.73a}$$

$$\mathbf{z}_t = \mathbf{h}(\mathbf{v}_t) + \boldsymbol{\epsilon}_t \tag{6.73b}$$

where $\mathbf{F}_t$ denotes the state transition matrix, $\mathbf{g}_t$ captures the process mismatch, and $\mathbf{w}_t$ is the additive noise. The nonlinear mapping $\mathbf{h}(\cdot)$ comes from conventional SCADA measurements. Values $\{(\mathbf{F}_t, \mathbf{g}_t)\}$ can be obtained in real time using, for example, Holt's system identification method [74]. Two common dynamic tracking approaches to cope with the nonlinearity in the measurement model of (6.73b) include the (extended or unscented) Kalman filters (EKF/UKF) and moving horizon estimators [66, 73, 75–77], and they are outlined in order next.

The *extended Kalman filter* (EKF) handles the nonlinearity by linearizing $\mathbf{h}(\mathbf{v})$ around the state predictor. To start, let $\hat{\mathbf{v}}_{t+1|t}$ stand for the predicted estimate at time $t+1$ given measurements $\{\mathbf{z}_\tau\}_{\tau=1}^{t}$ up to time t. Let also $\hat{\mathbf{v}}_{t+1|t+1}$ be the filtered estimate given measurements $\{\mathbf{z}_\tau\}_{\tau=1}^{t+1}$. If the noise terms $\boldsymbol{\omega}_t$ and $\boldsymbol{\epsilon}_t$ in (6.73) are assumed zero-mean Gaussian with known covariance matrices $\mathbf{Q}_t \succeq \mathbf{0}$ and $\mathbf{R}_t \succeq \mathbf{0}$, respectively, the EKF can be implemented with the following recursions:

$$\hat{\mathbf{v}}_{t+1|t+1} = \hat{\mathbf{v}}_{t+1|t} + \mathbf{K}_{t+1}\left[\mathbf{z}_{t+1} - \mathbf{h}\left(\hat{\mathbf{v}}_{t+1|t}\right)\right] \tag{6.74}$$

where the state predictor $\hat{\mathbf{v}}_{t+1|t}$ and the Kalman gain $\mathbf{K}_{t+1}$ are given by

$$\hat{\mathbf{v}}_{t+1|t} = \mathbf{F}_t\hat{\mathbf{v}}_{t|t} + \mathbf{g}_t \tag{6.75a}$$

$$\mathbf{K}_{t+1} = \mathbf{P}_{t+1|t}\mathbf{J}_{t+1}^{\mathcal{H}}\left(\mathbf{J}_{t+1}\mathbf{P}_{t+1|t}\mathbf{J}_{t+1}^{\mathcal{H}} + \mathbf{R}_{t+1}\right)^{-1} \tag{6.75b}$$

$$\mathbf{P}_{t+1|t+1} = \mathbf{P}_{t+1|t} - \mathbf{K}_{t+1}\mathbf{J}_{t+1}\mathbf{P}_{t+1|t} \tag{6.75c}$$

$$\mathbf{P}_{t+1|t} = \mathbf{F}_t\mathbf{P}_{t|t}\mathbf{F}_t^{\mathcal{H}} + \mathbf{Q}_t \tag{6.75d}$$

with $\mathbf{J}_{t+1}$ being the measurement Jacobian matrix of $\mathbf{h}$ evaluated at $\hat{\mathbf{v}}_{t+1|t}$ and $\mathbf{P}_{t+1|t+1} \succeq \mathbf{0}$ ($\mathbf{P}_{t+1|t} \succeq \mathbf{0}$) denoting the corrected (predicted) state estimation error covariance matrix at time $t+1$. To improve on the approximation accuracy of the EKF, UKF have been reported in [73]. Particle filtering may also be useful if its computational complexity can be supported during real-time power systems operations.

Because the EKF and UKF are known to diverge for highly nonlinear dynamics, *moving horizon estimation* (MHE) has been suggested as an accurate yet tractable alternative with proven robustness to bounded model errors [79]. Different from Kalman filtering, the initial state $\mathbf{v}_0$ and noises $\boldsymbol{\omega}_t$ and $\boldsymbol{\epsilon}_t$ in MHE are viewed as deterministic unknowns taking values from given bounded sets $\mathcal{S}$, $\mathcal{W}$, and $\mathcal{E}$, respectively. The sets $\mathcal{W}$ and $\mathcal{E}$ model disturbances with truncated densities [79].

The idea behind MHE is to perform PSSE by exploiting useful information present in a sliding window of the most recent observations. Consider here a sliding window of length $L+1$. Let $\hat{\mathbf{v}}_{t-L|t}$ denote the smoothed estimate at time $t-L$ given L past measurements, as well as the current one, namely, $\{\mathbf{z}_\tau\}_{\tau=t-L}^{t}$. MHE aims at obtaining the most recent L state estimates $\left\{\hat{\mathbf{v}}_{t-L+s|t}\right\}_{s=0}^{L}$ based on $\{\mathbf{z}_\tau\}_{t-L}^{t}$ and the available estimate $\breve{\mathbf{v}}_{t-L} := \hat{\mathbf{v}}_{t-L|t-1}$ from time $t-1$ and for $t \geq L$. A key simplification is that once $\hat{\mathbf{v}}_{t-L|t}$ becomes available, the other L recent estimates at time t can be recursively obtained through "noise-free" propagation based on the dynamic model (6.73a); that is,

$$\hat{\mathbf{v}}_{t-L+s|t} = \mathbf{F}_{t-L+s-1}\hat{\mathbf{v}}_{t-L+s-1|t} \tag{6.76}$$

for $s = 1, \ldots, L$. By relating all recent estimates to $\hat{\mathbf{v}}_{t-L|t}$ via successive multiplications of transition matrices, the update in (6.76) simplifies to

$$\hat{\mathbf{v}}_{t-L+s|t} = \mathbf{T}_{t-L+s}\hat{\mathbf{v}}_{t-L|t} \tag{6.77}$$

where $\mathbf{T}_{t-L+s} := \mathbf{F}_{t-L+s-1}\mathbf{T}_{t-L+s-1}$ for $s = 1, \ldots, L$, with $\mathbf{T}_{t-L} = \mathbf{I}$. The MHE-based state estimate $\hat{\mathbf{v}}_{t-L|t}$ is then given by

$$\hat{\mathbf{v}}_{t-L|t} := \arg\min_{\mathbf{v}} \sum_{s=0}^{L} \| \mathbf{z}_{t-L+s} - \mathbf{h}(\mathbf{T}_{t-L+s}\mathbf{v}) \|_2^2 + \lambda \, \| \mathbf{v} - \breve{\mathbf{v}}_{t-L} \|_2^2 \tag{6.78}$$

where $\lambda > 0$ can be tuned relying on our confidence in the state predictor $\breve{\mathbf{v}}_{t-L}$ and the measurements $\{\mathbf{z}_\tau\}_{t-L}^{t}$. Given the quadratic dependence of the SCADA measurements $\{\mathbf{h}(\mathbf{v}_t)\}$ and the state $\mathbf{v}$, the optimization problem in (6.78) is nonconvex.

Finding the MHE-based state estimates in real time entails online solutions of dynamic optimization problems. The MHE formulation can be convexified by exploiting the semidefinite relaxation: vector $\mathbf{v}$ is lifted to the matrix $\mathbf{V} := \mathbf{v}\mathbf{v}^{\mathcal{H}} \succeq \mathbf{0}$, and the mth entry of $\mathbf{h}(\mathbf{v}_{t-L+s})$ for $s = 0, \ldots, L$, is expressed as

$$h_m(\mathbf{T}_{t-L+s}\mathbf{v}) = \mathbf{v}^{\mathcal{H}}\mathbf{T}_{t-L+s}^{\mathcal{H}}\mathbf{H}_m\mathbf{T}_{t-L+s}\mathbf{v} = \mathrm{Tr}\left(\mathbf{T}_{t-L+s}^{\mathcal{H}}\mathbf{H}_m\mathbf{T}_{t-L+s}\mathbf{V}\right)$$

Upon dropping the nonconvex rank constraint rank($\mathbf{V}$) = 1, the SDP-based MHE yields

$$\hat{\mathbf{V}}_{t-L|t} := \arg\min_{\mathbf{V}\succeq\mathbf{0}} \sum_{s=0}^{L} \|\mathbf{z}_{t-L+s} - \mathrm{Tr}\left(\mathbf{T}_{t-L+s}^{\mathcal{H}}\mathbf{H}_m\mathbf{T}_{t-L+s}\mathbf{V}\right)\|_2^2 + \lambda \, \|\mathbf{v} - \breve{\mathbf{v}}_{t-L}\|_2^2$$

which can be solved in polynomial time using off-the-shelf toolboxes. Rank-one state estimates can be obtained again through eigendecomposition or randomization. The complexity of solving the last problem is rather high in its present form on the order of $N_{\mathrm{b}}^{4.5}$ [15]. Therefore, developing faster solvers for the SDP-based MHE by exploiting the rich sparsity structure in $\{\mathbf{H}_m\}$ matrices is worth investigating. Decentralized and localized MHE implementations are also timely and pertinent. Devising FPP-based solvers for the MHE in (6.78) constitutes another research direction.

6.7 DISCUSSION

This chapter has reviewed some of the recent advances in PSSE. After developing the CRLB, an SDP-based solver and its regularized counterpart were discussed. To overcome the high complexity involved, a scheme named feasible point pursuit relying on successive convex approximations was also advocated. A decentralized PSSE paradigm put forth provides the means for coping with the computationally intensive SDP formulations, it is tailored for the interconnected nature of modern grids, and it can also afford processing PMU data in a timely fashion. A better understanding of cyberattacks and disciplined ways for decentralized bad data processing were also provided. Finally, this chapter gave a fresh perspective to state tracking under model-free and model-based scenarios.

Nonetheless, there are still many technically challenging and practically pertinent grid monitoring issues to be addressed. Solving power grid data processing tasks *on the cloud* has been a major trend to alleviate data storage, communication, and interoperability costs for system operators and utilities. Moreover, with the current focus on low- and medium-voltage distribution grids, solvers for unbalanced and multiphase operating conditions are desirable. Smart meters and synchrophasor data from distribution grids (also known as micro-PMUs) call for new data processing solutions. Advances in machine learning and statistical signal processing, such as sparse and low-rank models, missing and incomplete data, tensor decompositions, deep learning, nonconvex and stochastic optimization tools, and (multi)kernel-based learning, to name a few, are currently providing novel paths to grid monitoring tasks while realizing the vision of smarter energy systems.

ACKNOWLEDGMENTS

G. Wang and G. B. Giannakis were supported in part by NSF grants 1711471 and 1901134. H. Zhu was supported in part by NSF grants 1610732 and 1653706.

6.A APPENDIX

Proof of Proposition 6.1. Consider the AGWN model (6.15) with $\epsilon \sim \mathcal{N}(\mathbf{0}, \mathrm{dg}(\{\sigma_m^2\}))$. The data likelihood function is

$$p(\mathbf{z}; \mathbf{v}) = \prod_{m=1}^{M} \frac{1}{\sqrt{2\pi\sigma_m^2}} \exp\left[-\frac{\left(z_m - \mathbf{v}^{\mathcal{H}}\mathbf{H}_m\mathbf{v}\right)^2}{2\sigma_m^2}\right] \tag{6.A1}$$

and the negative log-likelihood function denoted by $f(\mathbf{v}) = -\ln p(\mathbf{z}; \mathbf{x})$ is

$$f(\mathbf{v}) = \sum_{m=1}^{M} \left[\frac{1}{2\sigma_m^2}\left(z_m - \mathbf{v}^{\mathcal{H}}\mathbf{H}_m\mathbf{v}\right)^2 + \frac{1}{2}\ln\left(2\pi\sigma_m^2\right)\right] \tag{6.A2}$$

The Fisher information matrix (FIM) is defined as the Hessian of the real-valued function $f(\mathbf{v})$ with respect to the complex vector $\mathbf{v} \in \mathbb{C}^{N_b}$. Deriving the CRLB amounts to finding the Hessian of a real-valued function with respect to a complex-valued vector. *Wirtinger's calculus* confirms that $f(\mathbf{v})$ can be equivalently rewritten as $f(\mathbf{v}, \mathbf{v}^*)$; see, e.g. [8]. Upon introducing the conjugate coordinates $\left[\mathbf{v}^{\mathcal{T}} \; (\mathbf{v}^*)^{\mathcal{T}}\right]^{\mathcal{T}} \in \mathbb{C}^{2N_b}$, the *Wirtinger derivatives*, namely, the first-order partial differential operators of functions over complex domains, are given by [8]

$$\frac{\partial f}{\partial \mathbf{v}} := \left.\frac{\partial f(\mathbf{v}, \mathbf{v}^*)}{\partial \mathbf{v}^{\mathcal{T}}}\right|_{\text{constant } \mathbf{v}^*} = \left.\left[\frac{\partial f}{\partial \mathcal{V}_1} \cdots \frac{\partial f}{\partial \mathcal{V}_N}\right]\right|_{\text{constant } \mathbf{v}^*}$$

$$\frac{\partial f}{\partial \mathbf{v}^*} := \left.\frac{\partial f(\mathbf{v}, \mathbf{v}^*)}{\partial (\mathbf{v}^*)^{\mathcal{T}}}\right|_{\text{constant } \mathbf{v}} = \left.\left[\frac{\partial f}{\partial \mathcal{V}_1^*} \cdots \frac{\partial f}{\partial \mathcal{V}_N^*}\right]\right|_{\text{constant } \mathbf{v}}$$

These definitions follow the convention in multivariate calculus that derivatives are denoted by row vectors and gradients by column vectors. Define for notational brevity $\phi_m(\mathbf{v}, \mathbf{v}^*) := z_m - (\mathbf{v}^*)^{\mathcal{T}} \mathbf{H}_m \mathbf{v}$ for $m = 1, \ldots, M$. Accordingly, the Wirtinger derivatives of $f(\mathbf{v}, \mathbf{v}^*)$ in (6.A2) are obtained as

$$\frac{\partial f}{\partial \mathbf{v}} = \sum_{m=1}^{M} \frac{1}{\sigma_m^2} \phi_m \frac{\partial \phi_m}{\partial \mathbf{v}^{\mathcal{T}}} \quad \text{and} \quad \frac{\partial f}{\partial \mathbf{v}^*} = \sum_{m=1}^{L} \frac{1}{\sigma_m^2} \phi_m \frac{\partial \phi_m}{\partial (\mathbf{v}^*)^{\mathcal{T}}} \tag{6.A3}$$

and the Wirtinger derivatives of $\phi_m(\mathbf{v}, \mathbf{v}^*)$ can be found likewise

$$\frac{\partial \phi_m}{\partial \mathbf{v}^{\mathcal{T}}} = -(\mathbf{H}_m \mathbf{v})^{\mathcal{H}} \quad \text{and} \quad \frac{\partial \phi_m}{\partial (\mathbf{v}^*)^{\mathcal{T}}} = -(\mathbf{H}_m^* \mathbf{v}^*)^{\mathcal{H}} \tag{6.A4}$$

In the conjugate coordinate system, the complex Hessian of $f(\mathbf{v}, \mathbf{v}^*)$ with respect to the conjugate coordinates $\left[\mathbf{v}^{\mathcal{T}} \; (\mathbf{v}^*)^{\mathcal{T}}\right]^{\mathcal{T}}$ is defined as

$$\mathcal{H}(\mathbf{v}, \mathbf{v}^*) := \nabla^2 f(\mathbf{v}, \mathbf{v}^*) = \begin{bmatrix} \mathcal{H}_{\mathbf{vv}} & \mathcal{H}_{\mathbf{v}^*\mathbf{v}} \\ \mathcal{H}_{\mathbf{vv}^*} & \mathcal{H}_{\mathbf{v}^*\mathbf{v}^*} \end{bmatrix} \tag{6.A5}$$

whose blocks are given as

$$\mathcal{H}_{\mathbf{vv}} := \frac{\partial}{\partial \mathbf{v}^{\mathcal{T}}} \left(\frac{\partial f}{\partial \mathbf{v}}\right)^{\mathcal{H}}, \quad \mathcal{H}_{\mathbf{v}^*\mathbf{v}} := \frac{\partial}{\partial (\mathbf{v}^*)^{\mathcal{T}}} \left(\frac{\partial f}{\partial \mathbf{v}}\right)^{\mathcal{H}}$$

$$\mathcal{H}_{\mathbf{vv}^*} := \frac{\partial}{\partial \mathbf{v}^{\mathcal{T}}} \left(\frac{\partial f}{\partial \mathbf{v}^*}\right)^{\mathcal{H}}, \quad \mathcal{H}_{\mathbf{v}^*\mathbf{v}^*} := \frac{\partial}{\partial (\mathbf{v}^*)^{\mathcal{T}}} \left(\frac{\partial f}{\partial \mathbf{v}^*}\right)^{\mathcal{H}}$$

Substituting (6.A3) and (6.A4) into the last equations and after algebraic manipulations yields

$$\mathcal{H}_{\mathbf{vv}} = \sum_{m=1}^{M} \sigma_m^{-2} \left(\mathbf{H}_m \mathbf{v} (\mathbf{H}_m \mathbf{v})^{\mathcal{H}} - \phi_m \mathbf{H}_m\right) \tag{6.A6a}$$

$$\mathcal{H}_{\mathbf{v}^*\mathbf{v}} = \sum_{m=1}^{M} \sigma_m^{-2} \mathbf{H}_m \mathbf{v} \left(\mathbf{H}_m^* \mathbf{v}^*\right)^{\mathcal{H}} \tag{6.A6b}$$

$$\mathcal{H}_{\mathbf{v}\mathbf{v}^*} = \sum_{m=1}^{M} \sigma_m^{-2} \mathbf{H}_m^* \mathbf{v}^* (\mathbf{H}_m \mathbf{v})^{\mathcal{H}} \tag{6.A6c}$$

$$\mathcal{H}_{\mathbf{v}^*\mathbf{v}^*} = \sum_{m=1}^{M} \sigma_m^{-2} \left(\mathbf{H}_m^* \mathbf{v}^* \left(\mathbf{H}_m^* \mathbf{v}^*\right)^{\mathcal{H}} - \phi_m \mathbf{H}_m^*\right) \tag{6.A6d}$$

Evaluating the Hessian blocks of (6.A6) at the true value of $\mathbf{v}$, and taking the expectation with respect to $\boldsymbol{\epsilon}$, yields $\mathbb{E}[\phi_m] = 0$. Hence, the ϕ_m-related terms in (6.A6) disappear, and the FIM $\mathbf{F}(\mathbf{v}, \mathbf{v}^*) := \mathbb{E}[\mathcal{H}(\mathbf{v}, \mathbf{v}^*)]$ simplifies to the expression in (6.23); see also [80].

To show that the FIM is rank-deficient, define $\mathbf{g}_m := \left[(\mathbf{H}_m \mathbf{v})^{\mathcal{H}} \ \left(\mathbf{H}_m^* \mathbf{v}^*\right)^{\mathcal{H}}\right]^{\mathcal{H}}$ so that the FIM becomes $\mathbf{F} = \sum_{m=1}^{M} \sigma_m^{-2} \mathbf{g}_m \mathbf{g}_m^{\mathcal{H}}$. Observe now that the nonzero vector $\mathbf{d}(\mathbf{v}) := \left[\mathbf{v}^T - (\mathbf{v}^*)^T\right]^T$ is orthogonal to $\mathbf{g}_m$ for $m = 1, \ldots, M$; that is,

$$\mathbf{g}_m^{\mathcal{H}} \mathbf{d} = \mathbf{v}^{\mathcal{H}} \mathbf{H}_m \mathbf{v} - \left(\mathbf{v}^{\mathcal{H}} \mathbf{H}_m \mathbf{v}\right)^* = 0$$

Based on the latter, it is not hard to verify that $\mathbf{F}\mathbf{d} = \mathbf{0}$, which proves that the null space of $\mathbf{F}$ is non-empty.

REFERENCES

1. Schweppe, F.C., Wildes, J., and Rom, D. (Jan. 1970). Power system static state estimation: parts I, II, and III. *IEEE Trans. Power App. Syst.* 89: 120–135.
2. Abur, A. and Gómez-Expósito, A. (2004). *Power System State Estimation: Theory and Implementation*. New York: Marcel Dekker.
3. Monticelli, A. (Feb. 2000). Electric power system state estimation. *Proc. IEEE* 88 (2): 262–282.
4. Wood, A.J. and Wollenberg, B.F. (1996). *Power Generation, Operation, and Control*, 2e. New York: Wiley.
5. Phadke, A.G. and Thorp, J.S. (2008). *Synchronized Phasor Measurements and Their Applications*. New York: Springer.
6. Giannakis, G.B., Kekatos, V., Gatsis, N. et al. (Sep. 2013). Monitoring and optimization for power grids: a signal processing perspective. *IEEE Signal Process. Mag.* 30 (5): 107–128.
7. Kay, S.M. (1993). *Fundamentals of Statistical Signal Processing, Vol. I: Estimation Theory*. Prentice Hall.
8. Kreutz-Delgado, K. (2009). The complex gradient operator and the CR-calculus. *arXiv:0906.4835*.
9. Stoica, P. and Marzetta, T.L. (Jan. 2001). Parameter estimation problems with singular information matrices. *IEEE Trans. Signal Process.* 49 (1): 87–90.

10. Bertsekas, D.P. (1999). *Nonlinear Programming*, 2e. Belmont, MA: Athena Scientific.
11. Zhu, H. and Giannakis, G.B. (2011). Estimating the state of AC power systems using semidefinite programming. *Proceedings of the North American Power Symposium* (August 2011), Boston, MA.
12. Zhu, H. and Giannakis, G.B. (Dec. 2014). Power system nonlinear state estimation using distributed semidefinite programming. *IEEE J. Sel. Topics Signal Process.* 8 (6): 1039–1050.
13. Boyd, S. and Vandenberghe, L. (2004). *Convex Optimization*. New York: Cambridge University Press.
14. Goemans, M.X. and Williamson, D.P. (Nov. 1995). Improved approximation algorithms for maximum cut and satisfiability problems using semidefinite programming. *J. ACM* 42 (6): 1115–1145.
15. Luo, Z.-Q., Ma, W.-K., So, A.M.-C. et al. (May 2010). Semidefinite relaxation of quadratic optimization problems. *IEEE Signal Process. Mag.* 27 (3): 20–34.
16. Madani, R., Ashraphijuo, A., Lavaei, J., and Baldick, R. (2016). Power system state estimation with a limited number of measurements. *Proceedings of the IEEE Conference on Decision and Control* (December 2016), Las Vegas, NV.
17. Madani, R., Lavaei, J., and Baldick, R. (2015). Convexification of power flow problem over arbitrary networks. *Proceedings of the IEEE Conference on Decision and Control* (December 2015), Osaka, Japan.
18. Zhang, Y., Madani, R., and Lavaei, J. (Apr. 2017). Conic relaxations for power system state estimation with line measurements. *IEEE Trans. Control Netw. Syst.* 5 (3): 1193–205.
19. Mehanna, O., Huang, K., Gopalakrishnan, B. et al. (Nov. 2015). Feasible point pursuit and successive approximation of non-convex QCQPs. *IEEE Signal Process Lett.* 22 (7): 804–808.
20. Lipp, T. and Boyd, S. (Jun. 2016). Variations and extension of the convex–concave procedure. *Optim. Eng.* 17 (2): 263–287.
21. Zamzam, A.S., Sidiropoulos, N.D., and Dall'Anese, E. (Sept. 2018). Beyond relaxation and Newton–Raphson: Solving AC OPF for multi-phase systems with renewables. *IEEE Trans. Smart Grid* 9 (5): 3966–75.
22. Wang, G., Zamzam, A.S., Giannakis, G.B., and Sidiropoulos, N.D. (Mar. 2018). Power system state estimation via feasible point pursuit: Algorithms and Cramér-rao bound. *IEEE Trans. on Signal Process* 66 (6): 1649–1658.
23. Wang, G., Zamzam, A.S., Giannakis, G.B., and Sidiropoulos, N.D. (2016). Power system state estimation via feasible point pursuit. *Proceedings of the IEEE Global Conference on Signal and Information Processing* (November 2016), Washington, DC.
24. Pardalos, P.M. and Vavasis, S.A. (Mar. 1991). Quadratic programming with one negative eigenvalue is NP-hard. *J. Global Optim.* 1 (1): 15–22.
25. Power systems test case archive. University of Washington. https://labs.ece.uw.edu/pstca/ (accessed 11 September 2020).
26. Kekatos, V., Giannakis, G.B., and Wollenberg, B.F. (Aug. 2012). Optimal placement of phasor measurement units via convex relaxation. *IEEE Trans. Power Syst.* 27 (3): 1521–1530.
27. De La Ree, J., Centeno, V.A., Thorp, J., and Phadke, A. (Jun. 2010). Synchronized phasor measurement applications in power systems. *IEEE Trans. Smart Grid* 1 (1): 20–27.
28. Gómez-Expósito, A., Abur, A., de la Villa Jaén, A., and Gómez-Quiles, C. (Jun. 2011). A multilevel state estimation paradigm for smart grids. *Proc. IEEE* 99 (6): 952–976.

29. Cutsem, T.V. and Ribbens-Pavella, M. (Oct. 1983). Critical survey of hierarchical methods for state estimation of electric power systems. *IEEE Trans. Power App. Syst.* 102 (10): 3415–3424.
30. Iwamoto, S., Kusano, M., and Quintana, V.H. (Aug. 1989). Hierarchical state estimation using a fast rectangular-coordinate method. *IEEE Trans. Power Syst.* 4 (3): 870–880.
31. Korres, G.N. (Feb. 2011). A distributed multiarea state estimation. *IEEE Trans. Power Syst.* 26 (1): 73–84.
32. Zhao, L. and Abur, A. (May 2005). Multiarea state estimation using synchronized phasor measurements. *IEEE Trans. Power Syst.* 20 (2): 611–617.
33. Conejo, A.J., de la Torre, S., and Canas, M. (Feb. 2007). An optimization approach to multiarea state estimation. *IEEE Trans. Power Syst.* 22 (1): 213–221.
34. Lin, S.-Y. and Lin, C.-H. (Aug. 1994). An implementable distributed state estimator and distributed bad data processing schemes for electric power systems. *IEEE Trans. Power Syst.* 9 (3): 1277–1284.
35. Falcao, D.M., Wu, F.F., and Murphy, L. (May 1995). Parallel and distributed state estimation. *IEEE Trans. Power Syst.* 10 (2): 724–730.
36. Minot, A., Lu, Y.M., and Li, N. (Sep. 2016). A distributed Gauss–Newton method for power system state estimation. *IEEE Trans. Power Syst.* 31 (5): 3804–3815.
37. Ebrahimian, R. and Baldick, R. (Nov. 2000). State estimation distributed processing. *IEEE Trans. Power Syst.* 15 (4): 1240–1246.
38. Xie, L., Choi, D.-H., and Kar, S. (2011). Cooperative distributed state estimation: local observability relaxed. *Proceedings of the IEEE Power & Energy Society General Meeting* (July 2011), Detroit, MI.
39. Gómez-Expósito, A., de la Villa Jaén, A., Gómez-Quiles, C. et al. (Apr. 2011). A taxonomy of multi-area state estimation methods. *Electron. Power Syst. Res.* 81: 1060–1069.
40. Schizas, I.D., Ribeiro, A., and Giannakis, G.B. (Jan. 2008). Consensus in ad hoc WSNs with noisy links: part I: distributed estimation of deterministic signals. *IEEE Trans. Signal Process.* 56 (1): 350–364.
41. Zhu, H., Giannakis, G.B., and Cano, A. (Oct. 2009). Distributed in-network decoding. *IEEE Trans. Signal Process.* 57 (10): 3970–3983.
42. Boyd, S., Parikh, N., Chu, E. et al. (2010). Distributed optimization and statistical learning via the alternating direction method of multipliers. *Found. Trends Mach. Learn.* 3: 1–122.
43. Giannakis, G.B., Ling, Q., Mateos, G. et al. (2016). Decentralized learning for wireless communications and networking. In: *Splitting Methods in Communication, Imaging, Science, and Engineering*, 461–497. Springer.
44. Kekatos, V., Vlachos, E., Ampeliotis, D., et al. (2013). A decentralized approach to generalized power system state estimation. *Proceedings of the IEEE Workshop on Computational Advances in Multi-Sensor Adaptive Processing* (December 2013), Saint Martin.
45. Grone, R., Johnson, C.R., Sá, E.M., and Wolkowicz, H. (Apr. 1984). Positive definite completions of partial Hermitian matrices. *Linear Algebra Appl.* 58: 109–124.
46. Dall'Anese, E., Zhu, H., and Giannakis, G.B. (Sep. 2013). Distributed optimal power flow for smart microgrids. *IEEE Trans. Smart Grid* 4 (3): 1464–1475.
47. Jabr, R.A. (May 2012). Exploiting sparsity in SDP relaxations of the OPF problem. *IEEE Trans. Power Syst.* 27 (2): 1138–1139.
48. Lam, A.Y., Zhang, B., and David, T. (2012). Distributed algorithms for optimal power flow problem. *Proceedings of the IEEE Conference on Decision and Control* (December 2012), Maui, HI, pp. 430–437.

49. Kekatos, V. and Giannakis, G.B. (May 2013). Distributed robust power system state estimation. *IEEE Trans. Power Syst.* 28 (2): 1617–1626.
50. Duan, D., Yang, L., and Scharf, L.L. (2011). Phasor state estimation from PMU measurements with bad data. *Proceedings of the IEEE Workshop on Computational Advances in Multi-Sensor Adaptive Processing* (December 2011), San Juan, Puerto Rico.
51. Kekatos, V. and Giannakis, G.B. (Jul. 2011). From sparse signals to sparse residuals for robust sensing. *IEEE Trans. Signal Process.* 59 (7): 3355–3368.
52. Kosut, O., Jia, L., Thomas, J., and Tong, L. (Dec. 2011). Malicious data attacks on the smart grid. *IEEE Trans. Smart Grid* 2 (4): 645–658.
53. Celik, M.K. and Abur, A. (Feb. 1992). A robust WLAV state estimator using transformations. *IEEE Trans. Power Syst.* 7 (1): 106–113.
54. Haykin, S. (2002). *Adaptive Filter Theory*. New York: Prentice Hall.
55. Mili, L., Cheniae, M.G., and Rousseeuw, P.J. (May 1994). Robust state estimation of electric power systems. *IEEE Trans. Circuits Syst. I* 41 (5): 349–358.
56. Chen, S.S., Donoho, D.L., and Saunders, M.A. (Jul. 1998). Atomic decomposition by basis pursuit. *SIAM J. Sci. Comput.* 20: 33–61.
57. Wang, G., Giannakis, G.B., and Chen, J. (Feb. 2019). Robust and scalable power system state estimation via composite optimization. *IEEE Trans. Smart Grid* 10 (6): 6137–6147.
58. Maronna, R.A., Martin, R.D., and Yohai, V.J. (2006). *Robust Statistics: Theory and Methods*. Wiley.
59. Mangasarian, O.L. and Musicant, D.R. (Sep. 2000). Robust linear and support vector regression. *IEEE Trans. Pattern Anal. Mach. Intell.* 22 (9): 950–955.
60. Xu, W., Wang, M., Cai, J.F., and Tang, A. (Dec. 2013). Sparse error correction from nonlinear measurements with applications in bad data detection for power networks. *IEEE Trans. Signal Process.* 61 (24): 6175–6187.
61. Mo, B.Y., Kim, T.H.-J., Brancik, K. et al. (Jan. 2012). Cyber-physical security of a smart grid infrastructure. *Proc. IEEE* 100 (1): 195–209.
62. Zhu, H. and Giannakis, G.B. (2012). Robust power system state estimation for the nonlinear AC flow model. *Proceedings of the North American Power Symposium* (September 2012), Urbana-Champaign, IL.
63. Clements, K.A., Krumpholz, G.R., and Davis, P.W. (Jul. 1983). Power system state estimation with measurement deficiency: an observability/measurement placement algorithm. *IEEE Trans. Power App. Syst.* 102 (7): 2012–2020.
64. Monticelli, A. and Wu, F.F. (May 1985). Network observability: theory. *IEEE Trans. Power App. Syst.* 104 (5): 1042–1048.
65. Liu, Y., Ning, P., and Reiter, M.K. (May 2011). False data injection attacks against state estimation in electric power grids. *ACM Trans. Inf. Syst. Security* 14 (1): 13:1–13:33.
66. Huang, S.-J. and Shih, K.-R. (Nov. 2002). Dynamic-state-estimation scheme including nonlinear measurement function considerations. *IEE Proc. Generat. Transm. Distrib.* 149 (6): 673–678.
67. Shalev-Shwartz, S. (Mar. 2011). Online learning and online convex optimization. *Found. Trends Mach. Learn.* 4 (2): 107–194.
68. Kim, S.-J., Wang, G., and Giannakis, G.B. (2014). Online semidefinite programming for power system state estimation. *IEEE International Conference on Acoustics, Speech and Signal Processing* (May 2014), Florence, Italy, pp. 6024–6027.
69. Kim, S.-J. (2015). Online power system state estimation using alternating direction method of multipliers. *Proceedings of the IEEE PES Summer Meeting* (July 2015). Denver, CO, pp. 1–5.

70. Kekatos, V., Wang, G., Conejo, A.J., and Giannakis, G.B. (Nov. 2015). Stochastic reactive power management in microgrids with renewables. *IEEE Trans. Power Syst.* 30 (6): 3386–3395.
71. Kim, S.J., Giannakis, G.B., and Lee, K.Y. (2014). Online optimal power flow with renewables. *Asilomar Conference on Signals, Systems, and Computers* (November 2014), Pacific Grove, CA, pp. 355–360.
72. Wang, G., Kekatos, V., Conejo, A.J., and Giannakis, G.B. (Nov. 2016). Ergodic energy management leveraging resource variability in distribution grids. *IEEE Trans. Power Syst.* 31 (6): 4765–4775.
73. Valverde, G. and Terzija, V. (Jan. 2011). Unscented Kalman filter for power system dynamic state estimation. *IEE Proc. Generat. Transm. Distrib.* 5 (1): 29–37.
74. Da Silva, A.L., Do Coutto Filho, M., and De Queiroz, J. (Sep. 1983). State forecasting in electric power systems. *IEE Proc. Generat. Transm. Distrib.* 130 (5): 237–244.
75. Debs, A.S. and Larson, R.E. (Sep. 1970). A dynamic estimator for tracking the state of a power system. *IEEE Trans. Power App. Syst.* (7): 1670–1678.
76. Wang, G., Kim, S.-J., and Giannakis, G.B. (2014). Moving-horizon dynamic power system state estimation using semidefinite relaxation. *Proceedings of the IEEE PES General Meeting* (July 2014), Washington, DC, pp. 1–5.
77. Wang, S., Gao, W., and Meliopoulos, A.S. (May 2012). An alternative method for power system dynamic state estimation based on unscented transform. *IEEE Trans. Power Syst.* 27 (2): 942–950.
78. Zhao, J., Netto, M., and Mili, L. (2016). A robust iterated extended Kalman filter for power system dynamic state estimation. *IEEE Trans. Power Syst.* 32 (4): 3205–3216.
79. Rao, C.V., Rawlings, J.B., and Mayne, D.Q. (Feb. 2003). Constrained state estimation for nonlinear discrete-time systems: stability and moving horizon approximations. *IEEE Trans. Autom. Control.* 48 (2): 246–258.
80. Van den Bos, A. (Oct. 1994). A Cramér–Rao lower bound for complex parameters. *IEEE Trans. Signal Process.* 40 (10): 2859.

CHAPTER 7

ROBUST WIDE-AREA FAULT VISIBILITY AND STRUCTURAL OBSERVABILITY IN POWER SYSTEMS WITH SYNCHRONIZED MEASUREMENT UNITS

Mert Korkali

Lawrence Livermore National Laboratory, Livermore, CA, USA

7.1 INTRODUCTION

Operation of the existing power systems is rapidly going through major changes due to the widespread deployment of synchronized measurement systems. These systems provide unprecedented advantages in wide-area monitoring of power grids due to the availability of synchronization among measurements at geographically remote parts of the system. In order to estimate the system states, power system state estimator makes use of a set of available measurements. Given a set of measurements and their locations, the network observability analysis determines whether a unique estimate can be obtained for the system state [1]. This analysis is carried out offline during the initial phase of a state estimator installation in order to check the sufficiency of the existing measurement configuration. If the system is not found observable, then additional meters may have to be installed at certain locations. Observability analysis is executed online prior to performing the state estimator. It ensures that a state estimate can be obtained using the set of measurements at the last measurement scan. Telecommunication errors, telemetry failures, or changes in grid topology may at times result in cases where the state of the whole

Advances in Electric Power and Energy: Static State Estimation, First Edition.
Edited by Mohamed E. El-Hawary.

Published 2021 by John Wiley & Sons, Inc.

system cannot be estimated. Network observability test allows detection of such cases right before the execution of the state estimator. Observability of a given network is determined by the type and location of the available measurements as well as by the topology of the network. The concept of topological observability in power network state estimation is introduced as a partial requirement for state estimation solvability by [2] and further developed in [3]. To maintain topological network observability, installing a phasor measurement unit (PMU) at every bus in a wide-area power network is not economically justifiable; therefore, optimal deployment of such devices is extensively sought.

Widespread deployment of PMUs providing for both appropriate penetration and redundancy of synchronized measurements is a key enabling factor for system visibility. Such widespread deployment can be enhanced with the utilization of intelligent electronic devices (IEDs) such as microprocessor-based relays with synchronized measurement capability in order to enhance disturbance-recording capabilities of the power system. With a large volume of synchronized, high-resolution, and raw transient data collected through disbursed sensors deployed over a wide area, power system operators can obtain a coherent picture of the whole transient process and extract useful information that allows for pinpointing incipient problems and thus taking appropriate corrective actions. Sampled data obtained from around the entire network simultaneously can be used in forming a consistent picture of faults and other transient events as they occur on a power system. At present, in parallel with the availability of wide-area synchronized measurements, computational capabilities in substations allow the utilization of unconventional techniques, especially those based on traveling waves in the analysis of dynamic events such as faults. Moreover, recently developed signal-processing tools allow the analysis of sampled waveforms with localized transients.

The first part of this chapter delineates a methodology that enables synchronized measurement-based fault visibility in large-scale power systems. The approach taken employs traveling waves that propagate throughout the network after fault occurrence and requires capturing arrival times of fault-initiated traveling waves using synchronized sensors so as to localize the fault with the aid of the recorded times of arrival (ToAs) of these waves.

The second part of this chapter is devoted to optimization model for the deployment of PMUs paving the way for complete topological (structural) observability in power systems under various considerations, including PMU channel limits, zero-injection buses, and a single PMU failure.

7.2 ROBUST FAULT VISIBILITY USING STRATEGICALLY DEPLOYED SYNCHRONIZED MEASUREMENTS

With the ever-increasing deployment of synchronized wide-area measurement systems (WAMSs), novel control and protection functions are being investigated to improve system stability, protection, and reliability [4]. Postfault analysis requires accurate information from multiple transmission substations, where synchronized measurements are provided by highly accurate global positioning system (GPS)-synchronized PMUs. While such measurements are very useful for capturing events that occur in pseudosteady state, fast transients such as those caused by short-circuit faults will require capturing samples at much higher resolution, i.e., in the order of microseconds. In this work, it is assumed that such high-resolution raw samples of voltages are available from synchronized IEDs.

Further, with the availability of high-bandwidth optical instrument transformers, the new-generation PMUs are expected to provide synchronized point-on-waveform measurements at high sampling frequencies. Having access to these measurements can facilitate the acquisition of key information forming the system-wide picture of dynamic events through continuous monitoring of wide-area recordings of transient disturbances [5, 6].

In this section, we present a fault-location system in such a way that even with a relatively sparse penetration of measurements in an interconnection, the faults can be identified and observed at a wide-area system level. In particular, we propose a novel analytical and computational approach to fault location for large-scale power systems. The proposed framework involves an online and an off-line phase. The online phase is based solely on the utilization of the time-of-arrival (ToA) measurements of traveling waves propagating from the fault-occurrence point to synchronized wide-area fault-recording devices installed at strategically selected substations. The captured waveforms are processed together at the time of fault in order to identify the location of the fault under study. The applicability of the algorithm is independent of the fault type and can readily be extended to power grids of any size.

7.2.1 Wide-Area Synchronized Measurement-Based Fault Location Exploiting Traveling Waves

A power grid can be modeled as a weighted graph $\mathcal{G} = (\mathcal{V}, \mathcal{L})$ consisting of a set of $|\mathcal{V}| = N$ vertices (buses) and $|\mathcal{L}| = L$ edges (transmission lines), together with a distance measurement (transmission-line length), $\{\mathrm{d}_\ell\}_{\ell=1}^{L}$, associated with each edge. The waveform generated by a fault occurring on a transmission line can be decomposed into several electromagnetic-transient waveforms, each representing a mode and propagating throughout the network at its own propagation speed. For each mode, by recording the instants at which the wavefronts of the fault-generated waveform arrive at key buses in the system, the location of a fault can be identified.

Assume that ToAs, $\{T_k\}_{k=1}^{K}$, of a fault-originated electromagnetic wave are collected at optimally deployed K sensors in the grid. In the meantime, the set of propagation times, $\{D_\ell = \mathrm{d}_\ell/\nu_\ell\}_{\ell=1}^{L}$, where ν_ℓ is the wave-propagation speed for Line "ℓ," can be computed beforehand.

The shortest propagation delay from the fault on Line "ℓ" to Sensor "k" depends on the network topology, the propagation times, $\{D_\ell\}$, and the following unknown quantities:

1. The identity of the faulty line (say, "ℓ").
2. The location of the fault on the line (say, $\alpha^{(\ell)}D_\ell$, from a designated end of the line, so that $0 \le \alpha^{(\ell)} \le 1$).
3. The fault instant, $T_0^{(\ell)}$, associated with Line "ℓ."

Let us assume that the fault occurs on Line "ℓ." Then, the shortest propagation path from the fault to Sensor "k," whose corresponding time is denoted by $\zeta_{k,\ell}(\alpha^{(\ell)})$, will contain one of the two terminals of the faulty line. One of these terminals will be designated as the line *origin*, and likewise, the opposite endpoint will be called the line *terminus*. Since the endpoint that lies on the shortest propagation path from the fault to Sensor "k" is not known *a priori*, it is concluded that

$$\zeta_{k,\ell}\left(\alpha^{(\ell)}\right) = \min\left\{\mathcal{D}_{k,\ell}^{(o)} + \alpha^{(\ell)}D_\ell, \mathcal{D}_{k,\ell}^{(t)} + \left(1-\alpha^{(\ell)}\right)D_\ell\right\} \tag{7.1}$$

The delays, $\mathcal{D}_{k,\ell}^{(o)}$ and $\mathcal{D}_{k,\ell}^{(t)}$, can be determined in advance for every "k" and every "ℓ." The described approach is visualized in Figure 7.1.

At the breakeven point of the two possible paths in (7.1), we write

$$\mathcal{D}_{k,\ell}^{(o)} + \alpha^{(\ell)}D_\ell = \mathcal{D}_{k,\ell}^{(t)} + \left(1-\alpha^{(\ell)}\right)D_\ell \tag{7.2}$$

Let the corresponding value of $\alpha^{(\ell)}$ be called $\beta_{k,\ell}$, which will be given by

$$\beta_{k,\ell} = \frac{\mathcal{D}_{k,\ell}^{(t)} - \mathcal{D}_{k,\ell}^{(o)} + D_\ell}{2D_\ell} \tag{7.3}$$

Defining the vector of ToA measurements on sensors as

$$\mathbf{T} = [T_1\ T_2\ \cdots\ T_K]^{\mathrm{T}} \tag{7.4}$$

the overdetermined set of equations can be written in the form

$$\mathbf{T} - T_0^{(\ell)}\boldsymbol{\eta} = \boldsymbol{\zeta}_\ell\left(\alpha^{(\ell)}\right) \tag{7.5a}$$

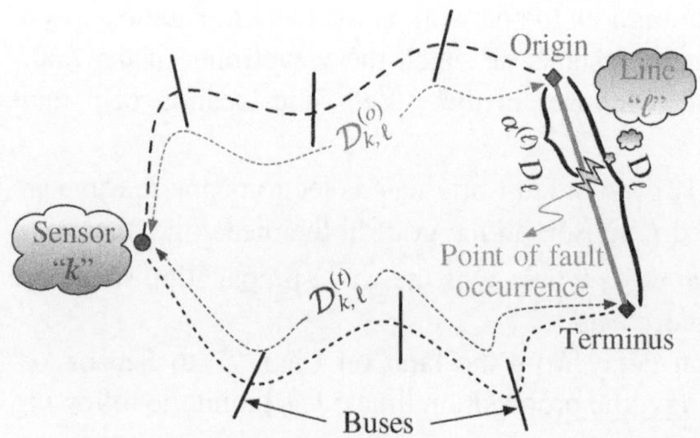

Figure 7.1 Shortest propagation paths from Sensor "k" to the fault on Line "ℓ."

where

$$\boldsymbol{\eta} = [1 \ \ 1 \ \ \cdots \ \ 1]^{\mathrm{T}}_{K \times 1} \tag{7.5b}$$

and

$$\boldsymbol{\zeta}_{\ell}\left(\boldsymbol{\alpha}^{(\ell)}\right) = \left[\zeta_{1,\ell}\left(\alpha^{(\ell)}\right) \ \ \zeta_{2,\ell}\left(\alpha^{(\ell)}\right) \ \ \cdots \ \ \zeta_{K,\ell}\left(\alpha^{(\ell)}\right)\right]^{\mathrm{T}} \tag{7.5c}$$

This system is linear in $T_0^{(\ell)}$, piecewise linear in $\alpha^{(\ell)}$, and highly nonlinear in the integer index "ℓ." The fault-location problem is then transformed into the following nonlinear constrained optimization problem:

$$\min_{\left\{\ell,\alpha^{(\ell)},T_0^{(\ell)}\right\}} \quad \| \ \mathbf{T} - T_0^{(\ell)}\boldsymbol{\eta} - \boldsymbol{\zeta}_{\ell}\left(\boldsymbol{\alpha}^{(\ell)}\right) \ \| \tag{7.6a}$$

$$\text{s.t.} \qquad 0 \leq \alpha^{(\ell)} \leq 1; \quad \ell \in \{1, 2, \ldots, L\} \tag{7.6b}$$

Note that one way to obtain closed-form expressions for $T_0^{(\ell)}$ and $\alpha^{(\ell)}$ is by "linearizing" the dependence of $\zeta_{k,\ell}(\alpha^{(\ell)})$ on the variable $\alpha^{(\ell)}$. This can be achieved by splitting the ℓth transmission line at the points defined by $\{\beta_{k,\ell}\}$. We first sort the set $\{\beta_{k,\ell}\}_{k=1}^{K}$ in ascending order, say,

$$0 \leq \beta_{k_1,\ell} \leq \beta_{k_2,\ell} \leq \cdots \leq \beta_{k_K,\ell} \leq 1 \tag{7.7}$$

and then introduce a fictitious node at each one of the points "$\beta_{k_i,\ell}D_\ell$" as depicted in Figure 7.2.

In this new graph, $\beta_{k,\ell} \in \{0, 1\}$ are the only possible values for every line segment. Now, $\zeta_{k,\ell}(\alpha^{(\ell)})$ is linear in $\alpha^{(\ell)}$ and can be compactly stated as

$$\zeta_{k,\ell}\left(\alpha^{(\ell)}\right) = \mathcal{D}_{k,\ell}^{(o)} + S_{k,\ell}\alpha^{(\ell)}D_\ell \tag{7.8}$$

where $S_{k,\ell} = 2\beta_{k,\ell} - 1 = \pm 1$.

Now, by letting $\psi^{(\ell)} = \alpha^{(\ell)}D_\ell$, our cost function becomes

$$\mathcal{J}_\ell = \ \| \ \boldsymbol{\mathcal{D}}_\ell + \psi^{(\ell)}\mathbf{S}_\ell - \mathbf{T} + T_0^{(\ell)}\boldsymbol{\eta}\|_2^2 \tag{7.9a}$$

where

$$\boldsymbol{\mathcal{D}}_\ell = \left[\mathcal{D}_{1,\ell}^{(o)} \ \ \mathcal{D}_{2,\ell}^{(o)} \ \ \cdots \ \ \mathcal{D}_{K,\ell}^{(o)}\right]^{\mathrm{T}} \tag{7.9b}$$

and

$$\mathbf{S}_\ell = [S_{1,\ell} \ \ S_{2,\ell} \ \ \cdots \ \ S_{K,\ell}]^{\mathrm{T}} \tag{7.9c}$$

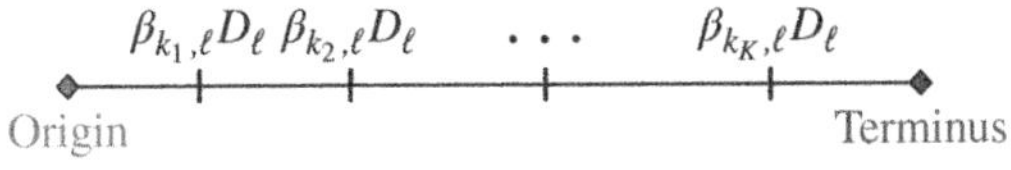

Figure 7.2 Virtual nodes generated at the points "$\beta_{k,\ell}D_\ell$" on Line "ℓ."

Taking first-order derivatives of $\mathcal{J}_\ell$ with respect to $\psi^{(\ell)}$ and $T_0^{(\ell)}$, closed-form expressions for $\psi^{(\ell)}$ and $T_0^{(\ell)}$ will be obtained as

$$\begin{pmatrix} K & \mathbf{S}_\ell^\top \boldsymbol{\eta} \\ \boldsymbol{\eta}^\top \mathbf{S}_\ell & K \end{pmatrix} \begin{pmatrix} \psi^{(\ell)} \\ T_0^{(\ell)} \end{pmatrix} = \begin{pmatrix} \mathbf{S}_\ell^\top (\mathbf{T} - \mathcal{D}_\ell) \\ \boldsymbol{\eta}^\top (\mathbf{T} - \mathcal{D}_\ell) \end{pmatrix} \tag{7.10}$$

where off-diagonal entries are equal and represent the following summations:

$$\mathbf{S}_\ell^\top \boldsymbol{\eta} = \boldsymbol{\eta}^\top \mathbf{S}_\ell = \sum_{k=1}^{K} S_{k,\ell} \tag{7.11}$$

By executing repeated solutions of (7.10) over a set of "ℓ" values and selecting the one yielding the minimum norm for (7.9a), faulted line as well as fault location can be determined.

Successful performance of the abovementioned methodology is well applicable when extracted ToA measurements are *free* of errors. However, fault-recording devices could be subject to various errors, which can threaten the reliability of the wide-area protection functions of the grid. In Section 7.2.3, gross measurement errors that can appear in such devices will be taken into consideration in identification of fault-occurrence points.

7.2.2 Optimal Deployment of Synchronized Sensors for Wide-Area Fault Visibility

As power system faults would theoretically occur at each bus and in every inch of each transmission line in a power system, it is highly desirable that an optimal measurement design ensures that all these faults be localizable. Thus, this section explains how these countably infinite number of faults in power systems can be made visible to system operators by taking advantage of optimally placed sensors with synchronized measurements. To that end, fault location in power systems is recast as an "observability" problem; thus, the ultimate objective becomes realizing optimal sensor selection that guarantees wide-scale coverage of faults.

In general, a system is regarded as "fault-observable" if any fault occurring in the system can be uniquely localized using the available set of measurements. It should be noted that fault-location observability is directly related to the number and location of the installed sensors and is independent of the captured ToAs. More precisely, optimal sensor deployment is an offline process, and placement strategy depends solely on the knowledge of grid topology and transmission-line lengths.

The values "–1" in the column vectors $\mathbf{S}_\ell$ in (7.9) imply that the shortest distances from the sensors at the corresponding buses to Line "ℓ" are from the *terminus* side, whereas the values "1" indicate that the sensors placed at those buses are closer to the same line from the *origin* than from the terminus side. Therefore, the line segments, for which the elements $S_{k,\ell}$ are all equal to "–1" or "1" in (7.11), will be regarded as unobservable (nonlocalizable) segments, or equivalently, *blind spots* from the viewpoint of fault location.

In order to design a sensor system that ensures full network fault observability, one simple solution will be to place a synchronized voltage sensor at every bus. While feasible, this will be a costly solution. A better and more systematic solution can be obtained by formulating the problem as a binary integer programming problem. The solver of this binary integer programming problem will then search for a reduced set of sensors that will accomplish the same task. Furthermore, as discussed earlier in Section 7.2.1, every line will be split into several segments by introducing virtual buses in order to simplify the optimization formulation. As a result, a large number (i.e., L_β [a number much larger than the number of lines in the original power grid topology]) of virtual branches will be created.

A matrix $\mathcal{S}$, which contains $2L_\beta$ rows and N_β columns, where N_β is the total number buses including the virtual ones created due to the splitting of lines, will then be built. Assuming that binary variables, x_j, represent (non)existence of sensors at Bus j, where a nonzero value indicates existence, the following constraint will be imposed for each Branch ℓ:

$$-K < S_{\ell,1}x_1 + S_{\ell,2}x_2 + \cdots + S_{\ell,N_\beta}x_{N_\beta} < K \tag{7.12}$$

The above two-sided inequality constraint can be expressed as two single-sided inequality constraints in two separate rows inside the matrix $\mathcal{S}$ as shown below:

$$\mathcal{S} = \left[\begin{array}{ccccc} S_{1,1} & S_{1,2} & \cdots & \cdots & S_{1,N_\beta} \\ -S_{1,1} & -S_{1,2} & \cdots & \cdots & -S_{1,N_\beta} \\ \hline S_{2,1} & S_{2,2} & \cdots & \cdots & S_{2,N_\beta} \\ -S_{2,1} & -S_{2,2} & \cdots & \cdots & -S_{2,N_\beta} \\ \hline \cdots & \cdots & \cdots & \cdots & \cdots \\ \cdots & \cdots & \cdots & \cdots & \cdots \\ \hline S_{L_\beta,1} & S_{L_\beta,2} & \cdots & \cdots & S_{L_\beta,N_\beta} \\ -S_{L_\beta,1} & -S_{L_\beta,2} & \cdots & \cdots & -S_{L_\beta,N_\beta} \end{array}\right] \begin{array}{l} \left.\vphantom{\begin{array}{c}S\\S\end{array}}\right\} \text{Branch } 1 \\ \left.\vphantom{\begin{array}{c}S\\S\end{array}}\right\} \text{Branch } 2 \\ \left.\vphantom{\begin{array}{c}S\\S\end{array}}\right\} \text{Branch } \ell \\ \left.\vphantom{\begin{array}{c}S\\S\end{array}}\right\} \text{Branch } L_\beta \end{array} \tag{7.13}$$

Doing so ensures that the sum of $S_{\ell,j}$ corresponding to those sensors placed at buses where $x_j = 1$ will not add up to K or $-K$, i.e., the corresponding virtual-branch faults will be observable (visible).

In light of these justifications, the optimization problem for sensor deployment can be explicitly formulated as the following integer programming problem:

$$\min \ \boldsymbol{wx} \tag{7.14a}$$

$$\text{s.t.} \ \mathcal{S}\boldsymbol{x} < \mathcal{K} \tag{7.14b}$$

$$\boldsymbol{x} = \begin{bmatrix} x_1 & x_2 & \cdots & x_{N_\beta} \end{bmatrix}^\top; \quad x_j \in \{0, 1\} \tag{7.14c}$$

$$\mathcal{K} = K \cdot \mathbf{1}_{2L_\beta \times 1}; \quad K \geq 0 \tag{7.14d}$$

$$w = \left[w_1 \cdot \mathbf{1}_{1\times N} \mid w_2 \cdot \mathbf{1}_{1\times (N_\beta - N)}\right] \tag{7.14e}$$

$$K = \sum_{j=1}^{N_\beta} x_j \tag{7.14f}$$

where $\mathbf{1}$ is the vector of ones; N is the number of (actual) buses in the system; $\boldsymbol{w}$ and $\boldsymbol{x}$ are the weight and the (binary) sensor placement vectors, respectively; and $\prec$ denotes the component-wise inequality. In vector $\boldsymbol{w}$, w_1, and w_2 are the weights assigned for actual and fictitious buses, respectively. The sum of nonzero elements in $\boldsymbol{x}$ gives the total sensor count required for system-wide fault visibility.

While the virtual buses are introduced in order to simplify the problem formulation, since they cannot really be used to place actual sensors, it is not desirable for the optimization algorithm to place sensors at such buses. Hence, the solution is forced to use such buses only as a last resort. This is accomplished by assigning $w_1 \ll w_2$ (e.g., $w_1 = 10^{-2}$ and $w_2 = 10^6$), hence making the optimization algorithm strongly favor placement of sensors at the "actual" buses over the fictitious ones (we refer the interested readers to [7] for detailed treatment of the concept of fault observability).

7.2.3 Application of Robust Estimation for Fault Location

Gross measurement errors can typically be invoked by sensor failures and inaccurate measurement scans. In addition to that, through remote access to smart meters from external locations, meter readings on substation IEDs can be manipulated by intruders into the power grid if proper security margins are not assured. Particularly, an intruder can tamper with IED data with the intent to jeopardize the intended protection functions of the grid. For instance, an adversary could create attack vectors by introducing time delays in synchronized measurements, which would, in turn, result in desynchronization of IEDs installed at multiple locations [8]. Therefore, countermeasures (e.g., bad data identification algorithms) need to be developed in order to prevent and mitigate the effect of such attacks on power grid operations; otherwise, reliability and security of the grid could be threatened to a great extent.

To motivate the discussion, consider the system of equations in (7.5), in which the state variables are related to the sensor measurements through the following model:

$$\Delta\mathbf{T}^{(\ell)} = h^{(\ell)}\left(\boldsymbol{\theta}^{(\ell)}\right) + \mathbf{e} \tag{7.15}$$

where $\Delta\mathbf{T}^{(\ell)} = \mathbf{T} - \mathcal{D}_\ell$ are the (modified) sensor measurements; $h^{(\ell)}\left(\boldsymbol{\theta}^{(\ell)}\right) = T_0^{(\ell)}\boldsymbol{\eta} + \psi^{(\ell)}\mathbf{S}_\ell$ are the measurement functions relating the state vectors $\boldsymbol{\theta}^{(\ell)} = \left[\psi^{(\ell)}\ T_0^{(\ell)}\right]^\mathrm{T}$ to the measurements $\Delta\mathbf{T}^{(\ell)}$; $\mathbf{e} = [e_1, e_2, \ldots, e_K]^T$, and $e_k \neq 0$ only if Sensor "k" is compromised.

The first-order derivatives of functions, $h^{(\ell)}(\boldsymbol{\theta}^{(\ell)})$, with respect to the state variables, $\psi^{(\ell)}$ and $T_0^{(\ell)}$, i.e.,

$$\frac{\partial h^{(\ell)}\left(\boldsymbol{\theta}^{(\ell)}\right)}{\partial \psi^{(\ell)}} = \mathbf{S}_\ell \quad \text{and} \quad \frac{\partial h^{(\ell)}\left(\boldsymbol{\theta}^{(\ell)}\right)}{\partial T_0^{(\ell)}} = \boldsymbol{\eta} \tag{7.16}$$

form the matrices $\mathcal{H}^{(\ell)} = [\mathbf{S}_\ell \quad \boldsymbol{\eta}]$. The optimization problem for the LAV estimator, which is a systematic way to identify the outliers, can be modeled as

$$\min \quad \mathcal{C}^\top \Theta^{(\ell)} \tag{7.17a}$$

$$\text{s.t.} \quad \mathcal{A}^{(\ell)} \Theta^{(\ell)} = \Delta \mathbf{T}^{(\ell)} \tag{7.17b}$$

$$\mathcal{A}^{(\ell)} = \left[\mathcal{H}^{(\ell)} - \mathcal{H}^{(\ell)} \quad \mathbf{I}_K - \mathbf{I}_K\right] \tag{7.17c}$$

$$\mathcal{C}^\top = [\mathbf{0}_{1 \times 4} \quad \mathbf{1}_{1 \times 2K}] \tag{7.17d}$$

$$\left[\Theta^{(\ell)}\right]^\top = \left[\left[\theta_{\mathbf{u}}^{(\ell)}\right]^\top \left[\theta_{\mathbf{v}}^{(\ell)}\right]^\top \left[\mathbf{u}^{(\ell)}\right]^\top \left[\mathbf{v}^{(\ell)}\right]^\top\right] \tag{7.17e}$$

$$\Theta^{(\ell)} \succeq \mathbf{0}_{(2K+4) \times 1} \tag{7.17f}$$

where $\mathbf{0}$ and $\mathbf{I}_K$ represent the vector of zeros and the identity matrix of order K, respectively.

7.2.4 Simulation Results on the Modified IEEE 118-Bus System

In the computational phase of the proposed fault-location algorithm [9], discrete wavelet transform has been applied to modal components of the three-phase transient-voltage measurements for the purpose of capturing the ToA values of traveling waves at key substations, where synchronized sensors are deployed.

Simulations are carried out in Alternative Transients Program (ATP) and MATLAB using a sampling frequency of 1 MHz. In all simulations, frequency-dependent line models are used. The fault-occurrence time is chosen to be 20 ms with respect to the simulation start time. In addition, the tower configuration and models of transmission lines used in the simulations are extracted from [9]. In this case, the traveling-wave speed of transients is approximated by evaluating it at the frequency corresponding to the midpoint of scale-1 of the utilized wavelet transform. This corresponds to the interval $[f_s/4 - f_s/2]$ (midpoint of which is $3f_s/8 = 375$ kHz), where f_s is the sampling frequency used in the transient simulations. This leads us to obtain an aerial-mode wave-propagation speed of 1.85885×10^5 mi/s. Identical tower configurations are assumed for all transmission lines in order to simplify the simulations without loss of generality. A lookup table for the transmission-line lengths and the corresponding wave-propagation times is created and used in the calculations (please refer to [10] for details).

TABLE 7.1 Synchronized sensor locations versus wave arrival times for the fault occurring at 99 miles away from Bus 63 on Line 63–64.

Buses	**1**	**2**	**4**	**6**	**10**	**14**	**20**	**29**
ToAs (ms)	*55.390*	*37.603*	*−8.240*	*47.452*	*47.112*	24.546	*16.477*	*41.190*
Buses	**35**	**39**	**41**	**46**	**53**	**55**	**57**	
ToAs (ms)	24.885	24.444	*44.255*	22.539	*51.150*	*9.242*	*13.801*	
Buses	**58**	**60**	**61**	**67**	**73**	**74**	**79**	
ToAs (ms)	*48.054*	*11.877*	*−0.761*	*11.410*	25.939	23.728	24.508	
Buses	**84**	**87**	**88**	**90**	**93**	**95**	**97**	
ToAs (ms)	*16.155*	*14.094*	*44.236*	*63.309*	*60.573*	*60.568*	24.476	
Buses	**99**	**101**	**104**	**106**	**107**	**109**	**111**	
ToAs (ms)	24.777	26.515	27.139	27.343	28.425	*43.264*	28.355	
Buses	**112**	**113**	**114**	**115**	**116**	**117**	**118**	
ToAs (ms)	*8.242*	25.116	*65.510*	26.574	24.992	*−4.207*	*12.455*	

We use the modified IEEE 118-bus system as done in [10]. To solve the optimization problems associated with wide-area fault location, synchronized sensor deployment, and robust fault estimation, IBM ILOG® CPLEX® Optimization Studio [11] is used. The optimally chosen locations for the 43 synchronized sensors in the modified IEEE 118-bus system are listed in Table 7.1.

Here, it is worth noting that the shortest propagation delay between each pair of buses is obtained by employing shortest-path algorithm since power grid is modeled as an undirected graph. Given the locations of optimally deployed sensors, the (newly partitioned) "pseudogrid" formed in this way involves 698 buses and 759 line segments. Obviously, the total number of buses and line segments constructed in this way is dependent upon the grid topology, the locations of the synchronized measurements, and the lengths of the transmission lines.

A fault, which occurs at 99 miles away from Bus 63 on a 111-mile-long Line 63–64, will be used to illustrate the robust fault-location approach using LAV. Note that in the studied faulty line, neither endpoint has a sensor deployed as an outcome of the sensor selection method described in Section 7.2.2. This is an advantage of the proposed sensor deployment strategy, in that faults occurring on a considerable number of such transmission lines are rendered visible even though neither end of those lines has a sensor. Meanwhile, the three-phase synchronized measurements of voltages are extracted at the time of fault occurrence. Then, the squares of aerial-mode wavelet-transform coefficients for each modal voltage are obtained following the decoupling of the phase quantities into the modal voltages.

Also, notice from Table 7.1 that out of 43 sensors where ToAs are captured following the fault occurrence, 26 of them (designated in italics) are contaminated with gross errors, which will later be suspected as contaminated measurements. For this case, the proposed LAV estimator can estimate the correct fault location using the remaining "uncompromised" measurements. In fact, LAV-based state estimator yields an optimal estimate of state vector, $\boldsymbol{\theta}^{(\ell)}$, for the suspected Line "ℓ," at which the objective function is minimized. As a result, one obtains

$$\boldsymbol{\theta}^{(314)} = \boldsymbol{\theta}_{\mathbf{u}}^{(314)} - \boldsymbol{\theta}_{\mathbf{v}}^{(314)} = \begin{pmatrix} \psi^{(314)} \\ T_0^{(314)} \end{pmatrix} = \begin{pmatrix} 0.1182 \\ 20.0001 \end{pmatrix} \text{ ms}$$

with $\boldsymbol{\theta}_{\mathbf{u}}^{(314)} = [0.1182, 20.0001]^{\mathrm{T}}$ and $\boldsymbol{\theta}_{\mathbf{v}}^{(314)} = [0, 0]^{\mathrm{T}}$. Accordingly, the resulting residual vector is found to be

$$\mathbf{r}_{\mathrm{LAV}}^{(314)} = \mathbf{u}^{(314)} - \mathbf{v}^{(314)} = \begin{pmatrix} \mathbf{28.8641} \\ \mathbf{10.7113} \\ \mathbf{-34.6851} \\ \mathbf{20.7107} \\ \mathbf{20.8595} \\ 0 \\ \vdots \\ 0.0005 \\ -0.0009 \\ \mathbf{-31.2000} \\ \mathbf{-12.0377} \end{pmatrix}$$

Note that among the elements of $\mathbf{r}_{\mathrm{LAV}}^{(314)}$, the residuals belonging to the corrupted measurements (e.g., Sensors 1–5, 42, and 43) have disproportionately large absolute values as shown in bold in the vector above.

Table 7.2 shows the varying values of $\| \mathbf{r}_{\mathrm{LAV}}^{(\ell)} \|_1$ (in ascending order) corresponding to selected transmission-line segments. Among the listed line segments, $\psi^{(314)}$ is chosen to be the optimum solution for the sought propagation delay associated with the fault. Indeed, this line has the minimum ℓ_1-norm value for the residual vector, $\mathbf{r}_{\mathrm{LAV}}^{(\ell)}$, satisfying the inequality, $0 \le \psi^{(\ell)} = \alpha^{(\ell)} D_\ell \le D_\ell$.

In Figure 7.3a, the location of the fault on Line 63–64 is displayed in terms of the propagation delay, $\psi^{(314)}$. Similar to computation of fault distance in Section 7.2.1, the distance to fault from Bus 63 is computed as follows:

$$\begin{aligned} \hat{d}_{\text{fault}} &= \left((5.972 - ((1.318 - 1.182) + 0.512)) \times 10^{-4} \text{ s}\right) \times \left(1.85885 \times 10^5 \text{ mi/s}\right) \\ &= 98.97 \approx 99 \text{ mi} \end{aligned}$$

To further enhance the fault-location accuracy, we first remove the sensors that are suspected to have carried erroneous measurements and then recompute the fault distance using the fault-location method presented in Section 7.2.1. This entails the elimination of faulty sensors, which, in turn, results in the creation of a

TABLE 7.2 Values of ℓ, $\| \mathbf{r}_{\mathrm{LAV}}^{(\ell)} \|_1$, $\psi^{(\ell)}$ (in ms), $T_0^{(\ell)}$ (in ms), and D_ℓ (in ms) for the fault occurring at 99 miles away from Bus 63 on Line 63–64.

ℓ	$\| \mathbf{r}_{\mathrm{LAV}}^{(\ell)} \|_1$	$\psi^{(\ell)}$	$T_0^{(\ell)}$	D_ℓ	$0 \leq \psi^{(\ell)} \leq D_\ell$?
680	496.9527	−5.1157	15.4384	0.1049	No
317	514.8469	5.6004	14.3306	0.2717	No
700	538.5804	0.7611	20.0001	0.0296	No
30	538.7489	−23.4815	41.2165	0.1049	No
302	539.8386	0.7799	19.8899	0.0484	No
314	**542.4916**	**0.1182**	**20.0001**	**0.1318**	**Yes**
157	547.0346	0.2042	19.8413	0.1345	No
330	552.2862	0.1720	21.7698	0.1722	Yes
188	553.2218	0.0352	20.9898	0.0890	Yes
618	566.6676	0.0021	17.0978	0.0027	Yes
⋮	⋮	⋮	⋮	⋮	⋮

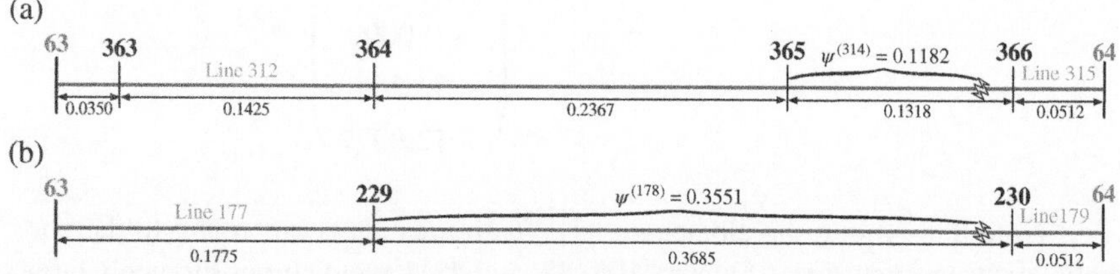

Figure 7.3 (a) Location of a fault occurring at 99 miles away from Bus 63 that is formed by means of optimally deployed sensors in the presence of gross errors (with 1-MHz sensors). (b) Reestimated location after elimination of sensors that are contaminated by gross errors.

virtual network consisting of 410 buses and 471 transmission-line segments. Figure 7.3b displays the newly created Line 63–64 and the propagation delay associated with the fault on Line 178. After the corrupted sensors are discarded, the remaining (true) measurements are used to identify the corrected distance to fault as follows:

$$\hat{d}_{\mathrm{fault_{corr}}} = \left((1.775 + 3.551) \times 10^{-4} \ \mathrm{s}\right) \times \left(1.85885 \times 10^5 \ \mathrm{mi/s}\right) = 99 \ \mathrm{mi}$$

In other words, the accuracy of the fault-location estimate is improved by $(99 - 98.97) \times 1609.34 = 48.28$ m.

The corrected vector of ToA measurements, $\mathbf{T}_{\mathrm{corr}}$, will then be computed by $\mathbf{T}_{\mathrm{corr}} = \mathbf{T} - \mathbf{r}_{\mathrm{LAV}}^{(\ell)} = \mathbf{T} - \mathbf{r}_{\mathrm{LAV}}^{(314)}$. The obtained values of corrected measurements are displayed in Table 7.3.

TABLE 7.3 Synchronized sensor locations versus wave arrival times (after correcting bad measurements) for the fault occurring at 99 miles away from Bus 63 on Line 63–64.

Buses	**1**	**2**	**4**	**6**	**10**	**14**	**20**	**29**
ToAs (ms)	26.526	26.892	26.445	26.741	26.252	24.546	25.901	25.842
Buses	**35**	**39**	**41**	**46**	**53**	**55**	**57**	
ToAs (ms)	24.886	24.444	23.260	22.539	22.938	21.237	22.211	
Buses	**58**	**60**	**61**	**67**	**73**	**74**	**79**	
ToAs (ms)	22.421	21.549	20.576	22.750	25.939	23.728	24.508	
Buses	**84**	**87**	**88**	**90**	**93**	**95**	**97**	
ToAs (ms)	25.197	28.995	26.402	27.489	26.590	26.053	24.476	
Buses	**99**	**101**	**104**	**106**	**107**	**109**	**111**	
ToAs (ms)	24.777	26.515	27.138	27.343	28.424	28.483	28.354	
Buses	**112**	**113**	**114**	**115**	**116**	**117**	**118**	
ToAs (ms)	28.714	25.116	625.810	26.574	24.993	26.994	24.492	

7.3 OPTIMAL PMU DEPLOYMENT FOR SYSTEM-WIDE STRUCTURAL OBSERVABILITY

In this section, the PMU deployment problem is revisited with the aim of relaxing the assumption that the bus voltage phasor and all current phasors along branches connected to that bus are available. To be specific, the problem is reformulated where the number of channels for each PMU can be varied, and the problem can be solved recurrently for each channel capacity to find the optimal number and locations of PMUs. Furthermore, the formulation takes into account any existing injection measurements, in particular the virtual measurements provided by the zero-injections of electrically passive buses. Afterward, the formulation is extended to account for loss of PMUs so that the final PMU measurement design remains robust against loss of a PMU due to device or communication-link failures. Another point of interest in this study is to recognize the effect of channel capacity of a given type of PMU on their optimal placement for network-wide observability and to develop an optimal solution to the PMU placement problem given a specified number of available channels for the candidate PMUs.

The approach taken here is one of exhaustive search among all possible combinations at a given bus for a given limit on the number of available channels. In formulating the problem, it is realized that for a given number of channel capacity, there will be a finite combination of possible assignments of incident branches to a given PMU placed at a bus. Hence, the choices will increase with the number of incident branches for a given bus while remaining bounded irrespective of the overall system size, thanks to the sparse interconnection of power system buses. In what follows, the developed measurement placement procedure will be outlined and illustrated by examples.

7.3.1 Optimal PMU Deployment Considering Channel Limits

Formulation of optimal PMU placement problem for the case of varying channels will be briefly reviewed first. Subsequently, revised formulations that account for zero-injection buses and a single PMU loss will be described.

Consider a PMU that has M channels and installed at Bus a as shown in Figure 7.4. Also, assume that Bus a is connected to N_a buses. Note that the actual number of channels may be three times more since the phasor measurements are usually the positive-sequence components derived from sampled waveforms of three-phase signals. Therefore, it should be noted that the number of channels refers to the number of positive-sequence phasor measurements that can be produced by the considered PMU.

If the number of channels, M, is larger than the number of neighbors, N_a, then a single PMU placed at the bus will provide phasor voltages at all its neighbor buses. Otherwise, there will be r_a combinations of possible channel assignments to branches incident at Bus a:

$$r_a = \begin{cases} {}^{N_a}C_M & \text{if } M < N_a \\ 1 & \text{if } N_a \leq M \end{cases} \tag{7.18}$$

where the number of possible combinations of M out of N_a branches is defined as

$${}^{N_a}C_M = \frac{N_a!}{(N_a - M)!M!} \tag{7.19}$$

Since a PMU is able to measure both the voltage phasor of the bus at which it is installed and the current phasors of all the lines in the neighborhood of this bus, the placement of PMUs becomes a problem with an objective of finding a minimal set of PMUs such that a bus must be observed *at least* once by the solution set of the PMUs. This leads us to define the binary connectivity matrix $\mathcal{B}$ consisting of all possible combinations at a given bus for a given limit on the number of available channels such that each Bus a will have r_a rows, each row containing $(M + 1)$ nonzeros for the bus itself and its neighbor buses. However, when $N_a < M$, i.e., the number of branches incident to Bus a is less than the channel limit of the PMUs,

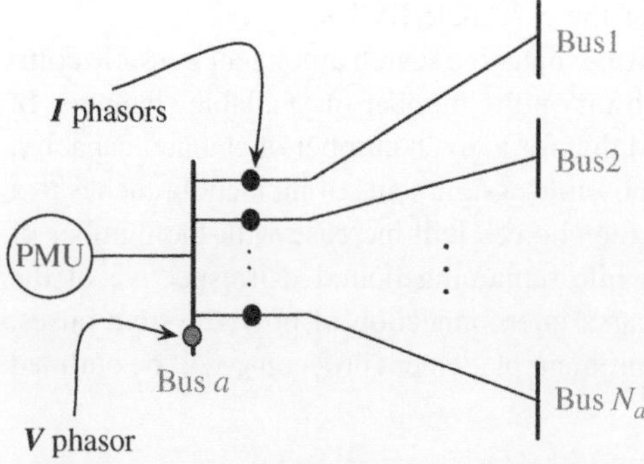

Figure 7.4 Phasor measurements provided by a PMU.

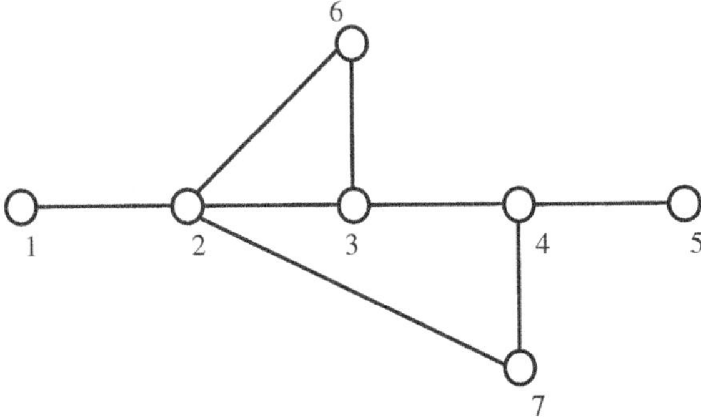

Figure 7.5 7-Bus system for illustration.

the associated row needs to be kept unchanged. The channel limit constraints can thus be imposed so that a PMU placed at a bus will observe its neighboring buses by selecting the appropriate combination(s) of M.

Description of the procedure can be illustrated using a 7-bus system example shown in Figure 7.5. Assuming a channel limit of 2 for the PMUs, the number of rows for each bus is found to be $r_1 = r_5 = r_6 = r_7 = 1$, $r_2 = 6$, and $r_3 = r_4 = 3$. In fact, consider the buses connected to Bus 2, which are Buses 1, 3, 6, and 7; therefore, there exist six combinations of two that can be formed with this set, i.e., $1-3$, $1-6$, $1-7$, $3-6$, $3-7$, and $6-7$. In this sense, each row associated with Bus 2 in matrix $\mathcal{B}$ will include a "1" corresponding to Bus 2 and its neighbor pairs. Accordingly, let $\mathcal{B}$ be defined as

$$\mathcal{B} = \left[\begin{array}{ccccccc}
1 & 1 & 0 & 0 & 0 & 0 & 0 \\ \hline
1 & 1 & 1 & 0 & 0 & 0 & 0 \\
1 & 1 & 0 & 0 & 0 & 1 & 0 \\
1 & 1 & 0 & 0 & 0 & 0 & 1 \\
0 & 1 & 1 & 0 & 0 & 1 & 0 \\
0 & 1 & 1 & 0 & 0 & 0 & 1 \\
0 & 1 & 0 & 0 & 0 & 1 & 1 \\ \hline
0 & 1 & 1 & 1 & 0 & 0 & 0 \\
0 & 1 & 1 & 0 & 0 & 1 & 0 \\
0 & 0 & 1 & 1 & 0 & 1 & 0 \\ \hline
0 & 0 & 1 & 1 & 1 & 0 & 0 \\
0 & 0 & 1 & 1 & 0 & 0 & 1 \\
0 & 0 & 0 & 1 & 1 & 0 & 1 \\ \hline
0 & 0 & 0 & 1 & 1 & 0 & 0 \\ \hline
0 & 1 & 1 & 0 & 0 & 1 & 0 \\ \hline
0 & 1 & 0 & 1 & 0 & 0 & 1
\end{array}\right]
\begin{array}{l}
\text{Bus 1} \\
\\
\\
\text{Bus 2} \\
\\
\\
\\
\\
\text{Bus 3} \\
\\
\\
\text{Bus 4} \\
\\
\text{Bus 5} \\
\text{Bus 6} \\
\text{Bus 7}
\end{array}$$

Using this approach, the relevant PMU placement problem can be formulated as

$$\min \sum_{i=1}^{s} \rho_i y_i \tag{7.20a}$$

$$\text{s.t.} \quad \mathcal{B}^{\top} \mathcal{Y} \succeq \mathbf{1}_{N \times 1} \tag{7.20b}$$

$$\mathcal{Y} = [y_1 \ y_2 \ \cdots \ y_s]^{\top}; \quad y_i \in \{0, 1\} \tag{7.20c}$$

where

$$s = \sum_{j=1}^{N} r_j \quad \text{and} \quad \boldsymbol{\rho} = \mathbf{1}_{1 \times s} \tag{7.20d}$$

Here, $\boldsymbol{\rho}$ represents the vector of installation cost of PMUs, elements of which are assumed to be uniform for simplicity; y_i are the binary variables for the PMU placement; and N denotes the number of buses in the system. In (7.20), vector $\mathcal{Y}$ represents the binary vector of all possible PMU channel assignments. Correspondingly, the nonzero entries in $\mathcal{Y}$ will point to the rows of associated buses, voltage angles of which can be observed by these PMU measurements.

The solution of the aforecited PMU placement problem, in which the specified channel limit for PMUs is 2, is given by

$$\mathcal{Y} = [0 \mid 0 \ 1 \ 1 \ 0 \ 0 \ 0 \mid 0 \ 0 \ 0 \mid 1 \ 0 \ 0 \mid 0 \mid 0 \mid 0]^{\top}$$

indicating deployment of a total of 3 PMUs, two of which are to be located at Bus 2 and one at Bus 4, enabling full network observability. Indeed, one PMU installed at Bus 2 measures the voltage phasor at Bus 2 as well as the current phasors for Branches 2 – 1 and 2 – 6; however, the other PMU at Bus 2 measures current phasors for Branches 2 – 1 and 2 – 7. Furthermore, a PMU located at Bus 4 will measure the voltage phasor at Bus 4 and current phasor on two of the incident branches, i.e., Branches 4 – 3 and 4 – 5 (see Figure 7.6).

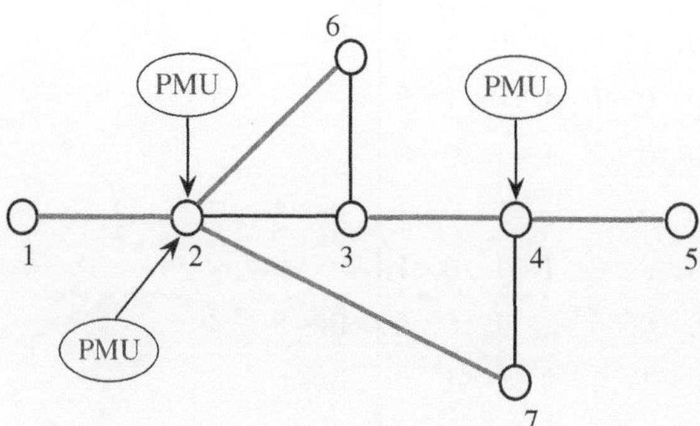

Figure 7.6 Placement of 3 two-channel PMUs in the 7-bus system.

7.3.2 Modeling Zero-Injection Buses

Zero-injection buses, which provide "free" measurements to the system, are incorporated into the optimization formulation as done in [12]. More specifically, zero-injection buses are buses where there is no generation or load, i.e., there exists no power injection into these buses. They act as transshipment or intermediate nodes in power systems. Therefore, zero injections can be used to reduce the number of required PMUs by *selectively* allowing some buses to be "unreachable" by the PMU measurements as long as these buses belong to a certain set of buses. This set is defined as the union of all zero-injection buses and their immediate neighbors.

For the sake of illustration, let us consider the four-bus system example shown in Figure 7.7. Figure 7.7a neglects information about the zero-injection buses; however, Figure 7.7b shows Bus 2 as a zero-injection bus. For the system in Figure 7.7a, it can be noticed that two PMUs are sufficient to make the system observable. For the system shown in Figure 7.7b, Bus 2 is assumed to be a zero-injection bus, and a single PMU will be sufficient to achieve observability. Indeed, Kirchhoff's current law implies that a current entering Node 2 will be equal to the current leaving that node; hence, the voltage phasor at Bus 4 can be calculated without placing a PMU at that bus. With a PMU at Bus 1, the current along Branch 2 – 4 can be computed since Bus 2 is a zero-injection bus, i.e., $I_{24} = I_{12}$. Hence, the voltage at Bus 4 can be computed by $V_4 = V_2 - I_{12}Z_{24}$, where Z_{24} is the impedance of Branches 2 – 4. This eliminates the need to place an extra PMU at Bus 4 in this system.

Now, let us define a set $\mathcal{N}_i$ as a set of buses including zero-injection Bus i and all its neighbors. Assuming "z_b" zero-injection buses to be present in the system, the following set can be defined:

$$\Phi = \bigcup_{i \in \mathcal{I}} \mathcal{N}_i = \mathcal{N}_{i_1} \cup \mathcal{N}_{i_2} \cup \cdots \cup \mathcal{N}_{i_{z_b}} \tag{7.21}$$

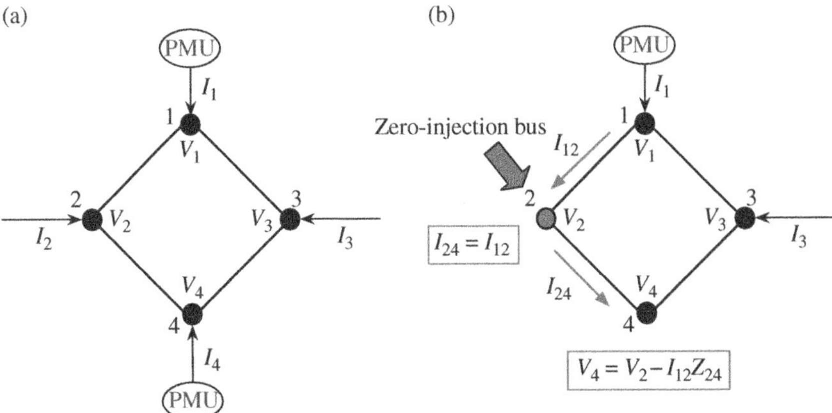

Figure 7.7 Optimal PMU deployment on a 4-bus system (a) ignoring zero-injection constraints and (b) considering Bus 2 as a zero-injection bus.

where $\mathcal{I} = \{i_1, i_2, \ldots, i_{z_b}\}$ designates the set of zero-injection buses.

Thus, the inequality constraints in (7.20) can be reestablished based on the above considerations as follows:

$$\left[\begin{array}{c|c} \mathcal{B}^{\mathrm{T}} & \mathcal{P} \\ \hline \mathbf{0} & \mathcal{Q} \end{array}\right] \begin{bmatrix} \mathcal{Y} \\ \mathbf{b} \end{bmatrix} \succeq \begin{bmatrix} \mathcal{R} \\ \mathbf{c} \end{bmatrix} \tag{7.22a}$$

where

$$\mathcal{P}_{j'k'} = \begin{cases} -1 & \text{if } j' \in \mathcal{N}_{\mathrm{i}} \text{ and } k' \in \{\Phi_{j'}\} \\ 0 & \text{otherwise} \end{cases} \tag{7.22b}$$

$$\mathcal{Q}_{j'k'} = \begin{cases} 1 & \text{if } j' \in \{\mathcal{I}_i\}_{i \in \mathcal{I}} \text{ and } k' \in \{\Phi_m\}_{m \in \mathcal{N}_i} \\ 0 & \text{otherwise} \end{cases} \tag{7.22c}$$

and $\mathcal{Y}$, $\mathcal{R}$, $\mathbf{b}$, and $\mathbf{c}$ are vectors of dimension s, N, $|\Phi|$, and $|\mathcal{I}|$, respectively, with $|\{\bullet\}|$ denoting the cardinality of a set, and matrices $\mathcal{B}$, $\mathcal{P}$, $\mathbf{0}$, and $\mathcal{Q}$ in (7.22) are of sizes $s \times N$, $N \times |\Phi|$, $|\mathcal{I}| \times s$, and $|\mathcal{I}| \times |\Phi|$, respectively. Also, $\mathbf{c}_i$ are equal to $(|\mathcal{N}_i| - 1)$ such that $i \in \mathcal{I}$.

In building matrices $\mathcal{P}$ and $\mathcal{Q}$, the set $\{\Phi_\epsilon\}_{\epsilon \in \mathcal{E}}$ is defined such that the elements of set Φ are indexed (labeled) by means of set $\mathcal{E}$. For the sake of convenience, the first row associated with the first partitioned matrix on the left-hand side of (7.22) splits up into two parts in such a way that the zero-injection buses and their neighbors are heaped together on the top of the new matrix. In this way, the elements of the matrix $\mathcal{Q}$ are clustered in the order of union set Φ. Moreover, the elements of vector $\mathcal{R}$ on the right-hand side of (7.22) comprise "0" for the variables associated with zero-injection buses and "1" for those of the remaining buses.

For the sake of illustration, consider the 7-bus system shown in Figure 7.8 where single-channel PMUs are used. Then, we can build the sets $\mathcal{N}_{\text{bus }4} = \{3,4,5,7\}$, $\mathcal{N}_{\text{bus }6} = \{2,3,6\}$, and $\Phi = \mathcal{N}_{\text{bus }4} \cup \mathcal{N}_{\text{bus }6} = \{2,3,4,5,6,7\}$ in light of the definitions above. In this context, (7.22) will take the following form:

$$\left[\begin{array}{c|cccccc} & -1 & & & & & \\ & & -1 & & & & \\ [\mathcal{B}_1^{\mathrm{T}}]_{6\times 16} & & & -1 & & & \\ & & & & -1 & & \\ & & & & & -1 & \\ & & & & & & -1 \\ \hline [\mathcal{B}_2^{\mathrm{T}}]_{1\times 16} & & & \mathbf{0}_{1\times 6} & & & \\ \hline \mathbf{0}_{2\times 16} & & 1 & 1 & 1 & & 1 \\ & 1 & 1 & & & 1 & \end{array}\right]_{9\times 22} \left[\begin{array}{c} \mathcal{Y} \\ \hline b_2 \\ b_3 \\ b_4 \\ b_5 \\ \boxed{b_6 = 0} \\ \boxed{b_7 = 0} \end{array}\right]_{22\times 1} \succeq \left[\begin{array}{c} 0 \\ 0 \\ 0 \\ 0 \\ 0 \\ 0 \\ \hline 1 \\ \hline 3 \\ 2 \end{array}\right]_{9\times 1}$$

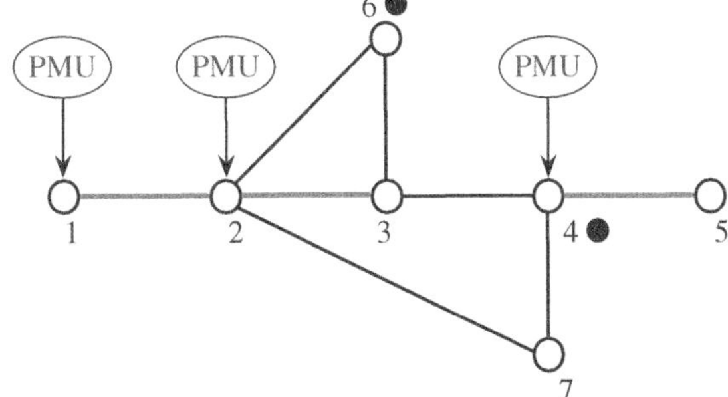

Figure 7.8 Configuration of 3 single-channel PMUs in the 7-bus system (the dots designate the zero-injection buses).

Inequalities $b_3 + b_4 + b_5 + b_7 \geq 3$ and $b_2 + b_3 + b_6 \geq 2$ with $b_i \in \{0, 1\}$ ensure that there is *at most* one unobservable bus in these sets provided that their observability is realized via the use of zero-injection buses in the corresponding sets. Indeed, $b_6 = b_7 = 0$, implying that Bus 6 is not directly reached by a PMU exploiting the fact that it is a zero-injection bus and Bus 7 is observed via the zero-injection Bus 4. Figure 7.8 also illustrates the deployed three PMUs along with the associated branches, through which the corresponding buses are observed.

7.3.3 Optimal Deployment Accounting for Single PMU Loss

In this subsection, the formulation introduced in Section 7.3.1 will be extended to account for loss of PMUs so that the final PMU measurement design remains robust against loss of a PMU due to device or communication-link failures. Thus, it is needed to guard against such unexpected failures of PMUs. In [13, 14], the primary set of PMUs is backed up by a secondary set, which is determined based on the same optimization formulation. In order to achieve a placement strategy for robust system observability, (7.20) can now be modified to ensure that each bus will be observed by *at least* two PMUs, as previously done in [12, 15, 16]. This ascertains that a PMU loss will not lead to loss of observability. In the binary integer programming framework, this can be easily achieved by multiplying $\mathcal{U}$ by 2, i.e., $\mathcal{U} = 2 \cdot \mathbf{1}_{1\times N}^{\top}$.

In this regard, the solution array for the PMU placement study done in Section 7.3.1, where the specified channel limit for PMUs is 2, is found to be

$$\mathcal{Y} = [0 \mid 0 \;\; 1 \;\; 1 \;\; 0 \;\; 0 \;\; 0 \mid 0 \;\; 0 \;\; 1 \mid 1 \;\; 0 \;\; 1 \mid 0 \mid 0 \mid 0]^{\top}$$

yielding a total of five PMUs, two of which are deployed at Buses 2 and 4 each and the remaining one at Bus 3, rendering the entire network observable even when the measurement from any one of the PMUs is lost (see Figure 7.9).

7.3.4 Consolidated PMU Deployment Results

To solve the binary integer programming problem devised for the optimal PMU placement, IBM ILOG® CPLEX® Optimization Studio [11] is utilized. Table 7.4 presents the optimum numbers of PMUs with respect to channel limits in the IEEE 30-, 57-, and 118-bus systems [17] for the following four cases: (i) conventional

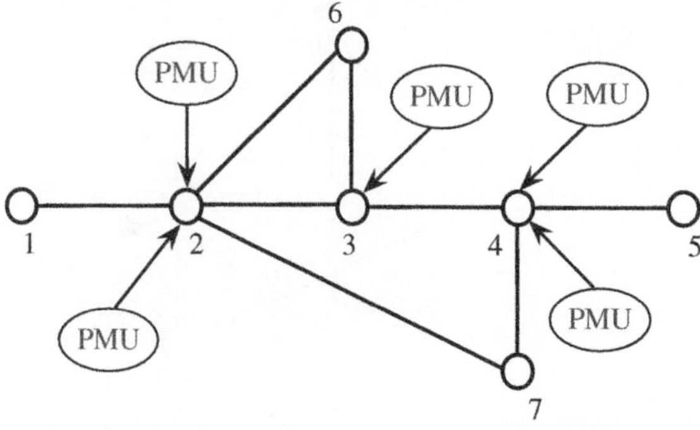

Figure 7.9 Placement of 5 two-channel PMUs in the 7-bus system.

TABLE 7.4 Optimal PMU deployment results for three IEEE test systems.

IEEE test system	Number of zero injections	Channel limit for the PMUs	Number of PMUs			
			Case 1	Case 2	Case 3	Case 4
30-bus	6	1	15	14	30	21
		2	11	9	22	14
		3	**10**	8	**20**	**13**
		4	10	**7**	20	13
		5	10	7	20	13
		6	10	7	20	13
		7	10	7	20	13
57-bus	15	1	29	23	57	34
		2	19	16	38	23
		3	**17**	**14**	34	**22**
		4	17	14	**33**	22
		5	17	14	33	22
		6	17	14	33	22
118-bus	10	1	61	57	121	103
		2	41	38	82	69
		3	33	32	**68**	**58**
		4	**32**	31	68	58
		5	32	**29**	68	58
		6	32	29	68	58
		7	32	29	68	58
		8	32	29	68	58
		9	32	29	68	58

placement ignoring zero injections (Case 1), (ii) conventional placement considering zero injections (Case 2), (iii) robust placement ignoring zero injections (Case 3), and (iv) robust placement considering zero injections (Case 4). In all cases, the upper channel limit of the PMUs is determined by the maximum number of branches incident to a certain bus in the corresponding test system. In most cases, having more than four (positive-sequence) channels does not reduce the required PMU count. Furthermore, through strategic placement of PMUs, a very reliable metering design can be achieved by deploying PMUs at less than 60% of the buses in the system. This number may be reduced significantly by taking advantage of zero-injection buses.

7.4 CONCLUSIONS

In this chapter, we consider the problem of robust, system-wide visibility of power system faults by means of synchronized measurement sensors that are optimally deployed across a large-scale power system. In connection with this, we present an optimal sensor deployment procedure that ensures unique localization of line faults appearing in power grids. Based on the topology of a power system and the distance measurements among system buses, we develop a fault-location technique that relies on ToA recordings of traveling waves at buses that are equipped with synchronized sensors.

This chapter also presents a formulation that is established to solve the problem of using PMUs with limited input capabilities to achieve complete structural observability in a power system. Moreover, zero-injection measurements are incorporated into the problem formulation in order to further minimize the required number of PMUs. We extend the results of conventional PMU placement to the case where the solution is expected to be robust against failure of any single PMU.

ACKNOWLEDGMENTS

The author gratefully acknowledges the intellectual guidance and technical assistance provided by Professors Ali Abur and Hanoch Lev-Ari during his graduate studies at Northeastern University, where this work was completed.

REFERENCES

1. Abur, A. and Expósito, A.G. (2004). *Power System State Estimation: Theory and Implementation*. New York: Marcel Dekker.
2. Clements, K.A. and Wollenberg, W.A. (1975). An algorithm for observability determination in power system state estimation. *IEEE Power Engineering Society Summer Meeting*, (20–25 July 1975). San Francisco, CA.
3. Krumpholz, G.R., Clements, K.A., and Davis, P.W. (1980). Power system observability: a practical algorithm using network topology. *IEEE Transactions on Power Apparatus and Systems* PAS-99 (4): 1534–1542.

4. Terzija, V., Valverde, G., Cai, D. et al. (2011). Wide-area monitoring, protection, and control of future electric power networks. *Proceedings of the IEEE* 99 (1): 80–93.
5. Adamiak, M.G., Apostolov, A.P., Begovic, M.M. et al. (2006). Wide area protection – technology and infrastructures. *IEEE Transactions on Power Delivery* 21 (2): 601–609.
6. Hu, Y., Moraes, R.M., Madani, V., and Novosel, D. (2007). Requirements of large-scale wide area monitoring, protection and control systems. *Proceedings of 10th Annual Fault Disturbance Analysis Conference* (30 April to 1 May 2007). Atlanta, GA, pp. 1–9.
7. Korkalı, M. and Abur, A. (2013). Optimal deployment of wide-area synchronized measurements for fault-location observability. *IEEE Transactions on Power Apparatus and Systems* 28 (1): 482–489.
8. Sridhar, S., Hahn, A., and Govindarasu, M. (2012). Cyber-physical system security for the electric power grid. *Proceedings of the IEEE* 100 (1): 210–224.
9. Korkalı, M., Lev-Ari, H., and Abur, A. (2012). Traveling-wave-based fault-location technique for transmission grids via wide-area synchronized voltage measurements. *IEEE Transactions on Power Apparatus and Systems* 27 (2): 1003–1011.
10. Korkalı, M. (2013). Robust and systemwide fault location in large-scale power networks via optimal deployment of synchronized measurements. PhD dissertation, Northeastern University.
11. IBM ILOG CPLEX V12.1 User's Manual for CPLEX, 2009.
12. Dua, D., Dambhare, S., Gajbhiye, R.K., and Soman, S.A. (2008). Optimal multistage scheduling of PMU placement: an ILP approach. *IEEE Transactions on Power Delivery* 23 (4): 1812–1820.
13. Xu, B. and Abur, A. (2004). Observability analysis and measurement placement for systems with PMUs. *Proceedings of the IEEE PES 2004 Power Systems Conference and Exposition* (10–13 October 2004). New York, NY, vol. 2, pp. 943–946.
14. Xu, B., Yoon, Y.J., and Abur, A. (2005). Optimal placement and utilization of phasor measurements for state estimation. *Power System Computation Conference* (22–26 August 2005). Liège, Belgium.
15. Chakrabarti, S., Kyriakides, E., and Eliades, D.G. (2009). Placement of synchronized measurements for power system observability. *IEEE Transactions on Power Delivery* 24 (1): 12–19.
16. Abbasy, N.H. and Ismail, H.M. (2009). A unified approach for the optimal PMU location for power system state estimation. *IEEE Transactions on Power Apparatus and Systems* 24 (2): 806–813.
17. Christie, R.D. (Aug. 1999). Power systems test case archive. http://www.ee.washington.edu/research/pstca (accessed 2 January 2018).

CHAPTER 8

A ROBUST HYBRID POWER SYSTEM STATE ESTIMATOR WITH UNKNOWN MEASUREMENT NOISE

Junbo Zhao[1], Lamine Mili[2], and Massimo La Scala[3]

[1]*Department of Electrical and Computer Engineering, Mississippi State University, Starkville, MS, USA*

[2]*Bradley Department of Electrical Computer Engineering, Virginia Polytechnic Institute and State University, Northern Virginia Center, Falls Church, VA, USA*

[3]*Dipartimento di Elettrotecnica ed Elettronica (DEE)-Politecnico di Bari, Bari, Italy*

8.1 INTRODUCTION

Accurate state information obtained by a state estimator (SE) is of vital importance for various power system applications, such as static security analysis, contingency analysis and optimal load dispatch, etc. [1]. The objective of a SE is to get the best estimate of the current system bus voltage magnitudes and phase angles given a set of redundant measurements and accurate network parameters [2]. Thus, the performance of an estimator highly depends on the accuracy of the measurements and the assumed estimation model. However, all the measurements are subject to noise or errors caused by the metering instruments, or even gross errors due to the instrument failures, impulsive communication noise, etc. The parameters of the network change with time due to the variations of environment temperature, yielding uncertainties to the assumed SE model. In addition, without enough field information or lack of calibration, erroneous zero injections may occur as well [3]. Therefore, the

Advances in Electric Power and Energy: Static State Estimation, First Edition.
Edited by Mohamed E. El-Hawary.

Published 2021 by John Wiley & Sons, Inc.

measurement noise that is used to account for all these uncertainties may be unknown in practice.

With the wide-area deployment of phasor measurement units (PMUs), more and more synchronized voltage and current phasor measurements are available. PMU measurements are time-stamped and have much higher accuracy than the traditional supervisory control and data acquisition (SCADA) measurements and thus are usually utilized to improve the accuracy of a SE as well as enhance bad data (BD) detection and identification capability. However, the bus voltage and branch current phasor measurements provided by PMUs are incompatible with conventional SCADA systems because of their significantly different sampling rates [4]. On the other hand, PMUs cannot be installed at every bus of a large power system due to economic constraints [5]. As a consequence, the incorporation of the limited number of PMUs to the conventional SCADA-based SE has been a major research subject in recent years. These approaches can be divided into two groups [1, 6]: a single SE, where PMU measurements are mixed with the traditional SCADA measurements [7–13], and a two-stage estimation scheme, where the state estimates obtained from the traditional SCADA measurements are used together with the PMU measurements for final state filtering [14–27]. The main idea of the former group is to augment SCADA and PMU measurement together and apply a single estimator for state estimates. Compared with SCADA measurements only SE, they improve the estimation accuracy thanks to the inclusion of better measurements provided by PMUs. While for the two-stage estimation scheme, two types of approaches have been proposed: (i) a PMU-based linear estimator is implemented at the first stage, processing only available PMU measurements; then a traditional nonlinear estimator that processes state estimates obtained at the first stage and all available SCADA measurements is used; (ii) a traditional SE using SCADA measurements is performed, and its state estimates are incorporated with PMU measurements for the second-stage linear state estimation. However, since there exists time skewness between the SCADA and PMU measurements, it is not always feasible to perform the two groups of approaches mentioned above [28, 29]. In addition, the temporal and spatial correlations among PMU measurements are neglected. To address this issue, the PMU measurements are buffered, and a hypothesis test is proposed to choose the optimal buffer length [30–32]. Then, the sample mean and sample covariance matrix of these buffered PMU measurements are used together with SCADA measurements for the state estimation. Note that the Gaussian assumption is used such that the sample mean and sample covariance matrix are reliable. However, this assumption does not hold true in practice. For example, an investigation conducted by Pacific Northwest National Laboratory (PNNL) reveals that PMU measurement noise follows a heavily tailed non-Gaussian distribution [33, 34]. Furthermore, the SCADA and PMU measurements are assumed to be free of gross errors when determining the optimal buffering length of the PMU measurements. However, this assumption might not be satisfied provided that the measurements can be with large errors due to the instrument failures, impulsive communication noise, etc.

In this chapter, we propose a robust hybrid power system SE to address the unknown non-Gaussian measurement noise and the time skewness problem between SCADA and PMU measurements in a systematic way. First of all, we resort to the robust statistical theory and adopt the robust Schweppe-type Huber generalized maximum likelihood (SHGM)-estimator to suppress BD in SCADA measurements. To address the time skewness problem between SCADA and PMU measurements as well as the unknown non-Gaussian noise in PMU measurements, robust Mahalanobis distances combined with a statistical test are proposed to determine a series of weighted PMU measurements for buffering. Finally, the weighted mean and robust covariance matrix of the buffered PMU measurements are used together with the SCADA measurements-based state estimates to yield the final state estimation.

The organization of this chapter is as follows: Section 8.2 shows the problem statement. Section 8.3 presents the proposed robust state estimation framework, where the modeling of non-Gaussian measurement noise, the determination of a series of weighted PMU measurements using robust Mahalanobis distances, and the robust hybrid state estimation are described. Section 8.4 presents and analyzes the simulation results on the IEEE 30-bus test system, and finally Section 8.5 concludes this chapter.

8.2 PROBLEM STATEMENT

For an N-bus power system using an AC power flow model, the relationship between the vector of measurements $z \in \mathbb{R}^m$ obtained from the SCADA system and the state vector $\boldsymbol{x} \in \mathbb{R}^n$, which contains the nodal voltage magnitudes and phase angles, yielding $n = 2N - 1 < m$, is given by

$$z = \boldsymbol{h}(\boldsymbol{x}) + \boldsymbol{e} \tag{8.1}$$

where $\boldsymbol{h}(\cdot) : \mathbb{R}^n \rightarrow \mathbb{R}^m$ is a vector-valued nonlinear function and $\boldsymbol{e} \in \mathbb{R}^m$ is the measurement error vector that is assumed to follow a Gaussian distribution with zero mean and covariance matrix $\boldsymbol{R} \in \mathbb{R}^{m \times m}$, i.e. $\boldsymbol{e} \sim \mathcal{N}(\boldsymbol{0}, \boldsymbol{R})$. The SE is solved by minimizing the weighted least squares (WLS) criterion, yielding

$$\hat{\boldsymbol{x}} = \arg \min_{\boldsymbol{x}} [z - \boldsymbol{h}(\boldsymbol{x})]^T \boldsymbol{R}^{-1} [z - \boldsymbol{h}(\boldsymbol{x})] \tag{8.2}$$

Let us apply the Gauss–Newton iterative algorithm to solve for the state vector. Formally, we have

$$\boldsymbol{x}^{k+1} = \boldsymbol{x}^k + \Delta \boldsymbol{x}^k, \quad k = 1, 2, \ldots \tag{8.3}$$

$$\Delta \boldsymbol{x}^k = \left(\boldsymbol{H}(\boldsymbol{x}^k)^T \boldsymbol{R}^{-1} \boldsymbol{H}(\boldsymbol{x}^k) \right)^{-1} \boldsymbol{H}(\boldsymbol{x}^k)^T \boldsymbol{R}^{-1} \left(z - \boldsymbol{h}(\boldsymbol{x}^k) \right) \tag{8.4}$$

where $\boldsymbol{H}(\boldsymbol{x}^k) = \partial \boldsymbol{h}(x) / \partial \boldsymbol{x}|_{\boldsymbol{x} = \boldsymbol{x}^k} \in \mathbb{R}^{m \times n}$ is the Jacobian matrix. The algorithm converges once the norm of $\Delta \boldsymbol{x}^k$ is smaller than a pre-specified threshold.

With the wide-area deployment of PMUs today, more and more accurate voltage and current phasors are available for performing power system online monitoring and controls. However, due to many reasons, there are still not enough PMUs installed that can ensure the observability of a system. Thus, the traditional low-sampling-rate SCADA measurements have to be used together with PMU measurements for power system monitoring through a SE. When the synchronized PMU measurements are integrated with SCADA measurements to improve the SE performance, time skewness problem occurs as the sampling rate of PMU is much higher than SCADA measurements. Recently, the authors in [30–32] propose to buffer the PMU measurements during the waiting period of two SCADA measurement scans and use a hypothesis test to choose the optimal buffer length. Note that all these processes are based on the Gaussian assumption and the PMU measurements are assumed to be cleaned without any BD, which are frequently violated in practice. For example, investigations performed by PNNL [33, 34] reveal that the PMU measurement noise follows a non-Gaussian distribution; the statistical BD is very difficult to be cleaned through simple data pre-processing approach. Under those conditions, the performances of the approaches in [30–32] can deteriorate significantly. In this chapter, we will develop a robust SE to address both the non-Gaussian measurement noise and the time skewness problem between SCADA and PMU measurements in a systematic way.

8.3 PROPOSED FRAMEWORK FOR ROBUST HYBRID STATE ESTIMATION

The schematic of the proposed robust hybrid state estimation framework is shown in Figure 8.1. During the waiting period of the two SCADA measurement scans, only the time series PMU data is processed. The robust Mahalanobis distances are proposed to detect BD or large system changes that violate the stationary of the static SE assumption. Then, the optimal buffer length of the PMU measurements

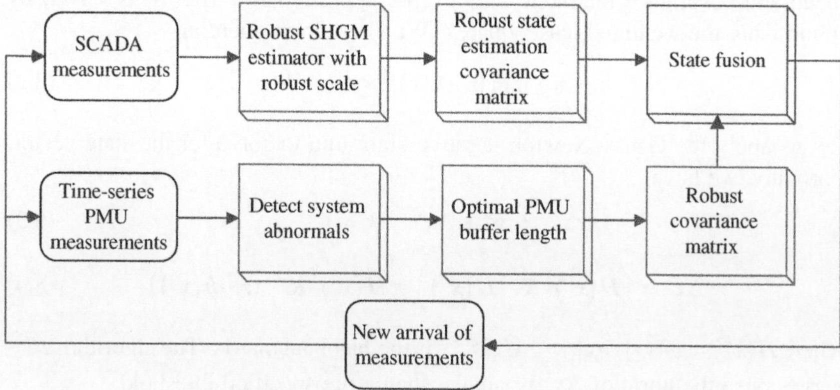

Figure 8.1 Schematic of the proposed robust hybrid state estimation framework.

is determined. The statistical information of these buffered PMU data is further represented by its weighted mean and a robust covariance matrix. Next, when SCADA measurements arrive, the robust SHGM-estimator is utilized to obtain the state estimates and its covariance matrix. Finally, these state estimates are combined with the buffered PMU measurements to yield final state estimation. In what follows, we present the detailed descriptions of the proposed framework.

8.3.1 Statistical Model of Noises

Due to the communication channel noises, GPS synchronization process, changing environment temperature, and different operating conditions of the system, the statistics of the PMU and SCADA measurement noises may be unknown to the control center, and they may not follow Gaussian distributions. There are several models that can be used to model deviations from Gaussian assumption. Among them, the Gaussian sum approach is widely used since any non-Gaussian noise distribution $p(\boldsymbol{x})$ can be expressed as, or approximated sufficiently well by, a finite sum of known Gaussian densities according to the Wiener approximation theorem [35], i.e.

$$p(\boldsymbol{x}) = \sum_{i=1}^{N_A} a_i \mathcal{N}(\overline{\boldsymbol{x}}_i, \Sigma_i) \tag{8.5}$$

where a_i is the weight and $\sum_{i=1}^{N_A} a_i = 1$; N_A is the number of Gaussian components; the mean and covariance matrix associated with the ith Gaussian component are $\overline{\boldsymbol{x}}_i$ and Σ_i, respectively.

Define sets $\mathcal{S}$ and $\mathcal{Q}$ as the Gaussian components with large weights and small weights, respectively. The model in (8.5) can be further rewritten as

$$p(\boldsymbol{x}) = \sum_{i \in \mathcal{S}} a_i \mathcal{N}(\overline{\boldsymbol{x}}_i, \Sigma_i) + \sum_{j \in \mathcal{Q}} a_j \mathcal{N}\left(\overline{\boldsymbol{x}}_j, \Sigma_j\right) \tag{8.6}$$

The functional form of (8.6) can be expressed as

$$G = (1 - \epsilon)\Phi + \epsilon K \tag{8.7}$$

where G is the cumulative probability distribution of $p(\mathbf{x})$ and Φ and K are the functional forms of the terms $\sum_{i \in \mathcal{S}} a_i \mathcal{N}(\overline{\boldsymbol{x}}_i, \Sigma_i)$ and $\sum_{j \in \mathcal{Q}} a_j \mathcal{N}\left(\overline{\boldsymbol{x}}_j, \Sigma_j\right)$, respectively. In practical system, Φ is usually assumed to be Gaussian because according to the central limit theorem, the distribution of a random variable will converge to Gaussian asymptotically. Thus, we anticipate that majority of the data follow roughly Gaussian. K is an unknown distribution, which can be a heavy-tailed density, such as Laplacian density, or Gaussian noise with large unknown variance; $\epsilon\varepsilon \in [0\ \ 0.5)$ regulates the contribution of the non-Gaussian component, e.g. for small ϵ, this model indicates a large fraction of the errors follows Gaussian distribution while maintaining a small fraction of non-Gaussian errors; if $\epsilon = 0$, the noise follows a Gaussian distribution. Note that as long as $\epsilon \neq 0$, G is a non-Gaussian distribution, which is usually the case for practical power systems.

8.3.2 Determining Optimal Buffered PMU Data and Its Robust Covariance Matrix

The static SE typically runs every three to five seconds according to the scan rate of SCADA measurements. In this chapter, we assume the update rate of the static SE is five seconds and the sampling rate of PMU measurements is 30 samples/second. The PMU installed at a bus measures its bus voltage phasor and the current phasors that are adjacent to this bus. During the two SCADA measurement scans, i.e. $N_t =$ 5 seconds, there are $N = 150$ PMU snapshots, yielding a time series measurement matrix $\mathbf{Z}$. If $\mathbf{Z}$ is divided into N_t smaller intervals, we obtain $\mathbf{Z} = [\mathbf{Y}_1\ \ \mathbf{Y}_2\ \ \mathbf{Y}_3\ \ \mathbf{Y}_4\ \ \mathbf{Y}_5]$, where each $\mathbf{Y}_i$ is with $n_t = 30$ PMU snapshots. The reason for such matrix partition is to reduce the synchronization error since the global positioning system radio clock of PMUs sends one pulse synchronization signal per second [36]. Furthermore, the system is assumed to be stationary for a short period [30–32]. In this case, the longer the buffer length of the PMU is, the more efficient the sample mean is since it is the maximum likelihood estimator of the measurement mean with independent, identically distributed (i.i.d) Gaussian noise [30, 31]. However, with more and more intermittent, large- and small-scale renewable-based distributed energy resources (DER) integration to the power grid, the probability of sudden changes in complex bus voltage phasors within a small time frame increases significantly [37]. These abrupt changes are mainly driven by the changes in active power injections (stochastic DER or load). In the meantime, once BD occurs in the PMU measurements, or the PMU measurement noise is not Gaussian, the sample mean and sample covariance matrix would induce large bias. Thus, to determine the optimal PMU data buffer length, we need to detect the system abnormality first. This is done by using the robust Mahalanobis distances.

8.3.2.1 Detect System Abnormality by Proposed Robust Mahalanobis Distances

Given a cloud of l points in κ dimensions denoted by $\mathbf{y}_i$, $i = 1, \ldots, l$, the Mahalanobis distances are calculated through

$$\mathcal{D}_i = \sqrt{(\mathbf{y}_i - \bar{\mathbf{y}})^T \mathbf{C}^{-1} (\mathbf{y}_i - \bar{\mathbf{y}})} \tag{8.8}$$

where $\bar{\mathbf{y}} = \frac{1}{l}\sum_{i=1}^{l} \mathbf{y}_i$ and $\mathbf{C} = \frac{1}{l-1}\sum_{i=1}^{l} (\mathbf{y}_i - \bar{\mathbf{y}})(\mathbf{y}_i - \bar{\mathbf{y}})^T$ are the sample mean and sample covariance matrix, respectively. Equation (8.8) can also be expressed as the solution to the following maximization form:

$$\mathcal{D}_i = \max_{\|\mathbf{v}\|=1} \frac{\left| \mathbf{y}_i^T \mathbf{v} - \frac{1}{l} \sum_{j=1}^{l} \mathbf{y}_j^T \mathbf{v} \right|}{\sqrt{\frac{1}{l-1} \sum_{k=1}^{l} \left(\mathbf{y}_k^T \mathbf{v} - \frac{1}{l} \sum_{j=1}^{l} \mathbf{y}_j^T \mathbf{v} \right)^2}} \tag{8.9}$$

The Mahalanobis distances represent the surface of an κ dimensional ellipsoid centered at the sample mean $\bar{\mathbf{y}}$. It can be shown that they follow a χ^2 distribution with κ

degree of freedom if $\boldsymbol{y}_i$ is a Gaussian random variable. However, it is well known that the sample mean and sample covariance are sensitive to the cluster of outliers [38, 39]. A way to robustify the Mahalanobis distances is to replace the sample mean and covariance by sample median and the median absolute deviation (MAD) from the median, i.e.

$$\mathcal{P}_i = \max_{\|\boldsymbol{v}\|=1} \frac{\left|\boldsymbol{y}_i^T\boldsymbol{v} - \text{median}\left(\boldsymbol{y}_j^T\boldsymbol{v}\right)\right|}{\lambda\cdot\text{median}\left(\left|\boldsymbol{y}_k^T\boldsymbol{v} - \text{median}\left(\boldsymbol{y}_j^T\boldsymbol{v}\right)\right|\right)} \tag{8.10}$$

where $\mathcal{P}_i$ are the robust Mahalanobis distances and λ is a correction factor to ensure unbiasedness for a given distribution and is determined by $\lambda = 1/\Phi^{-1}(0.75)$, where $\Phi^{-1}(\cdot)$ is the inverse of the cumulative distribution function for a given distribution. This is because as the number of samples tends to be very large, the MAD asymptotically tends to the 75th percentile of the given distribution, i.e.

$$\text{MAD}/\sigma \rightarrow \Phi^{-1}(0.75) \text{ as } \kappa \rightarrow \infty \tag{8.11}$$

where σ is the true scale of the given distribution. Thus, to compensate the bias, $\sigma \leftarrow 1/\Phi^{-1}(0.75)\cdot\text{MAD} = \lambda\cdot\text{MAD}$. For example, this value is 1.4826 for Gaussian distribution, 1 for Cauchy distribution with mean 0 and scale 1, 1.4427 for Laplace distribution with mean 0 and scale 1, etc.

Remark: *It should be noted that there are several differences between the proposed robust Mahalanobis distances and the robust Mahalanobis distances (also called projection statistics–PS) in [40]. The aim of [40] is to apply PS to the system sparse Jacobian matrix for identifying leverage outliers, and the weights of the measurements are determined under Gaussian noise assumption. However, the aim of this paper is to modify PS with analytical correction factor to strategically constructed matrix for detecting system abnormalities.*

To detect the abnormalities in the time series PMU data, the following algorithm is proposed:

- Step 1: Calculate the median of time series measurement matrix, yielding $\boldsymbol{Y}' = \text{median}(\mathbf{Z}) = \left[y_1'\ y_2'\ y_3'\ y_4'\ y_5'\right]$, where $\boldsymbol{y}_i'$ is a column vector.
- Step 2: For the latest measurement set $\boldsymbol{Y}_5$, the M-statistics for change-point detection technique [41] is used to detect whether the change of system states occur, i.e. the system is not operating under steady-state conditions. If change is detected, no PMU measurements are buffered, and the recent PMU and SCADA measurements will be used for state estimation; otherwise, $\boldsymbol{Y}_5$ would be included in the buffered PMU measurement set, and go to step 3.
- Step 3: Apply the proposed robust Mahalanobis distances to the matrices, $\eta_1 = \left[y_5'\ y_4'\right]$, $\eta_2 = \left[y_4'\ y_3'\right], \ldots, \eta_4 = \left[y_2'\ y_1'\right]$ for detecting system changes. If all the calculated $\mathcal{P}_i$ values for $\boldsymbol{\eta}_1$ satisfy $\mathcal{P}_i < \xi$, where ξ is the detection

threshold and will be discussed later, no system change is declared, and $\boldsymbol{Y}_4$ will be included into the PMU measurement buffering set; otherwise, only $\boldsymbol{Y}_5$ is buffered, and the algorithm is terminated. Continue to test $\eta_2,\ldots,\eta_4$ and finally end up with the PMU measurement buffering set, $\boldsymbol{Z}'$. It should be noted that the median is used in the test to determine the optimal PMU buffering length because the median is very robust to outliers so that the detected abnormality is guaranteed to be system change.

- Step 4: After determining the optimal PMU buffering length, the next process is to determine the statistical information of the buffered PMU measurements, i.e. weighted mean and $\overline{\boldsymbol{h}}$ its associated covariance matrix $\overline{\boldsymbol{C}}$. Since outliers may still exist in the buffered PMU measurements, the robust estimation is required. This is done by applying proposed robust Mahalanobis distances to $\boldsymbol{Z}'$, yielding

$$\overline{\boldsymbol{h}} = \frac{\sum_{i=1}^{\alpha} \omega_i \boldsymbol{h}_i}{\sum_{i=1}^{\alpha} \omega_i} \tag{8.12}$$

$$\overline{\boldsymbol{C}} = \frac{\sum_{i=1}^{\alpha} \left(\omega_i \boldsymbol{h}_i - \overline{\boldsymbol{h}}\right)\left(\omega_i \boldsymbol{h}_i - \overline{\boldsymbol{h}}\right)^T}{\sum_{i=1}^{\alpha} \omega_i - 1} \tag{8.13}$$

where ω_i are weights and determined by

$$\omega_i = \min\left[1, \ \frac{\xi}{\mathcal{P}_i^2}\right] \tag{8.14}$$

where ξ is the cutoff value of the proposed robust Mahalanobis distances.

Under Gaussian noise, the robust Mahalanobis distances asymptotically approximate to the conventional Mahalanobis distances that follow a χ^2 distribution [42]. The cutoff value for the $\mathcal{P}_i$ values in this condition can be set as similar to the conventional Mahalanobis distance-based outliers detection, i.e. $\mathcal{P}_i > \chi^2_{\nu,\beta}$, where ν is the degrees of freedom and β is the confidence level (for example, $\beta = 0.975$). However, the distribution of the robust Mahalanobis distances with non-Gaussian noise is not clear.

In this chapter, we use extensive Monte Carlo simulations and QQ-plots to investigate the probability distribution of robust Mahalanobis distances under non-Gaussian measurement noise using the model in (8.7). To conduct the test, consider two random variables $\boldsymbol{v}_1$ and $\boldsymbol{v}_2$ that are independent and identically distributed according to a given non-Gaussian distribution, i.e. Gaussian mixture distribution with two components, i.e. mean 0 and variance 0.01 with weight 0.85 and mean 0 and variance 0.1 with weight 0.15. We generate 1000 realizations of these two random variables to get 1000 two-dimensional data points, $[\boldsymbol{v}_1\ \boldsymbol{v}_2]$. All the experiments are repeated for 100 times. Finally, the medians and the interquartile ranges of the empirical robust Mahalanobis distances quartiles are plotted versus the corresponding quartiles of the chi-square distribution with the appropriate degree of freedom, $\nu = 2$ in this case. The results reveal that the robust Mahalanobis distances follow a χ^2 distribution with 2 degree of freedom. As a result, the cutoff value of the proposed robust Mahalanobis distances is determined as $\xi = \chi^2_{\alpha,\beta}$, where α is the number of columns of the vector $\boldsymbol{Z}'$.

8.3.3 Proposed Robust Nonlinear State Estimation using SCADA Measurements

Once SCADA measurements are received, a robust nonlinear state estimation is performed to filter out unknown non-Gaussian measurement noise and suppress bad SCADA measurements. It aims to minimize the following objective function [43, 44]:

$$J = \sum_{i=1}^{m} \varpi_i^2 \rho(r_{S_i}) \tag{8.15}$$

where ϖ_i is the weight used to bound the influence of leverage point and is calculated by applying the projection statistics [40] to the Jacobian matrix $\boldsymbol{H} = \partial \boldsymbol{h} / \partial \boldsymbol{x}$ and $\rho(\cdot)$ is the Huber convex cost function defined as

$$\rho(r_{S_i}) = \begin{cases} \frac{1}{2} r_{S_i}^2, & \text{for } |r_{S_i}| < c \\ c|r_{S_i}| - c^2/2, & \text{elesewhere} \end{cases} \tag{8.16}$$

where $r_{S_i} = r_i / \sigma_i \varpi_i$ is the standardized residual, $r_i = z_i - h_i(x)$ is the residual associated with the ith measurement, and σ_i is the known standard deviation of the ith measurement. Note that due to the uncertainty of the measurement devices and the aging process of the meter devices, σ_i may not be constant or can be unknown. On the other hand, according to the processes of calculating real reactive power injections and power flows using PT and CT signals in [45], we can conclude that the accuracy of the SCADA measurements is in the same order. Therefore, in this chapter, we propose to use a robust scale estimation to approximate σ_i, which distinguishes with the existing GM-estimators [3, 40, 46, 47]. A common scale estimation in robust statistics is the MAD as it has fast computation efficiency and good robustness. Its definition is

$$s = 1.4826 \cdot b_m \cdot \text{median}_i \left| r_i - \text{median}_j (r_j) \right| \tag{8.17}$$

where b_m is a correction factor for unbiasedness at a given probability distribution. It should be noted that s^2 will asymptotically tend to the true measurement variance σ^2 if the number of samples for estimation tends to infinity [42]. However, MAD is aimed at symmetric distributions, and it has low (37%) statistical efficiency under Gaussian noise. To improve its statistical efficiency, we propose to use the following robust scale estimation [42]:

$$s = 1.1926 \cdot f_m \cdot \underset{i=1,\ldots,m}{\text{lomed}} \; \underset{j \neq i}{\text{lomed}} \left| r_i - r_j \right| \tag{8.18}$$

where f_m is a correction factor; the outer lomed operator is a low median (that is, the $[(m+1)/2]$-th order statistic out of m numbers), and the inner lomed is a high median (the $([m/2]+1)$-th order statistic out of m numbers); [] is the integer operator; this scale estimator has better finite-sample efficiencies than the MAD even at very asymmetric non-Gaussian distributions while maintaining good robustness.

To minimize (8.15), one takes its partial derivative and sets it equal to zero, yielding

$$\frac{\partial J}{\partial \boldsymbol{x}} = \sum_{i=1}^{m} - \frac{\varpi_i \boldsymbol{a}_i}{s_i^2} \psi(r_{S_i}) = \boldsymbol{0} \tag{8.19}$$

where $\psi(r_{S_i}) = \partial \rho(r_{S_i}) / \partial r_{S_i}$ and $\boldsymbol{a}_i$ is the ith row of the Jacobian matrix $\boldsymbol{H}$. By dividing and multiplying standardized residual to both sides of (8.19), we obtain the following matrix form:

$$\boldsymbol{H}^T \boldsymbol{Q} \hat{\boldsymbol{R}}^{-1} (\boldsymbol{z} - \boldsymbol{h}(\boldsymbol{x})) = \boldsymbol{0} \tag{8.20}$$

where $\boldsymbol{Q} = \text{diag}(q(r_{S_i}))$, $q(r_{S_i}) = \psi(r_{S_i}) / r_{S_i}$, and $\hat{\boldsymbol{R}} = \text{diag}(s_i^2)$. Using the iteratively reweighted least squares (IRLS) algorithm [40] and the first-order Taylor series expansion, the state correction at the k iteration is

$$\left(\boldsymbol{H}^T \boldsymbol{Q}_k \hat{\boldsymbol{R}}^{-1} \boldsymbol{H} \right) \Delta \boldsymbol{x}_k = \boldsymbol{H}^T \boldsymbol{Q}_k \hat{\boldsymbol{R}}^{-1} (\boldsymbol{z} - \boldsymbol{h}(\boldsymbol{x}_k)) \tag{8.21}$$

with the convergence condition

$$\|\Delta \boldsymbol{x}_k\|_\infty = \|\boldsymbol{x}_{k+1} - \boldsymbol{x}_k\|_\infty \leq 10^{-2} \tag{8.22}$$

8.3.3.1 Asymptomatic Normality of Estimation Error Covariance Matrix

After the convergence of the IRLS algorithm, the estimation error covariance matrix needs to be calculated so that the state fusion that simultaneously processes the state estimates by SCADA measurements and the buffered PMU measurements can be performed. Consider the ϵ-contamination model $G = (1 - \epsilon)\Phi + \epsilon\Delta_r$, where G and Φ are the contaminated and the true cumulative probability distribution function of the residual, respectively, and Δ_r is the point mass to model outliers. Inspired by Fernholz [48] and Hampel et al. [49], the error covariance matrix is updated based on the following theorem:

Theorem 8.1 *Let $T(\cdot)$ be the functional form of the GM-estimator with a bounded $\psi(\cdot)$ function and Φ_m be the empirical cumulative probability distribution function, then*

$$\sqrt{m}(\boldsymbol{T}(\Phi_m) - \boldsymbol{T}(\Phi)) \xrightarrow{d} \mathcal{N}(\boldsymbol{0}, \Sigma) \tag{8.23}$$

where $\xrightarrow{d}$ *means convergence in probability distribution;* $\Sigma = \mathbb{E}[\boldsymbol{IF}(\boldsymbol{x};\Phi,\mathbf{T})\cdot \boldsymbol{IF}(\boldsymbol{x};\Phi,\boldsymbol{T})^T]$ *with the total influence function* $\boldsymbol{IF}(\boldsymbol{x};\ \Phi,\ \boldsymbol{T})$ *evaluated at* Φ.

Proof. By taking a von Mises expansion of the functional form of the estimator $\boldsymbol{T}$ with respect to Φ, we get

$$\boldsymbol{T}(\Phi_m) = \boldsymbol{T}(\Phi) + \boldsymbol{T}(\Phi_m - \Phi) + \text{Rem}(\Phi_m - \Phi) \tag{8.24}$$

which can be reorganized into the following form by multiplying $\sqrt{m}$ on both sides of the equality:

$$\sqrt{m}(\boldsymbol{T}(\Phi_m) - \boldsymbol{T}(\Phi)) = \sqrt{m}\boldsymbol{T}'(\Phi_m - \Phi) + \sqrt{m}\text{Rem}(\Phi_m - \Phi) \tag{8.25}$$

$$= \sqrt{m}\int \boldsymbol{IF}(\boldsymbol{x};\Phi,\boldsymbol{T})\text{d}(\Phi_m - \Phi) + \sqrt{m}\text{Rem}(\Phi_m - \Phi) \tag{8.26}$$

$$= \sqrt{m}\int \boldsymbol{IF}(\boldsymbol{x};\Phi,\boldsymbol{T})\text{d}\Phi_m + \sqrt{m}\text{Rem}(\Phi_m - \Phi) \tag{8.27}$$

$$= \frac{1}{\sqrt{m}}\sum_{i=1}^{m} \boldsymbol{IF}(x_i;\Phi,\boldsymbol{T}) + \sqrt{m}\text{Rem}(\Phi_m - \Phi) \tag{8.28}$$

where the definition of the influence function is applied to yield (8.25) and (8.26). By virtues of Fisher consistency at the distribution Φ, that is, $\int \boldsymbol{IF}(\boldsymbol{x};\ \Phi,\ \boldsymbol{T})\text{d}\Phi = \boldsymbol{0}$, (8.26) reduces to (8.27). Finally, by using the property of the empirical cumulative probability distribution function, we have

$$\int \boldsymbol{IF}(\boldsymbol{x};\Phi,\boldsymbol{T})\text{d}\Phi_m = \frac{1}{m}\sum_{i=1}^{m} \boldsymbol{IF}(x_i;\Phi,\boldsymbol{T}) \tag{8.29}$$

yielding (8.27) and (8.28).

Following the work of Fernholz [48], we can show that

$$\sqrt{m}\text{Rem}(\Phi_m - \Phi) \xrightarrow{p} 0 \tag{8.30}$$

where $\xrightarrow{p}$ means probability convergence. Therefore, by applying the central limit theorem and Slutsky's lemma to (8.28), it follows that

$$\sqrt{m}(\boldsymbol{T}(\Phi_m) - \boldsymbol{T}(\Phi)) \xrightarrow{d} \mathcal{N}(\boldsymbol{0},\Sigma) \tag{8.31}$$

where $\Sigma = \mathbb{E}\left[\boldsymbol{IF}(\boldsymbol{x};\Phi,\boldsymbol{T})\cdot\boldsymbol{IF}(\boldsymbol{x};\Phi,\boldsymbol{T})^T\right]$. □

8.3.3.2 Total Influence Function of the GM-estimator

Let the cumulative probability distribution of the residual vector $\boldsymbol{r} = \boldsymbol{z} - \boldsymbol{h}(\boldsymbol{x})$ be $\Phi(\boldsymbol{r})$. The GM-estimator provides an estimate of the state by processing the redundant observation vector $\boldsymbol{z}$ and solving the following implicit equation [50]:

$$\sum_{i=1}^{m} \varpi_i \frac{\partial \boldsymbol{h}(\boldsymbol{x})}{\partial \boldsymbol{x}} \psi(r_{S_i}) = \mathbf{0} \tag{8.32}$$

which, by virtue of the Glivenko–Cantelli theorem, asymptotically tends to

$$\int \lambda(\boldsymbol{r}, \boldsymbol{T})\mathrm{d}G = \mathbf{0} \tag{8.33}$$

Here, the GM-estimator $\hat{\boldsymbol{x}}$ at G has been put in a functional form, $\boldsymbol{T}(G)$. The asymptotic total influence function of $\boldsymbol{T}(G)$ is given by

$$\boldsymbol{IF}(\boldsymbol{r}, \Phi) = \left.\frac{\partial \boldsymbol{T}(G)}{\partial \epsilon}\right|_{\epsilon = 0} = \lim_{\Delta\epsilon \to 0} \frac{\boldsymbol{T}((1-\epsilon)\Phi + \epsilon\Delta r) - \boldsymbol{T}(\Phi)}{\epsilon} \tag{8.34}$$

Substituting G into (8.33) yields

$$\int \lambda(\boldsymbol{r}, \boldsymbol{T}(G))\mathrm{d}\Phi + \epsilon \int \lambda(\boldsymbol{r}, \boldsymbol{T}(G))\mathrm{d}(\Delta_r - \Phi) = \mathbf{0} \tag{8.35}$$

Taking the differentiation with respect to ϵ and evaluating it at $\epsilon = 0$, assuming regularity conditions and Fisher consistency at Φ, given by $\int \lambda(\boldsymbol{r}, \boldsymbol{T}(\Phi))\mathrm{d}\Phi = \mathbf{0}$, yields

$$\frac{\partial}{\partial \varepsilon} \int \lambda(\boldsymbol{r}, \boldsymbol{T}(G))\mathrm{d}\Phi|_{\epsilon = 0} + \left.\int \lambda(\boldsymbol{r}, \boldsymbol{T}(G))\mathrm{d}(\Delta_r)\right|_{\epsilon = 0} = \mathbf{0} \tag{8.36}$$

After applying the sifting property of the Dirac impulse to the second term, we obtain

$$\frac{\partial}{\partial \epsilon} \int \lambda(\boldsymbol{r}, \boldsymbol{T}(G))\mathrm{d}\Phi|_{\epsilon = 0} + \lambda(\boldsymbol{r}, \boldsymbol{T}(\Phi)) = \mathbf{0} \tag{8.37}$$

Assuming $\lambda(\cdot)$ is continuous and measurable and $\lambda'(\cdot)$ is measurable, we can apply the interchangeability of differentiation and integration theorem to the first term in (8.37), yielding

$$\int \left.\frac{\partial \lambda(\boldsymbol{r}, \boldsymbol{T}(G))}{\partial \epsilon}\right|_{\boldsymbol{T}(\Phi)} \cdot \left.\frac{\partial \boldsymbol{T}(G)}{\partial \epsilon}\right|_{\epsilon = 0} \mathrm{d}\Phi + \lambda(\boldsymbol{r}, \boldsymbol{T}(\Phi)) = \mathbf{0} \tag{8.38}$$

Thus, $\boldsymbol{IF}(\boldsymbol{r}, \Phi)$ is expressed as

$$\begin{aligned} \boldsymbol{IF}(\boldsymbol{r}, \Phi) &= \left.\frac{\partial \boldsymbol{T}(G)}{\partial \epsilon}\right|_{\epsilon = 0} \\ &= -\left[\int \left.\frac{\partial \lambda(\boldsymbol{r}, \boldsymbol{T}(G))}{\partial \epsilon}\right|_{\boldsymbol{T}(\Phi)} \mathrm{d}\Phi\right]^{-1} \lambda(\boldsymbol{r}, \boldsymbol{T}(\Phi)) \end{aligned} \tag{8.39}$$

Taking the derivative of $\lambda(\cdot)$ with respect to $\boldsymbol{x}$ and assuming that ϖ and s are independent of $\boldsymbol{x}$, we obtain

$$\frac{\partial \lambda(\boldsymbol{r})}{\partial \boldsymbol{x}} = \varpi\psi(r_{S_i})\left[\frac{\partial^2 \boldsymbol{h}(\boldsymbol{x})}{\partial x_i \partial x_j}\right] + h\left[\frac{\partial \psi(r_{S_i})}{\partial \boldsymbol{x}}\right]\left[\frac{\partial \boldsymbol{h}(\boldsymbol{x})}{\partial \boldsymbol{x}}\right]^T \tag{8.40}$$

By neglecting the second-order term on the right-hand side of (8.40), we get

$$\frac{\partial \lambda(\boldsymbol{r})}{\partial \boldsymbol{x}} = \varpi\left[\frac{\partial \psi(r_{S_i})}{\partial \boldsymbol{x}}\right]\left[\frac{\partial \boldsymbol{h}(\boldsymbol{x})}{\partial \boldsymbol{x}}\right]^T \tag{8.41}$$

Applying the chain rule to the derivative of $\psi(r_{S_i})$ yields

$$\frac{\partial \lambda(\boldsymbol{r})}{\partial \boldsymbol{x}} = -\frac{1}{s}\psi'(r_{S_i})\left[\frac{\partial \boldsymbol{h}(\boldsymbol{x})}{\partial \boldsymbol{x}}\right]\left[\frac{\partial \boldsymbol{h}(\boldsymbol{x})}{\partial \boldsymbol{x}}\right]^T \tag{8.42}$$

By substituting (8.32) and (8.42) into (8.39), we get

$$\boldsymbol{IF}(\boldsymbol{r}, \Phi) = \frac{\psi(r_S)}{(\mathbb{E}_\Phi[\psi'(r_s)])^2}\left(\boldsymbol{H}^T\boldsymbol{H}\right)^{-1}\boldsymbol{a}\varpi \tag{8.43}$$

Therefore, the asymptotic covariance matrix Σ is [50]

$$\begin{aligned}\Sigma &= \mathbb{E}\left[\boldsymbol{IF}\cdot\boldsymbol{IF}^T\right] \\ &= \frac{\mathbb{E}_\Phi[\psi^2(r_S)]}{\{\mathbb{E}_\Phi[\psi'(r_S)]\}^2}\left(\boldsymbol{H}^T\boldsymbol{H}\right)^{-1}\left(\boldsymbol{H}^T\boldsymbol{Q}_\varpi\boldsymbol{H}\right)\left(\boldsymbol{H}^T\boldsymbol{H}\right)^{-1}\end{aligned} \tag{8.44}$$

where $\boldsymbol{H}$ is the Jacobian matrix; $\boldsymbol{Q}_\varpi = \text{diag}\left(\varpi_i^2\right)$.

Remark: *It is shown by Theorem 8.1 that the state estimates provided by the SHGM-estimator follow a Gaussian distribution with mean $\hat{\boldsymbol{x}}$ and covariance matrix Σ. Thanks to this nice asymptotical normality, the state estimates using SCADA measurements can be easily integrated with the buffered PMU measurements for the final state estimation.*

8.3.4 State Fusion: Robust Linear State Estimation Incorporating Buffered PMU Measurements

In Section 8.3.2.1, the statistical information of the buffered PMU measurements is represented by its weighted mean $\overline{\boldsymbol{h}}$ and the robust covariance matrix $\overline{\boldsymbol{C}}$. In this section, they are augmented with the results from the SCADA measurements-based state estimates $\hat{\boldsymbol{x}}$ for the final state estimation. Formally, we have

$$z_{\text{aug}} = \boldsymbol{A}\boldsymbol{x} + \varepsilon \tag{8.45}$$

where $z_{\text{aug}} = \left[\hat{\boldsymbol{x}}\ \overline{\boldsymbol{h}}\right]^T$ is the augmented measurement vector, $\boldsymbol{A} = [\boldsymbol{I}\ \ \boldsymbol{M}]^T$ is a constant matrix, $\boldsymbol{I}$ is an identity matrix, $\boldsymbol{M}$ is a constant matrix that represents the linear relationship between the PMU measurements and the state vector $\boldsymbol{x}$, and ε is the error vector with zero mean and covariance matrix

$$\boldsymbol{P} = \mathbb{E}\left[\varepsilon\varepsilon^{T}\right] = \begin{bmatrix} \Sigma & \boldsymbol{0} \\ \boldsymbol{0} & \overline{\boldsymbol{C}} \end{bmatrix} = \boldsymbol{S}\boldsymbol{S}^{T} \tag{8.46}$$

where $\boldsymbol{S}$ is calculated through the Cholesky decomposition. It should be noted that after the SCADA measurements-based state estimation, the error covariance matrix Σ is no longer a diagonal matrix due to the nonlinear measurement function; furthermore, as demonstrated in [31, 32] that the time series PMU measurements have both temporal and spatial correlations, then the covariance matrix of the buffered PMU measurements is not a diagonal matrix as well. As a result, (8.45) is not the widely used linear regression model. In fact, it is a generalized regression model with the consideration of measurement correlations. To uncorrelate the error vector and obtain the simple linear regression model, we multiply $\boldsymbol{S}^{-1}$ to both sides of (8.45), yielding

$$\boldsymbol{y}_{\text{aug}} = \boldsymbol{G}\boldsymbol{x} + \boldsymbol{\xi} \tag{8.47}$$

where $\mathbb{E}\left[\boldsymbol{\xi}\boldsymbol{\xi}^{T}\right] = \boldsymbol{I}$.

It should be noted that if the buffered PMU measurements do not follow a Gaussian distribution, its weighted mean $\overline{\boldsymbol{h}}$ will not be Gaussian. Instead, it will follow an unknown thick-tailed distribution with majority of the data clustered that has a covariance matrix $\overline{\boldsymbol{C}}$. As a result, $\boldsymbol{\xi}$ does not follow a Gaussian distribution but with its majority of the data associated with a Gaussian distribution, which can be described by the model shown in (8.7). Therefore, the WLS estimator cannot be applied. To filter out non-Gaussian noise and suppress BD in PMU measurements, the SHGM-estimator is used again to obtain the final state estimates.

8.3.5 Implementation Issues

The PS values can be calculated off-line from the Jacobian matrix evaluated at the flat voltage profile, and they do not need to be updated if no measurement configuration or topology changes occur [3, 40]. However, flat voltage start is not recommended for the initialization of the GM-estimator since most of the measurement residuals will be greater than the threshold of the Huber function and will be strongly downweighted at the first iteration. Although good measurements will be re-assigned with weights 1 and bad measurements remained downweighted after several iterations, it is not necessary to increase the number of iterations of the algorithm. Therefore, it is preferred to perform flat voltage start for PS calculation, while utilizing the WLS for the first iteration and then switch to the IRLS algorithm. To obtain linear relationship between PMU measurements and system states, the rectangular coordinate is used for the voltage and current phasors.

8.4 NUMERICAL RESULTS

In this section, the proposed method is tested on the IEEE 30-bus test system with various types of BD and non-Gaussian measurement noise. The measurement configuration of the test system is the same as that in [51], where 38 pairs of power measurements including 15 pairs of real and reactive power injections and 23 pairs of real and reactive power flow measurements are considered; the real and reactive power injections on buses 11, 12, 24, 27, and 30 and the real and reactive power flows 24–23, 25–26, and 30–27 are critical measurements without considering the PMU measurements; 6 PMUs are deployed on buses 8, 9, 12, 24, 25, and 26 to increase the redundancy of the test system and make some critical measurements non-critical. The PMUs can measure the bus voltage magnitudes and phase angles as well as the current phasors in all branches that are adjacent to the bus. In the simulation, the parameters for the proposed robust estimator are set as $c = 1.5$, the maximal iteration is 20 and the convergence threshold is 10^{-2}; the non-robust two-stage estimator using both SCADA and PMU measurements [51] termed as two-stage WLS is used to make comparisons. The absolute estimation error representing the difference between the estimated state and the true state is used to assess the performance of the two estimators.

8.4.1 Gaussian Measurement Noise

In this scenario, Gaussian random variables with means 0 and variances 10^{-4} and 10^{-6} are assumed for the errors of SCADA and PMU measurements, respectively. Figures 8.2 and 8.3 show the comparison results of the two-stage WLS estimator and the proposed robust estimator. It can be found that most of the estimation results obtained by the proposed method is better than the two-stage WLS estimator. But, in general, the state estimation errors of both two estimators are very small.

8.4.2 Non-Gaussian Measurement Noise

Two types of non-Gaussian measurement noise are considered and tested in this section. The Gaussian mixture model with two Gaussian components, which are represented by zero means and variances 10^{-6} and 10^{-4} with weights 0.9 and 0.1, respectively, are assumed. Note that the sum of these two Gaussian components results in non-Gaussian measurement noise. In the second scenario, the Laplace distribution with zero mean and scale parameter 10^{-3} is used for the noises of both SCADA and PMU measurements.

Figures 8.4 and 8.5 present the test results of two SEs with mixture model-based non-Gaussian measurement noises. While the results with Laplace measurement noises are displayed in Figures 8.6 and 8.7. Compared with the results obtained under Gaussian assumption, the estimation errors of the two-stage WLS estimator increase significantly in the presence of non-Gaussian

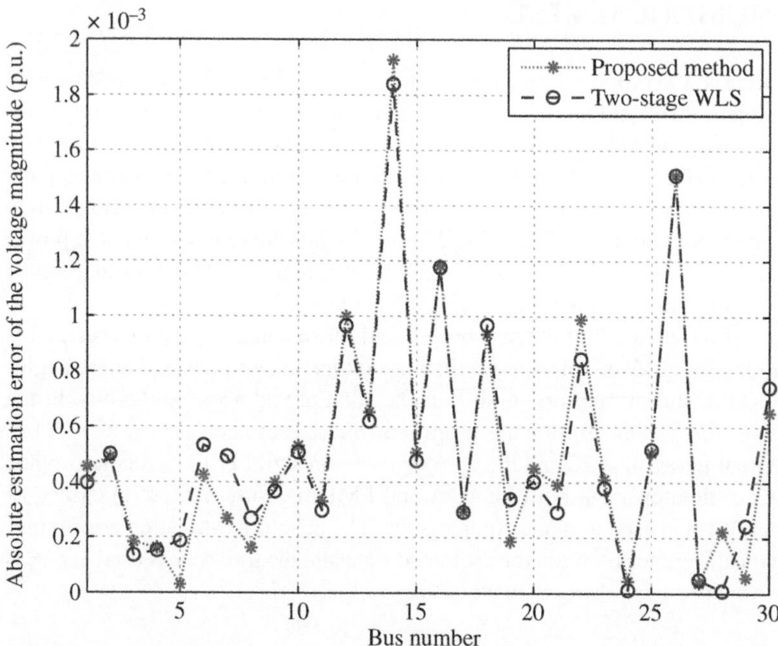

Figure 8.2 The absolute estimation error of the bus voltage magnitude with Gaussian measurement noise in IEEE 30-bus system.

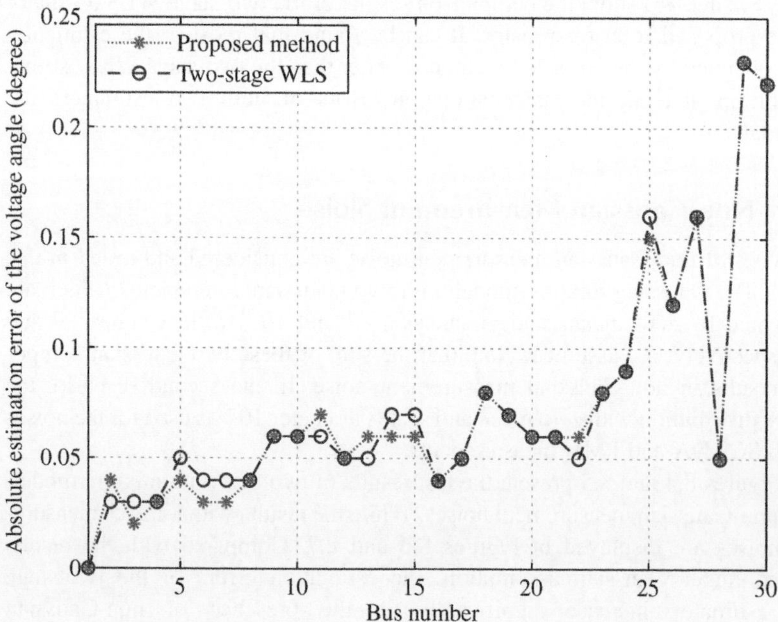

Figure 8.3 The absolute estimation error of the bus voltage angles with Gaussian measurement noise in IEEE 30-bus system.

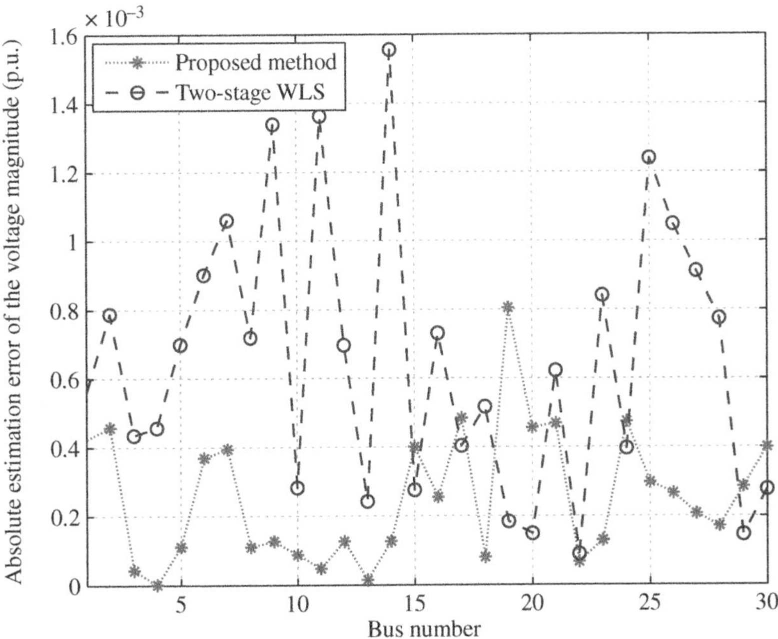

Figure 8.4 The absolute estimation error of the bus voltage magnitude with Gaussian measurement noise simulated by Gaussian mixture model in IEEE 30-bus system.

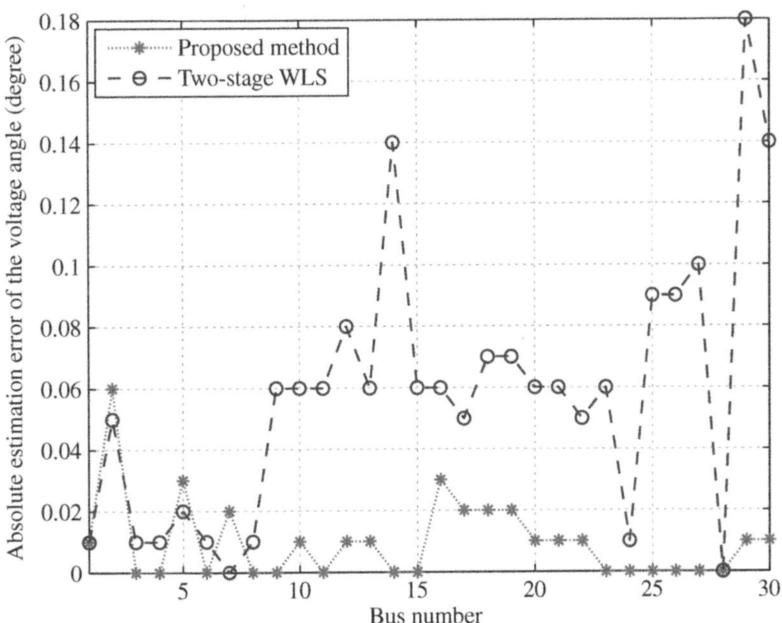

Figure 8.5 The absolute estimation error of the bus voltage angles with non-Gaussian measurement noise simulated by Gaussian mixture model in IEEE 30-bus system.

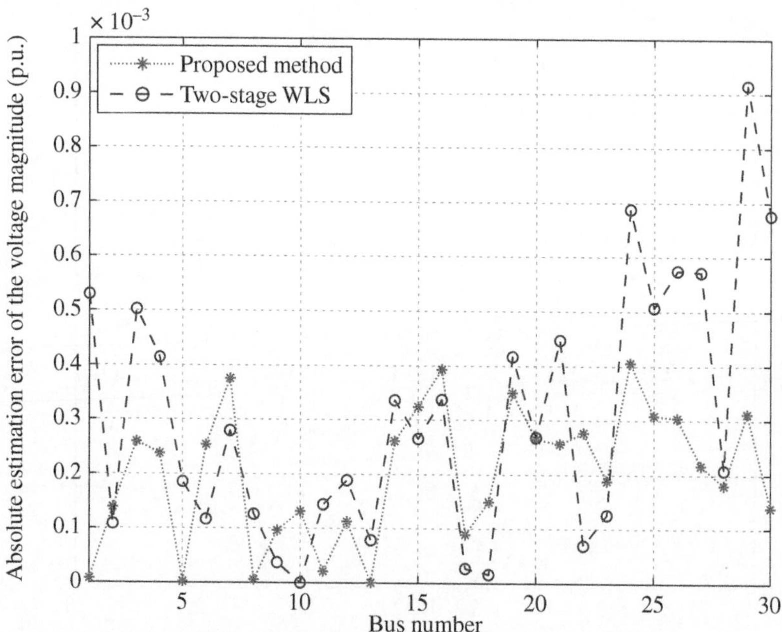

Figure 8.6 The absolute estimation error of the bus voltage magnitude with Laplace measurement noise in IEEE 30-bus system.

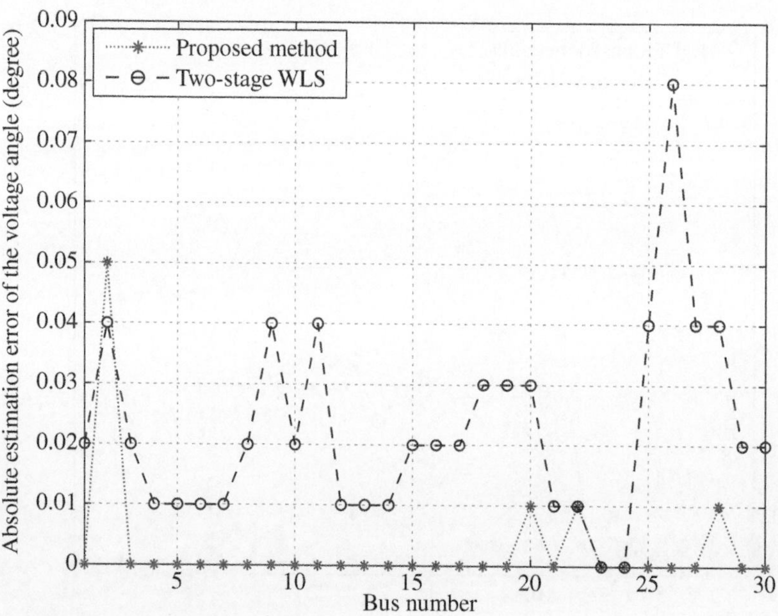

Figure 8.7 The absolute estimation error of the bus voltage angles with Laplace measurement noise in IEEE 30-bus system.

measurement noises. By contrast, the proposed GM-estimator is able to filter out non-Gaussian noise while maintaining a good estimation statistical efficiency.

8.4.3 Robustness to Various Types of Bad Data

To show the robustness of the proposed estimator to vertical BD, bad leverage points, erroneous zero injection, and the bad critical measurements, the following cases are considered:

- Case 1: Single BD, where real power injection P_2 is changed to 1.
- Case 2: Bad leverage points, where P_{19} and P_{19-20} are both changed to 0.1.
- Case 3: Bad critical measurement, where P_{11} is changed to 0.2.
- Case 4: Bad zero injection, where P_6 is changed to 0.2.

Figure 8.8 shows the simulation results with different types of BD. It is clear that, thanks to the robustness of projection statistics and the Huber M-estimator, the influences of various types of BD are bounded, yielding very small estimation biases. Please note that bad critical measurement cannot be identified by any BD processing method if no redundant measurements are included. In Case 3, with the inclusion of additional PMU measurements, the local redundancy of P_{11} is increased. Then, the SHGM-estimator is able to bound its influence on the state estimates.

8.5 CONCLUSIONS

In this chapter, we resort to the robust statistical theory and propose a robust SE framework to tackle the non-Gaussian measurement noise and measurement skewness issue. In the framework, the SHGM-estimator is advocated to filter non-Gaussian SCADA measurement noise and suppress BD. We show that the state estimates provided by the SHGM-estimator follow roughly a Gaussian distribution. This allows us to effectively combine it with the buffered PMU measurements for final state estimation. Specifically, the robust Mahalanobis distances are proposed to detect BD and assign appropriate weights to buffered PMU measurements. Those weights are further utilized by the SHGM-estimator to suppress outliers and filter out unknown non-Gaussian measurement noise. Extensive simulation results carried out on the IEEE 30-bus test system demonstrate the effectiveness and robustness of the proposed method.

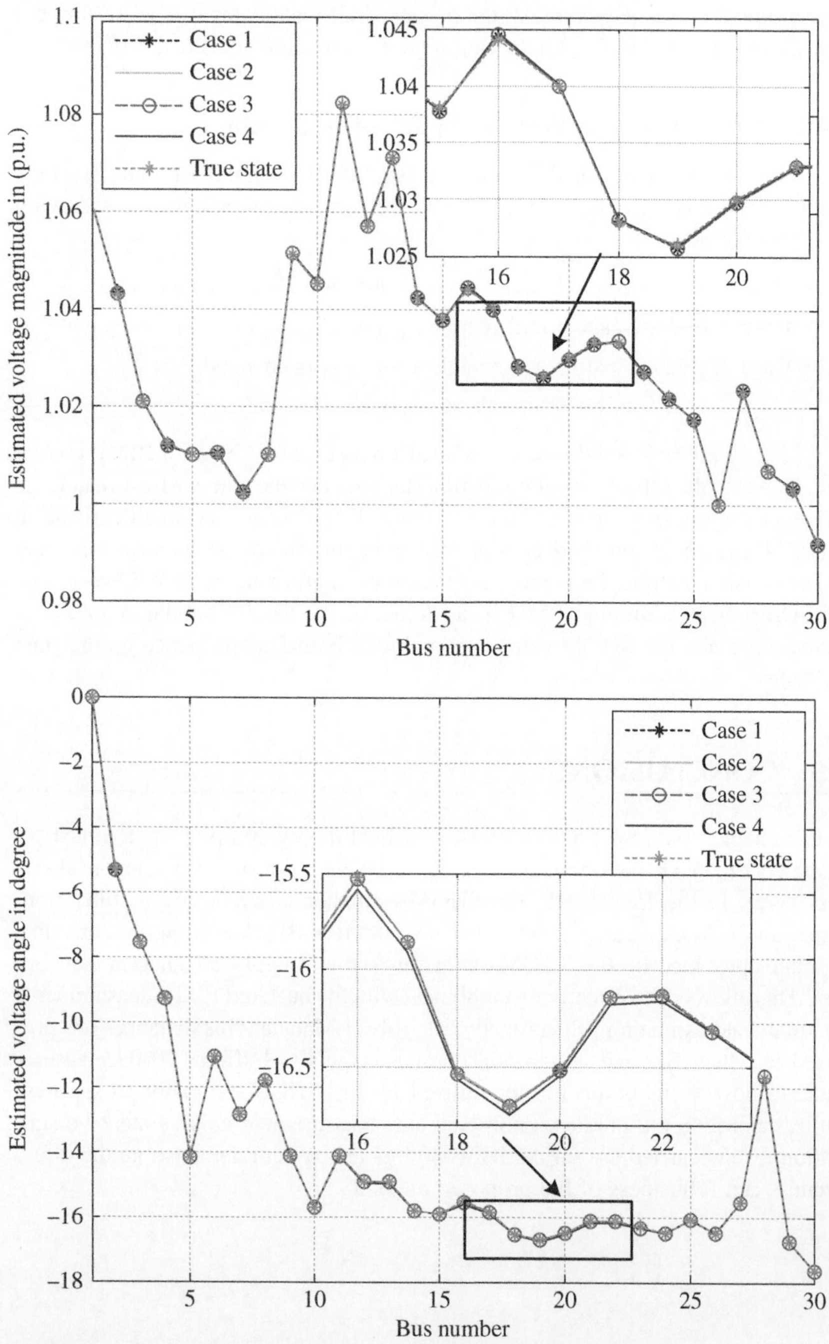

Figure 8.8 The estimated states of the proposed method under different bad data conditions.

REFERENCES

1. Huang, Y.F., Werner, S.F., Huang, J. et al. (2012). State estimation in electric power grids: meeting new challenges presented by the requirements of the future grid. *IEEE Signal Process. Mag.* 29 (5): 33–43.
2. Abur, A. and Gómez-Expósito, A. (2004). *Power System State Estimation-Theory and Implementation*. New York: Marcel Dekker.
3. Pires, R.C., Mili, L., and Lemos, F. (May 2014). Constrained robust estimation of power system state variables and transformer tap positions under erroneous zero-injections. *IEEE Trans. Power Syst.* 29 (3): 1144–1152.
4. Phadke, A.G., Thorp, J.S., Nuqui, R.F., and Zhou, M. (2009). Recent developments in state estimation with phasor measurements. *Proceedings of the IEEE/PES Power Systems Conference and Exposition* (March 2009), Seattle, WA.
5. Phadke, A.G. and Thorp, J.S. (2008). *Synchronized Phasor Measurements and Their Applications*. Springer.
6. Gómez-Expósito, A., Abur, A., de la Villa Jaén, A., and Gómez-Quiles, C. (Jun. 2011). A multilevel state estimation paradigm for smart grids. *Proc. IEEE* 99 (6): 952–976.
7. Korres, G.N. and Manousakis, N.M. (Jul. 2011). State estimation and bad data processing for systems including PMU and SCADA measurements. *Electr. Power Syst. Res.* 81 (7): 1514–1524.
8. Chakrabarti, S., Kyriakides, E., Ledwich, G., and Ghosh, A. (Otc. 2010). Inclusion of PMU current phasor measurements in a power system state estimator. *IET Gen. Transm. Distrib.* 4 (10): 1104–1115.
9. Bruno, C., Candia, C., Franchi, L. et al. (2009). Possibility of enhancing classical weighted least squares state estimation with linear PMU measurements. *IEEE PowerTech*, Bucharest (2009), pp. 1–6.
10. Ga, Z.H. (2006). A new state estimation model of utilizing PMU measurements. *International Conference on Power System Technology*, Chongqing, China (2006), pp. 1–5.
11. Chen, F., Han, X., Pan, Z., and Han, L. (2008). State estimation model and algorithm including PMU. *Third International Conference on Electric Utility Deregulation and Restructuring and Power Technologies*, Nanjing, China (April 2008), pp. 1097–1102.
12. Xue, H., Jia, Q., Wang, N. et al. (2007). A dynamic state estimation method with PMU and SCADA measurement for power systems. *International Power Engineering Conference*, Singapore (2007), pp. 848–853.
13. Jain, A. and Shivakumar, N.R. (2008). Impact of PMU in dynamic state estimation of power systems. *40th North American Power Symposium*, Calgary, AB, Canada (2008), pp. 1–8.
14. Korres, G.N. and Manousakis, N.M. (Sep. 2012). State estimation and observability analysis for phasor measurement unit measured systems. *IET Gen. Transm. Distrib.* 6 (9): 902–913.
15. Zhao, L. and Abur, A. (May 2005). Multi-area state estimation using synchronized phasor measurements. *IEEE Trans. Power Syst.* 20 (2): 611–617.
16. Zhou, M., Centeno, V.A., Thorp, J.S., and Phadke, A.G. (Nov. 2006). An alternative for including phasor measurements in state estimators. *IEEE Trans. Power Syst.* 21 (4): 1930–1937.
17. Vanfretti, L., Chow, J.H., Sarawgi, S., and Fardanesh, B. (Feb. 2011). A phasor data based state estimation incorporating phase bias correction. *IEEE Trans. Power Syst.* 26 (1): 111–119.

18. Chen, J. and Abur, A. (Nov. 2006). Placement of PMUs to enable bad data detection in state estimation. *IEEE Trans. Power Syst.* 21 (4): 1608–1615.
19. Gou, B. and Kavasseri, R.G. (Nov. 2014). Unified PMU placement for observability and bad data detection in state estimation. *IEEE Trans. Power Syst.* 29 (6): 2573–2580.
20. Do Coutto Filho, M.B., Stacchini de Souza, J.C., and Guimaraens, R. (2014). Enhanced bad data processing by phasor-aided state estimation. *IEEE Trans. Power Syst.* 29 (5): 1–10.
21. Zhao, J.B., Zhang, G.X., Das, K. et al. (2016). Power system real-time monitoring by using PMU-based robust state estimation method. *IEEE Trans. Smart Grid* 7 (1): 300–309.
22. Zhao, J.B., Zhang, G.X., La Scala, M., and Zhang, J.H. (2015). Multistage phasoraided bad data detection and identification. *Proceedings of the IEEE Power and Energy Society General Meeting*, Denver, CO, USA (26–30 July 2015).
23. Jiang, W., Vittal, V., and Heydt, G.T. (May 2007). A distributed state estimator utilizing synchronized phasor measurements. *IEEE Trans. Power Syst.* 22 (2): 563–571.
24. Farantatos, E., Stefopoulos, G.K., Cokkinides, G.J., and Meliopoulos, A.P. (2009). PMU based dynamic state estimation for electric power systems. *IEEE Power and Energy Society General Meeting*, Calgary, AB, Canada (January 2009), pp. 1–8.
25. Simoes Costa, A., Albuquerque, A., and Bez, D. (May 2013). An estimation fusion method for including phasor measurements into power system real time modelling. *IEEE Trans. Power Syst.* 28 (2): 1910–1920.
26. Nuqui, R.F. (2001). State estimation and voltage security monitoring using synchronized phasor measurements. Ph.D. dissertation. Virginia Poly-technic Institute and State University.
27. Gómez-Expósito, A., de la Villa Jaén, A., Gómez-Quiles, C. et al. (Apr. 2011). A taxonomy of multi-area state estimation methods. *Electr. Power Syst. Res.* 81 (4): 1060–1069.
28. Zhang, Q., Vittal, V., Heydt, G. et al. (2012). The time skew problem in PMU measurements. *Proceedings of the IEEE Power and Energy Society General Meeting*, San Diego, CA (July 2012).
29. Yang, P., Tan, Z., Wiesel, A., and Nehora, A. (Nov. 2013). Power system state estimation using PMUs with imperfect synchronization. *IEEE Trans. Power Syst.* 28 (4): 4162–4172.
30. Zhang, Q., Chakhchoukh, Y., Vittal, V. et al. (May 2013). Impact of PMU measurement buffer length on state estimation and its optimization. *IEEE Trans. Power Syst.* 28 (2): 1657–1665.
31. Chakhchoukh, Y., Vittal, V., and Heydt, G.T. (Mar. 2014). PMU based state estimation by integrating correlation. *IEEE Trans. Power Syst.* 29 (2): 617–626.
32. Murugesan, V., Chakhchoukh, Y., Vittal, V. et al. (2015). PMU data buffering for power system state estimators. *IEEE Power Energy Technol. Syst. J.* 2 (3): 94–102.
33. Zhou, N., Huang, Z., and Meng, D. (2014). Capturing dynamics in the power grid: formulation of dynamic state estimation through data assimilation. Technical report PNNL-23213. Pacific Northwest National Laboratory.
34. Huang, Z., Zhou, N., Diao, R. et al. (2015). Capturing real-time power system dynamics: opportunities and challenges. *Proceedings of the IEEE Power and Energy Society General Meeting*, Denver, CO, USA (July 2015).
35. Arasaratnam, I., Haykin, S., and Elliott, R.J. (2007). Discrete-time nonlinear filtering algorithms using Gauss–Hermite quadrature. *Proc. IEEE* 95 (5): 953–977.

36. Hart, D.G., Uy, D., Gharpure, V. et al. (2001). *PMUs–A New Approach to Power Network Monitoring*. ABB.
37. Hassanzadeh, M., Evrenosoglu, C.Y., and Mili, L. (2016). A short-term nodal voltage phasor forecasting method using temporal and spatial correlation. *IEEE Trans. Power Syst.* 31 (5): 3881–3890.
38. Rousseeuw, P.J. and van Zomeren, B.C. (1990). Unmasking multivariate outliers and leverage points. *J. Am. Stat. Assoc.* 85 (411): 633–639.
39. Rousseeuw, P.J. and van Zomeren, B.C. (1991). Robust distances: simulations and cutoff values. In: *Directions in Robust Statistics and Diagnostics, Part II, Vol. 34, The IMA Volumes in Mathematics and Its Applications* (eds. W. Stahel and S. Weisberg), 195–203. New York: Springer-Verlag.
40. Mili, L., Cheniae, M., Vichare, N., and Rousseeuw, P. (1996). Robust state estimation based on projection statistics. *IEEE Trans. Power Syst.* 11 (2): 1118–1127.
41. Li, S., Xie, Y., Dai, H., and Song, L. (2015). M-statistic for kernel change-point detection. In: *Advances in Neural Information Processing Systems*, 3366–3374. MIT Press.
42. Croux, C. and Rousseeuw, P.J. (1992). Time-efficient algorithms for two highly robust estimators of scale. In: *Computational Statistics*, vol. 1 (eds. Y. Dodge and J. Whittaker), 411–428. Heidelberg: Physica-Verlag.
43. Thomas, L. and Mili, L. (2007). A robust GM-estimator for the automated detection of external defects on barked hardwood logs and stems. *IEEE Trans. Signal Process.* 55 (7): 3568–3576.
44. Gandhi, M. and Mili, L. (2010). Robust Kalman filter based on a generalized maximum-likelihood-type estimator. *IEEE Trans. Signal Process.* 58 (5): 2509–2520.
45. Caro, E. and Valverde, G. (2014). Impact of transformer correlations in state estimation using the unscented transformation. *IEEE Trans. Power Syst.* 1 (29): 368–376.
46. Pires, R., Simoes-Costa, A., and Mili, L. (1999). Iteratively reweighted least-squares state estimation through Givens rotations. *IEEE Trans. Power Syst.* 14 (4): 1499–1505.
47. Jabr, R. (2005). Power system Huber M-estimation with equality and inequality constraints. *Electr. Power Syst. Res.* 74: 239–246.
48. Fernholz, L. (1983). *Von Mises Calculus for Statistical Functionals*,". Lecture Notes in Statistics, vol. 19. New York: Springer-Verlag.
49. Hampel, F.R., Ronchetti, E.M., Rousseeuw, P.J., and Stahel, W.A. (1986). *Robust Statistics: The Approach Based on Influence Functions*. New York: Wiley.
50. Zhao, J.B., Netto, M., and Mili, L. (2016). A robust iterated extended Kalman filter for power system dynamic state estimation. *IEEE Trans. Power Syst.* https://doi.org/10.1109/TPWRS.2016.2628344.
51. Tarali, A. and Abur, A. (2012). Bad data detection in two-stage state estimation using phasor measurements. *Proceedings of the 3rd IEEE PES Innovative Smart Grid Technologies Conference* (2012), Berlin, Germany, pp. 1–8.

CHAPTER 9

LEAST-TRIMMED-ABSOLUTE-VALUE STATE ESTIMATOR

Ibrahim Omar Habiballah and Yuanhai Xia

King Fahd University of Petroleum & Minerals, Dhahran, Saudi Arabia

Many estimation algorithms have been proposed to improve the reliability and efficiency of the estimation process. Estimation algorithms such as least-absolute-value (LAV) and least-mean-square (LMS), least-median-of-squares (LMedS), least measurement rejection (LMR), and maximum-exponential-square (MES) are able to suppress bad data without any extra bad data detection and identification loop, and at the same time they are computationally fast [1]. Based on the alternating direction method of multipliers, Reference [2] considers local and overall robustness of power system in distributed state estimation (DSE). Moreover, the popular genetic evolution algorithms, especially neural networks, are used in very large and complex power system [3–5], but the results still need improvement.

The main objective of the chapter is to introduce a new robust estimator known as least-trimmed-absolute-value estimator. The algorithm evolves from the two estimators, LAV and least-trimmed-square (LTS), and benefits the merits of both. It can detect and eliminate both single and multiple bad data more efficiently.

Numerical experiments are conducted on 6-bus system and IEEE 14-bus system first, then these two systems, plus IEEE 30-bus system, are used to conduct AC estimation. Various types of bad data are simulated to evaluate the performance of the proposed robust estimator. MATLAB is used as the implementation platform.

Advances in Electric Power and Energy: Static State Estimation, First Edition.
Edited by Mohamed E. El-Hawary.

Published 2021 by John Wiley & Sons, Inc.

9.1 BAD DATA DETECTION AND ROBUST ESTIMATORS

9.1.1 Bad Data Detection

In any state estimator, detection and removal of bad -data are essential in order to reduce the residual error. Considering a practical raw data set for estimation, many scenarios may lead to both small and large errors. Generally there are three different interpretations: small departures for all data points, large departures for a small number of data points, or both situations occur simultaneously, which is more common in practice. Situations with large errors lead to the notion of outlier points, which are generally the most stressful part for statistical procedures. In fact, a single outlier can completely spoil the least squares estimation, causing it to break down.

Some obvious bad data, such as error with several order difference, negative-voltage magnitude, and large disparity between input/output power flow, are easy to detect by simple plausible check. But most of time those outliers are far away from the true value, so estimators must have the ability to correct those bad data themselves.

The general idea of detecting bad data is that removal of these corresponding measurements should not interfere the observability of the system. It is based on the results derived in the theory of network observability [6–8]. One of the most widely used methods for detecting bad data is the χ^2 test.

9.1.1.1 χ^2 Distribution

For a data set $x = \{x_1, x_2, \ldots, x_m\}$, where each x_i is from standard normal distribution $x_i \sim N(0, 1)$; a new variable Y is defined by

$$Y = \sum_{i=1}^{m} x_i^2 \tag{9.1}$$

Then Y is said to have an χ^2 distribution with m degrees of freedom, i.e. $Y \sim \chi_m^2$. Where m is the number of independent variables in the summation of square. It can be proved that the χ_m^2 distribution has mean m and variance $2m$.

When the m random variables are constrained by n independent equations, Y will have an χ_2 distribution with utmost $m - n$ degrees of freedom. This is the m measurements and n state variables in power system state estimation.

According to central limit theorem, the sum of a large number of random variables following any distribution with bounded variance approximates a normal distribution. Figure 9.1 illustrates χ_2 probability density function with 8 degrees of freedom. Since $\chi_{8,0.95}^2 = 15.5$, i.e. Prob($\chi_2 > 15.5$) = 0.05, the region on the right of the vertical dash line that represents the probability of x is larger than a certain threshold $x_t = 15.5$ with probability error 0.05. If the measured value of x exceeds this threshold, then with 0.95 probability, the measured x will not have an χ_2 distribution, i.e. presence of bad data will be suspected.

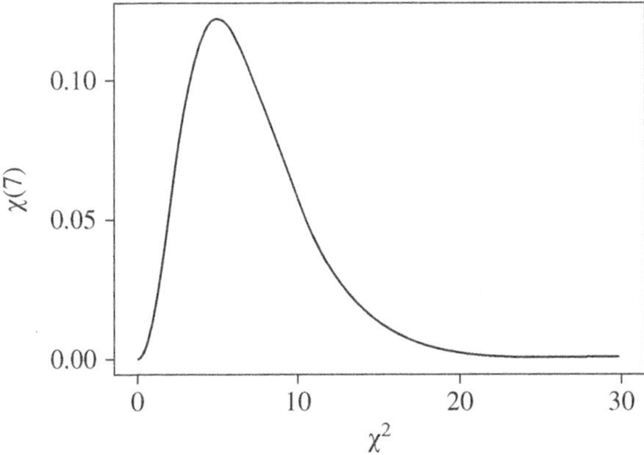

Figure 9.1 χ^2 probability density function.

9.1.1.2 χ^2 Test for Bad-Data Detection in WLS Estimator

The WLS objective function $J(x)$ can be used for bad-data detection. The details are given as follows:

- Perform WLS estimator and calculate the objective function:

$$J(\hat{x}) = \sum_{i=1}^{m} \frac{(z_i - h_i(\hat{x}_i))^2}{\sigma_i^2} \tag{9.2}$$

where $\hat{x}$ is the estimated state vector.

- Look up the value from the χ_2 distribution table corresponding to a detection confidence with probability p and $(m - n)$ degrees of freedom, let the value be $\chi^2_{(m-n),p}$.
- Test if $J(\hat{x}) > \chi^2_{(m-n),p}$. If yes, bad data will be suspected and trimmed; else, the measurements will be assumed to be free of bad data.

9.1.2 Robust Estimators

The term "robust" was coined in statistics by G.E.P. Box in 1953. When referring to a statistical estimator, it means "insensitive to small departures from the idealized assumptions for which the estimator is optimized" [9].

There are two parameters to evaluate the robustness of an estimator:

- **Breakdown point**: The breakdown point of an estimator for finite sample is the fraction of data that can be given arbitrarily large or small extreme values without yielding a bad result.
- **Leverage point**: The leverage point is a small amount of change that causes a small or large change in system behavior.

Statisticians have developed various sorts of robust statistical estimators; most of them can be grouped in one of three categories:

- **M-estimator**: Maximum-likelihood arguments are usually the most relevant class for model-fitting, i.e. estimation of parameters.
- **L-estimator**: "Linear combinations of order statistics." These are most applicable to estimations of central value and tendency, though they can occasionally be applied to some problems in estimation of parameters. Two "typical" L-estimates are the median and Tukey's trimean.
- **R-estimator**: Each residual (Section 9.1.2.3) is weighted by a score based on selected criteria.

Besides these three main categories, there are some other kinds of robust techniques from the fields of optimal control and filtering rather than from the field of mathematical statistics. In this section, the basic weighted-least-square (WLS) algorithm and the three main categories are reviewed.

9.1.2.1 M-Estimator

M-estimators, introduced by Huber [9], are a broad class of estimators. M-estimation is maximum-likelihood estimation obtained by minimizing an objective function, which is expressed as a function of the measurement residual $\rho(r)$, subject to the constraints given by the measurement equations:

$$\min J(r) = \sum_{i=1}^{m} \rho(r_i) \tag{9.3}$$

$$\text{subject to } z = h(x) + r$$

where

$\rho(r_i)$ is a function of the measurement residual r_i,

z is the measurement vector,

x is the state vector,

$h(x)$ is the measurement function related to the state vector.

The function $\rho(r_i)$ plays an important role in generating desirable results. It must be chosen with the following properties:

- $\rho(r) > 0$ for $r = 0$
- $\rho(r) \geq 0$ for any r
- $\rho(r)$ is monotonically increasing in both $+r$ and $-r$ directions
- $\rho(r) = \rho(-r)$, i.e. symmetric with respect to $r = 0$

According to [10, 11], the expression of $\rho(r)$ can be

- Quadratic-Constant (QC) [10]

$$\rho(r) = \begin{cases} r_i^2/\sigma_i^2, & |r_i/\sigma_i| \le a \\ a^2/\sigma_i^2, & \text{otherwise} \end{cases} \tag{9.4}$$

- Quadratic–Linear (QL) [10]

$$\rho(r) = \begin{cases} r_i^2/\sigma_i^2, & |r_i/\sigma_i| \le a \\ 2a\sigma_i|r_i| - a^2\sigma_i^2, & \text{otherwise} \end{cases} \tag{9.5}$$

- Square Root (SR) [10]

$$\rho(r) = \begin{cases} r_i^2/\sigma_i^2, & |r_i/\sigma_i| \le a \\ 4a^{3/2}\sqrt{r_i/\sigma_i} - 3a^2, & \text{otherwise} \end{cases} \tag{9.6}$$

- Schweppe–Huber Generalized-M (SHGM) [11]

$$\rho(r) = \begin{cases} \frac{1}{2} r_i^2/\sigma_i^2, & |r_i/\sigma_i| \le a \\ a w_i |r_i/\sigma_i| - \frac{1}{2} a^2 w_i^2, & \text{otherwise} \end{cases} \tag{9.7}$$

where a is a turning parameter specified according to different situations and w_i is an iteratively modified weighting factor.

With all the parameters settled, there are many methods to solve this problem, such as Newton's method. These methods require computation of the first and second derivatives of ρ.

9.1.2.2 L-Estimator

L-estimator [12] is a linear combination of order statistics of the measurements. The main benefit of L-estimators is that they are often extremely simple and robust. With sorted data, they are very easy to calculate and interpret and are often resistant to outliers. Though L-estimators are inefficient for whole estimation, in many circumstances they are reasonably efficient and adequate for initial estimation.

L-estimators are based on a definition of quintiles as follows:

$$\varphi(r_j) = \begin{cases} p-1, & r_j < 0 \\ p, & \text{otherwise} \end{cases} \tag{9.8}$$

where $r_j = x_j - z$ is the signed residual from the jth estimated data, x_j, to the measurement value z.

The objective function is

$$\min J = \sum_{j=1}^{N} \rho(r_j) \tag{9.9}$$

where

$$\rho(r_j) = r_j\phi(r_j) = \begin{cases} r_j(p-1), & r_j < 0 \\ r_j p, & \text{otherwise} \end{cases} \tag{9.10}$$

It is easy to show that when $p = 1/2$, the half-quintile, $\phi_p(r_j) = \frac{1}{2}\,\text{sgn}\,(r_j)$, corresponds to the sample median.

9.1.2.3 R-Estimator

R-estimator [13] involves ranking residuals; each residual is weighted by a score based on its rank with following objective:

$$\min J = \sum_{j=1}^{m} a_m(R_j) r_j \tag{9.11}$$

where R_j is the rank of the jth residuals in $\{r_1, \ldots, r_m\}$ and a_m is a non-decreasing score function following $\sum_k a_m(k) = 0$.

As proposed in the same paper, Jaeckel's estimator focuses on finding the coefficient that minimizes J:

$$\min J = \sum_{j=1}^{m} \left(R_j - \frac{m+1}{2}\right) r_j \tag{9.12}$$

Like L-estimators, the major inconvenience in R-estimators is that they are low efficient and not easy for optimization, and the definition of the score function implicitly needs prior information about the noise contamination rate.

As for the three main categories stated above, M-estimators are preferred in modern robust statistics though they are more complex for computation. But some L-estimator and R-estimator features are used in combination with M-estimator for better robustness.

9.1.3 Existing Robust Estimators

In this section, major robust estimation algorithms will be presented. Besides the original weighted-least-square (WLS), least-absolute-value (LAV) [14], which belongs to M-estimator, least-median-square (LMS) [15, 16], which belongs to L-estimator, least-trimmed-square (LTS) [15], and least-measurements-rejected (LMR) [17], which belongs to R-estimator, will also be implemented.

For a general expression, the objective function is set up upon the classical linear regression model presented in Eq. (9.9). Figure 9.2 shows the $\rho(r/\sigma)$ of the basic least-square algorithm.

9.1.3.1 Least-Absolute-Value (LAV)

Least-absolute-value (LAV) is a mathematical optimization technique similar to the popular least-squares algorithm that attempted to find a function, which closely

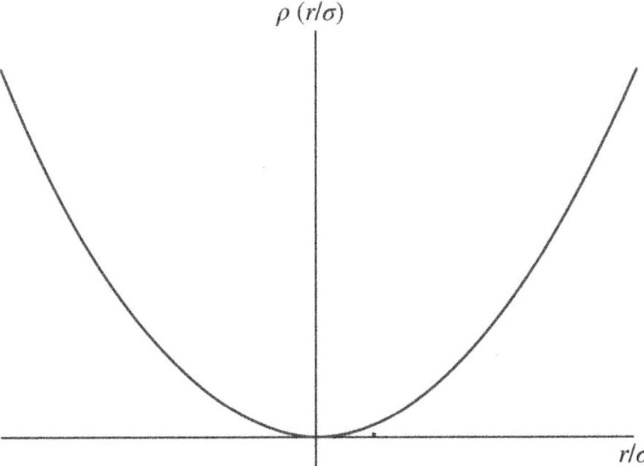

Figure 9.2 Least square objective function.

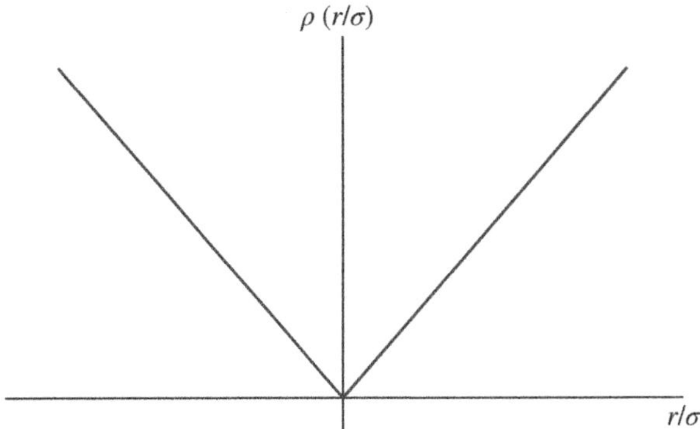

Figure 9.3 Least-absolute-value objective function.

approximates a set of data; the method minimizes the sum of absolute errors. Figure 9.3 shows the $\rho(r/\sigma)$ of the basic least-absolute-value algorithm.

LAV estimator minimizes the objective function:

$$\min J = \sum_{j=1}^{m} |r_j| \tag{9.13}$$

where $r_j = z - h_j(x)$ is the jth residue.

Assuming an initial solution x_o for the state and using the first-order approximation of $h_j(x)$ around x_o, the above equation can be written around the operating point x_o as the following:

$$\Delta z = H(x_o \Delta x + r) \tag{9.14}$$

where $\Delta z = z - h(x_o)$, $\Delta x = x - x_o$, and $H(x_o) = \dfrac{\partial h(x_o)}{\partial x}$

The problem can be transformed into a successive set of linear programming (LP) problems; each minimizes the objective function:

$$\min J\left(x^k\right) = \sum_{j=1}^{m} u_i^k + v_i^k \tag{9.15}$$

where u_i^k and v_i^k are both positive and $u_i^k - v_i^k = z - h\left(x^k\right) - H\left(x^k\right)\Delta x = \Delta z^k - H\left(x^k\right)\Delta x$ is the measurement residual vector at the kth iteration. For the kth iteration, the optimization problem to be solved can be formulated as follows (k are dropped for simplicity):

$$\min \sum_{j=1}^{m} u_i^k + v_i^k \tag{9.16}$$

subject to

$$H\Delta x_u - H\Delta x_v + u - v = \Delta z$$

$$\Delta x_u, \Delta x_v, u, v \geq 0$$

where $\Delta x = \Delta x_u - \Delta x_v$.

This LP problem can be solved using the simplex method. The overall state estimation solution will be obtained by successively solving these LP problems until $|\Delta x|$ is less than a chosen threshold.

9.1.3.2 *Least-Median-of-Squares (LMedS)*

Unlike WLS and LAV, which have a breakdown point of zero, least-median-square is an estimator with high breakdown point. They may yield the correct solution even if up to half the redundant measurements are bad leverage points.

The LMedS estimator minimizes the median of squared residues instead of the regular summation, as its name hints. The objective function for minimization is

$$\min J = r_{k(v)}^2 \tag{9.17}$$

where $v = \frac{m}{2}$, the largest integer that is smaller or equal to $\frac{m}{2}$, and $r_{k(v)}^2$, mean of the vth ordered squared residue at the kth iteration, $r_{k(i)}^2$, are ordered increasingly:

$$r_{k(1)}^2 \leq \ldots \leq r_{k(m)}^2$$

The objective function of LMedS estimator is that it seeks a regression that minimizes the value of a tolerance t, whereby the majority of the measurements fall within tolerance. The implementation can be formed using the following equation:

$$\min_{x,k,t} t \tag{9.18}$$

subject to

$$b - t - Mk \leq Ax \leq b + t + Mk$$

$$k_1 + k_2 + \cdots + k_m \leq K$$

where k is an unknown binary integer vector (either 0 or 1), t is an unknown scalar tolerance, M is a specified arbitrary large positive scalar, and K is $\frac{m}{2}$.

In this case, if M is sufficiently large, one can eliminate those large outliers and keep other measurements within the smallest tolerance.

9.1.3.3 Least-Measurement-Rejected (LMR)

Least-measurement-rejected algorithm is a variant of LMS, compared with LMS, LMR minimizes the number of rejected values with fixed tolerance, given by

$$\min_{x,k} k_i \tag{9.19}$$

subject to

$$b - t - Mk \leq Ax \leq b + t + Mk$$

$$k_1 + k_2 + \cdots + k_m \leq K$$

where t is given fixed tolerance and other parameters are the same as in LMS algorithm.

9.1.3.4 Least-Trimmed-Square (LTS)

Least-trimmed-square is an approach that simply ignores or trims the largest residuals by applicable analysis and computation. Figure 9.4 shows the $\rho(r/\sigma)$ of the basic least-trimmed-square algorithm.

Assume there are n points, and $r_1, \ldots, r_n$ represent the residuals corresponding to some choices for the slope and intercept. Then, order the squared residuals and label them as $r_1^2 \leq \ldots \leq r_n^2$; r_1^2 is the smallest squared residual, and r_n^2 is the largest.

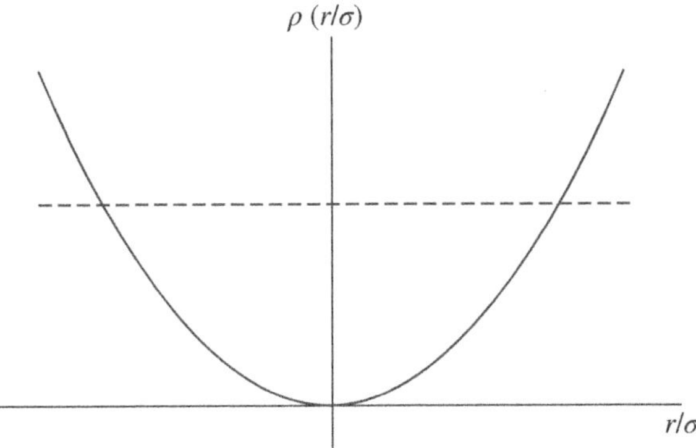

Figure 9.4 Least trimmed square objective function.

The minimization function can be formulated as follows:

$$\min_{x,k,r} \sum r^T r \tag{9.20}$$

subject to

$$b - Mk \leq Ax - r \leq b + Mk$$

$$k_1 + k_2 + \cdots + k_m \leq K$$

where K is a specified number of measurements that should be rejected, r is residue, and the others are same as in LMS.

9.1.4 Proposed LTAV Robust Estimators

Among robust estimators proposed above, LAV and LTS are two excellent representatives. LAV, a variant of WLS by changing the objective function $\rho(r)$ in Section 9.1.2.1, is proved to be a good estimator. LTS, by adding a trimming strategy to WLS estimator, performs well in correcting large outliers.

Inspired by these two algorithms, least-trimmed-absolute-value (LTAV) estimator uses the fitting function of LAV and trims the given largest residues in each iteration during the estimation process. Figure 9.5 shows the $\rho(r/\sigma)$ of the proposed least-trimmed-absolute-value algorithm.

The minimization problem can be written as follows:

$$\min \sum_{i=1}^{m-K} |\hat{r}_i| \tag{9.21}$$

where K is the number of trimmed data and $|\hat{r}_i|$ is sorted from smallest to largest.

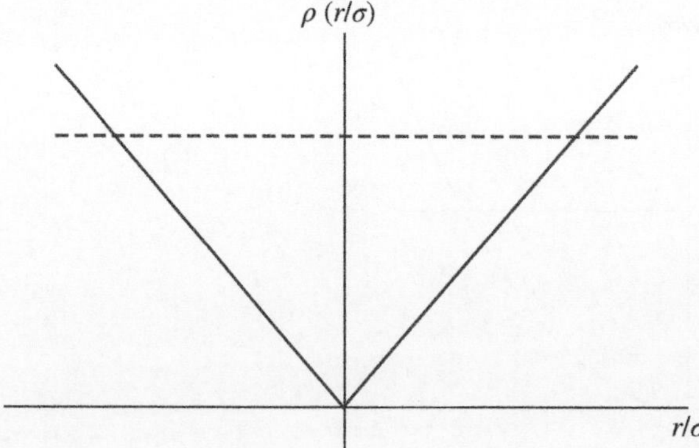

Figure 9.5 Least-trimmed-absolute-value objective function.

9.1.4.1 General Steps of LTAV Estimator

The general steps to use LTAV estimator are the following:

1. **Sampling of measurements**: Randomly select a specified k observable sample sets from 75% of the m measurements. For each selected sample set, estimate the state variables by the LAV estimator.
2. **Summation of least-trimmed-absolute-value error**: Find the state candidates having the least summation of trimmed absolute value among all candidates in step 1.
3. **Bad data removal**: Based on the residual generated by the state candidate, eliminate the data according to a certain threshold.
4. **Re-estimation**: Re-estimate the states using the remaining data.

9.1.4.2 Mixed Integer Linear Programming Implementation

LTAV can be formulated as a mixed integer linear programming (MILP) problem, which in turn can be solved by applying one of the well-developed MILP solution methods. The MILP model is built as follows:

$$\min_{x,k,u,v} \sum (u+v) \tag{9.22}$$

subject to

$$b - Mk \leq Ax - u + v \leq b + Mk$$

$$k_1 + k_2 + \cdots + k_m \leq K$$

Equation (9.22) can be transformed into the basic MILP problem format as follows:

$$\min\ c^T \cdot Y \tag{9.23}$$

subject to

$$A \cdot Y \leq B$$

where

$$c^T = [0_m, 0_n, 1_m, 1_n]$$

$$A = \begin{bmatrix} H & -M & -I & I \\ -H & -M & I & -I \\ 0m & 0n & 1m & 1n \end{bmatrix}$$

$$B^T = [b, -b, K], \qquad Y^T = [x, k, u, v]$$

M = diag(m), I = diag(1), $b > 0$, and $k = 0$ or 1.

The K here is different from LTS estimator, whose K is a fixed number, copied from the idea of LMS and LMR. K in LTAV is changeable with the maximum integer smaller than $m/2$, where m is the number of measurements.

9.2 RESULTS AND DISCUSSION

In this section, the proposed Least-Trimmed-Absolute-Value (LTAV) is presented and compared with other estimators. Three power systems: 6-bus, IEEE 14-bus, and IEEE 30-bus are used [19]. DC and AC systems are used to evaluate the performance of the proposed estimator. Various kinds of bad data configurations are simulated for testing the performance of the proposed estimator. LP solver is used to solve the mixed integer linear programming [18]. All the codes are written in MATLAB platform.

9.2.1 6-Bus System

For the 6-bus system both DC and AC state estimation are carried out to test the performance of the proposed LTAV and compared with other existing robust estimators.

9.2.1.1 6-Bus DC State Estimation

The line and bus data of this system are presented in Appendix 9.A.1. Four major cases are conducted to demonstrate the efficiency of estimators: base case with normal random error, single bad-data case, multiple noninteracting bad-data case, and multiple interacting bad-data case.

The parameters used in the estimation are the following:

1. Tolerance of iterative loop for all estimators: 0.0001.
2. Fixed trimmed number for LTS: $K = 2$.
3. M for LMS: 3; M for LMR: 500; and M for LTAV: 3.

For many cases, the results have the same trends though with different number. In these cases, comments will be omitted. Only cases with new type of results are carefully analyzed.

Table 9.1 shows the actual values of real power injection and flow in the system from load flow. Random noise are generated using normal distribution, with $\sigma = 0.2$, and added to the actual value to simulate the measurements.

TABLE 9.1 Simulated measurements for 6-bus DC system.

Meter	Actual	Meas.	Meter	Actual	Meas.	Meter	Actual	Meas.
P1	1.0	1.0108	**P1–2**	0.1875	0.1788	**P2–6**	0.2875	0.302
P2	0.5	0.5367	**P1–4**	0.425	0.4319	**P3–5**	0.2375	0.2362
P3	0.6	0.5548	**P1–5**	0.3875	0.4591	**P3–6**	0.325	0.3393
P4	−0.7	−0.6828	**P2–3**	−0.0375	0.0179	**P4–5**	−0.0375	−0.0416
P5	−0.7	−0.6936	**P2–4**	0.2375	0.2105	**P5–6**	0.0875	0.085
P6	−0.7	−0.7262	**P2–5**	0.2	0.2607			

The performance of the proposed LTAV estimator is compared with the following estimators: WLS, LAV, and LTS. Full redundancy is considered.

9.2.1.1.1 Case 0: Basic Case with Only Random Noise In this case, the measured values in Table 9.1 are used. Table 9.2 shows the estimated values. Table 9.3 shows the error between each estimator and actual value.

As illustrated in Table 9.2, the results for this case show that all estimators generate good results. The results of LTS is worse than WLS, because there are no outliers; the data with the largest error lay on the same bus, making the bus in less coverage by true value, and induce large estimation error.

Table 9.3 shows the absolute difference between the estimated and actual values for the selected estimators. The last raw in the table is the summation of these errors. This summation is used as an indicator to observe the performance of the proposed estimator against the other ones. It can be observed that LTAV is among the best estimators for this base case.

9.2.1.1.2 Case 1: Single Bad Data In this case, a single bad data is presented by reversing the polarity of the measured meter to simulate improper connection of the meter. This type of error is considered in general throughout this section. Based on Table 9.1, the power flow between bus 2 and bus 4 is changed to −0.2105. Table 9.4 shows the estimated values. Table 9.5 shows the error between each estimator and actual value. The selected bad data is highlighted in both tables.

TABLE 9.2 Estimation results of case 0, 6-bus DC system.

Meter	Actual	Meas.	WLS	LAV	LTS	LTAV
P1	1.0	1.0108	1.0236	1.0108	1.102	1.0108
P2	0.5	0.5367	0.5475	0.5367	0.5539	0.5367
P3	0.6	0.5548	0.5527	0.5549	0.5613	0.5548
P4	−0.7	−0.6828	−0.682	−0.6828	−0.6678	−0.6828
P5	−0.7	−0.6936	−0.713	−0.6936	−0.829	−0.6936
P6	−0.7	−0.7262	−0.7289	−0.726	−0.7203	−0.7259
P1–2	0.1875	0.1788	0.1936	0.1912	0.2145	0.1912
P1–4	0.425	0.4319	0.4264	0.4234	0.4424	0.4234
P1–5	0.3875	0.4591	0.4036	0.3962	0.445	0.3962
P2–3	−0.0375	0.0179	−0.0111	−0.0148	−0.0052	−0.0148
P2–4	0.2375	0.2105	0.2328	0.2322	0.2279	0.2322
P2–5	0.2	0.2607	0.2101	0.2051	0.2305	0.2051
P2–6	0.2875	0.302	0.3093	0.3054	0.3152	0.3054
P3–5	0.2375	0.2362	0.2212	0.2199	0.2357	0.2198
P3–6	0.325	0.3393	0.3204	0.3202	0.3204	0.3202
P4–5	−0.0375	−0.0416	−0.0228	−0.0272	0.0026	−0.0272
P5–6	0.0875	0.085	0.0992	0.1004	0.0847	0.1003

TABLE 9.3 Estimation errors of case 0, 6-bus DC system.

Meter	Meas.	WLS	LAV	LTS	LTAV
P1	0.0108	0.0236	0.0108	0.102	0.0108
P2	0.0367	0.0475	0.0367	0.0539	0.0367
P3	0.0452	0.0473	0.0451	0.0387	0.0452
P4	0.0172	0.018	0.0172	0.0322	0.0172
P5	0.0064	0.013	0.0064	0.129	0.0064
P6	0.0262	0.0289	0.026	0.0203	0.0259
P1–2	0.0087	0.0061	0.0037	0.027	0.0037
P1–4	0.0069	0.0014	0.0016	0.0174	0.0016
P1–5	0.0716	0.0161	0.0087	0.0575	0.0087
P2–3	0.0554	0.0264	0.0227	0.0323	0.0227
P2–4	0.027	0.0047	0.0053	0.0096	0.0053
P2–5	0.0607	0.0101	0.0051	0.0305	0.0051
P2–6	0.0145	0.0218	0.0179	0.0277	0.0179
P3–5	0.0013	0.0163	0.0176	0.0018	0.0177
P3–6	0.0143	0.0046	0.0048	0.0046	0.0048
P4–5	0.0041	0.0147	0.0103	0.0401	0.0103
P5–6	0.0025	0.0117	0.0129	0.0028	0.0128
Sum	**0.4095**	**0.3121**	**0.2527**	**0.6276**	**0.2528**

TABLE 9.4 Estimation results of case 1, 6-bus DC system.

Meter	Actual	Meas.	WLS	LAV	LTS	LTAV
P1	1.0	1.0108	1.0316	1.0108	1.0095	1.0108
P2	0.5	0.5367	0.4874	0.5367	0.5513	0.5367
P3	0.6	0.5548	0.5327	0.5549	0.5569	0.5551
P4	−0.7	−0.6828	−0.5898	−0.6828	−0.689	−0.6828
P5	−0.7	−0.6936	−0.713	−0.6936	−0.7041	−0.6936
P6	−0.7	−0.7262	−0.7489	−0.726	−0.7247	−0.7262
P1–2	0.1875	0.1788	0.2131	0.1912	0.1878	0.1912
P1–4	0.425	0.4319	0.4054	0.4234	0.4246	0.4234
P1–5	0.3875	0.4591	0.4132	0.3962	0.3971	0.3962
P2–3	−0.0375	0.0179	−0.0061	−0.0148	−0.0136	−0.0149
P2–4	**0.2375**	**−0.2105**	**0.1923**	**0.2322**	**0.2368**	**0.2322**
P2–5	0.2	0.2607	0.2001	0.2051	0.2092	0.2051
P2–6	0.2875	0.302	0.3143	0.3054	0.3068	0.3055
P3–5	0.2375	0.2362	0.2062	0.2199	0.2229	0.2199
P3–6	0.325	0.3393	0.3204	0.3202	0.3204	0.3203
P4–5	−0.0375	−0.0416	0.0078	−0.0272	−0.0276	−0.0272
P5–6	0.0875	0.085	0.1142	0.1004	0.0975	0.1004

TABLE 9.5 Estimation errors of case 1, 6-bus DC system.

Meter	Meas.	WLS	LAV	LTS	LTAV
P1	0.0108	0.0316	0.0108	0.0095	0.0108
P2	0.0367	0.0126	0.0367	0.0513	0.0367
P3	0.0452	0.0673	0.0451	0.0431	0.0449
P4	0.0172	0.1102	0.0172	0.011	0.0172
P5	0.0064	0.013	0.0064	0.0041	0.0064
P6	0.0262	0.0489	0.026	0.0247	0.0262
P1–2	0.0087	0.0256	0.0037	0.0003	0.0037
P1–4	0.0069	0.0196	0.0016	0.0004	0.0016
P1–5	0.0716	0.0257	0.0087	0.0096	0.0087
P2–3	0.0554	0.0314	0.0227	0.0239	0.0226
P2–4	**0.448**	**0.0452**	**0.0053**	**0.0007**	**0.0053**
P2–5	0.0607	0.0001	0.0051	0.0092	0.0051
P2–6	0.0145	0.0268	0.0179	0.0193	0.018
P3–5	0.0013	0.0313	0.0176	0.0146	0.0176
P3–6	0.0143	0.0046	0.0048	0.0046	0.0047
P4–5	0.0041	0.0453	0.0103	0.0099	0.0103
P5–6	0.0025	0.0267	0.0129	0.01	0.0129
Sum	**0.8305**	**0.5661**	**0.2527**	**0.2463**	**0.2526**

As illustrated in Table 9.4, the results for this case show that all estimators, except WLS, successfully rejected the simulated bad data and generated good estimates. The results of WLS is the worst estimator due to the presence of the single bad data P2–4. Table 9.5 shows the indicator of each estimator. It confirms the outcome observed in Table 9.4. It can be observed that the LAV, LTS, and LTAV estimators are almost performing identically.

9.2.1.1.3 Case 2: Multiple Noninteracting Bad Data For actual large system, multiple bad-data case is more common. In this case, two noninteracting bad data are generated fir simulation. Since the connection between each bus of the 6-bus system is high, it is very hard to pick up buses with no connection with each other. Measurements with weak connection are selected.

Based on the raw data used in the single reverse bad-data case, a weakly connected power flow P3–6 is selected beside P2–4. The reading of these two meters are reversed in polarity to simulate the presence of two noninteracting bad measurements. Table 9.6 shows the estimated values. Table 9.7 shows the error between each estimator and actual value. The two selected bad data are highlighted in both tables.

As illustrated in Table 9.6, all estimators generate better results, but WLS fails to present good estimates. The results of WLS are getting worse as the number of bad data increases. The LAV and LTAV estimators are performing identically the same and give better indicator than the LTS estimator as presented in Table 9.7. Since LAV generated good results, LTAV did not trim any data.

TABLE 9.6 Estimation results of case 2, 6-bus DC system.

Meter	Actual	Meas.	WLS	LAV	LTS	LTAV
P1	1.0	1.0108	1.0316	1.0108	1.0231	1.0108
P2	0.5	0.5367	0.4874	0.5367	0.5511	0.5367
P3	0.6	0.5548	0.3969	0.5548	0.5497	0.5548
P4	−0.7	−0.6828	−0.5898	−0.6828	−0.6874	−0.6828
P5	−0.7	−0.6936	−0.713	−0.6936	−0.713	−0.6936
P6	−0.7	−0.7262	−0.6132	−0.7259	−0.7235	−0.7259
P1–2	0.1875	0.1788	0.2131	0.1912	0.1924	0.1912
P1–4	0.425	0.4319	0.4054	0.4234	0.4276	0.4234
P1–5	0.3875	0.4591	0.4132	0.3962	0.4031	0.3962
P2–3	−0.0375	0.0179	0.0278	−0.0148	−0.0104	−0.0148
P2–4	**0.2375**	**−0.2105**	**0.1923**	**0.2322**	**0.2352**	**0.2322**
P2–5	0.2	0.2607	0.2001	0.2051	0.2107	0.2051
P2–6	0.2875	0.302	0.2804	0.3054	0.3079	0.3054
P3–5	0.2375	0.2362	0.1722	0.2198	0.221	0.2198
P3–6	**0.325**	**−0.3393**	**0.2525**	**0.3202**	**0.3183**	**0.3202**
P4–5	−0.0375	−0.0416	0.0078	−0.0272	−0.0246	−0.0272
P5–6	0.0875	0.085	0.0803	0.1003	0.0973	0.1003

TABLE 9.7 Estimation errors of case 2, 6-bus DC system.

Meter	Meas.	WLS	LAV	LTS	LTAV
P1	0.0108	0.0316	0.0108	0.0231	0.0108
P2	0.0367	0.0126	0.0367	0.0511	0.0367
P3	0.0452	0.2031	0.0452	0.0503	0.0452
P4	0.0172	0.1102	0.0172	0.0126	0.0172
P5	0.0064	0.013	0.0064	0.013	0.0064
P6	0.0262	0.0868	0.0259	0.0235	0.0259
P1–2	0.0087	0.0256	0.0037	0.0049	0.0037
P1–4	0.0069	0.0196	0.0016	0.0026	0.0016
P1–5	0.0716	0.0257	0.0087	0.0156	0.0087
P2–3	0.0554	0.0653	0.0227	0.0271	0.0227
P2–4	0.448	0.0452	0.0053	0.0023	0.0053
P2–5	0.0607	0.0001	0.0051	0.0107	0.0051
P2–6	0.0145	0.0071	0.0179	0.0204	0.0179
P3–5	0.0013	0.0653	0.0177	0.0165	0.0177
P3–6	0.6643	0.0725	0.0048	0.0067	0.0048
P4–5	0.0041	0.0453	0.0103	0.0129	0.0103
P5–6	0.0025	0.0072	0.0128	0.0098	0.0128
Sum	**1.4805**	**0.8362**	**0.2528**	**0.3031**	**0.2528**

9.2.1.1.4 Case 3: Multiple Interacting Bad Data In case 2, the two selected bad data were not interacting with each other. In this case, two bad data that are interacting with each other: P2–4 and P2–5. Thus for bus 2, only the power injection is available for estimation; it should be harder than noninteracting case.

The reading of these two meters are reversed in polarity to simulate the presence of two interacting bad measurements. Table 9.8 shows the estimated values. Table 9.9 shows the error between each estimator and actual value. The two selected bad data are highlighted in both tables.

As illustrated in Table 9.8, all estimators generate better results, but WLS fails to present good estimates. The results of WLS is getting worse as the number of bad-data increases. The LAV and LTAV estimators are performing identically the same, and give better indicator than the LTS estimator as presented in Table 9.9. Since LAV generated good results, LTAV did not trim any data.

9.2.1.1.5 Summary Table 9.10 presents the comparison between each estimator.

As illustrated in Table 9.10 in all cases, LAV and LTAV estimators generate the best estimation, while LTS estimator is a little fall behind. WLS estimator results are not satisfied, especially for the bad-data cases. Some cases share the same results, especially LTS and LTAV; the reason is that they trim the same data and then have the same raw data for estimation.

TABLE 9.8 Estimation results of case 3, 6-bus DC system.

Meter	Actual	Meas.	WLS	LAV	LTS	LTAV
P1	1.0108	1.0316	1.0108	1.0232	1.0108	1.0108
P2	0.5367	0.4129	0.5367	0.5434	0.5367	0.5367
P3	0.5548	0.5327	0.5549	0.5538	0.5551	0.5548
P4	−0.6828	−0.5898	−0.6828	−0.6871	−0.6828	−0.6828
P5	−0.6936	−0.6385	−0.6936	−0.7054	−0.6936	−0.6936
P6	−0.7262	−0.7489	−0.726	−0.7278	−0.7262	−0.7262
P1–2	0.1788	0.2255	0.1912	0.1937	0.1912	0.1788
P1–4	0.4319	0.4054	0.4234	0.4276	0.4234	0.4319
P1–5	0.4591	0.4008	0.3962	0.4019	0.3962	0.4591
P2–3	**0.0179**	**−0.0185**	**−0.0148**	**−0.0126**	**−0.0149**	**0.0179**
P2–4	**−0.2105**	**0.1798**	**0.2322**	**0.2338**	**0.2322**	**−0.2105**
P2–5	−0.2607	0.1752	0.205	0.2081	0.205	−0.2607
P2–6	0.302	0.3019	0.3054	0.3078	0.3055	0.302
P3–5	0.2362	0.1937	0.2199	0.2208	0.2199	0.2362
P3–6	0.3393	0.3204	0.3202	0.3204	0.3203	0.3393
P4–5	−0.0416	−0.0046	−0.0272	−0.0257	−0.0272	−0.0416
P5–6	0.085	0.1266	0.1004	0.0996	0.1004	0.085

TABLE 9.9 Estimation errors of case 3, 6-bus DC system.

Meter	Meas.	WLS	LAV	LTS	LTAV
P1	0.0108	0.0316	0.0108	0.0232	0.0108
P2	0.0367	0.0871	0.0367	0.0434	0.0367
P3	0.0452	0.0673	0.0451	0.0462	0.0449
P4	0.0172	0.1102	0.0172	0.0129	0.0172
P5	0.0064	0.0615	0.0064	0.0054	0.0064
P6	0.0262	0.0489	0.026	0.0278	0.0262
P1–2	0.0087	0.038	0.0037	0.0062	0.0037
P1–4	0.0069	0.0196	0.0016	0.0026	0.0016
P1–5	0.0716	0.0133	0.0087	0.0144	0.0087
P2–3	**0.0554**	**0.019**	**0.0227**	**0.0249**	**0.0226**
P2–4	**0.448**	**0.0577**	**0.0053**	**0.0037**	**0.0053**
P2–5	0.4607	0.0248	0.005	0.0081	0.005
P2–6	0.0145	0.0144	0.0179	0.0203	0.018
P3–5	0.0013	0.0438	0.0176	0.0167	0.0176
P3–6	0.0143	0.0046	0.0048	0.0046	0.0047
P4–5	0.0041	0.0329	0.0103	0.0118	0.0103
P5–6	0.0025	0.0391	0.0129	0.0121	0.0129
SUM	**1.2305**	**0.7138**	**0.2527**	**0.2841**	**0.2526**

TABLE 9.10 Comparison of 6-bus indicators.

Case no.	Meas.	WLS	LAV	LTS	LTAV
Case 0	0.4095	0.3121	0.2527	0.6276	0.2528
Case 1	0.8305	0.5661	0.2527	0.2463	0.2526
Case 2	1.4805	0.8362	0.2528	0.3031	0.2528
Case 3	1.2305	0.7138	0.2527	0.2841	0.2526

9.2.1.2 5.1.2 6-Bus AC State Estimation

The line and bus data of the 6-bus AC system are presented in Appendix 9.A.2. For consistency, the same four different cases are conducted as in DC system with the same kind of error at the same location for the four estimators: WLS, LAV, LTS, and LTAV.

Table 9.11 shows the actual values of real power injection and flow from load flow as obtained from [19]. Random noise of normal distribution, with $\sigma = 0.016\,65$ for voltage measurements and $\sigma = 0.05$ for power measurements, is generated and added to the actual value to simulate the measurements [19]. Full redundancy is considered.

The performance of the four estimators will be shown using the estimated values and the indicators.

TABLE 9.11 Simulated measurements for 6-bus AC system.

Meter	Actual	Meas.	Meter	Actual	Meas.	Meter	Actual	Meas.
V1	1.0500	1.0365	**P2–1**	–0.2780	–0.3490	**Q1–5**	0.1130	0.0940
V2	1.0500	1.0339	**P2–4**	0.3310	0.3280	**Q2–1**	0.1280	0.0970
V3	1.0700	1.0900	**P2–5**	0.1550	0.1740	**Q2–4**	0.4610	0.3830
V4	0.9896	0.9813	**P2–6**	0.2620	0.2230	**Q2–5**	0.1540	0.2200
V5	0.9857	0.9791	**P2–3**	0.0290	0.0860	**Q2–6**	0.1240	0.1500
V6	1.0043	0.9952	**P3–2**	–0.0290	–0.0210	**Q2–3**	–0.1230	–0.1190
P1	1.0790	1.1310	**P3–5**	0.1910	0.1770	**Q3–2**	0.0570	0.1020
P2	0.5000	0.4840	**P3–6**	0.4380	0.4330	**Q3–5**	0.2320	0.2390
P3	0.6000	0.5510	**P4–1**	–0.4250	–0.4010	**Q3–6**	0.6070	0.5830
P4	–0.7000	–0.7180	**P4–2**	–0.3160	–0.2980	**Q4–1**	–0.1990	–0.1430
P5	–0.7000	–0.7200	**P4–5**	0.0410	0.0070	**Q4–2**	–0.4510	–0.4430
P6	–0.7000	–0.7230	**P5–4**	–0.0400	–0.0210	**Q4–5**	–0.0490	–0.1740
Q1	0.1600	0.2020	**P5–1**	–0.3450	–0.3660	**Q5–4**	–0.0280	–0.0150
Q2	0.7440	0.7190	**P5–2**	–0.1500	–0.1170	**Q5–1**	–0.1350	–0.1750
Q3	0.8960	0.9060	**P5–3**	–0.1800	–0.2510	**Q5–2**	–0.1800	–0.2220
Q4	–0.7000	–0.7190	**P5–6**	0.0160	–0.0210	**Q5–3**	–0.2610	–0.1990
Q5	–0.7000	–0.6770	**P6–5**	–0.0160	0.0100	**Q5–6**	–0.0970	–0.0080
Q6	–0.7000	–0.6090	**P6–2**	0.2570	–0.1960	**Q6–5**	0.0390	0.0290
P1–2	0.2870	0.3150	**P6–3**	–0.4280	–0.4680	**Q6–2**	–0.1600	–0.2230
P1–4	0.4360	0.3890	**Q1–2**	–0.1540	–0.1320	**Q6–3**	–0.5790	–0.5110
P1–5	0.3560	0.3570	**Q1–4**	0.2010	0.2120			

9.2.1.2.1 Case 0: Basic Case with Only Random Noise In this case, the measured values in Table 9.11 are used. Table 9.12 shows the estimated value for each estimator.

Table 9.12 shows that all four estimators generate better overall results than measurements, and the difference between each estimator are really small. Since there is no bad data simulated in this 0-case, no bad data is detected and eliminated.

9.2.1.2.2 Case 1: Single Bad Data Based on the raw data in Table 9.11, the reverse of meter P2–4 is considered as bad data. Table 9.13 shows the estimated value for each estimator.

As shown in Table 9.13, all estimators generate much closer results to actual value when compared with the measurement meters. The estimation from LAV and LTAV are much better. Since LAV can successfully detect the bad data, LTAV generates the best estimation after the bad data is removed.

9.2.1.2.3 Case 2: Multiple Noninteracting Bad Data In this case, the two noninteracting P2–4 and P3–6 m are used to simulate bad data by reversing the polarity of the original measured values. Table 9.14 shows the estimated value for each estimator.

TABLE 9.12 Estimation errors of case 0, 6-bus AC system.

Meter	Actual	Meas.	WLS	LAV	LTS	LTAV
V1	1.0500	1.0365	1.0462	1.0442	1.0462	1.0462
V2	1.0500	1.0339	1.0423	1.0397	1.0423	1.0423
V3	1.0700	1.0900	1.0626	1.0591	1.0626	1.0626
V4	0.9896	0.9813	0.9827	0.9789	0.9827	0.9827
V5	0.9857	0.9791	0.9794	0.9764	0.9794	0.9794
V6	1.0043	0.9952	0.9991	0.9952	0.9991	0.9991
P1	1.0790	1.1310	1.1189	1.1310	1.1189	1.1189
P2	0.5000	0.4840	0.4750	0.4840	0.4750	0.4750
P3	0.6000	0.5510	0.5970	0.5907	0.5970	0.5970
P4	–0.7000	–0.7180	–0.7021	–0.7104	–0.7021	–0.7021
P5	–0.7000	–0.7200	–0.7196	–0.7200	–0.7196	–0.7196
P6	–0.7000	–0.7230	–0.6896	–0.6940	–0.6896	–0.6896
Q1	0.1600	0.2020	0.1894	0.2020	0.1894	0.1894
Q2	0.7440	0.7190	0.7017	0.7190	0.7017	0.7017
Q3	0.8960	0.9060	0.8647	0.8595	0.8647	0.8647
Q4	–0.7000	–0.7190	–0.7001	–0.7158	–0.7001	–0.7001
Q5	–0.7000	–0.6770	–0.6828	–0.6770	–0.6828	–0.6828
Q6	–0.7000	–0.6090	–0.6641	–0.6706	–0.6641	–0.6641
P1–2	0.2870	0.3150	0.3032	0.3071	0.3032	0.3032
P1–4	0.4360	0.3890	0.4476	0.4521	0.4476	0.4476
P1–5	0.3560	0.3570	0.3681	0.3717	0.3681	0.3681
P2–1	–0.2780	–0.3490	–0.2935	–0.2972	–0.2935	–0.2935
P2–4	0.3310	0.3280	0.3236	0.3280	0.3236	0.3236
P2–5	0.1550	0.1740	0.1566	0.1578	0.1566	0.1566
P2–6	0.2620	0.2230	0.2589	0.2630	0.2589	0.2589
P2–3	0.0290	0.0860	0.0293	0.0325	0.0293	0.0293
P3–2	–0.0290	–0.0210	–0.0289	–0.0321	–0.0289	–0.0289
P3–5	0.1910	0.1770	0.1920	0.1898	0.1920	0.1920
P3–6	0.4380	0.4330	0.4338	0.4330	0.4338	0.4338
P4–1	–0.4250	–0.4010	–0.4359	–0.4401	–0.4359	–0.4359
P4–2	–0.3160	–0.2980	–0.3090	–0.3128	–0.3090	–0.3090
P4–5	0.0410	0.0070	0.0428	0.0425	0.0428	0.0428
P5–4	–0.0400	–0.0210	–0.0424	–0.0421	–0.0424	–0.0424
P5–1	–0.3450	–0.3660	–0.3566	–0.3599	–0.3566	–0.3566
P5–2	–0.1500	–0.1170	–0.1518	–0.1528	–0.1518	–0.1518
P5–3	–0.1800	–0.2510	–0.1814	–0.1792	–0.1814	–0.1814
P5–6	0.0160	–0.0210	0.0125	0.0140	0.0125	0.0125
P6–5	–0.0160	0.0100	–0.0120	–0.0135	–0.0120	–0.0120
P6–2	0.2570	–0.1960	–0.2533	–0.2572	–0.2533	–0.2533
P6–3	–0.4280	–0.4680	–0.4243	–0.4234	–0.4243	–0.4243
Q1–2	–0.1540	–0.1320	–0.1413	–0.1396	–0.1413	–0.1413
Q1–4	0.2010	0.2120	0.2131	0.2214	0.2131	0.2131

TABLE 9.12 *(continued)*

Meter	Actual	Meas.	WLS	LAV	LTS	LTAV
Q1–5	0.1130	0.0940	0.1175	0.1202	0.1175	0.1175
Q2–1	0.1280	0.0970	0.1171	0.1160	0.1171	0.1171
Q2–4	0.4610	0.3830	0.4496	0.4585	0.4496	0.4496
Q2–5	0.1540	0.2200	0.1461	0.1467	0.1461	0.1461
Q2–6	0.1240	0.1500	0.1119	0.1168	0.1119	0.1119
Q2–3	−0.1230	−0.1190	−0.1229	−0.1190	−0.1229	−0.1229
Q3–2	0.0570	0.1020	0.0585	0.0549	0.0585	0.0585
Q3–5	0.2320	0.2390	0.2240	0.2216	0.2240	0.2240
Q3–6	0.6070	0.5830	0.5822	0.5830	0.5822	0.5822
Q4–1	−0.1990	−0.1430	−0.2076	−0.2140	−0.2076	−0.2076
Q4–2	−0.4510	−0.4430	−0.4410	−0.4486	−0.4410	−0.4410
Q4–5	−0.0490	−0.1740	−0.0515	−0.0531	−0.0515	−0.0515
Q5–4	−0.0280	−0.0150	−0.0246	−0.0225	−0.0246	−0.0246
Q5–1	−0.1350	−0.1750	−0.1358	−0.1371	−0.1358	−0.1358
Q5–2	−0.1800	−0.2220	−0.1725	−0.1726	−0.1725	−0.1725
Q5–3	−0.2610	−0.1990	−0.2531	−0.2507	−0.2531	−0.2531
Q5–6	−0.0970	−0.0080	−0.0969	−0.0942	−0.0969	−0.0969
Q6–5	0.0390	0.0290	0.0397	0.0373	0.0397	0.0397
Q6–2	−0.1600	−0.2230	−0.1481	−0.1519	−0.1481	−0.1481
Q6–3	−0.5790	−0.5110	−0.5556	−0.5559	−0.5556	−0.5556

TABLE 9.13 Estimation errors of case 1, 6-bus AC system.

Meter	Actual	Meas.	WLS	LAV	LTS	LTAV
V1	1.0500	1.0365	1.0476	1.0448	1.0476	1.0462
V2	1.0500	1.0339	1.0413	1.0401	1.0413	1.0423
V3	1.0700	1.0900	1.0608	1.0591	1.0608	1.0626
V4	0.9896	0.9813	0.9876	0.9798	0.9876	0.9827
V5	0.9857	0.9791	0.9790	0.9768	0.9790	0.9794
V6	1.0043	0.9952	0.9969	0.9952	0.9969	0.9991
P1	1.0790	1.1310	1.1345	1.1310	1.1345	1.1190
P2	0.5000	0.4840	0.3686	0.4840	0.3686	0.4741
P3	0.6000	0.5510	0.5648	0.5783	0.5648	0.5967
P4	−0.7000	−0.7180	−0.5370	−0.6857	−0.5370	−0.7007
P5	−0.7000	−0.7200	−0.7298	−0.7200	−0.7298	−0.7197
P6	−0.7000	−0.7230	−0.7228	−0.7066	−0.7228	−0.6800
Q1	0.1600	0.2020	0.1816	0.2020	0.1816	0.1893
Q2	0.7440	0.7190	0.6958	0.7190	0.6958	0.7017
Q3	0.8960	0.9060	0.8669	0.8603	0.8669	0.8647
Q4	−0.7000	−0.7190	−0.6926	−0.7188	−0.6926	−0.7000
Q5	−0.7000	−0.6770	−0.6824	−0.6770	−0.6824	−0.6828

(continued)

TABLE 9.13 *(continued)*

Meter	Actual	Meas.	WLS	LAV	LTS	LTAV
Q6	–0.7000	–0.6090	–0.6632	–0.6688	–0.6632	–0.6641
P1–2	0.2870	0.3150	0.3314	0.3096	0.3314	0.3034
P1–4	0.4360	0.3890	0.4176	0.4465	0.4176	0.4473
P1–5	0.3560	0.3570	0.3855	0.3749	0.3855	0.3683
P2–1	–0.2780	–0.3490	–0.3201	–0.2996	–0.3201	–0.2937
P2–4	**0.3310**	**–0.3280**	**0.2099**	**0.3128**	**0.2099**	**0.3226**
P2–5	0.1550	0.1740	0.1547	0.1592	0.1547	0.1566
P2–6	0.2620	0.2230	0.2782	0.2717	0.2782	0.2591
P2–3	0.0290	0.0860	0.0459	0.0400	0.0459	0.0295
P3–2	–0.0290	–0.0210	–0.0454	–0.0396	–0.0454	–0.0291
P3–5	0.1910	0.1770	0.1758	0.1850	0.1758	0.1919
P3–6	0.4380	0.4330	0.4344	0.4330	0.4344	0.4338
P4–1	–0.4250	–0.4010	–0.4074	–0.4347	–0.4074	–0.4357
P4–2	–0.3160	–0.2980	–0.1984	–0.2980	–0.1984	–0.3081
P4–5	0.0410	0.0070	0.0687	0.0470	0.0687	0.0430
P5–4	–0.0400	–0.0210	–0.0677	–0.0465	–0.0677	–0.0426
P5–1	–0.3450	–0.3660	–0.3730	–0.3628	–0.3730	–0.3567
P5–2	–0.1500	–0.1170	–0.1499	–0.1542	–0.1499	–0.1518
P5–3	–0.1800	–0.2510	–0.1657	–0.1746	–0.1657	–0.1812
P5–6	0.0160	–0.0210	0.0266	0.0181	0.0266	0.0126
P6–5	–0.0160	0.0100	–0.0260	–0.0176	–0.0260	–0.0121
P6–2	0.2570	–0.1960	–0.2720	–0.2656	–0.2720	–0.2535
P6–3	–0.4280	–0.4680	–0.4248	–0.4234	–0.4248	–0.4243
Q1–2	–0.1540	–0.1320	–0.1404	–0.1393	–0.1404	–0.1413
Q1–4	0.2010	0.2120	0.2006	0.2207	0.2006	0.2130
Q1–5	0.1130	0.0940	0.1214	0.1207	0.1214	0.1176
Q2–1	0.1280	0.0970	0.1193	0.1160	0.1193	0.1171
Q2–4	0.4610	0.3830	0.4429	0.4597	0.4429	0.4495
Q2–5	0.1540	0.2200	0.1445	0.1463	0.1445	0.1461
Q2–6	0.1240	0.1500	0.1114	0.1160	0.1114	0.1119
Q2–3	–0.1230	–0.1190	–0.1224	–0.1190	–0.1224	–0.1229
Q3–2	0.0570	0.1020	0.0585	0.0550	0.0585	0.0585
Q3–5	0.2320	0.2390	0.2249	0.2223	0.2249	0.2240
Q3–6	0.6070	0.5830	0.5836	0.5830	0.5836	0.5822
Q4–1	–0.1990	–0.1430	–0.2013	–0.2144	–0.2013	–0.2076
Q4–2	–0.4510	–0.4430	–0.4405	–0.4506	–0.4405	–0.4410
Q4–5	–0.0490	–0.1740	–0.0509	–0.0538	–0.0509	–0.0515
Q5–4	–0.0280	–0.0150	–0.0245	–0.0217	–0.0245	–0.0246
Q5–1	–0.1350	–0.1750	–0.1359	–0.1370	–0.1359	–0.1358
Q5–2	–0.1800	–0.2220	–0.1711	–0.1722	–0.1711	–0.1725
Q5–3	–0.2610	–0.1990	–0.2550	–0.2517	–0.2550	–0.2531
Q5–6	–0.0970	–0.0080	–0.0959	–0.0944	–0.0959	–0.0969
Q6–5	0.0390	0.0290	0.0390	0.0375	0.0390	0.0397
Q6–2	–0.1600	–0.2230	–0.1456	–0.1504	–0.1456	–0.1481
Q6–3	–0.5790	–0.5110	–0.5566	–0.5559	–0.5566	–0.5556

As shown in Table 9.14, LAV and LTAV estimators generate good results to actual value when compared with the measurement meters. The estimation from LTAV is slightly better. Both estimators successfully detected and eliminated the two noninteracting bad data. On the other hand, WLS and LTS estimators generate bad overall estimation. The reason is that LTS relays on WLS and does not trim all the two bad data, so the estimation is not as good as the other estimators.

TABLE 9.14 Estimation errors of case 2, 6-bus AC system.

Meter	Actual	Meas.	WLS	LAV	LTS	LTAV
V1	1.0500	1.0365	1.0477	1.0447	1.0477	1.0462
V2	1.0500	1.0339	1.0415	1.0400	1.0415	1.0423
V3	1.0700	1.0900	1.0595	1.0588	1.0595	1.0626
V4	0.9896	0.9813	0.9878	0.9798	0.9878	0.9827
V5	0.9857	0.9791	0.9788	0.9767	0.9788	0.9794
V6	1.0043	0.9952	0.9986	0.9952	0.9986	0.9990
P1	1.0790	1.1310	1.1349	1.1310	1.1349	1.1190
P2	0.5000	0.4840	0.3726	0.4840	0.3726	0.4741
P3	0.6000	0.5510	0.3664	0.5510	0.3664	0.5969
P4	−0.7000	−0.7180	−0.5355	−0.6855	−0.5355	−0.7007
P5	−0.7000	−0.7200	−0.7360	−0.7200	−0.7360	−0.7197
P6	−0.7000	−0.7230	−0.5275	−0.6799	−0.5275	−0.6901
Q1	0.1600	0.2020	0.1818	0.2020	0.1818	0.1893
Q2	0.7440	0.7190	0.6987	0.7190	0.6987	0.7017
Q3	0.8960	0.9060	0.8661	0.8610	0.8661	0.8648
Q4	−0.7000	−0.7190	−0.6920	−0.7182	−0.6920	−0.7001
Q5	−0.7000	−0.6770	−0.6877	−0.6770	−0.6877	−0.6828
Q6	−0.7000	−0.6090	−0.6737	−0.6720	−0.6737	−0.6640
P1–2	0.2870	0.3150	0.3304	0.3096	0.3304	0.3034
P1–4	0.4360	0.3890	0.4170	0.4465	0.4170	0.4473
P1–5	0.3560	0.3570	0.3876	0.3750	0.3876	0.3683
P2–1	−0.2780	−0.3490	−0.3192	−0.2995	−0.3192	−0.2937
P2–4	**0.3310**	**−0.3280**	**0.2106**	**0.3127**	**0.2106**	**0.3226**
P2–5	0.1550	0.1740	0.1574	0.1593	0.1574	0.1566
P2–6	0.2620	0.2230	0.2464	0.2672	0.2464	0.2591
P2–3	0.0290	0.0860	0.0774	0.0443	0.0774	0.0294
P3–2	−0.0290	−0.0210	−0.0767	−0.0438	−0.0767	−0.0290
P3–5	0.1910	0.1770	0.1527	0.1816	0.1527	0.1919
P3–6	**0.4380**	**−0.4330**	**0.2904**	**0.4132**	**0.2904**	**0.4340**
P4–1	−0.4250	−0.4010	−0.4068	−0.4346	−0.4068	−0.4357
P4–2	−0.3160	−0.2980	−0.1990	−0.2980	−0.1990	−0.3081
P4–5	0.0410	0.0070	0.0704	0.0471	0.0704	0.0430

(*continued*)

TABLE 9.14 *(continued)*

Meter	Actual	Meas.	WLS	LAV	LTS	LTAV
P5–4	−0.0400	−0.0210	−0.0694	−0.0466	−0.0694	−0.0426
P5–1	−0.3450	−0.3660	−0.3749	−0.3629	−0.3749	−0.3567
P5–2	−0.1500	−0.1170	−0.1526	−0.1543	−0.1526	−0.1518
P5–3	−0.1800	−0.2510	−0.1431	−0.1713	−0.1431	−0.1813
P5–6	0.0160	−0.0210	0.0039	0.0152	0.0039	0.0126
P6–5	−0.0160	0.0100	−0.0035	−0.0147	−0.0035	−0.0121
P6–2	0.2570	−0.1960	−0.2412	−0.2613	−0.2412	−0.2535
P6–3	−0.4280	−0.4680	−0.2828	−0.4039	−0.2828	−0.4244
Q1–2	−0.1540	−0.1320	−0.1405	−0.1392	−0.1405	−0.1413
Q1–4	0.2010	0.2120	0.2003	0.2205	0.2003	0.2130
Q1–5	0.1130	0.0940	0.1220	0.1207	0.1220	0.1176
Q2–1	0.1280	0.0970	0.1192	0.1158	0.1192	0.1171
Q2–4	0.4610	0.3830	0.4426	0.4592	0.4426	0.4495
Q2–5	0.1540	0.2200	0.1449	0.1462	0.1449	0.1461
Q2–6	0.1240	0.1500	0.1138	0.1167	0.1138	0.1119
Q2–3	−0.1230	−0.1190	−0.1219	−0.1190	−0.1219	−0.1229
Q3–2	0.0570	0.1020	0.0589	0.0551	0.0589	0.0585
Q3–5	0.2320	0.2390	0.2302	0.2229	0.2302	0.2240
Q3–6	0.6070	0.5830	0.5770	0.5830	0.5770	0.5822
Q4–1	−0.1990	−0.1430	−0.2010	−0.2143	−0.2010	−0.2076
Q4–2	−0.4510	−0.4430	−0.4402	−0.4502	−0.4402	−0.4410
Q4–5	−0.0490	−0.1740	−0.0508	−0.0538	−0.0508	−0.0515
Q5–4	−0.0280	−0.0150	−0.0245	−0.0217	−0.0245	−0.0246
Q5–1	−0.1350	−0.1750	−0.1361	−0.1369	−0.1361	−0.1358
Q5–2	−0.1800	−0.2220	−0.1713	−0.1721	−0.1713	−0.1725
Q5–3	−0.2610	−0.1990	−0.2614	−0.2525	−0.2614	−0.2531
Q5–6	−0.0970	−0.0080	−0.0945	−0.0938	−0.0945	−0.0969
Q6–5	0.0390	0.0290	0.0372	0.0368	0.0372	0.0397
Q6–2	−0.1600	−0.2230	−0.1509	−0.1515	−0.1509	−0.1481
Q6–3	−0.5790	−0.5110	−0.5599	−0.5574	−0.5599	−0.5556

9.2.1.2.4 Case 3: Multiple Interacting Bad Data In this case, the two bad data that are interacting with each other P2–4 and P2–5 are used. The reading of these two meters are reversed in polarity to simulate the presence of two interacting bad measurements. Table 9.15 shows the estimated values.

As shown in Table 9.15, LAV and LTAV estimators generate good results to actual value when compared with the measurement meters. Similar to case 2, the estimation from LTAV is slightly better. Both estimators successfully detected and eliminated the two interacting bad data. On the other hand, WLS and LTS estimators generate bad overall estimation.

TABLE 9.15 Estimation errors of case 3, 6-bus AC system.

Meter	Actual	Meas.	WLS	LAV	LTS	LTAV
V1	1.0500	1.0365	1.0474	1.0448	1.0474	1.0462
V2	1.0500	1.0339	1.0408	1.0401	1.0408	1.0423
V3	1.0700	1.0900	1.0609	1.0591	1.0609	1.0626
V4	0.9896	0.9813	0.9871	0.9798	0.9871	0.9827
V5	0.9857	0.9791	0.9799	0.9768	0.9799	0.9795
V6	1.0043	0.9952	0.9971	0.9952	0.9971	0.9991
P1	1.0790	1.1310	1.1311	1.1310	1.1311	1.1188
P2	0.5000	0.4840	0.3366	0.4840	0.3366	0.4724
P3	0.6000	0.5510	0.5644	0.5783	0.5644	0.5967
P4	–0.7000	–0.7180	–0.5473	–0.6857	–0.5473	–0.7013
P5	–0.7000	–0.7200	–0.6837	–0.7200	–0.6837	–0.7173
P6	–0.7000	–0.7230	–0.7240	–0.7066	–0.7240	–0.6899
Q1	0.1600	0.2020	0.1795	0.2020	0.1795	0.1892
Q2	0.7440	0.7190	0.6949	0.7190	0.6949	0.7016
Q3	0.8960	0.9060	0.8685	0.8603	0.8685	0.8648
Q4	–0.7000	–0.7190	–0.6935	–0.7188	–0.6935	–0.7001
Q5	–0.7000	–0.6770	–0.6842	–0.6770	–0.6842	–0.6829
Q6	–0.7000	–0.6090	–0.6632	–0.6688	–0.6632	–0.6641
P1–2	0.2870	0.3150	0.3351	0.3096	0.3351	0.3036
P1–4	0.4360	0.3890	0.4208	0.4465	0.4208	0.4475
P1–5	0.3560	0.3570	0.3752	0.3749	0.3752	0.3678
P2–1	–0.2780	–0.3490	–0.3236	–0.2996	–0.3236	–0.2939
P2–4	**0.3310**	**–0.3280**	**0.2088**	**0.3128**	**0.2088**	**0.3226**
P2–5	**0.1550**	**–0.1740**	**0.1419**	**0.1592**	**0.1419**	**0.1560**
P2–6	0.2620	0.2230	0.2706	0.2717	0.2706	0.2587
P2–3	0.0290	0.0860	0.0389	0.0400	0.0389	0.0291
P3–2	–0.0290	–0.0210	–0.0385	–0.0396	–0.0385	–0.0287
P3–5	0.1910	0.1770	0.1674	0.1850	0.1674	0.1915
P3–6	0.4380	0.4330	0.4355	0.4330	0.4355	0.4339
P4–1	–0.4250	–0.4010	–0.4104	–0.4347	–0.4104	–0.4358
P4–2	–0.3160	–0.2980	–0.1973	–0.2980	–0.1973	–0.3080
P4–5	0.0410	0.0070	0.0603	0.0470	0.0603	0.0426
P5–4	–0.0400	–0.0210	–0.0596	–0.0465	–0.0596	–0.0422
P5–1	–0.3450	–0.3660	–0.3633	–0.3628	–0.3633	–0.3562
P5–2	–0.1500	–0.1170	–0.1375	–0.1542	–0.1375	–0.1511
P5–3	–0.1800	–0.2510	–0.1575	–0.1746	–0.1575	–0.1808
P5–6	0.0160	–0.0210	0.0341	0.0181	0.0341	0.0130
P6–5	–0.0160	0.0100	–0.0335	–0.0176	–0.0335	–0.0125
P6–2	0.2570	–0.1960	–0.2646	–0.2656	–0.2646	–0.2531
P6–3	–0.4280	–0.4680	–0.4258	–0.4234	–0.4258	–0.4243

(*continued*)

TABLE 9.15 *(continued)*

Meter	Actual	Meas.	WLS	LAV	LTS	LTAV
Q1–2	–0.1540	–0.1320	–0.1409	–0.1393	–0.1409	–0.1413
Q1–4	0.2010	0.2120	0.2014	0.2207	0.2014	0.2130
Q1–5	0.1130	0.0940	0.1189	0.1207	0.1189	0.1174
Q2–1	0.1280	0.0970	0.1203	0.1160	0.1203	0.1171
Q2–4	0.4610	0.3830	0.4440	0.4597	0.4440	0.4496
Q2–5	0.1540	0.2200	0.1434	0.1463	0.1434	0.1460
Q2–6	0.1240	0.1500	0.1106	0.1160	0.1106	0.1119
Q2–3	–0.1230	–0.1190	–0.1235	–0.1190	–0.1235	–0.1230
Q3–2	0.0570	0.1020	0.0595	0.0550	0.0595	0.0586
Q3–5	0.2320	0.2390	0.2254	0.2223	0.2254	0.2240
Q3–6	0.6070	0.5830	0.5836	0.5830	0.5836	0.5822
Q4–1	–0.1990	–0.1430	–0.2015	–0.2144	–0.2015	–0.2076
Q4–2	–0.4510	–0.4430	–0.4415	–0.4506	–0.4415	–0.4410
Q4–5	–0.0490	–0.1740	–0.0506	–0.0538	–0.0506	–0.0515
Q5–4	–0.0280	–0.0150	–0.0253	–0.0217	–0.0253	–0.0247
Q5–1	–0.1350	–0.1750	–0.1358	–0.1370	–0.1358	–0.1358
Q5–2	–0.1800	–0.2220	–0.1712	–0.1722	–0.1712	–0.1725
Q5–3	–0.2610	–0.1990	–0.2562	–0.2517	–0.2562	–0.2531
Q5–6	–0.0970	–0.0080	–0.0957	–0.0944	–0.0957	–0.0969
Q6–5	0.0390	0.0290	0.0388	0.0375	0.0388	0.0397
Q6–2	–0.1600	–0.2230	–0.1456	–0.1504	–0.1456	–0.1481
Q6–3	–0.5790	–0.5110	–0.5565	–0.5559	–0.5565	–0.5556

TABLE 9.16 Comparison of indicators for 6-bus AC system.

Case no.	Meas.	WLS	LAV	LTS	LTAV
Case 0	2.4385	1.1079	1.1898	1.1079	1.1079
Case 1	3.0945	1.9207	1.2904	1.9207	1.1096
Case 2	3.9605	2.6214	1.3695	2.6214	1.1088
Case 3	3.4045	1.9087	1.2904	1.9087	1.1044

9.2.1.2.5 Summary Table 9.16 presents the performance comparison of the four estimators for the four cases.

As illustrated in Table 9.16 in cases 1, 2, and 3, LAV and LTAV estimators generate the best estimation. WLS and LTS estimators are not satisfactory when bad-data cases are introduced.

9.2.2 14-Bus System

The estimation results for the IEEE 14-bus system are presented in this section. AC state estimation is conducted with full- and median-redundancy for various scenarios. The line and bus data are listed in Appendix 9.B. Table 9.17 shows the actual values of voltage magnitudes, real- and reactive-power injections, and real- and reactive- power flows from load flow.

Random noises of normal distribution, with $\sigma = 0.016\,65$ for voltage measurements and $\sigma = 0.05$ for power measurements, are generated and added to the actual value to simulate the measurements. Table 9.18 shows the simulated measured values for the full-redundancy scenario. The selected meters along with their measurements for the median redundancy scenario are given in Table 9.19.

Due to the huge amount of estimated results, the performance of the four estimators will be shown using the indicators only. The following cases are considered:

- Case 0: Normal noise and no bad data.
- Case 1: Single bad data by reversal of meter P4.
- Case 2: Multiple noninteracting bad data by reversal of meters P4, P9–10, and P13–14.
- Case 3: Multiple interacting bad data by reversal of meters P4, P4–7, and P7–9.

Table 9.20 presents the performance of each estimator for full and median-redundancy cases, respectively.

It can be observed from Tables 9.20 and 9.21 that WLS and LTS generate the worst results for almost all cases but the overall estimation is still better than measurements meters. LAV's performance is very good and stable by detecting different cases of bad data. LTAV estimator generated better results in almost all cases. LTS estimator, though it generates good estimation in some cases, but its performance is not stable since it is based on bad-data detection of WLS.

The performance of full-redundancy cases is more stable when compared with median cases. Full-redundancy cases have almost 2.5 times rawer data than median cases. In the presence of bad data, the local-redundancy of full-redundancy cases is larger than local-redundancy of median redundancy cases, which further induce estimators' stability issue.

9.2.3 30-Bus System

The estimation results for the IEEE 30-bus system are presented in this section. AC state estimation is conducted with only median redundancy for various scenarios. The line and bus data are listed in Appendix 9.C.

Random noises of normal distribution, with $\sigma = 0.016\,65$ for voltage measurements and $\sigma = 0.05$ for power measurements, are generated and added to the actual value to simulate the measurements. The median-redundancy rate is selected to be 1.9, i.e. the number of measurement is equal to $1.9 \times N$, where Ns is the number of state variable and $N = 59$ in this section. The number of measurements is $m = 113$.

TABLE 9.17 Actual values for 14-bus AC system.

Meter	Actual	Meter	Actual	Meter	Actual	Meter	Actual	Meter	Actual
V1	1.0600	**P12**	–0.0610	**P4–9**	0.1608	**P13–6**	–0.1754	**Q9–14**	0.0361
V2	1.0450	**P13**	–0.1350	**P5–6**	0.4409	**P8–7**	0.0000	**Q10–11**	–0.0162
V3	1.0100	**P14**	–0.1490	**P6–11**	0.0735	**P9–7**	–0.2807	**Q12–13**	0.0075
V4	1.0180	**Q1**	–0.1655	**P6–12**	0.0779	**P10–9**	–0.0521	**Q13–14**	0.0175
V5	1.0200	**Q2**	0.3086	**P6–13**	0.1775	**P14–9**	–0.0931	**Q2–1**	0.2768
V6	1.0700	**Q3**	0.0608	**P7–8**	0.0000	**P11–10**	0.0380	**Q5–1**	0.0223
V7	1.0620	**Q4**	0.0390	**P7–9**	0.2807	**P13–12**	–0.0161	**Q3–2**	0.0160
V8	1.0900	**Q5**	–0.0160	**P9–10**	0.0523	**P14–13**	–0.0559	**Q4–2**	0.0302
V9	1.0560	**Q6**	0.0523	**P9–14**	0.0943	**Q1–2**	–0.2040	**Q5–2**	–0.0210
V10	1.0510	**Q7**	0.0000	**P10–11**	–0.0379	**Q1–5**	0.0385	**Q4–3**	–0.0484
V11	1.0570	**Q8**	0.1762	**P12–13**	0.0161	**Q2–3**	0.0356	**Q5–4**	0.1420
V12	1.0550	**Q9**	–0.1660	**P13–14**	0.0564	**Q2–4**	–0.0155	**Q7–4**	0.1138
V13	1.0500	**Q10**	–0.0580	**P2–1**	–1.5259	**Q2–5**	0.0117	**Q9–4**	0.0173
V14	1.0360	**Q11**	–0.0180	**P5–1**	–0.7275	**Q3–4**	0.0447	**Q6–5**	–0.0805
P1	2.3239	**Q12**	–0.0160	**P3–2**	–0.7091	**Q4–5**	0.1582	**Q11–6**	–0.0344
P2	0.1830	**Q13**	–0.0580	**P4–2**	–0.5445	**Q4–7**	–0.0968	**Q12–6**	–0.0235
P3	–0.9420	**Q14**	–0.0500	**P5–2**	–0.4061	**Q4–9**	–0.0043	**Q13–6**	–0.0680
P4	–0.4780	**P1–2**	1.5688	**P4–3**	0.2366	**Q5–6**	0.1247	**Q8–7**	0.1762
P5	–0.0760	**P1–5**	0.7551	**P5–4**	0.6167	**Q6–11**	0.0356	**Q9–7**	–0.0498
P6	–0.1120	**P2–3**	0.7324	**P7–4**	–0.2807	**Q6–12**	0.0250	**Q10–9**	–0.0418
P7	0.0000	**P2–4**	0.5613	**P9–4**	–0.1608	**Q6–13**	0.0722	**Q14–9**	–0.0336
P8	0.0000	**P2–5**	0.4152	**P6–5**	–0.4409	**Q7–8**	–0.1716	**Q11–10**	0.0164
P9	–0.2950	**P3–4**	–0.2329	**P11–6**	–0.0730	**Q7–9**	0.0578	**Q13–12**	–0.0075
P10	–0.0900	**P4–5**	–0.6116	**P12–6**	–0.0771	**Q9–10**	0.0422	**Q14–13**	–0.0164
P11	–0.0350	**P4–7**	0.2807						

TABLE 9.18 Measured meters for 14-bus AC system, full redundancy.

Meter	Meas.	Meter	Meas.	Meter	Meas.	Meter	Meas.	Meter	Meas.
V1	1.0686	**P12**	0.0907	**P4–9**	0.1771	**P13–6**	–0.2299	**Q9–14**	–0.0041
V2	1.0743	**P13**	–0.0987	**P5–6**	0.4032	**P8–7**	0.0016	**Q10–11**	0.0186
V3	0.9739	**P14**	–0.1522	**P6–11**	0.1420	**P9–7**	–0.2531	**Q12–13**	0.0493
V4	1.0318	**Q1**	–0.1298	**P6–12**	–0.0077	**P10–9**	0.0029	**Q13–14**	0.0053
V5	1.0251	**Q2**	0.2984	**P6–13**	0.1724	**P14–9**	–0.0159	**Q2–1**	0.2876
V6	1.0491	**Q3**	0.0546	**P7–8**	–0.0121	**P11–10**	0.0423	**Q5–1**	–0.0360
V7	1.0551	**Q4**	0.1135	**P7–9**	0.2967	**P13–12**	–0.0907	**Q3–2**	–0.0414
V8	1.0955	**Q5**	0.0545	**P9–10**	0.0679	**P14–13**	–0.0930	**Q4–2**	0.0354
V9	1.1131	**Q6**	0.1232	**P9–14**	0.0511	**Q1–2**	–0.2571	**Q5–2**	0.0151
V10	1.0952	**Q7**	0.0336	**P10–11**	–0.0394	**Q1–5**	0.1560	**Q4–3**	0.0809
V11	1.0355	**Q8**	0.1158	**P12–13**	0.0079	**Q2–3**	0.0048	**Q5–4**	0.1087
V12	1.1034	**Q9**	–0.1301	**P13–14**	0.0878	**Q2–4**	0.0219	**Q7–4**	0.1232
V13	1.0616	**Q10**	0.0235	**P2–1**	–1.4712	**Q2–5**	0.0021	**Q9–4**	0.0132
V14	1.0350	**Q11**	0.0064	**P5–1**	–0.6720	**Q3–4**	0.0891	**Q6–5**	–0.1772
P1	2.3508	**Q12**	0.0357	**P3–2**	–0.7523	**Q4–5**	0.1200	**Q11–6**	–0.0563
P2	0.2747	**Q13**	–0.0217	**P4–2**	–0.5406	**Q4–7**	–0.1669	**Q12–6**	–0.1132
P3	–1.0549	**Q14**	–0.0652	**P5–2**	–0.4668	**Q4–9**	–0.0754	**Q13–6**	–0.0260
P4	–0.4349	**P1–2**	1.5835	**P4–3**	0.1809	**Q5–6**	0.1491	**Q8–7**	0.1318
P5	–0.0601	**P1–5**	0.7157	**P5–4**	0.6164	**Q6–11**	0.0267	**Q9–7**	–0.0448
P6	–0.1774	**P2–3**	0.7768	**P7–4**	–0.2041	**Q6–12**	0.0152	**Q10–9**	–0.0690
P7	–0.0217	**P2–4**	0.5039	**P9–4**	–0.1993	**Q6–13**	0.1432	**Q14–9**	–0.0184
P8	0.0171	**P2–5**	0.3618	**P6–5**	–0.4223	**Q7–8**	–0.1570	**Q11–10**	–0.0136
P9	–0.1161	**P3–4**	–0.2734	**P11–6**	–0.0843	**Q7–9**	0.0677	**Q13–12**	0.0170
P10	0.0485	**P4–5**	–0.7588	**P12–6**	–0.0212	**Q9–10**	0.1216	**Q14–13**	0.0206
P11	–0.1025	**P4–7**	0.3526						

TABLE 9.19 Selected measured meters for 14-bus AC system, median redundancy.

Meter	Meas.	Meter	Meas.	Meter	Meas.
V1	1.0686	Q4	0.1135	P7–8	–0.0121
P1	2.3508	Q5	0.0545	P7–9	0.2967
P2	0.2747	Q6	0.1232	P9–10	0.0679
P3	–1.0549	Q7	0.0336	P10–11	–0.0394
P4	–0.4349	Q8	0.1158	P13–14	0.0878
P5	–0.0601	Q9	–0.1301	Q1–2	–0.2571
P6	–0.1774	Q10	0.0235	Q1–5	0.156
P7	–0.0217	Q11	0.0064	Q2–3	0.0048
P8	0.0171	Q12	0.0357	Q3–4	0.0891
P9	–0.1161	Q13	–0.0217	Q4–7	–0.1669
P10	0.0485	Q14	–0.0652	Q5–6	0.1491
P11	–0.1025	P1–2	1.5835	Q6–12	0.0152
P12	0.0907	P1–5	0.7157	Q7–8	–0.157
P13	–0.0987	P2–3	0.7768	Q7–9	0.0677
P14	–0.1522	P3–4	–0.2734	Q9–10	0.1216
Q1	–0.1298	P4–7	0.3526	Q10–11	0.0186
Q2	0.2984	P5–6	0.4032	Q13–14	0.0053
Q3	0.0546	P6–12	–0.0077		

TABLE 9.20 Full redundancy of 14-bus AC system.

Case no.	WLS	LAV	LTS	LTAV
Case 0	2.1408	2.0361	2.1408	1.9526
Case 1	3.6465	2.6437	2.2273	2.0061
Case 2	3.6727	2.7602	2.2691	2.1637
Case 3	4.0497	2.2664	3.2721	2.1690

TABLE 9.21 Median redundancy of 14-bus AC system.

Case no.	WLS	LAV	LTS	LTAV
Case 0	2.5633	2.9987	2.5633	3.0435
Case 1	4.7656	3.3894	4.7656	3.1157
Case 2	4.7477	3.6854	4.7477	3.2348
Case 3	5.981	3.5747	10.7832	3.4798

Real- and reactive-power injection meters are placed on each bus. The involvement of power injections ensures the observability of the system. Real- and reactive- power flow meters are placed on 26 lines. A voltage magnitude meter is placed on the slack-bus "bus 1."

Table 9.22 shows the selected actual and measured values of voltage magnitude, real- and reactive-power injection, and real- and reactive- power flow meters for a median redundancy scenario.

TABLE 9.22 Actual and measured values for 30-bus AC system.

Meter	Actual	Meas.	Meter	Actual	Meas.	Meter	Actual	Meas.
V1	1.0600	1.0690	Q8	0.0611	0.0240	P10–20	0.0903	0.1383
P1	2.6096	2.6540	Q9	0.0000	−0.0531	P10–17	0.0533	0.0595
P2	0.1830	0.1256	Q10	−0.0200	0.0975	P10–21	0.1579	0.2297
P3	−0.0240	−0.0774	Q11	0.1606	0.1298	P21–22	−0.0183	−0.0282
P4	−0.0760	−0.1165	Q12	−0.0750	−0.0376	P22–24	0.0574	0.2028
P5	−0.9420	−1.0892	Q13	0.1045	0.0949	P23–24	0.0180	0.0593
P6	0.0000	0.0719	Q14	−0.0160	0.0284	P25–27	−0.0476	−0.0710
P7	−0.2280	−0.2117	Q15	−0.0250	−0.0632	P28–27	0.1807	0.1671
P8	−0.3000	−0.3377	Q16	−0.0180	−0.0881	P27–30	0.0709	0.0570
P9	0.0000	0.0685	Q17	−0.0580	−0.1291	P29–30	0.0370	0.0721
P10	−0.0580	−0.1436	Q18	−0.0090	0.0154	P6–28	0.1867	0.1690
P11	0.0000	−0.0051	Q19	−0.0340	−0.0429	Q1–3	0.0428	0.0595
P12	−0.1120	−0.1241	Q20	−0.0070	−0.0168	Q2–4	0.0475	0.0671
P13	0.0000	0.0160	Q21	−0.1120	−0.0410	Q3–4	−0.0385	−0.0159
P14	−0.0620	−0.0464	Q22	0.0000	0.0146	Q2–5	0.0278	0.0213
P15	−0.0820	−0.1252	Q23	−0.0160	−0.0061	Q4–6	−0.1591	−0.1829
P16	−0.0350	−0.0365	Q24	−0.0670	0.0124	Q5–7	0.1149	0.1580
P17	−0.0900	−0.0982	Q25	0.0000	−0.0402	Q6–7	−0.0278	−0.0959
P18	−0.0320	−0.0006	Q26	−0.0230	0.0118	Q6–9	−0.0809	−0.1233
P19	−0.0950	−0.0403	Q27	0.0000	0.0418	Q9–10	0.0588	0.1108
P20	−0.0220	0.0335	Q28	0.0000	−0.0122	Q12–14	0.0240	0.0570
P21	−0.1750	−0.2182	Q29	−0.0090	0.0018	Q12–15	0.0679	0.0645
P22	0.0000	0.0039	Q30	−0.0190	−0.0773	Q16–17	0.0144	−0.0008
P23	−0.0320	−0.0927	P1–3	0.8765	0.8817	Q15–18	0.0160	0.0172
P24	−0.0870	−0.1427	P2–4	0.4365	0.4726	Q18–19	0.0062	0.0088
P25	0.0000	−0.0003	P3–4	0.8214	0.9507	Q19–20	−0.0279	0.0134
P26	−0.0350	0.0416	P2–5	0.8236	0.7903	Q10–20	0.0371	0.1134
P27	0.0000	−0.0385	P4–6	0.7213	0.7172	Q10–17	0.0443	0.0676

(continued)

TABLE 9.22 *(continued)*

Meter	Actual	Meas.	Meter	Actual	Meas.	Meter	Actual	Meas.
P28	0.0000	0.0186	P5–7	−0.1478	−0.2445	Q10–21	0.1001	0.0896
P29	−0.0240	−0.0353	P6–7	0.3813	0.3594	Q21–22	−0.0143	−0.0051
P30	−0.1060	−0.0501	P6–9	0.2772	0.3192	Q22–24	0.0306	0.0781
Q1	−0.2042	−0.2587	P9–10	0.2772	0.2500	Q23–24	0.0124	0.0278
Q2	0.4337	0.4353	P12–14	0.0786	0.1031	Q25–27	−0.0037	0.0094
Q3	−0.0120	0.0156	P12–15	0.1789	0.2159	Q28–27	0.0504	0.0033
Q4	−0.0160	0.039	P16–17	0.0369	−0.0700	Q27–30	0.0166	0.0093
Q5	0.1666	0.2438	P15–18	0.0602	0.0182	Q29–30	−0.0061	0.0205
Q6	0.0000	0.0043	P18–19	0.0278	0.0955	Q6–28	0.0011	−0.0427
Q7	−0.1090	−0.1836	P19–20	−0.0673	−0.1209			

The performance of the four estimators will be shown using the indicators only. The following cases are considered:

- Case 0: Normal noise and no bad data.
- Case 1: Single bad data by reversal of meter P2.
- Case 2: Multiple noninteracting bad data by reversal of meters P2, P10–21, P12–14, and P25–27.
- Case 3: Multiple interacting bad data by reversal of meters P2, P2–4, P4–6, and P6–9.

Table 9.23 presents the performance of each estimator for all cases.

It can be observed from Table 9.23 that LTAV estimator generates better overall estimation than the other estimators. In the presence of multiple interacting bad data, WLS and LTS estimators did not perform well when compared with LAV and LTAV.

TABLE 9.23 Median redundancy of 30-bus AC system.

Case no.	WLS	LAV	LTS	LTAV
Case 0	6.5405	7.8971	6.5405	6.5405
Case 1	7.0127	8.4977	7.0127	7.0127
Case 2	7.9108	8.3359	7.9108	7.1024
Case 3	15.5771	9.2069	17.8626	7.1332

9.2.4 Section Summary

In this section, 6-bus power system is used for both DC and AC state estimations, and full-redundancy scenario. IEEE 14-bus system is used for AC state estimation and full- and median- redundancy scenarios. IEEE 30-bus power system is used for AC median-redundancy state estimation. In all systems, the performance of WLS, LAV, LTS, and LTAV estimators are investigated for four different cases: normal noise, presence of single bad data, multiple noninteracting bad data, and multiple interacting bad data.

Among all cases, overall performance of LTAV estimator is better than other estimators. LAV has performed very good in most of the cases and sometimes match LTAV estimator. LTS performed, in general, better than WLS. However, in certain cases it gives worse results than WLS. Full-redundancy cases generate better estimation due to the high local redundancy.

9.3 CONCLUSIONS

Robust estimator plays a key role in power system state estimation. Various estimators are used to improve the efficiency in detecting different types of bad data and estimating system's state variables.

Least-trimmed-absolute-value, inspired by LAV and LTS, is proposed to improve the estimation process. LTAV is introduced by combining the merits of both LAV and LTS estimators.

The performance of four estimators WLS, LAV, LTS, and LTAV are investigated using three power systems: 6-bus, IEEE 14-bus, and 30-bus system. The 6-bus power system is used for both DC and AC state estimation, and full-redundancy scenario. IEEE 14-bus system is used for AC state estimation, and full- and median-redundancy scenarios. IEEE 30-bus power system is used for AC median-redundancy state estimation.

In all systems, the performance of WLS, LAV, LTS, and LTAV estimators are investigated for four different cases: normal noise, presence of single bad data, multiple noninteracting bad data, and multiple interacting bad data.

System data redundancy plays a very important role in state estimation. If overall and local-redundancy are large, estimators can generate better estimation. The larger the redundancy is, the estimation are more accurate and reliable. If redundancy cannot be guaranteed or large redundancy are not realistic, robust estimator helps a lot in state estimation with available data.

9.A.1 6-BUS DC SYSTEM

TABLE 9.A.1 Line data of 6-bus DC system.

Line no.	From bus	To bus	R (p.u.)	X (p.u.)	$B/2$ (p.u.)	Tap
1	1	2	0	1	0	1
2	1	4	0	1	0	1
3	1	5	0	1	0	1
4	2	3	0	1	0	1
5	2	4	0	1	0	1
6	2	5	0	1	0	1
7	2	6	0	1	0	1
8	3	5	0	1	0	1
9	3	6	0	1	0	1
10	4	5	0	1	0	1
11	5	6	0	1	0	1

TABLE 9.A.2 Bus data of 6-bus DC system.

		Voltage		Load		Generator				
Bus	Type	Mag.	Ang.	Real	Reac.	Real	Reac.	Q-min	Q-max	Q-inj
1	1	1	0	0	0	0	0	0	0	0
2	2	1	0	0	0	0.5	0	0	0	0
3	2	1	0	0	0	0.6	0	0	0	0
4	0	1	0	0.7	0	0	0	0	0	0
5	0	1	0	0.7	0	0	0	0	0	0
6	0	1	0	0.7	0	0	0	0	0	0

9.A.2 6-BUS AC SYSTEM

TABLE 9.A.3 Line data of 6-bus AC system.

Line no.	From bus	To bus	R (p.u.)	X (p.u.)	$B/2$ (p.u.)	Tap
1	1	2	0.1	0.2	0.02	1
2	1	4	0.05	0.2	0.02	1
3	1	5	0.08	0.3	0.03	1
4	2	3	0.05	0.25	0.03	1
5	2	4	0.05	0.1	0.01	1
6	2	5	0.1	0.3	0.02	1

TABLE 9.A.3 (*continued*)

Line no.	From bus	To bus	R (p.u.)	X (p.u.)	$B/2$ (p.u.)	Tap
7	2	6	0.07	0.2	0.025	1
8	3	5	0.12	0.26	0.025	1
9	3	6	0.02	0.1	0.01	1
10	4	5	0.2	0.4	0.04	1
11	5	6	0.1	0.3	0.03	1

TABLE 9.A.4 Bus data of 6-bus AC system.

		Voltage		Load		Generator				
Bus	Type	Mag.	Ang.	Real	Reac.	Real	Reac.	Q-min	Q-max	Q-inj
1	1	1.05	0	0	0	0	0	0	0	0
2	2	1.05	0	0	0	0.5	0	0	0	0
3	2	1.07	0	0	0	0.6	0	0	0	0
4	0	1	0	0.7	0.7	0	0	0	0	0
5	0	1	0	0.7	0.7	0	0	0	0	0
6	0	1	0	0.7	0.7	0	0	0	0	0

9.B 14-BUS AC SYSTEM

TABLE 9.B.1 Line data of 14-bus AC system.

Line no.	From bus	To bus	R (p.u.)	X (p.u.)	$B/2$ (p.u.)	Tap
1	1	2	0.019 38	0.059 17	0.0264	1
2	1	5	0.054 03	0.223 04	0.0246	1
3	2	3	0.046 99	0.197 97	0.0219	1
4	2	4	0.058 11	0.176 32	0.017	1
5	2	5	0.056 95	0.173 88	0.0173	1
6	3	4	0.067 01	0.171 03	0.0064	1
7	4	5	0.013 35	0.042 11	0	1
8	4	7	0	0.209 12	0	0.978
9	4	9	0	0.556 18	0	0.969

(*continued*)

TABLE 9.B.1 *(continued)*

Line no.	From bus	To bus	R (p.u.)	X (p.u.)	$B/2$ (p.u.)	Tap
10	5	6	0	0.252 02	0	0.932
11	6	11	0.094 98	0.1989	0	1
12	6	12	0.122 91	0.255 81	0	1
13	6	13	0.066 15	0.130 27	0	1
14	7	8	0	0.176 15	0	1
15	7	9	0	0.110 01	0	1
16	9	10	0.031 81	0.0845	0	1
17	9	14	0.127 11	0.270 38	0	1
18	10	11	0.082 05	0.192 07	0	1
19	12	13	0.220 92	0.199 88	0	1
20	13	14	0.170 93	0.348 02	0	1

TABLE 9.B.2 Bus data of 14-bus AC system.

		Voltage		Load		Generator				
Bus	Type	Mag.	Ang.	Real	Reac.	Real	Reac.	Q-min	Q-max	Q-inj
1	1	1.06	0	0	0	2.3239	−0.1655	0	0	0
2	2	1.045	−4.983	0.217	0.127	0.4	0.4356	−0.4	0.5	0
3	2	1.01	−12.725	0.942	0.19	0	0.2508	0	0.4	0
4	0	1.018	−10.313	0.478	−0.039	0	0	0	0	0
5	0	1.02	−8.774	0.076	0.016	0	0	0	0	0
6	2	1.07	−14.221	0.112	0.075	0	0.1273	−0.06	0.24	0
7	0	1.062	−13.36	0	0	0	0	0	0	0
8	2	1.09	−13.36	0	0	0	0.1762	−0.06	0.24	0
9	0	1.056	−14.939	0.295	0.166	0	0	0	0	0.19
10	0	1.051	−15.097	0.09	0.058	0	0	0	0	0
11	0	1.057	−14.791	0.035	0.018	0	0	0	0	0
12	0	1.055	−15.076	0.061	0.016	0	0	0	0	0
13	0	1.05	−15.156	0.135	0.058	0	0	0	0	0
14	0	1.036	−16.034	0.149	0.05	0	0	0	0	0

9.C 30-BUS AC SYSTEM

TABLE 9.C.1 Line data of 30-bus AC system.

No.	From bus	To bus	*R* (p.u.)	*X* (p.u.)	*B*/2(p.u.)	Tap
1	1	2	0.0192	0.0575	0.0264	1
2	1	3	0.0452	0.1652	0.0204	1
3	2	4	0.057	0.1737	0.0184	1
4	3	4	0.0132	0.0379	0.0042	1
5	2	5	0.0472	0.1983	0.0209	1
6	2	6	0.0581	0.1763	0.0187	1
7	4	6	0.0119	0.0414	0.0045	1
8	5	7	0.046	0.116	0.0102	1
9	6	7	0.0267	0.082	0.0085	1
10	6	8	0.012	0.042	0.0045	1
11	6	9	0	0.208	0	0.978
12	6	10	0	0.556	0	0.969
13	9	11	0	0.208	0	1
14	9	10	0	0.11	0	1
15	4	12	0	0.256	0	0.932
16	12	13	0	0.14	0	1
17	12	14	0.1231	0.2559	0	1
18	12	15	0.0662	0.1304	0	1
19	12	16	0.0945	0.1987	0	1
20	14	15	0.221	0.1997	0	1
21	16	17	0.0524	0.1923	0	1
22	15	18	0.1073	0.2185	0	1
23	18	19	0.0639	0.1292	0	1
24	19	20	0.034	0.068	0	1
25	10	20	0.0936	0.209	0	1
26	10	17	0.0324	0.0845	0	1
27	10	21	0.0348	0.0749	0	1
28	10	22	0.0727	0.1499	0	1
29	21	22	0.0116	0.0236	0	1
30	15	23	0.1	0.202	0	1
31	22	24	0.115	0.179	0	1

(continued)

TABLE 9.C.1 *(continued)*

No.	From bus	To bus	R (p.u.)	X (p.u.)	$B/2$(p.u.)	Tap
32	23	24	0.132	0.27	0	1
33	24	25	0.1885	0.3292	0	1
34	25	26	0.2544	0.38	0	1
35	25	27	0.1093	0.2087	0	1
36	28	27	0	0.396	0	0.968
37	27	29	0.2198	0.4153	0	1
38	27	30	0.3202	0.6027	0	1
39	29	30	0.2399	0.4533	0	1
40	8	28	0.0636	0.2	0.0214	1
41	6	28	0.0169	0.0599	0.0065	1

TABLE 9.C.2 Bus data of 30-bus AC system.

		Voltage		Load		Generator				
Bus	Type	Mag.	Ang.	Real	Reac.	Real	Reac.	Q-min	Q-max	Q-inj
1	1	1.06	0	0	0	2.6096	−0.2042	0	0	0
2	2	1.045	−5.378	0.217	0.127	0.4	0.5607	−0.4	0.5	0
3	0	1	−7.520	0.024	0.012	0	0	0	0	0
4	0	1.012	−0.270	0.076	0.016	0	0	0	0	0
5	2	1.01	−14.140	0.042	0.10	0	0.3566	−0.4	0.4	0
6	0	1.011	−11.055	0	0	0	0	0	0	0
7	0	1	−12.852	0.228	0.109	0	0	0	0	0
8	2	1.01	−11.707	0.3	0.3	0	0.3611	−0.1	0.4	0
9	0	1.051	−14.008	0	0	0	0	0	0	0
10	0	1.045	−15.688	0.058	0.02	0	0	0	0	0.19
11	2	1.082	−14.008	0	0	0	0.1606	−0.06	0.24	0
12	0	1.057	−14.033	0.112	0.075	0	0	0	0	0
13	2	1.071	−14.033	0	0	0	0.1045	−0.06	0.24	0
14	0	1.043	−15.825	0.062	0.016	0	0	0	0	0
15	0	1.038	−15.016	0.082	0.025	0	0	0	0	0
16	0	1.045	−15.515	0.035	0.018	0	0	0	0	0
17	0	1.04	−15.85	0.00	0.058	0	0	0	0	0
18	0	1.028	−16.53	0.032	0.009	0	0	0	0	0
10	0	1.026	−16.704	0.005	0.034	0	0	0	0	0
20	0	1.03	−16.507	0.022	0.007	0	0	0	0	0
21	0	1.033	−16.131	0.175	0.112	0	0	0	0	0
22	0	1.034	−16.116	0	0	0	0	0	0	0
23	0	1.027	−16.307	0.032	0.016	0	0	0	0	0

TABLE 9.C.2 *(continued)*

		Voltage		Load		Generator				
Bus	Type	Mag.	Ang.	Real	Reac.	Real	Reac.	Q-min	Q-max	Q-inj
24	0	1.022	−16.483	0.087	0.067	0	0	0	0	0.043
25	0	1.018	−16.055	0	0	0	0	0	0	0
26	0	1	−16.474	0.035	0.023	0	0	0	0	0
27	0	1.024	−15.53	0	0	0	0	0	0	0
28	0	1.007	−11.677	0	0	0	0	0	0	0
20	0	1.004	−16.750	0.024	0.009	0	0	0	0	0
30	0	0.002	−17.642	0.106	0.019	0	0	0	0	0

REFERENCES

1. Wu, W., Guo, Y., Zhang, B. et al. (2011). Robust state estimation method based on maximum exponential square. *IET Generation, Transmission and Distribution* 5 (11): 1165–1172.
2. Kekatos, V. and Giannakis, G. (2013). Distributed robust power system state estimation. *IEEE Transactions on Power Systems* 28 (2): 1617–1626.
3. Bernieri, A., Betta, G., Liguori, C., and Losi, A. (1996). Neural networks and pseudo-measurements for real-time monitoring of distribution systems. *IEEE Transactions on Instrumentation and Measurement* 45 (2): 645–650.
4. Glazunova, A. (2010). Forecasting power system state variables on the basis of dynamic state estimation and artificial neural networks. 2010 IEEE Region 8 International Conference on Computational Technologies in Electrical and Electronics Engineering (SIBIRCON), pp. 470–475.
5. Manitsas, E., Singh, R., Pal, B., and Strbac, G. (2012). Distribution system state estimation using an artificial neural network approach for pseudo measurement modeling. *IEEE Transactions on Power Systems* 27 (4): 1888–1896.
6. Wu, F. and Monticelli, A. (1985). Network observability: theory. *IEEE Transactions on Power Apparatus and Systems* PAS-104 (5): 1042–1048.
7. Monticelli, A. and Wu, F. (1985). Network observability: identification of observable islands and measurement placement. *IEEE Transactions on Power Apparatus and Systems* PAS-104 (5): 1035–1041.
8. Gomez-Exposito, A. and Abur, A. (1998). Generalized observability analysis and measurement classification. *IEEE Transactions on Power Systems* 13 (3): 1090–1095.
9. Huber, P.J. (1981). *Robust Statistics*, Wiley Series in Probability and Mathematics Statistics. Wiley.
10. Handschin, E., Schweppe, F., Kohlas, J., and Fiechter, A. (1975). Bad data analysis for power system state estimation. *IEEE Transactions on Power Apparatus and Systems* 94 (2): 329–337.
11. Mili, L., Cheniae, M., Vichare, N., and Rousseeuw, P. (1996). Robust state estimation based on projection statistics [of power systems]. *IEEE Transactions on Power Systems* 11 (2): 1118–1127.

12. Koenker, R. and Bassett, G. Jr. (1978). Regression quantiles. *Econometrica: Journal of the Econometric Society* 46: 33–50.
13. Jaeckel, L.A. (1972). Estimating regression coefficients by minimizing the dispersion of the residuals. *The Annals of Mathematical Statistics* 43: 1449–1458.
14. Singh, H. and Alvarado, F. (1993). "*Fast approximations to LAV solutions for state estimation of power systems. Proceedings of PSCC Avignon* (30 August–3 September 1993), France.
15. Rousseeuw, P.J. (1984). Least median of squares regression. *Journal of the American Statistical Association* 79 (388): 871–880.
16. Mili, L., Phaniraj, V., and Rousseeuw, P. (1991). Least median of squares estimation in power systems. *IEEE Transactions on Power Systems* 6 (2): 511–523.
17. Irving, M. (2008). Robust state estimation using mixed integer programming. *IEEE Transactions on Power Systems* 23 (3): 1519–1520.
18. Berkelaar, M., Eikland, K., and Notebaert, P. (2004). lp solve 5.5, open source (mixed-integer) linear programming system. *Software*, May 1, 2004. http://lpsolve.sourceforge.net/5.5 (accessed 18 December 2009).
19. Wood, A.J. and Wollenberg, B.F. (1996). *Power Generation, Operation, and Control.* Wiley.

PART III

STATE ESTIMATION FOR DISTRIBUTION SYSTEMS

CHAPTER 10

PROBABILISTIC STATE ESTIMATION IN DISTRIBUTION NETWORKS

Bernd Brinkmann and Michael Negnevitsky

Centre for Renewable Energy and Power Systems, University of Tasmania, Hobart, Australia

10.1 INTRODUCTION

State estimation is a procedure that can be used to obtain an estimate of the network state by processing the available set of measurements. The state of a network is commonly defined as the voltage magnitude and angle at every bus. Other parameters in the network such as power flows and currents can be calculated from the network state. The information provided by the state estimation is used to assess the network security, to analyze contingencies, and to make decisions on required control actions. The concept of applying state estimation to power systems was developed around 1970 [1]. Since then state estimation has become a routine task in transmission systems. Over time a large number of redundant measurement devices were installed throughout transmission networks, and accurate network models have been developed. As a result, it is usually possible to estimate the state of a transmission network with a high degree of accuracy.

Historically state estimation has not been applied to distribution networks. In recent years, however, smart grid technologies such as demand response and energy storage have emerged. These technologies can be used to make distribution networks more flexible and efficient and increase the maximum amount of renewable generation connected to the distribution systems. In order to implement and manage these technologies in an optimal manner, information about the network

Advances in Electric Power and Energy: Static State Estimation, First Edition.
Edited by Mohamed E. El-Hawary.

Published 2021 by John Wiley & Sons, Inc.

state has to be available at the distribution level. However, in distribution networks the number of available real-time measurements is usually very limited compared with transmission networks [2]. Therefore, pseudo-measurements, which are forecasts of the load and/or generation at specific buses, are often used as additional measurements. It is, however, not possible to accurately forecast the load and generation based on historical data. Therefore, if a large number of pseudo-measurements are used in the state estimation process, the estimated network state could contain a significant amount of uncertainty [3, 4]. This makes it difficult to implement the traditional approach to state estimation in distribution networks.

Before the state estimation can be performed, it has to be determined if a unique estimate of the network state can be obtained from the available set of measurements. This is done by the observability analysis. In order to calculate the state of a network, a number of linearly independent measurements at least, but really more than equal to the number of elements in the state vector are required. A network that fulfills this criterion is considered to be observable [5]. If the number of linearly independent measurements is lower than the number of elements in the state vector, the network is considered to be unobservable. This means that it is not possible to obtain a unique solution for the network state from the available set of measurements. However, a lack of measurements can always be compensated by the use of pseudo-measurements, irrespective of their accuracy. Therefore, if a network is made observable by using a large number of pseudo-measurements, the estimated state may not be accurate enough to be practical even if the network is classified as observable. This represents a limitation of the traditional approach to observability.

This chapter presents a new probabilistic approach to state estimation in distribution networks based on confidence. This probabilistic approach is then also extended to observability, resulting in a new observability assessment that focuses on the accuracy of the state estimation results.

10.2 STATE ESTIMATION IN DISTRIBUTION NETWORKS

State estimation refers to the process of obtaining the state of a network from the set of redundant measurements. The state of a network with a number of n buses is uniquely identified by $2n$ parameters. These are the voltage magnitude and angle at every bus. The voltage angle at one bus is used as a reference and commonly set to zero ($\theta_1 = 0$) [6]. Hence, the state vector has $2n - 1$ elements and is defined by

$$x^T = [\theta_2, \ldots, \theta_n, V_1, \ldots, V_n] \tag{10.1}$$

where the voltage angle is given by θ and the voltage magnitude by V. The relationship between the state vector described in Eq. (10.1) and the vector, z, which contains all measurements that are used in the state estimation process, is given by

$$z = h(x) + e \tag{10.2}$$

where e refers to the vector of measurement errors and h represents the nonlinear function that relates the state vector, x, to the measurement vector, z.

A number of state estimation methods have been developed for an application in transmission networks. The most popular method is the weighted least squares method that is an iterative process and minimizes the following objective function [7]:

$$J(x) = [z - h(x)]^T R^{-1} [z - h(x)] \tag{10.3}$$

$$R = \text{diag}\left[\sigma_1^2, \sigma_2^2, \ldots, \sigma_m^2\right] \tag{10.4}$$

where R is the measurement error covariance matrix, σ^2 is the measurement variance, and m is the number of measurements used in the state estimation. The weighted least squares method is an iterative process given by

$$x^{k+1} = x^k + G(x^k)^{-1} H^T(x^k) R^{-1} [z - h(x)] \tag{10.5}$$

$$H(x) = \left[\frac{\partial h(x)}{\partial x}\right] \tag{10.6}$$

$$G(x) = H^T(x) R^{-1} H(x) \tag{10.7}$$

where x^k represents the network state at iteration k, H is the Jacobian matrix, and G is the Gain matrix.

A variety of state estimation methods have been developed specifically for distribution networks. One of the first papers to consider state estimation in distribution networks was [8]. Here the fact that the number of available measurements is very limited was acknowledged and the proposed method filled missing information with statistical data. A three-phase state estimation method based on the weighted least squares approach was presented in [9–11]. Since distribution networks can be very large, computationally efficient state estimation methods have been proposed in [12–14]. These methods use, among others, a constant gain matrix in order to increase the computational efficiency of the state estimation in distribution networks. This is achieved by using a current-based approach in which power measurements are converted to equivalent current measurements. In order to further decrease the computation time for large distribution networks, multi-area state estimation methods have been proposed in [15–18]. These methods can perform the state estimation in parallel for the individual network areas and combine the results afterward by using reference buses that are located at the boundaries between the areas and are equipped with measurements.

Current distribution network state estimation methods predominantly aim at estimating the expected value of the state vector elements. However, the estimated network state can be significantly different from the true state if a large proportion of the measurements are pseudo-measurements. For this reason, additional information about the accuracy of the estimated state is required in order to make an objective assessment of the network state. The idea of considering the statistical properties of the state estimation result has been discussed in [3]. In [19] an efficient method that estimates the network state as well as the statistical properties of the estimated parameters is presented. However, the main output of the presented state estimators is still the expected value of the estimated parameters.

10.2.1 A Confidence-Based Approach to State Estimation in Distribution Networks

The following method represents a significant change in the way how information about the network state is provided to the operator. Previous methods focused on providing the expected values of the estimated parameters to the network operator. However, considering that the main concern of the distribution network operator is to maintain a network state that is within its constraints, it can be argued that the network operator does not need to know the exact value of the estimated parameters but rather if these parameters are within their constraints. This also implies that the expected value alone may not be practical for the network operation, since it is not possible to determine if an estimated parameter is within its constraints. This is due to the typically high uncertainty in distribution network state estimation as discussed in [3]. Therefore, if a high uncertainty is present in a state estimation result, the result can only be interpreted if the expected values of the estimated parameters are considered together with information about the accuracy of the estimates [4]. However, assessing these two pieces of information can be difficult, especially considering the usually large size of distribution networks. Instead of providing the expected value of the estimated parameters, the approach presented in this section focuses on the confidence that the estimated parameters are within their respective constraints [20]. Using this confidence value makes it possible to combine the expected value and the state estimation accuracy into a single number, which makes an interpretation of the results straightforward.

10.2.2 State Estimation Accuracy

The accuracy of an estimated parameter expresses how close an estimated value is to its true value. However, the true value of an estimated parameter is never known. For this reason, the accuracy of a parameter estimate expresses the maximum expected difference between a parameter estimate and the true value of this parameter. This maximum expected difference is given by the endpoints of the confidence interval of the estimated parameter. A confidence interval expresses the range of values in which the true value is expected to lie in with a predefined level of confidence. Since the risk that the true value is outside of the confidence interval can never be zero (except if the interval ranges from minus infinity to plus infinity, the endpoints are equal to physical limitations of the underlying system, or a combination of the two), it is required to define a confidence level that represents the risk that the true value lies outside of the confidence interval. Hence, a high confidence level results in a low risk that the true value is outside of the confidence interval and vice versa. The confidence intervals of a Gaussian-distributed

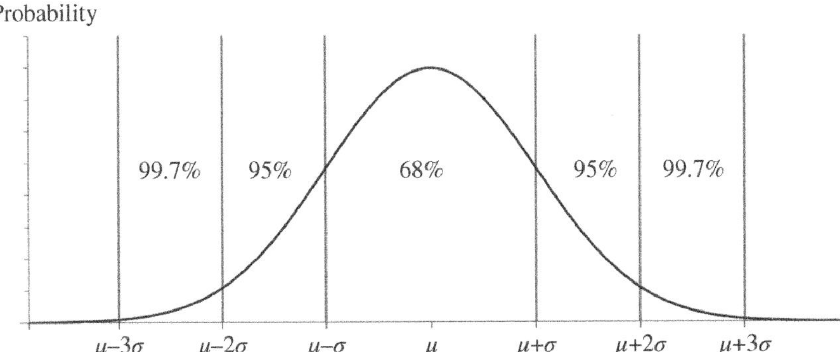

Figure 10.1 The 68, 95, and 99.7% confidence interval of a Gaussian-distributed probability density function.

probability density function (PDF) for the confidence values of 68, 95, and 99.7% are illustrated in Figure 10.1. Here μ refers to the mean value and σ to the standard deviation of the PDF.

The confidence that the true value is within a specified range is equal to the area under the PDF within this range. Therefore, as a first step, the PDFs of all estimated parameters have to be calculated in order to assess the state estimation accuracy. A number of methods are available for this purpose [19, 21–23]. A common method is to use the gain matrix to calculate the variance of the estimated parameter under the assumption of Gaussian-distributed measurement errors [24].

$$\mathrm{cov}(x) = \left[H^T(x)R^{-1}H(x)\right]^{-1} \tag{10.8}$$

where R is the measurement error covariance matrix with $R = \mathrm{diag}\left[\sigma_1^2, \sigma_2^2, \ldots, \sigma_m^2\right]$ and σ^2 represents the measurement variance. The diagonal elements of cov(x) represent the variance of the state vector entries. If the variance of a Gaussian-distributed parameter is known, its PDF is given by

$$\mathrm{PDF}_i(a_i) = \frac{1}{\sqrt{2\pi\sigma_i{}^2}}\, e^{-\frac{(a_i - E_i)^2}{2\sigma_i{}^2}} \tag{10.9}$$

where an estimated parameter is denoted by i for $i = [1, 2, \ldots, i, \ldots, I]$, I is the total number of estimated parameters, the estimated value of the ith estimated parameter is given by E_i, the variance of the estimate E_i is represented by σ_i^2, and a_i is a possible value of the parameter i.

In [22] this approach was expanded to non-Gaussian-distributed inputs by using Gaussian mixture models to represent the PDF of the state estimation inputs. This method performs a state estimation for each combination of the Gaussian components used to represent the PDFs of the state estimation inputs. The PDF of the estimated parameters is then constructed by combining the Gaussian distributions of the individual estimates that are weighted according to the product of the weights of the Gaussian components used in each state estimation. Another approach is to use Monte Carlo simulations to calculate the PDF of the estimated parameters. The advantages of the Monte Carlo method are its flexibility and straightforward implementation. However, since a large number of simulations are required for convergence, the computation time can be significant. Many other methods are also available, and the most appropriate method has to be chosen depending on the individual network. At this point it is important to note that the confidence-based state estimation approach does not depend on a particular method to calculate the PDF.

For the confidence-based state estimation approach, the confidence that the true value of an estimated parameter is within its constraints needs to be obtained. The confidence value of the ith estimated parameter is defined by the area under the PDF within its constraints ($i_{\min}$ and $i_{\max}$), calculated by

$$\text{Conf}_i = \int_{i_{\min}}^{i_{\max}} \text{PDF}_i(a_i)da_i \tag{10.10}$$

where Conf_i is the probability or confidence that the true value of the estimated parameter is within its constraints, $i_{\min}$ and $i_{\max}$ represent the minimum and maximum value of the ith estimated parameter, a_i is a possible value of the ith estimated parameter, and the probability density function of the estimated parameter i is given by PDF_i. An increase in Conf_i represents a decrease in risk that the true value of the estimated parameter i is outside its constraints, and a decrease of Conf_i implies the opposite. The constraints are defined by either the physical limitations of the network such as the current-carrying capacity of lines or regulations such as the voltage compliance range.

10.2.3 Computation Time and System Integration

The computation time of the presented confidence-based approach to state estimation mainly depends on the method chosen to obtain the PDF of the estimated parameters and therefore can vary widely depending on the exact method used. For instance, two common methods to calculate the PDF are the gain matrix approach and the Monte Carlo simulation. If the weighted least squares method is used to estimate the network state, the gain matrix can be employed to obtain the PDF of the estimated parameters with nearly no increase in computation time. This is due to the fact that the gain matrix is a by-product of the weighted least squares state estimation method. Under some circumstances it may, however, not be possible to use this method due to specific network topologies and measurement configurations. Especially nonlinear network elements such as voltage regulators with unmonitored tap positions can make it difficult to apply the gain matrix approach. The Monte Carlo method repeatedly performs state estimations with randomly generated inputs until the histograms of the estimated parameters converge.

The resulting histograms then represent the PDFs of the estimated parameters. The computation time of this method can be significant due to the large number of simulations required for convergence. Hence, the computation time is often reduced by, for instance, the use of additional processors, which then perform the Monte Carlo simulations in parallel. The advantages of this method are its straightforward implementation, easy to interpret results, and the fact that this method can generally be applied to any network or feeder that is classified as observable by traditional observability analysis methods. These two methods were taken as an example since they represent the extremes with a low and high amount of additional computational time. As mentioned earlier, several methods have been proposed in the literature for the purpose of calculating the PDFs of the estimated parameters. The best method has to be chosen depending on the network topology, measurement configuration, and requirements in terms of computation time and accuracy.

The presented approach is intended to deliver situational awareness to the distribution network operator. This is achieved by providing the confidence values that the estimated parameters are within their respective constraints to the network operator. The confidence value can also be easily visualized by using diagrams such as heat maps as shown in Figure 10.10. If the operator requires additional information about a particular estimated parameter, its PDF and expected value are available on request. As proposed in [25], it is also possible to use the calculated confidence values to implement an alarm system that informs the operator about a potential constraint violation. This system would be triggered if the confidence that an estimated parameter is within its constraints drops below a predefined threshold.

Furthermore, the confidence value of the estimated parameters could be used to implement control methods such as the voltage control approach presented in [26]. Here a method was presented to adjust the on-load tap-changer setting at the substation depending on the confidence values of the estimated voltage magnitudes. The objective of this method is to increase the maximum amount of distributed generation that could be connected to a distribution feeder if the limiting factor is a potential voltage constraint violation.

10.2.4 Case Studies

Distribution networks consist of a number of individual feeders and can be very large in size. However, a large system would not be suitable to illustrate the presented method due to the large number of results that would have to be provided. Therefore, the case studies in this chapter only consider individual feeders. However, all methods presented here can be applied to larger networks as well.

The confidence-based approach to state estimation in distribution networks is demonstrated on a real 13-bus feeder and a real 148-bus feeder and compared with the traditional approach. The 13-bus feeder is a low-voltage feeder with a nominal voltage magnitude of 230 V. Voltage magnitude measurements are located at buses 1, 3, 8, and 13 as designated by the arrows in Figure 10.2. A current measurement is also available at bus 1. The accuracy of these measurements is ±1%

of the measured value. All loads are represented as pseudo-measurements that are obtained from historical data and have the accuracy of ±100%.

This feeder has a voltage compliance range equal to ±6% of the nominal voltage magnitude as specified in the electricity code that is relevant for this feeder [27]. Three different cases are considered. For case 1 the feeder is studied under high load, and the state estimation is performed without the voltage magnitude measurements at buses 3, 8, and 13. The measurements have been recorded over a period of two weeks, and the instant with the highest recorded load during the two-week period is used as the high-load scenario. In order to demonstrate how the state estimation results can change with the measurement configuration, an additional voltage magnitude measurement at bus 13 is used for case 2 while everything else is identical to the previous case. In case 3 the feeder is considered under low load while using the same measurement configuration as case 2. For the low-load scenario, a typical night load that was recorded around midnight is used.

The estimated voltage magnitudes for case 1 are shown in Figure 10.3. Considering that real measurements are only available at bus 1, it is clear that the estimate could contain a large amount of uncertainty and therefore can be significantly different from the actual state. However, it is also apparent that the results provided in Figure 10.3 contain no information about the accuracy of the estimated state.

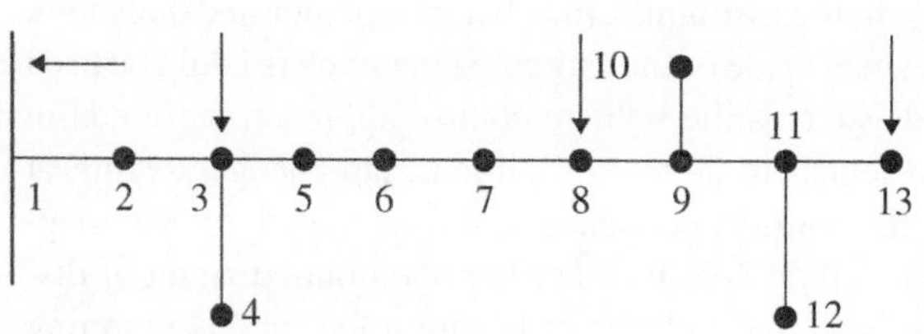

Figure 10.2 The 13-bus feeder.

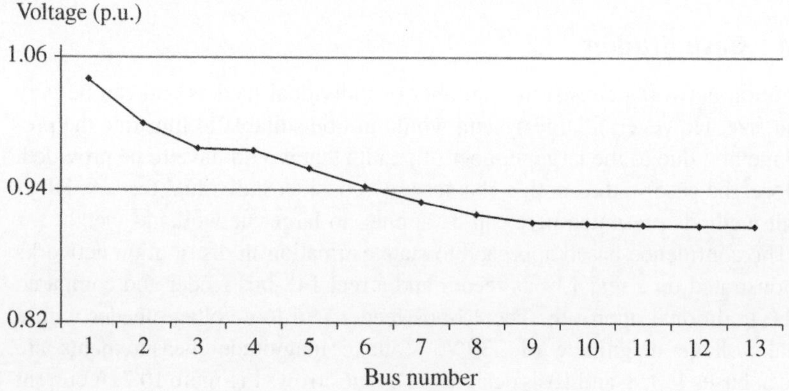

Figure 10.3 The voltage profile of the 13-bus feeder (case 1).

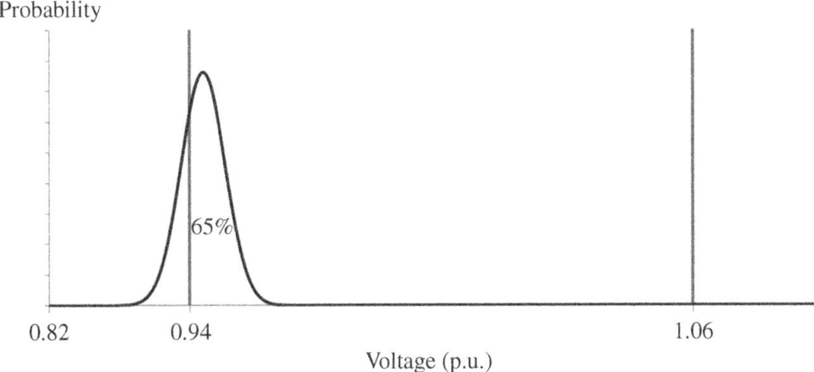

Figure 10.4 PDF of the voltage estimate at bus 6 of the 13-bus feeder (case 1).

Figure 10.4 provides the PDF of the estimated voltage magnitude at bus 6. The vertical bars represent the voltage constraints and the black curve, the PDF provided by the estimated value, and its standard deviation. The area under the curve within the voltage constraints is equal to 0.65 that corresponds to a 65% confidence that the true value is within the constraints. This shows that even if an estimated value is within its constraints, the confidence that the true value is within the constraints can still be low. This is especially the case if estimates are in proximity to their constraints and if the accuracy of the estimate is low.

The confidence-based approach is able to provide more practical information about the state of a distribution network in the presence of uncertain state estimation inputs. Table 10.1 provides the estimated voltage magnitudes as well as information about the statistical properties of the estimates in the form of the standard deviation.

TABLE 10.1 Example simulation results with confidence values.

Bus number	V (p.u.)	Standard deviation	Confidence (%)
1	1.040	0.0030	100
2	1.000	0.0030	100
3	0.978	0.0031	100
4	0.976	0.0033	100
5	0.959	0.0041	100
6	0.943	0.0052	65
7	0.929	0.0065	9
8	0.918	0.0078	2
9	0.911	0.0089	1
10	0.910	0.0090	1
11	0.908	0.0093	0
12	0.908	0.0094	0
13	0.908	0.0094	0

These two pieces of information represent the proximity of the estimate to its constraints as well as the accuracy of the estimate. However, especially for larger networks it would be difficult to make decisions on the network operation based on this information since it is difficult to interpret in this form. After combining these two pieces of information into a confidence value, it is much easier to interpret the state estimation results and to identify critical buses that may need attention.

The results of the confidence-based state estimation approach are shown in Figure 10.5 where the bars indicate the 95% confidence interval. The confidence that the true values of the parameters are within their respective constraints is shown as percentage value above the respective estimates. The estimated voltages have been compared with the actual voltage magnitude measurements in order to get an indication of how close this estimate is to its true value. This resulted in an average error of 1.8% and a maximum error of 3.4% at bus 13.

Figure 10.6 provides the estimated state for case 2. Here the average error between the measurements and the estimated state is reduced to 0.4% with the

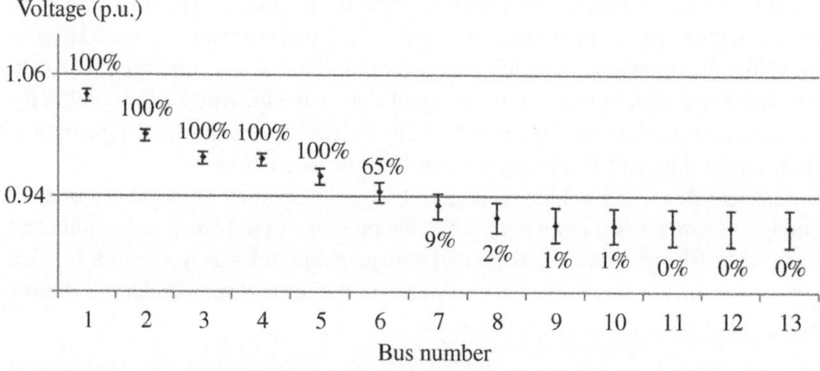

Figure 10.5 The voltage profile of the 13-bus feeder including confidence values (case 1).

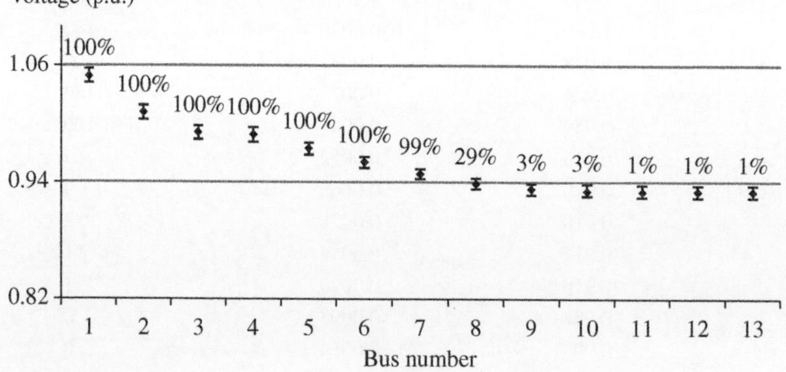

Figure 10.6 Voltage profile of the 13-bus feeder confidence values (case 2).

maximum error equal to 0.8% at bus 3. The reduced error is also reflected by the 95% confidence intervals of the estimated voltage magnitudes, which are much smaller compared with case 1. The change in the confidence values due to the added voltage magnitude measurement is especially apparent at buses 6 and 7, since the estimated values are close to their constraint. For instance, the confidence at bus 6 was 65% for case 1, and after increasing the accuracy by adding an additional measurement at bus 13, the confidence at bus 6 increased to 100%. As a general rule, if the state estimation accuracy increases, the respective confidence values will move toward the extreme values (very low or very high confidence) since it becomes easier to determine if an estimated parameter is within its constraints.

In case 3 the load was changed to a typical night load while keeping the metering configuration identical to case 2. For this simulation, the average error between the measured voltages and the estimates is equal to 0.4%, and the maximum error was determined to be at bus 13 with 0.7% (Figure 10.7).

How the proposed approach can be implemented in a larger feeder is demonstrated on a real 145-bus feeder. The nominal voltage of this feeder is 11 kV, and the topology is outlined in Figure 10.8. Here bus 1, which is also the slack bus, is marked by a square, arrows indicate power flow measurements, and a voltage regulator is identified by a circle. Since this feeder only has a very small number of real-time measurements compared with the number of parameters that have to be estimated, pseudo-measurements are added at each bus. These pseudo-measurements represent the power injection at the respective buses and are modeled by using the number of customers connected to each bus in combination with an average customer load profile, which was provided by the distribution network operator. The accuracy of these pseudo-measurements is assumed to be ±50%. The power flow measurements indicated by the arrows have the accuracy of ±10%, and the voltage magnitude at bus 1 was recorded with the accuracy of ±1%. For this feeder all measurement errors are assumed to be normally distributed, and the voltage compliance range is ±5% of the nominal feeder voltage as specified in the electricity code that is relevant for this feeder [28].

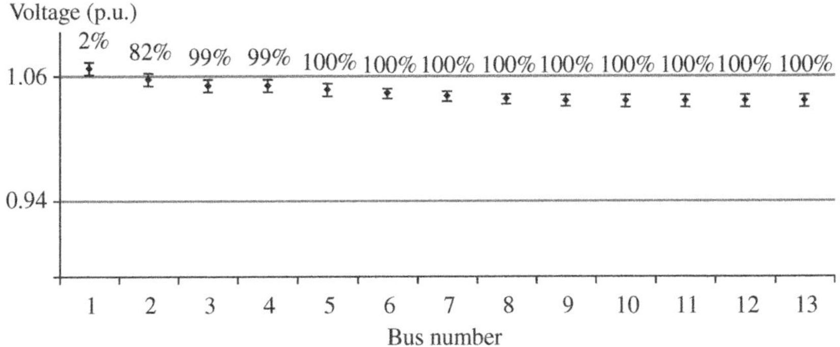

Figure 10.7 Voltage profile of the 13-bus feeder confidence values (case 3).

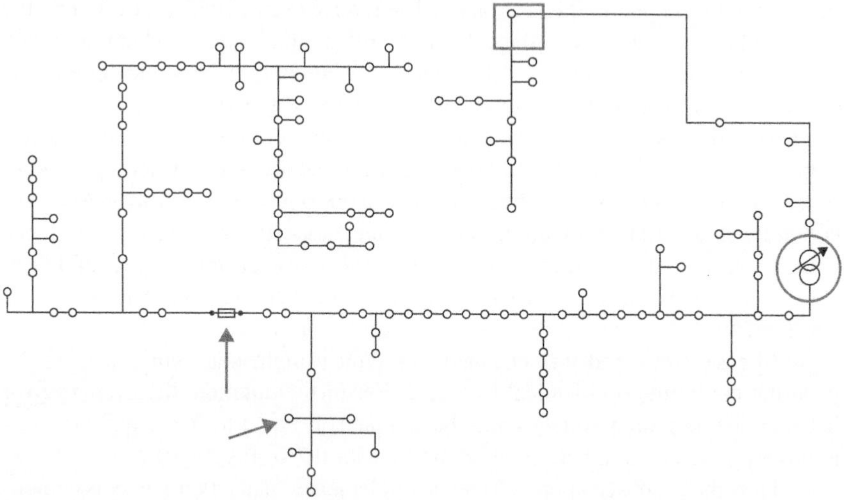

Figure 10.8 The 145-bus test feeder.

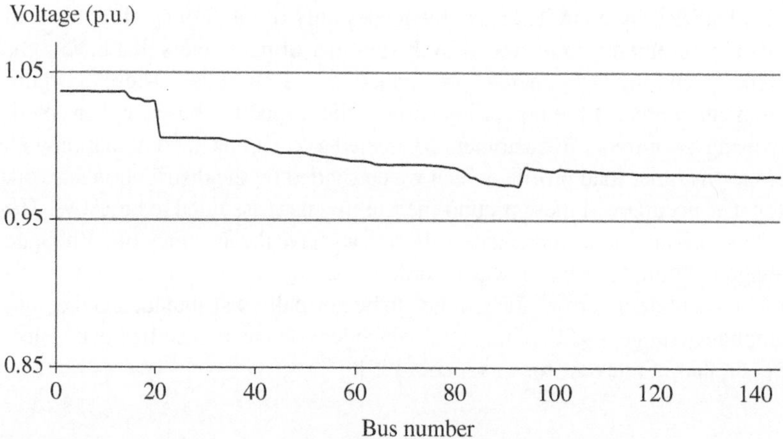

Figure 10.9 The voltage profile of the 145-bus feeder.

The 145-bus feeder is considered under the maximum forecasted load, which according to the average load profile occurs at 7:30 p.m. When considering the location of the feeder, this time is after sunset, which means that no distributed generation from photovoltaics is present in the feeder at the simulated time. The resulting voltage profile is shown in Figure 10.9.

The estimated values are all within the compliance range. However, as demonstrated earlier, the confidence that the true values are within their respective constraints can still be relatively low [3]. This further illustrates that an interpretation of state estimation results is difficult without additional information about the accuracy.

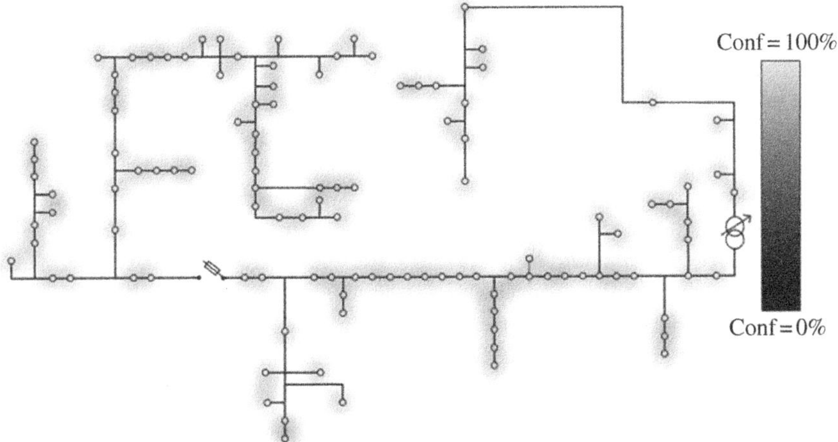

Figure 10.10 The voltage of the 145-bus feeder with a heat map overlay.

However, even if information about the accuracy or the estimate is available, it is difficult to present this information to the operator in a manner that is easy to interpret for a network with a large number of buses. The confidence-based approach makes it straightforward to provide this information since the expected value and information about the estimation accuracy are combined into a single number. An example of how the confidence value could be used to quickly identify problem areas in a network is given in Figure 10.10 by using a heat map overlay over the networks one line diagram of a network. From Figure 10.10 it is immediately apparent that the confidence of the whole network operating within its constraints is high. Therefore, the network is in a normal state, and no control actions are required. This conclusion could not be drawn from information provided in Figure 10.9 alone.

10.3 IMPROVING OBSERVABILITY IN DISTRIBUTION NETWORKS

The observability analysis is performed before the state estimation and determines whether it is possible to estimate the network state with the available set of measurements or not. If a network is defined as observable, a unique solution to the state estimation problem exists. If the observability analysis method defines a network as unobservable (that is under-determined), additional pseudo–measurements have to be used in order to perform the state estimation. In transmission network this approach works well since the number of available real-time measurements is relatively high. Hence, even if a small number of pseudo-measurements have to be used to make a network observable, the accuracy of the estimate is usually still high. Compared with transmission networks, the number of buses in distribution networks is usually significantly larger. Furthermore, the number of real-time measurements available for state estimation is often very limited due to economic constraints. Therefore, distribution networks are nearly always classified as unobservable. In order to achieve observability, pseudo-measurements, which are forecasted values for the connected load and/or generation, are used in the absence of real measurements [29, 30]. Pseudo-measurements are based on historical data such as billing information and generally make up a large portion of the measurement set. However, since it is impossible to accurately predict loads and distributed generation from historical data, pseudo-measurements have a large amount of uncertainty associated with them if compared with real-time measurements. Hence, if a large portion of the measurements used in a state estimation process are pseudo-measurements, the estimated network state might differ significantly from the true state of the network, even if the network is classified as observable [3].

Two main observability analysis methods can be identified in the literature. Namely, the numerical and the topological observability analysis [31]. The numerical observability analysis is based on the inverse function theorem and resolves around analyzing the Jacobian matrix. In the numerical observability analysis, the relationship between the network state and the measurement values is considered without measurement errors given by

$$z = h(x) \tag{10.11}$$

This simplification is justified by the fact that the observability analysis only determines if a network state can be estimated and not how accurate the resulting estimate might be. By linearizing Eq. (10.11), the mismatch between the measured values and the calculated values at an estimated state x_0 is given by

$$\Delta z = H\Delta x \tag{10.12}$$

where Δz is the difference between the measurement vector and the calculated values at the estimated state x_0 calculated by $\Delta z = z - h(x_0)$, Δx is given by $\Delta x = x - x_0$, and H is the Jacobian matrix. The Jacobian matrix has the dimensions $m \times (2n - 1)$, with m representing the number of measurements used for the state estimation and n being the number of buses in the system. The minus one signifies that one bus is used as reference and the angle at this bus is set to zero. Consequently this angle does not have to be estimated since it is already fixed. According to the inverse function theory, a unique solution for the network state can be obtained if the determinant of the Jacobian matrix is not equal to zero ($\det(H) \neq 0$) or if the column rank of the Jacobian matrix is of full rank. If the column rank of the Jacobian matrix is not full or the determinant is equal to zero, the network is considered to be unobservable, and a unique solution to the state estimation problem does not exist. In this case, pseudo-measurements have to be introduced in order to make the network observable.

The topological observability analysis is based on the graph theory and attempts to build a spanning tree of full rank. This approach only relies on logical operations and therefore does not use any floating point calculations [32–34]. For this reason, the values of the state estimation inputs as well as the measurement accuracy are not considered in the topological observability analysis. The topological method classifies a network as observable if it is possible to build a spanning tree of full rank. The tree of a network is the set of its loop free and connected branches. If a tree contains every branch in a network, it is referred to as a spanning tree. Finally, a tree is considered to be of full rank if a measurement can be uniquely assigned to every branch in the network [35].

Since the concept of state estimation has been introduced to power systems, significant research has been conducted in order to improve numerical performance of observability analysis methods. Another focus of this research was to adapt existing methods to new developments in state estimation such as the use of more detailed network models and phasor measurement units (PMUs).

For instance, under some circumstances, the Jacobian matrix can become ill-conditioned. Possible reasons for this could be if a very long line is connected to a very short line, if a large number of injection measurements are used, or if very low/high weights are used to represent pseudo-measurements or zero injection buses. In [36] a method based on the orthogonal transformation approach was proposed in order to address this problem. The proposed approach is numerically more stable and can potentially alleviate the problem of ill-conditioning. The numerical observability analysis was also modified in [34, 37, 38] to incorporate more detailed network models that included zero impedance branches to account for equipment such as switches and circuit breakers. By adding zero impedance branches, the power flow through these branches is added to the state vector x. The standard numerical observability analysis was expanded in [37, 38] to account for this generalized state estimation approach and [34] did this for the topological observability analysis. In [39–41] the observability analysis was extended to incorporate PMU measurements. PMUs use the Global Positioning System (GPS) as a time reference to provide synchronized measurements of the voltage magnitude and phase angle.

A main focus of the research on distribution network observability is the problem of unbalanced distribution networks. Since transmission systems are assumed to be balanced, observability analysis methods were developed to analyze a single-phase representation of a network. Distribution networks, on the other hand, often consist of long untransposed feeders that serve three-, two-, and single-phase loads. For this reason, three-phase state estimators have been introduced, which estimate the network state for each phase individually. Therefore, an observability analysis that can handle three-phase network models and measurement sets is discussed in [42–44]. Another feature of distribution networks is the low X/R ratio of distribution lines. In transmission systems, lines are predominantly inductive (X >> R), and therefore, the line resistance is generally neglected. In this case it can be observed that the phase angle depends predominantly on the active power and the system voltage is mainly affected by the reactive power. This is a key assumption in the numerical observability analysis for transmission systems. In [5] a method has been described, which is based on the numerical observability analysis and uses the orthogonal linear rotational transformation to account for the low X\R ratio of the distribution lines.

A lack of measurements in the state estimation process can always be compensated by adding pseudo-measurements. However, these additional measurements do not have to be accurate in order to make a network observable as shown in Eq. (10.11). It is obvious that if a large number of pseudo-measurements are used in the state estimation, the estimated state could contain a significant amount of uncertainty. Hence, the estimated state may not be accurate enough to be practical even if the network is classified as observable. This represents a significant limitation of the traditional approach to observability if applied to distribution networks.

10.3.1 Probabilistic Approach to Observability

In this section a probabilistic approach to assess the observability of distribution networks is introduced. Considering that the aim of a distribution network operator is to maintain a network state that does not violate any network constraints, it is important to know if the estimated parameters such as voltages and currents are within their respective constraints. In terms of state estimation accuracy, this implies that the network state has to be estimated with an accuracy that is at least high enough to determine if the estimated parameters are within their respective constraints under normal network operating conditions. A network is considered to operate normally if all connected loads are being supplied, without violating any constraints [7].

Traditionally a network has been considered observable if it is possible to calculate the network state based on the available set of measurements. The method discussed in this section, however, takes a different approach. It defines a network as observable if it is possible to estimate the network state with a degree of accuracy that is sufficient to determine if the true value of the calculated parameters is within their respective constraints under normal operating conditions. From here on this approach is referred to as probabilistic observability assessment [45].

As explained earlier, due to the random nature of the measurement errors, it can never be guaranteed that the true value of an estimated parameter is within a specific range. For this reason, a confidence interval has to be used, which represents the range of values that contains the true value of the estimated parameter with a predefined level of confidence. Hence, for an estimated parameter i, the endpoints of its confidence interval have to fulfill the following equations for a predefined confidence level:

$$\int_{E_i - a_{i,\min}}^{E_i} \mathrm{PDF}_i(a_i)\,da_i = \mathrm{CL}/2 \tag{10.13}$$

$$\int_{E_i}^{E_i + a_{i,\max}} \mathrm{PDF}_i(a_i)\,da_i = \mathrm{CL}/2 \tag{10.14}$$

where CL is the predefined confidence level and the accuracy of the estimate E_i is given by $a_{i,\min}$ and $a_{i,\max}$, which represent the maximum expected difference between E_i and its true value for values below and above E_i, respectively. In case of a Gaussian-distributed PDF, the confidence interval can be obtained by multiplying its standard deviation with a coverage factor that relates to the predefined confidence level. If the PDFs are not Gaussian distributed, other well-known methods such as bootstrapping could be used to obtain the confidence interval [46].

Now that it is defined how the accuracy of an estimated parameter can be calculated, it has to be quantified how much accuracy is required for each estimated parameter. As stated earlier, the accuracy of an estimated network state has to be high enough to determine if the estimated parameters are within their respective constraints under normal operating conditions. Therefore, the required accuracy for a parameter estimate is defined as the difference between the estimated value of this parameter and its constraints.

$$MA_{i,\min} = -\left[i_{\min} - E_i\right] \tag{10.15}$$

$$MA_{i,\max} = \left[i_{\max} - E_i\right] \tag{10.16}$$

where the minimum and maximum values of the estimated parameter i are given by $i_{\min}$ and $i_{\max}$, respectively, and the required accuracy for values below or above E_i are given by $MA_{i,\min}$ and $MA_{i,\max}$, respectively.

Since the minimum required accuracy calculated by Eqs. (10.10) and (10.11) changes depending on the estimated value, every possible network state would have to be evaluated in order to determine if a specific network complies with the minimum required accuracy. However, the network states where it is most likely for constraints to be violated are also likely to be the states with the highest required accuracy. Furthermore, in [47] it has been shown that the highest deviation of the estimated state from the true network state is likely to occur during the worst-case scenarios. This is due to the common assumption that the measurement error variance is proportional to the measured value. Hence, it is possible to reduce the infinite number of possible network states that have to be evaluated to the number of worst-case scenarios in which the network constraints are most likely to be violated. Generally, two worst-case scenarios have to be considered [48]:

- Maximum load and minimum generation.
- Minimum load and maximum generation.

The worst-case scenarios of a network should be known by the distribution network operator since these are commonly used during the network planning phase [49–51].

10.3.2 Probabilistic Observability Assessment Algorithm

In order to classify a network as observable, an observability criterion has to be satisfied by every possible estimate of a network state and each worst-case scenario. A network state estimate satisfies the observability criterion if the accuracy of every estimated parameter is equal to or higher than their associated minimum accuracy value given by Eqs. (10.10) and (10.11). This is defined by

$$a_{i,\min} \leq MA_{i,\min} \wedge a_{i,\max} \leq MA_{i,\max}. \tag{10.17}$$

The following steps can be used to determine if an estimated parameter satisfies the observability criterion:

Step 1: Specify the desired CL value.

Step 2: Obtain the PDF of the estimated parameter.

Step 3: Calculate the confidence interval that satisfies Eqs. (10.8) and (10.9).

Step 4: Use Eqs. (10.10) and (10.11) to calculate the required accuracy values.

Step 5: Determine if the accuracy of the parameter estimate is equal to or higher than the required accuracy values as defined in Eq. (10.17).

It is, however, not feasible to evaluate if every possible network state estimate fulfills the observability criterion due to the infinite number of possible network state estimates. To solve this problem, a subset of possible estimates is used to represent the infinite number of possible estimates. For this purpose a number of Monte Carlo simulations have to be performed. In each simulation, a possible set of state estimation inputs is randomly generated according to the accuracy and expected value of the individual measurements. The generated set of inputs is then used to perform a state estimation. For each simulation it has to be determined if the estimated parameters meet the observability criterion defined in Eq. (10.17). This is done by the steps outlined above. Once a sufficiently large number of simulations have been performed, the ratio between the number of estimates that satisfied Eq. (10.17) and the total number of performed simulations is calculated for each estimated parameter. From here onward this ratio is referred to as compliance ratio. For example, if 1 out of 100 simulations does not satisfy Eq. (10.17), the compliance ratio is equal to 99%. This value has to be equal to or higher than a desired value in order to classify an estimated parameter as observable. Hence, if the value of the compliance ratio is equal to or higher than a desired value for every estimated parameter and each worst-case scenario, the network is classified as observable. The advantages of using the compliance ratio are twofold. First, the computation time is kept realistic by using a limited number of estimates as a representation of the infinite number of possible estimates. Second, it accounts for the random nature of the measurement errors, which implies that it is never impossible that an estimated network state does not satisfy the observability criterion. This means that if the probabilistic observability assessment is repeated several times for the same network with a desired compliance ratio equal to one, the results may be inconsistent, and the network could be classified observable at one time and unobservable at another. By using a desired compliance ratio that is lower than one, it is possible to account for the random variation and therefore improve the repeatability of the method. The flowchart in Figure 10.11 illustrates how the probabilistic observability assessment is performed.

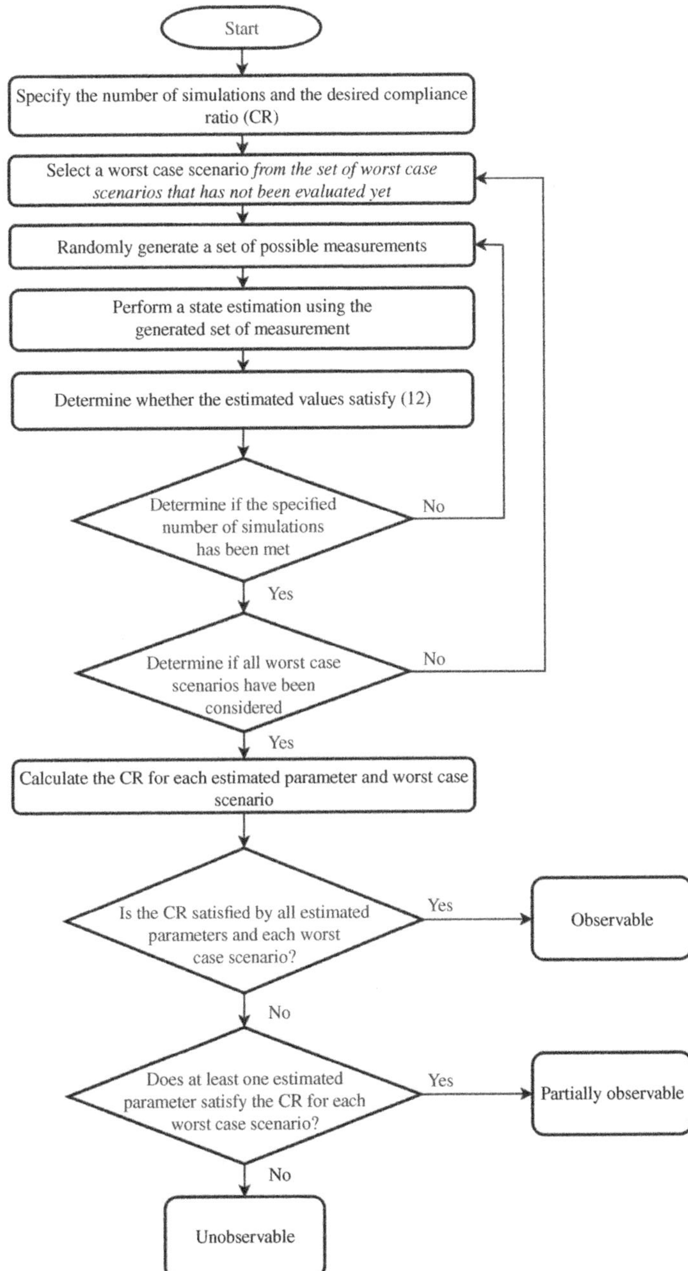

Figure 10.11 Flowchart of the probabilistic observability assessment.

10.3.3 Scalability

The probabilistic observability assessment only needs to be repeated if the network topology, the set of available measurements, or the worst-case scenarios change. This means that the computation time for the probabilistic observability assessment is not critical since it is performed offline. However, any method used in distribution networks has to be scalable. In distribution networks scalability is an important topic due to the usually large number of buses. The computation time of the presented observability assessment can vary significantly with the method chosen to calculate the network states and the PDF of the estimated parameters. If considering the methods chosen in the case studies, the computation time does increase with the size of the network, and therefore, every simulation will take more time. However, the computation time does not increase exponentially with the size of the network [52]. Furthermore, the number of simulations that are required for convergence is independent of the number of buses in the network. This is due to the fact that the compliance ratio is calculated for each bus individually and not the system as a whole. Also, reducing the computation time is straightforward since the probabilistic observability assessment is highly suitable for parallel processing, which means that it is possible to significantly reduce the computation time by performing the calculations in parallel using additional processors.

10.3.4 Case Studies

The two case studies in this section are performed on a modified version of the IEEE 34-bus test feeder. The first case study considers the feeder without voltage control devices. The second case study considers the same feeder with voltage control devices at two buses in order to demonstrate how the network configuration can impact the outcome of the probabilistic observability assessment. For the sake of brevity, only voltage magnitudes are considered in the following case studies. The methodology illustrated in this section can, however, be applied to any parameter that is estimated by the state estimation process.

Two worst-case scenarios are evaluated in each case studies. The first worst-case scenario considers the network under maximum load and no distributed generation, and the second worst-case scenario considers the network under minimum load and maximum distributed generation. For the first case, the maximum load is equal to one third of the loading specified in [53], and the distributed generation is equal to zero. For the second worst-case scenario the distributed generation is equal to 20% of the maximum load used in worst-case one, and the minimum load is equal to 30% of the maximum load used in worst-case one.

The desired confidence level and compliance ratio are set to 95% and 99%, respectively. For an actual application of the method, these values would have to be defined by the distribution network operator depending on the individual application. As a general rule, the higher the desired confidence level, the higher the accuracy required to determine if the estimated parameters are within their respective constraints. The desired compliance ratio should be chosen small enough to account for the random variations in the probabilistic observability assessment results, and at the same time it has to be as large as possible to avoid false positive results. For instance, a desired compliance ratio of 1 means that a network is not classified as observable if one out of all simulations is not compliant with Eq. (10.17) by chance. However, if the compliance ratio is chosen too small, the risk of a false positive result increases.

Figure 10.12 shows the test feeder used for the first case study. For simplicity, the connected loads are assumed to be three-phase balanced, and the voltage control devices have been removed. The original data of the feeder is given in [53].

A number of assumptions are made in regard to the measurements used in the state estimation:

- All measurement errors are normally distributed.
- The only real-time measurement in the feeder is located at bus 1, which measures the voltage magnitude with the accuracy of ±1%.
- All loads are modeled as pseudo-measurements, which have the accuracy of ±50%.
- Distributed generation is assumed to be connected at every bus and is proportional to the load that is connected to the same bus.
- The distributed generation is represented in the state estimation by pseudo-measurements that have the measurement accuracy of ±100%.

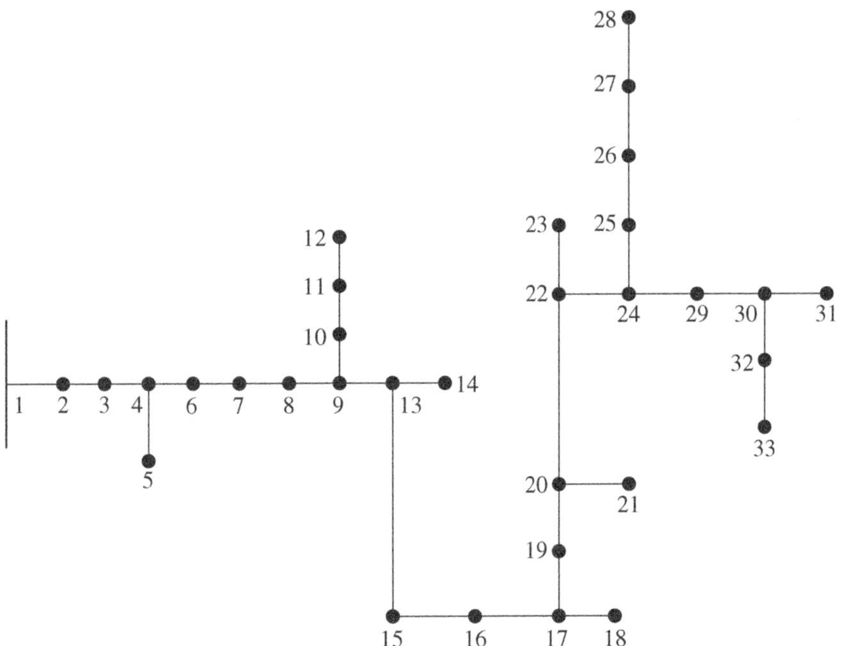

Figure 10.12 Modified IEEE test feeder.

The results of the state estimation for the two worst-case scenarios are shown in Figures 10.13 and 10.14. From these voltage profiles it is clear that the accuracy required to determine if the voltage magnitudes are within their constraints changes with the network state and depends on the proximity of the estimated values to their constraints. For instance, the voltage estimate at bus 7 is likely to be within its constraints. The voltage estimate at bus 33, however, requires a much higher accuracy in order to determine if the voltage at this bus is within its constraints compared with the estimate at bus 7.

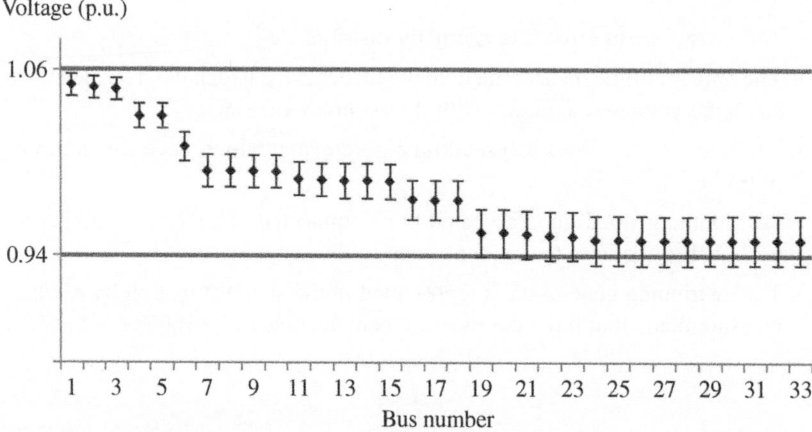

Figure 10.13 Voltage profile for worst-case scenario one (maximum load and no distributed generation).

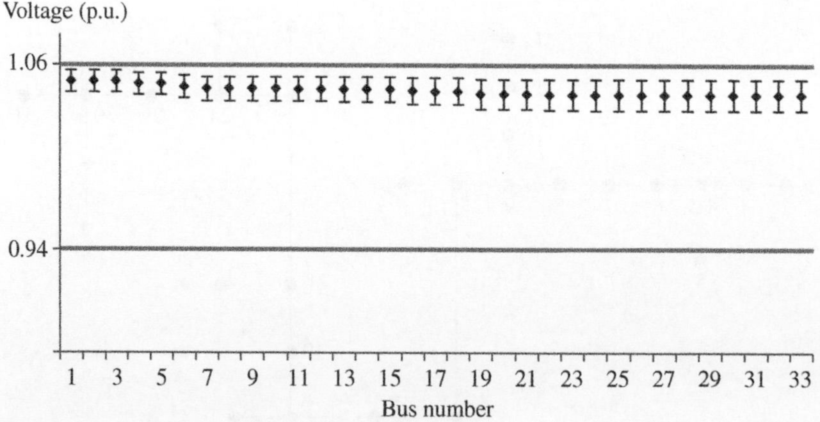

Figure 10.14 Voltage profile for worst-case scenario two (minimum load and maximum distributed generation).

The bars shown in Figures 10.13 and 10.14 indicate the 95% confidence interval of the estimated voltages. The confidence interval was obtained by multiplying the standard deviation of the estimated voltages with the corresponding coverage factor. This is possible due to the assumption of Gaussian-distributed measurement errors. The horizontal lines represent the voltage constraints at ±6% of the nominal voltage [27]. In order to illustrate how the compliance of an estimated state is calculated, an example is given based on the worst-case scenario one. The values for the estimated voltage magnitudes and the corresponding standard deviations are shown in Table 10.2.

After calculating the standard deviation using Eq. (10.8), the PDF is given by Eq. (10.9). Next, the required minimum accuracy for values above and below the estimated value has to be determined. Using bus 33 as an example, the minimum accuracy is calculated by Eqs. (10.15) and (10.16) for values below and above the estimated value, respectively. This results in

$$MA_{i,\min} = -[i_{\min} - E_i] = -[0.94 - 0.949] = 0.009$$

$$MA_{i,\max} = [i_{\max} - E_i] = [1.06 - 0.949] = 0.111$$

Now the accuracy of the state estimation has to be calculated. As specified in Eqs. (10.13) and (10.14), the accuracy is represented by the endpoints of the confidence interval that correspond to the specified confidence level. Since it is assumed that the measurement errors are Gaussian distributed, it is possible to use a coverage factor that corresponds to the desired confidence level in order

TABLE 10.2 Example simulation results (voltages and standard deviations).

Bus number	*V* (p.u.)	Standard deviation	Bus number	*V* (p.u.)	Standard deviation
1	1.050	0.0035	18	0.975	0.0060
2	1.049	0.0035	19	0.954	0.0072
3	1.048	0.0035	20	0.954	0.0072
4	1.030	0.0038	21	0.953	0.0073
5	1.030	0.0038	22	0.952	0.0073
6	1.010	0.0044	23	0.952	0.0073
7	0.995	0.0050	24	0.950	0.0075
8	0.995	0.0050	25	0.950	0.0075
9	0.994	0.0050	26	0.949	0.0075
10	0.994	0.0050	27	0.949	0.0075
11	0.989	0.0051	28	0.949	0.0075
12	0.989	0.0051	29	0.950	0.0075
13	0.988	0.0053	30	0.949	0.0075
14	0.988	0.0053	31	0.949	0.0075
15	0.988	0.0053	32	0.949	0.0075
16	0.976	0.0060	33	0.949	0.0075
17	0.975	0.0060			

to calculate the confidence interval. In case of the specified 95% confidence interval, the corresponding coverage factor is equal to 1.96. Multiplying this factor with the standard deviation shown in Table 10.2 results in the maximum expected difference of 0.015 between the estimate and the true value with a confidence of 95%, as shown in the computation below. Since the Gaussian distribution is symmetrical, the accuracy value is identical for values above and below the estimated value.

$$a_{33,\min} = a_{33,\max} = 0.0075 \times 1.96 = 0.015$$

The results for the minimum required accuracy and the state estimation accuracy for all buses are shown in Table 10.3. Since the required as well as the actual state estimation accuracy are now known, the observability criterion defined in Eq. (10.17) can be assessed. The result for bus 33 is the following:

$$a_{33,\min} \leq MA_{33,\min} \wedge a_{33,\max} \leq MA_{33,\max} = 0.015 \leq 0.009 \wedge 0.015 \leq 0.111$$

It can be seen that this particular estimate of the voltage magnitude at bus 33 does not satisfy the observability criterion.

From Table 10.3 it is apparent that buses 21–33 do not comply with the observability criterion for this particular estimated state. It is therefore not possible to determine if the true voltage magnitude at these buses is within constraints with a certainty of at least 95%. Hence a higher accuracy is required at these buses according to this simulation. However, since this is only one possible estimate, we have to

TABLE 10.3 Example simulation results (accuracy and minimum accuracy values).

Bus number	$MA_{i,\min}$	$MA_{i,\max}$	$a_{i,\min}$	$a_{i,\max}$	Bus number	$MA_{i,\min}$	$MA_{i,\max}$	$a_{i,\min}$	$a_{i,\max}$
1	0.110	0.010	0.007	0.007	18	0.035	0.085	0.012	0.012
2	0.109	0.011	0.007	0.007	19	0.014	0.106	0.014	0.014
3	0.108	0.012	0.007	0.007	20	0.014	0.106	0.014	0.014
4	0.090	0.030	0.007	0.007	21	0.013	0.107	0.014	0.014
5	0.090	0.030	0.007	0.007	22	0.012	0.108	0.014	0.014
6	0.070	0.050	0.009	0.009	23	0.012	0.108	0.014	0.014
7	0.055	0.065	0.010	0.010	24	0.010	0.110	0.015	0.015
8	0.055	0.065	0.010	0.010	25	0.010	0.110	0.015	0.015
9	0.054	0.066	0.010	0.010	26	0.009	0.111	0.015	0.015
10	0.054	0.066	0.010	0.010	27	0.009	0.111	0.015	0.015
11	0.049	0.071	0.010	0.010	28	0.009	0.111	0.015	0.015
12	0.049	0.071	0.010	0.010	29	0.010	0.110	0.015	0.015
13	0.048	0.072	0.010	0.010	30	0.009	0.111	0.015	0.015
14	0.048	0.072	0.010	0.010	31	0.009	0.111	0.015	0.015
15	0.048	0.072	0.010	0.010	32	0.009	0.111	0.015	0.015
16	0.036	0.084	0.012	0.012	33	0.009	0.111	0.015	0.015
17	0.035	0.085	0.012	0.012					

perform a number of simulations in order to determine how likely it is that an estimated parameter does not satisfy Eq. (10.17). A number of 5000 simulations are used in this case study. The convergence behavior of bus 33 for this particular network under maximum load and no distributed generation can be seen in Figure 10.15.

The results of the probabilistic observability assessment for both worst-case scenarios are shown in Table 10.4.

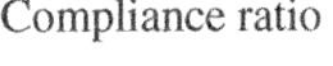

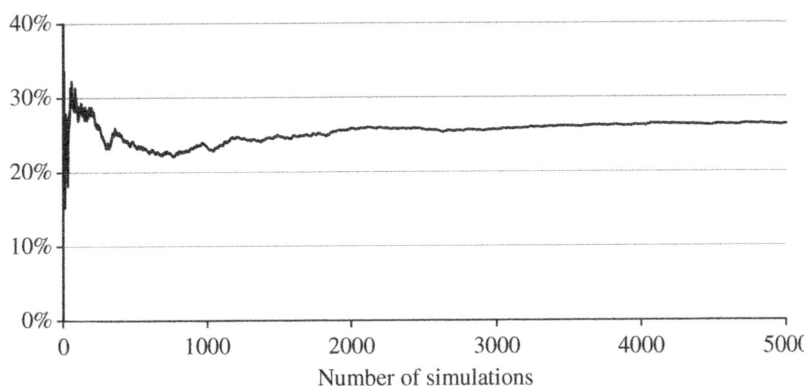

Figure 10.15 The convergence behavior of the compliance ratio at bus 33.

TABLE 10.4 Compliance ratios.

Bus number	Worst case 1 (%)	Worst case 2 (%)	Bus number	Worst case 1 (%)	Worst case 2 (%)
1	82	81	18	100	97
2	90	81	19	53	97
3	94	82	20	53	97
4	100	90	21	48	97
5	100	90	22	42	97
6	100	94	23	42	97
7	100	96	24	29	97
8	100	96	25	29	97
9	100	96	26	27	97
10	100	96	27	26	97
11	100	97	28	26	97
12	100	97	29	28	97
13	100	97	30	27	97
14	100	97	31	27	97
15	100	97	32	27	97
16	100	97	33	27	97
17	100	97			

It can be seen that the desired compliance ratio of 99% is not met by any of the buses for both worst-case scenarios. Hence, the network is classified as unobservable according to the probabilistic observability assessment. This means that it is not always possible to determine if the values of the estimated parameters are within their respective constraints during normal operation of the network. This is highlighted by the fact that only a single measurement device is installed in the network. In combination with the widely unknown loads, it is clear that the state of the feeder cannot be estimated with a high degree of accuracy and that the estimate will contain a large amount of uncertainty. However, this information is not available from the traditional observability analysis methods, which classify the feeder as observable due to the use of pseudo-measurements. This underlines the importance of assessing the accuracy of a state estimation if a large proportion of pseudo-measurements are used for the state estimation, as is commonly the case in distribution networks.

The second case study considers the same feeder as in case study one but with added voltage control devices at buses 8 and 20 as shown in Figure 10.16.

For the following simulations, the voltage control devices are assumed to operate normally and are able to maintain a voltage magnitude at buses 8 and 20 within the bandwidth specified in [53]. The voltage control devices are modeled by keeping the voltage magnitude at the voltage controlled buses constant at the set points. In order to account for variations introduced by inexact voltage control,

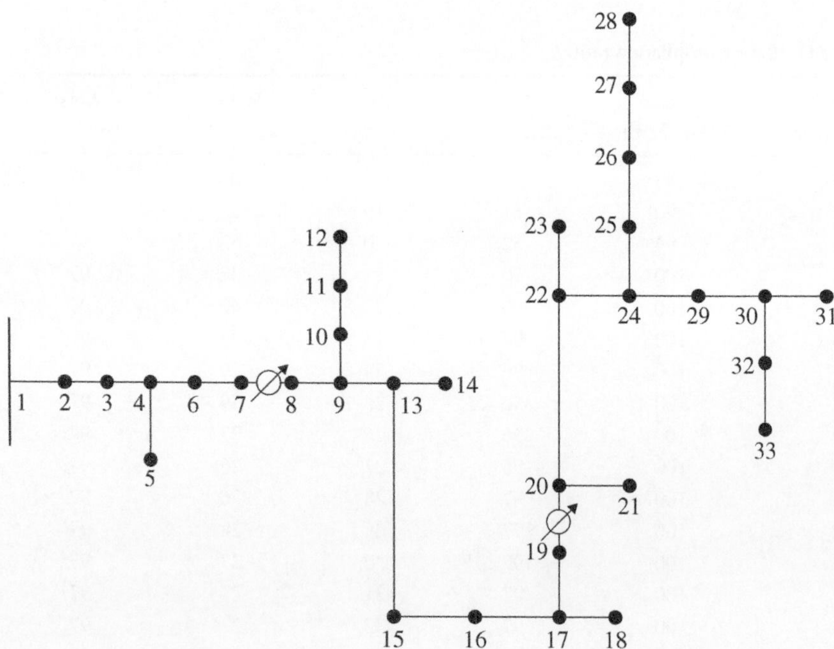

Figure 10.16 Modified IEEE test network with the voltage control devices.

voltage magnitude measurements are added at the voltage controlled buses. These additional measurements have the accuracy of ±1%, which corresponds to the specified bandwidth. It is also assumed that the tap positions of the control devices are not monitored. For this reason it is not possible for the additional measurements to provide any information about voltage magnitudes at buses that are hierarchically higher than the respective voltage control device (bus 1 is assumed to be at the top of the hierarchy). In order to increase the accuracy of the simulations, the feeder is divided into sections at buses 8 and 20. The state estimation is then performed sequentially from the hierarchically lowest section to the highest. The estimated value and PDF of the power flow through the branch connecting to the next section are used as an input for the next section.

The estimated voltage profiles for the two worst-case scenarios are presented in Figures 10.17 and 10.18.

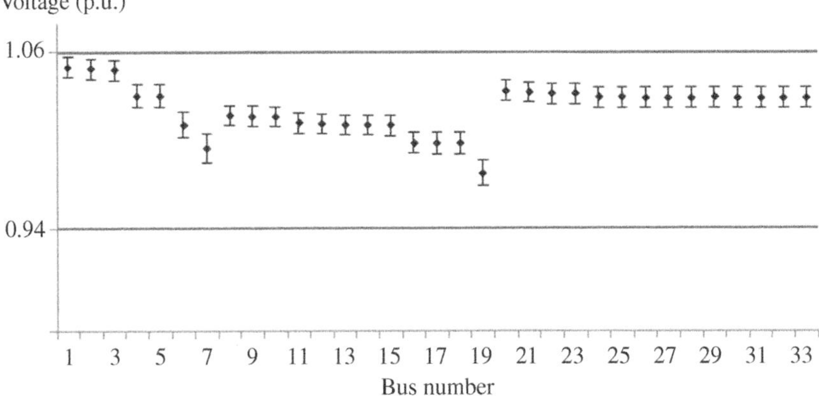

Figure 10.17 Voltage profile for worst-case one (maximum load and no distributed generation).

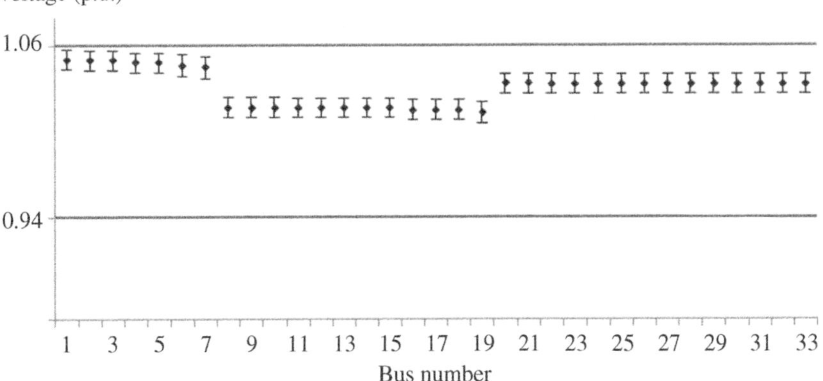

Figure 10.18 Voltage profile for worst-case two (minimum load and maximum distributed generation).

TABLE 10.5 Compliance ratios (case study 2).

Bus number	Worst case 1 (%)	Worst case 2 (%)	Bus number	Worst case 1 (%)	Worst case 2 (%)
1	81	81	18	100	100
2	90	82	19	100	100
3	94	83	20	100	100
4	100	91	21	100	100
5	100	96	22	100	100
6	100	97	23	100	100
7	100	97	24	100	100
8	100	100	25	100	100
9	100	100	26	100	100
10	100	100	27	100	100
11	100	100	28	100	100
12	100	100	29	100	100
13	100	100	30	100	100
14	100	100	31	100	100
15	100	100	32	100	100
16	100	100	33	100	100
17	100	100			

The results of the probabilistic observability assessment are shown in Table 10.5. It can be seen that buses 1–7 do not satisfy the desired compliance ratio of 99% for each worst-case scenario. As a result the feeder is classified as partially observable.

Two options exist to make the network observable. First, placing additional measurement devices in order to increase the state estimation accuracy and second, decreasing the minimum required accuracy by increasing the distance between the estimated parameters and their respective constraints. This could be done by, for example, adjusting the set point of the voltage control devices as described in [26].

10.4 CONCLUSION

The traditional approach to state estimation only provides the estimated values to the network operator without any information about the accuracy of the estimates. This works well in transmission networks where a large number of redundant measurements are generally available. However, due to economic constraints, the number of available real-time measurements is usually low in distribution networks, which can lead to a significant amount of uncertainty in the state estimation result. This makes it difficult to adapt the traditional state estimation approach to distribution networks. Therefore, a new probabilistic approach to state estimation in distribution networks based on confidence has been discussed in this chapter. This approach uses the confidence that the estimated parameters are within their constraints as a primary output of the estimator. By using the confidence value, it is possible to combine information about the estimated value as well as the accuracy of the estimate into a single number. This approach has been demonstrated in case studies using real 13-bus and 145-bus feeders. The results of the case studies show that even if a large amount of uncertainty is present in the state estimation result, the proposed approach is able to provide practical information about the network state in a form that is easy to interpret.

A probabilistic observability assessment was also presented, which uses a similar probabilistic approach. The traditional approach to observability in distribution networks has the limitation that even if a network is classified as observable, the state estimation result could be completely decoupled from reality. The presented method on the other hand determines if the state of a distribution network can be estimated with a degree of accuracy that is sufficient to evaluate if the true value of the estimated parameters is within their respective constraints. This method is able to identify critical arrears in a distribution network that have a low accuracy or violate the constraints under worst-case consideration. The implementation of the probabilistic observability assessment has been outlined in detail, and several case studies have been performed to illustrate this method.

REFERENCES

1. Schweppe, F.C. and Wildes, J. (Jan 1970). Power system static-state estimation, part I: exact model. *IEEE Transactions on Power Apparatus and Systems* PAS-89 (1): 120–125.
2. Colome, D.G. and Vega, C.E.D.L. (2008). Voltage quality control at distribution systems applying a fuzzy state estimator. *PES T&D LA*, pp. 1–7.
3. Ke, L. (May 1996). State estimation for power distribution system and measurement impacts. *IEEE Transactions on Power Systems* 11 (2): 911–916.
4. Brinkmann, B., Negnevitsky, M., Yee, T., and Nguyen, T. (2015). An observability index for distribution networks using information entropy. *Proceedings of the Australasian Universities Power Engineering Conference (AUPEC)*, Wollongong, NSW, pp. 1–6.
5. Gelagaev, R., Vermeyen, P., Vandewalle, J., and Driesen, J. (2010). Numerical observability analysis of distribution systems. *Proceedings of the 14th International Conference on Harmonics and Quality of Power (ICHQP)*, Bergamo, Italy, pp. 1–6.
6. Grainger, J.J. and Stevenson, W.D. (1994). *Power System Analysis*. New York: McGraw-Hill, International ed.
7. Abur, A. and Expósito, A.G. (2004). *Power System State Estimation Theory and Implementation*. New York: Marcel Dekker, Inc.
8. Roytelman, I. and Shahidehpour, S.M. (1993). State estimation for electric power distribution systems in quasi real-time conditions. *IEEE Transactions on Power Delivery* 8 (4): 2009–2015.

9. Baran, M.E. and Kelley, A.W. (1994). State estimation for real-time monitoring of distribution systems. *IEEE Transactions on Power Systems* 9 (3): 1601–1609.
10. Hansen, C.W. and Debs, A.S. (1995). Power system state estimation using three-phase models. *IEEE Transactions on Power Systems* 10 (2): 818–824.
11. Haughton, D.A. and Heydt, G.T. (2013). A linear State estimation formulation for smart distribution systems. *IEEE Transactions on Power Systems* 28 (2): 1187–1195.
12. Baran, M.E. and Kelley, A.W. (1995). A branch-current-based state estimation method for distribution systems. *IEEE Transactions on Power Systems* 10 (1): 483–491.
13. Lin, W.M. and Teng, J.H. (1996). Distribution fast decoupled state estimation by measurement pairing. *IEE Proceedings - Generation, Transmission and Distribution* 143 (1): 43–48.
14. Whei-Min, L., Jen-Hao, T., and Shi-Jaw, C. (2001). A highly efficient algorithm in treating current measurements for the branch-current-based distribution state estimation. *IEEE Transactions on Power Delivery* 16 (3): 433–439.
15. Muscas, C., Pau, M., Pegoraro, P.A., et al. (2014). Two-step procedures for wide-area distribution system state estimation. *Proceedings of the 2014 IEEE International Instrumentation and Measurement Technology Conference (I2MTC) Proceedings*, Montevideo, Uruguay, pp. 1517–1522.
16. Muscas, C., Pau, M., Pegoraro, P.A. et al. (2015). Multiarea distribution system state estimation. *IEEE Transactions on Instrumentation and Measurement* 64 (5): 1140–1148.
17. Muscas, C., Pegoraro, P.A., Sulis, S., et al. (2016). Fast multi-area approach for distribution system state estimation. *Proceedings of the 2016 IEEE International Instrumentation and Measurement Technology Conference Proceedings*, Taipei, Taiwan, pp. 1–6.
18. Liang, Z. and Abur, A. (2005). Multi area state estimation using synchronized phasor measurements. *IEEE Transactions on Power Systems* 20 (2): 611–617.
19. Ghosh, A.K., Lubkeman, D.L., Downey, M.J., and Jones, R.H. (1997). Distribution circuit state estimation using a probabilistic approach. *IEEE Transactions on Power Systems* 12 (1): 45–51.
20. Brinkmann, B. and Negnevisky, M. (2016). Robust state estimation in distribution networks. *Proceedings of the 2016 Australasian Universities Power Engineering Conference (AUPEC)*, Brisbane, QLD, Australia, pp. 1–5.
21. Strelec, M., Janecek, P., Georgiev, D., et al. (2015). Backward/forward probabilistic network state estimation tool and its real world validation. *Proceedings of the 56th International Scientific Conference on Power and Electrical Engineering of Riga Technical University (RTUCON)*, Riga, Latvia, pp. 1–6.
22. Valverde, G., Saric, A.T., and Terzija, V. (2013). Stochastic monitoring of distribution networks including correlated input variables. *IEEE Transactions on Power Systems* 28 (1): 246–255.
23. Guang, L., Ning, Z., Ferryman, T., and Tuffner, F. (2011). Uncertainty quantification in state estimation using the probabilistic collocation method. *Proceedings of the IEEE/PES Power Systems Conference and Exposition (PSCE)*, Phoenix, AZ, pp. 1–8.
24. Mathews, G.M. (2012). The accuracy of factored nonlinear weighted least squares state estimation. *Proceedings of the IEEE International Energy Conference and Exhibition (ENERGYCON)*, Florence, pp. 860–866.
25. Lubkeman, D.L., Jianzhong, Z., Ghosh, A.K., and Jones, R.H. (2000). Field results for a distribution circuit state estimator implementation. *IEEE Transactions on Power Delivery* 15 (1): 399–406.

26. Hird, C. M., Leite, H., Jenkins, N., and Li, H. (2004). Network voltage controller for distributed generation. *Proceedings of the Generation, Transmission and Distribution* (March 2004), pp. 150–156.
27. Office of the Tasmanian Economic Regulator (28 Apr. 2015). *Tasmanian Electricity Code*. http://www.energyregulator.tas.gov.au/domino/otter.nsf/elect-v/003 (accessed 28 April 2015).
28. The Queensland Department of Natural Resources, Mines and Energy (DNRME) (25 Feb. 2016). *Queensland Electricity Regulation*. https://www.legislation.qld.gov.au/LEGISLTN/CURRENT/E/ElectricR06.pdf (accessed 25 February 2016).
29. Krumpholz, G.R., Clements, K.A., and Davis, P.W. (Jul. 1980). Power system observability: a practical algorithm using network topology. *IEEE Transactions on Power Apparatus and Systems* PAS-99 (4): 1534–1542.
30. Manitsas, E., Singh, R., Pal, B.C., and Strbac, G. (Apr. 2012). Distribution system state estimation using an artificial neural network approach for pseudo measurement modeling. *IEEE Transactions on Power Systems* 27 (4): 1888–1896.
31. Gomez-Exposito, A. and Abur, A. (Aug. 1998). Generalized observability analysis and measurement classification. *IEEE Transactions on Power Systems* 13 (3): 1090–1095.
32. Clanents, K.A., Krutnpholz, G.R., and Davis, P.W. (1983). Power system state estimation with measurement deficiency: an observability/measurement placement algorithm. *IEEE Transactions on Power Apparatus and Systems* PAS-102 (7): 2012–2020.
33. Mori, H. and Tsuzuki, S. (1991). A fast method for topological observability analysis using a minimum spanning tree technique. *IEEE Transactions on Power Systems* 6 (2): 491–500.
34. Costa, A.S., Lourenco, E.M., and Clements, K.A. (2002). Power system topological observability analysis including switching branches. *IEEE Transactions on Power Systems* 17 (2): 250–256.
35. Mori, H. (1994). A GA-based method for optimizing topological observability index in electric power networks. *Proceedings of the IEEE World Congress on Computational Intelligence*, Orlando, FL, pp. 565–568.
36. Monticelli, A. and Wu, F.F. (1986). Observability analysis for orthogonal transformation based state estimation. *IEEE Transactions on Power Systems* 1 (1): 201–206.
37. Monticelli, A. (1993). The impact of modeling short circuit branches in state estimation. *IEEE Transactions on Power Systems* 8 (1): 364–370.
38. Katsikas, P.J. and Korres, G.N. (2003). Unified observability analysis and measurement placement in generalized state estimation. *IEEE Transactions on Power Systems* 18 (1): 324–333.
39. Baldwin, T.L., Mili, L., Boisen, M.B., and Adapa, R. (1993). Power system observability with minimal phasor measurement placement. *IEEE Transactions on Power Systems* 8 (2): 707–715.
40. Göl, M. and Abur, A. (2013). Observability and criticality analyses for power systems measured by phasor measurements. *IEEE Transactions on Power Systems* 28 (3): 3319–3326.
41. Korres, G.N. and Manousakis, N.M. (2012). State estimation and observability analysis for phasor measurement unit measured systems. *IET Generation, Transmission and Distribution* 6 (9): 902–913.
42. Jerome, J. (2001). Network observability and bad data processing algorithm for distribution networks. 2001 Power Engineering Society Summer Meeting, pp. 1692–1697.

43. Toyoshima, D., Castillo, M.R.C., Fantin, C.A., and London, J.B.A. (2012). Observability and measurement redundancy analysis on three-phase state estimation. *Proceedings of the 2012 IEEE Power and Energy Society General Meeting*, San Diego, CA, pp. 1–8.
44. Magnago, F., Zhang, L., and Celik, M.K. (2016). Multiphase observability analysis in distribution systems state estimation. *Proceedings of the 2016 Power Systems Computation Conference (PSCC)*, Genoa, Italy, pp. 1–7.
45. Brinkmann, B. and Negnevitsky, M. (2016). A probabilistic approach to observability of distribution networks. *IEEE Transactions on Power Systems* 32: 1–10.
46. Crawley, M.J. (2005). *Statistics an Introduction Using R*. Wiley.
47. Damavandi, M.G., Krishnamurthy, V., and Martí, J.R. (2015). Robust meter placement for state estimation in active distribution systems. *IEEE Transactions on Smart Grid* 6 (4): 1972–1982.
48. Grond, M.O.W., Luong, H.N., Morren, J., et al. (2014). Practice-oriented optimization of distribution network planning using metaheuristic algorithms. *Proceedings of the Power Syst. Computation Conf. (PSCC)*, Wroclaw, pp. 1–8.
49. Kulmala, A., Repo, S., and Jarventausta, P. (2011). Using statistical distribution network planning for voltage control method selection. *Proceedings of the Renewable Power Generation Conference*, Edinburgh, pp. 1–6.
50. Grond, M.O.W., Morren, J., and Slootweg, J.G. (2013). Integrating smart grid solutions into distribution network planning. *Proceedings of the PowerTech (POWERTECH), 2013 IEEE*, Grenoble, pp. 1–6.
51. Celli, G., Pilo, F., Soma, G.G., et al. (2013). A comparison of distribution network planning solutions: traditional reinforcement versus integration of distributed energy storage. *Proceedings of the IEEE PowerTech (POWERTECH)*, Grenoble, pp. 1–6.
52. Dejun, M., Yong, W., Xiuxia, T. (2015). Least squares class power system state estimation algorithm efficiency analysis. *2015 Third International Conference on Proceedings of the Technological Advances in Electrical, Electronics and Computer Engineering (TAEECE)*, Beirut, Lebanon, pp. 17–21.
53. IEEE PES AMPS DSAS Test Feeder Working Group (5 Aug. 2013). *IEEE PES Radial Distribution Test Feeders*. http://ewh.ieee.org/soc/pes/dsacom/testfeeders/ (accessed 5 August 2013).

CHAPTER 11

ADVANCED DISTRIBUTION SYSTEM STATE ESTIMATION IN MULTI-AREA ARCHITECTURES

Marco Pau[1], *Paolo Attilio Pegoraro*[2], *Ferdinanda Ponci*[1], *and Sara Sulis*[2]

[1]*Institute for Automation of Complex Power Systems, RWTH Aachen University, Aachen, Germany*

[2]*Department of Electrical and Electronic Engineering, University of Cagliari, Cagliari, Italy*

11.1 ISSUES AND CHALLENGES OF DISTRIBUTION SYSTEM STATE ESTIMATION

11.1.1 Distribution Grid Peculiarities

The power systems were designed to transport energy from a limited number of traditional power plants toward the consumers, with a unidirectional energy flow and a unidirectional information flow. The systems were hierarchical and divided into four parts: the generation system, the transmission and the distribution networks, and the utilization system. In particular, the electrical energy is transferred to final customers by means of the distribution systems.

Distribution grids are usually characterized by open-loop topologies, with radial or weakly meshed structures (rare reconfiguration possibilities, generally only in case of faults), with many feeders and laterals, a very large number of nodes, with varied load, and different voltage levels. Distribution networks are defined by medium/low voltage levels (MV/LV). A typical distribution grid consists of high-to-medium voltage (HV/MV) transformation centers, the MV grid, MV/LV transformation centers, and the LV grid.

Advances in Electric Power and Energy: Static State Estimation, First Edition.
Edited by Mohamed E. El-Hawary.

Published 2021 by John Wiley & Sons, Inc.

These grids have not only three-phase but also two-/single-phase configurations, and several types of nonsymmetrical loads, drawing nonsymmetrical electrical quantities, are connected to the grids. As a consequence, different degrees of unbalance can occur. Another relevant difference compared with transmission systems is that power distribution lines present low reactance to resistance (X/R) ratios that can be equal to few units or even equal to unity.

In recent years, this consolidated infrastructure has been affected by significant changes. The actors in the distribution system scenario are changing along with management and control policies. The European requirements (the well-known 20-20-20 target and the Strategic Research Agenda 2035 [1]) have called for a rising amount of electric power produced downstream the transmission level.

In Europe, the energy market deregulation is now completed, and there is an explosive growth of distributed generation (DG), mainly renewable energy sources (RESs), and new distributed energy resources (DERs), as electric vehicles, are rapidly gaining ground. These changes have transformed the traditional passive grid into a highly complex active network with possible bidirectional power flows, system reliability problems, and significant power quality (PQ) issues. Moreover, one of the aforementioned typical characteristics of the distribution system, the possibly high level of unbalance of the network, is now exacerbated by a high penetration level of DG and DERs that is taking place without appropriate planning.

Distribution systems are operated by distribution system operators, DSOs, which have a monopoly on well-defined parts of the power grid and are monitored only by the national regulator. It is worth highlighting that, in the liberalized market, operation, maintenance, and development of adjacent parts of distribution networks can be addressed by different DSOs, which can have different needs and constraints.

In any case, the management and control of this kind of network can be problematic, mainly because of the very low number of measurement devices on the field.

The network operation plan is traditionally updated starting from an "idea" of the actual operating conditions of the grid obtained in the HV/MV substations. DSO makes decisions based on the measurement devices available on the field, which are currently very few, the past behavior of the grid, and the forecast for the load and generation. Moreover, DSO commonly works in a conservative manner, in order to protect the grid infrastructure. Nevertheless, it is now recognized that such traditional approach cannot satisfy an efficient operation of modern networks.

The distribution networks are typically very large networks, with a high number of nodes (up to thousands) that are becoming, more and more often and at an increasing rate, active nodes, which can inject power. Most DG and DERs work in an intermittent or substantially unpredictable way, depending on weather conditions and other unforeseeable factors. Furthermore, grid stability is based on inertia of large rotating machinery, but most "green" energy sources have short-term inertia or are inertia-less generators. Therefore, unexpected local critical operating conditions (e.g. local power surplus) may occur. Besides, DERs commonly have power electronics interfaces, and this, together with the nonlinear loads, makes possible local PQ issues increasingly relevant.

In this perspective, traditional management and control systems will be no more able to face untimely protection activations and avoid out-of-service configurations, while novel approaches to network monitoring that serve such higher-level applications appear to be essential.

11.1.2 Future Scenarios

The future scenarios of the electric grids are now calling for more effective management and control systems. To address this need, a new generation of networks has been conceived, the so-called smart grids (SGs) [2, 3].

Despite the meaning of the expression "smart grid" is not globally shared, what is commonly recognized is that SGs aim at fulfilling consumers' demand while avoiding infrastructure collapses and poor quality in energy supply. In the SG scenario, the variations of the electrical quantities during the operation can be impressively rapid and significant. Automatic management, without human presence, is necessary to design advanced management and control functions, so that the network is handled effectively during real-time operation. To this aim, accurate data with appropriate reporting rates are required to obtain an accurate knowledge of the status of the grid. Therefore, in the SG framework, a suitable distributed measurement system, integrated with an efficient and future-proof information and communication system (ICT), is needed as the basis for every management and control application. In particular, a scalable ICT infrastructure is required in order to follow the increasing amount of data and possibly reassign grid functions and data in a context-aware manner.

The accurate knowledge on the actual operative conditions would allow the DSO to perform prompt and successful regulations in case, for example, the voltage level is approaching the boundary condition set by the regulatory framework, or to determine if some assets are in a stress condition, e.g. in case of overload, meanwhile avoiding untimely or overly conservative control actions.

The problem is how an accurate evaluation of the status of these wide-area networks can be obtained with such a limited number of measurement devices. In recent times, the number of devices in the distribution networks has been increasing, but redundancy of measured quantities is still far to reach.

An important help to improve awareness on the operating condition of the network comes from the progressively adopted state estimation (SE) methodologies.

SE can be used not only to determine the quantities of interest at unmonitored locations of a considered system but also to improve the accuracy of the information obtained from measurements in nodes equipped with measurement devices. SE is based on the mathematical relations between system state variables and available measurements. The state variables represent a minimum set of quantities that allows deriving a complete description of network status. Commonly, the considered state variables are the node voltages, but it is also possible to choose the branch currents, expressed in polar or rectangular terms, depending on the grid characteristics and the available monitoring system. SE techniques were proposed in the early 1970s and nowadays are broadly exploited as support to make secure control operations in transmission systems.

11.1.3 Distribution System State Estimation, DSSE

It is now expected that estimation techniques will also spread in distribution systems, where operating conditions, in terms of node voltages, branch currents, and powers, can be obtained by means of numerical procedures, the distribution system state estimation (DSSE) techniques, that are specifically designed for distribution grids.

It is worth noting that well-established methodologies that have been used to perform SE for several decades in transmission systems cannot be applied directly to distribution systems due to the aforementioned peculiarities of this kind of grids. Research activities on the topic of DSSE started in the 1990s, when the idea to automate some management operations began to spread and utilities planned to install supervisory control and data acquisition (SCADA) systems. In particular, DSSE methodologies are designed to exploit the few real measurements available on the field and any other kind of available information about the network. The number of MV/LV substations is about two orders of magnitude larger than the number of HV/MV substations, and each of them has several LV lines. Therefore, the cost required for a suitable coverage of distribution systems with monitoring units is a crucial factor and a major obstacle, and thus, detailed information on LV loads is commonly missing. The observability of the network is thus obtained by exploiting a high number of derived measurements. Historical and forecast data about load power consumption and/or generation at substations are used to evaluate these derived measurements, the so-called pseudo-measurements. Prior data concerning generation and load demand are commonly available with a time resolution of 15–30 minutes. The uncertainty associated with this kind of information is very high, and commonly it is assumed in the order of 50%. Obviously, this poor accuracy, along with the high quantity of prior data to use, significantly affects the accuracy of the estimated state and, consequently, can impact on the efficiency of the distribution network operation [4].

Moreover, due to the presence of nonsymmetrical three-phase electrical quantities, the simplification given by the single-phase equivalent scheme or direct sequence, commonly assumed in power systems, cannot be applied. Therefore, the estimators for distribution networks should generally be developed according to a three-phase model. In addition, the physical characteristics of the lines lead to the impossibility of adopting simplifications frequently used in the estimators developed for transmission systems, as, for example, neglecting resistances because of dominant inductive terms. As a further consequence, decoupled versions of the estimators are not so easily obtained. Finally, in these types of grids, the knowledge about the actual values of line parameters can present quite a large uncertainty, due to network aging and lack of accurate measurement campaigns.

Considering the challenges offered by the presented scenario, several DSSE methods have been proposed in literature in the last decades, with different approaches and considering different measurement systems. In particular, different families of DSSE algorithms exist, with static and dynamic models. Static estimators include the widespread weighted least squares (WLS) estimators [5–7] and Bayesian methods [8], while dynamic estimators often rely on Kalman filter (KF) application [9].

11.1.4 Open Issues of DSSE

Each estimator designed for distribution systems must cope with or address the following open issues:

- The lack of measurement redundancy that affects the estimator uncertainty and robustness (for instance, with respect to possible device malfunctions or accuracy degradation) and the possibility to implement auxiliary functions such as bad data detection, topology error identification, and parameter check.
- The high number of nodes that, along with the need of developing three-phase estimators, leads to tackle very large systems, resulting in the explosion in the execution times of the algorithms and in strong requirements for data storage within the calculation center of the network.
- The geographical dimensions of these systems that cause technical problems, concerning the collection (need of reliable communication systems) and the comparison of the available measurements (need of accurate processing methodologies).
- The constant evolution of distribution networks toward active distribution networks and SGs that leads to increasingly complicated operating conditions, requiring newly designed estimating strategies and a new generation of measurement devices.

A way to deal with these problems is to apply multi-area state estimation (MASE) approaches. The MASE process can be seen as a distributed or hierarchic complex monitoring system that, relying on the available measurements and integrating them with all the available information, yields an accurate view of the actual operating conditions of the overall network.

In the following sections of this chapter, different aspects of MASE DSSE implementations are introduced and discussed. The measurement devices that are the building blocks of the DSSE architecture are presented, and the communication infrastructure requirements are concisely addressed. Then, the basic principles of MASE approach are explained along with the latest algorithms proposed in the literature and examples of application.

11.1.5 Sensors and Smart Metering

SGs are expected to be equipped with new generations of smart devices able to generate vast amounts of data about the network status, which will be key components for the efficient, dynamic, and prompt management of such complex grids. In this regard, it is worth noting that a sort of time tag, stating the reference time of the measurements considered in the estimation process, would be necessary as for the coordination and comparison of the measured data.

Focusing on the new generations of measurement devices required for the estimation of the actual operating conditions of the grids, it is possible to cite SRA2035 [1], according to which "Usage of measurement devices as PMU … opens new system awareness and even an early warning approach for the control systems at all levels in the grid."

Phasor measurement units, PMUs, are devices able to measure synchronized voltage and current phasors (the so-called synchrophasors) with a time accuracy of about 1 μs (or less) with respect to a common coordinated universal time (UTC) reference and with reporting rates that can vary from 10 to 100 measurements per second, when the fundamental frequency is 50 Hz. Frequency and rate of change of frequency (ROCOF) have also to be measured, respecting given accuracy limits fixed by standards. The IEEE Standard C37.118.1-2 for PMUs was released in 2011 by means of two documents [10] and [11], with measurement and data communication specifications, respectively. An amendment, released in 2014 [12], modifies or suspends some of the performance requirements specified in [10]. IEEE Standard C37.118.1 introduces, for the first time, accuracy limits to be respected under dynamic conditions, thus drawing attention to a more complete characterization of PMU measurement accuracy. Standardization is in progress, and, in particular, it is worth noting that an IEEE/IEC working group published a joint synchrophasor standard, the IEC/IEEE 60255-118-1, in December 2018 [13]. The main metric to evaluate the accuracy of PMUs is the total vector error (TVE), which represents the relative absolute vector error between measured and reference phasor. However, such concise metric was designed without considering possible SG scenarios that can ask for a more detailed specification of PMU performance under conditions that may become particularly interesting in future networks. The signals presented in [10] were thought for transmission systems, and PMU employment in distribution networks will require specific characterization processes [14].

Nowadays, PMUs are the most accurate instruments for power system monitoring, but accurate measurement of synchrophasors is still a major challenge. Several dynamic algorithms for synchrophasor estimation have been presented in literature [15–18], and this kind of study is still in progress.

Challenges are open also because of the need for reliable communication of measured data. PMUs typically communicate measurement data to phasor data concentrators (PDCs). PDCs should be able to retransmit data received by several PMUs, aligned by timestamp, and, eventually, archive the data for post event analysis. Protocols have been developed to set standards for the requirements of these devices. However, PDCs still do not have an intelligent brain able to operate locally crucial analyses on the received data. Thus, a new generation PDC should be designed to rapidly detect possible critical situations and solve them by means of pseudo-real-time analysis [19].

The PMUs are very accurate measurement devices, but their price and, above all, their installation costs do not yet allow a widespread deployment in the distribution grids. In this regard, it is possible to note that a new generation of energy meters, the so-called smart meters, SMs, is spreading.

SMs are expected to be smart devices designed to process several measurements and to communicate accurate data by means of suitable communication protocols. Several types of commercial energy meters exist, but measured data are not easily comparable and suitably characterized considering SG requirements, despite the growing importance of such kind of information. It is worth noting that SMs can make several types of measurements. However, a clear standardization does not exist. The European Directive 2014/32/EU [20], Measuring Instruments Directive (MID), considers the requirements of different types of measurement devices, including meters for active energy. MID-approved instruments passed specific conformity assessment procedures, and the aim is to create a single market in measurement instruments. However, energy meters are usually based on data processed by means of different measurement algorithms and proprietary software, on which there is no regulatory requirement. For instance, information on the time windows assumed for the measurements is commonly missing. Nevertheless, new tariff structures and incentives for energy saving and home automation call new studies on the possibility to use effectively data measured by SMs.

11.1.6 Automation and Communication Requirements

The full achievement of a SG also passes through an efficient and future-proof information and communication system able to keep up with the needs of SG operator [21, 22].

A reliable data updating architecture is needed for SG management, because the monitoring systems have measurement points located on wide areas, while decisional points are instead commonly concentrated in the substations. The communication infrastructure has to be designed considering different aspects: latency, transmission reliability, two-way communication, quality of service, installation costs, security, and data management. Without going into details of all these factors, it is important to recall here that, while transmission systems have dedicated communication infrastructure (the implementation of modern wide area measurement systems, WAMS, can be kept in mind), the situation of distribution systems is more complicated. DSO can operate at local, regional, or national scale and can be of different size, so that it can support different costs of the ICT infrastructure. For this reason, it is foreseeable that DSO will choose also public-shared networks to build their architecture, considering the annual fees of the service instead of maintaining a private infrastructure. It is also important to recall that in distribution networks all the information available for DSSE should be exploited, thus meaning that all the existing measurement devices both traditional and modern should be integrated with the monitoring architecture. It is not possible to conceive the measurement system as a new infrastructure to be deployed from scratch, without considering the constraints imposed by the already available devices and connections. It is more likely that DSOs will build their architecture in an incremental way, upgrading and expanding their systems over the years.

In this context, it is worth noting that in the last decade, Internet of Things (IoT, [23]) paradigm has received increasing attention and appears promising in the SG domain. Interest is related specifically to the possibility of creating cyber-physical worlds where everything can be measured, used, controlled, composed, and updated. One major building block behind the success of the IoT paradigm is the virtualization [23, 24], in which counterparts in the cyber world are created for any real entity, for example, measurement device, and its functionalities. Virtualization permits to extend and to share the functionalities and the features of the devices, making heterogeneous objects interoperable using semantic descriptions. These aspects are among those envisioned in the specific context of SGs. The SRA2035 [1] identified the need of a successful interfacing of heterogeneous grid equipment to ensure interoperability with scalable solutions (i.e. suitable for the needs ranging from those of small utilities to those of large utilities and industrial users).

In particular, Cloud-based solutions can be used to address the nontrivial tasks related to storage, manipulation, and management of a large amount of data. Indeed, the Cloud can ensure a reliable environment for DSSE and every application relying on it, with massive computational and storage capabilities. Moreover, it can elastically react to critical situations in which the need for resources dynamically changes [25, 26].

The use of MASE architecture, besides all the advantages that have been mentioned and will be detailed in the following, answers to the need of lighter communication infrastructure, thus showing how the DSSE algorithm can be studied together with ICT issues to limit bandwidth, losses, costs, and delays.

11.1.7 Multi-Area State Estimation

The MASE process can be seen as a distributed or hierarchic complex monitoring system that, relying on the available measurements and integrating them with all the available information, yields an accurate view of the actual operating conditions of the overall network.

Multi-area techniques have been proposed to enable wide-area management and control in power systems. In a multi-area framework, the whole grid is divided into several zones that can be processed independently: this allows simplifying and speeding up the overall estimation process. Furthermore, a MASE approach is sometimes required also to handle portions of the grid having different features (e.g. different voltage levels) or being operated by different system operators.

Several MASE approaches have been proposed in the literature for transmission systems. They rely on the high redundancy of measurements available in these systems and are based on different criteria for the network partition, the computing architecture, and the coordination scheme [27–30]. The common goal is to exploit SEs performed on different sub-networks to deal with a large network that can belong to different transmission system operators.

On the other hand, to date, there has been a limited research activity dealing with multi-area DSSE. Because of the peculiarities of the distribution networks, the approaches applied in transmission systems are not directly applicable, and, thus, ad hoc MASE algorithms designed for DSSE are required.

In this case, the main aim is to decompose the DSSE problem to handle smaller systems and then recombine local estimations to get an overall picture of the network status. Nevertheless, several issues prevent an easy and straightforward implementation.

Once again, the most crucial problem is the limited number of measurement instruments installed on the field. The possibility to split the entire grid into limited areas can be strictly conditioned by the availability of measurements allowing the observability, or a minimum redundancy, for each sub-network. For this reason, the design of the areas must consider duly the configuration of the available measurement system, or, in a perspective of a measurement system upgrade, an appropriate meter placement.

In general, the design of a multi-area approach requires a trade-off among accuracy, efficiency, and communication and storage requirements. In the following, different ways to address the MASE problem are presented to highlight strengths and weaknesses of each method. Then, a proposal specifically conceived for the distribution networks is discussed.

11.1.8 Multi-Area Approaches

The multi-area proposals can be classified according to several criteria, like the level of area overlapping, the timing of the estimation processes, the computing architecture, or the adopted solution methodology [27]. The knowledge of strengths and weaknesses of the different solutions is useful to identify the approach that is more suited to the features and requirements of the given distribution network.

11.1.8.1 Level of Area Overlapping

The approaches to the MASE problem can be distinguished depending on the way the network is partitioned. The typologies of multi-area partition can be classified in three main groups: no overlapping, minimum overlapping, and extended overlapping, as shown in Figure 11.1.

1. No overlapping: no node is shared among areas.
   In this case each node belongs to one sub-network, and the size of each area is minimized. With a state vector composed of node voltages, this solution leads to autonomous estimation solutions for the different sub-networks. Afterward, integration procedures can be applied to improve

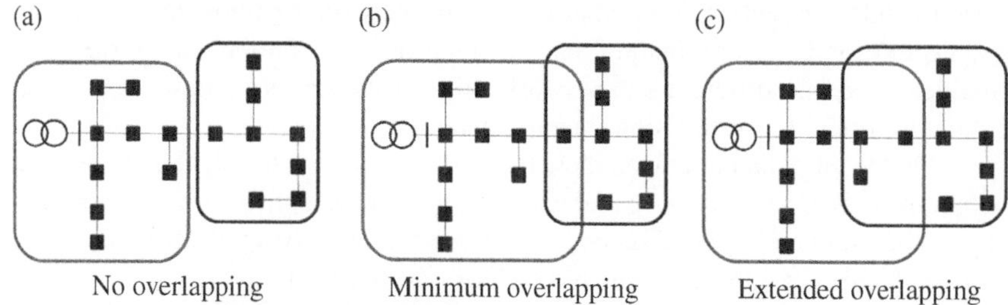

Figure 11.1 Multi-area partition strategies.

the estimation results of each area by including the border information of the neighboring areas (the voltage estimations at the border nodes and/or the powers at the branches connecting the considered area to the adjacent ones) [30, 31].

2. Minimum overlapping: only one node is shared among adjacent areas.
   In this case, some measurements can be shared among local decision points of contiguous areas. Each local estimator can update its results using information coming from the adjacent areas. In particular, the voltage estimations obtained by the different estimators on the shared nodes can be duly exploited to refine estimation results [28, 32].
3. Extended overlapping: more nodes are shared among different areas.
   In this case, the amount of shared measurements is larger than the previous scheme. Some advantages can be achieved, but a larger size for the subnetworks must also be considered, with a possible impact on the MASE efficiency (in terms of overall execution times) [29, 33].

11.1.8.2 Execution of the Estimation Processes

The timing for the execution of the local application is an essential element characterizing MASE approaches. In transmission systems, usually, all the proposals refer to a parallel running of the local estimations.

In general, two options are possible: in-series or in-parallel estimation processes, showed in Figure 11.2 in a schematic overview.

11.1.8.2.1 In-Series Implementation In this case, local estimators ("SE *i*", in Figure 11.2, where i is the area index, $i = 1, \ldots, n$) are performed sequentially, and each area can exploit the estimation results obtained by the neighboring upstream areas. This solution allows a flexible partition of the network and permits to create areas that become observable thanks to the information provided by the upstream areas. This can be particularly interesting in the distribution system scenario, considering the lack of measurement devices. Moreover, each area can store only its own data plus few information about the border nodes and the connecting

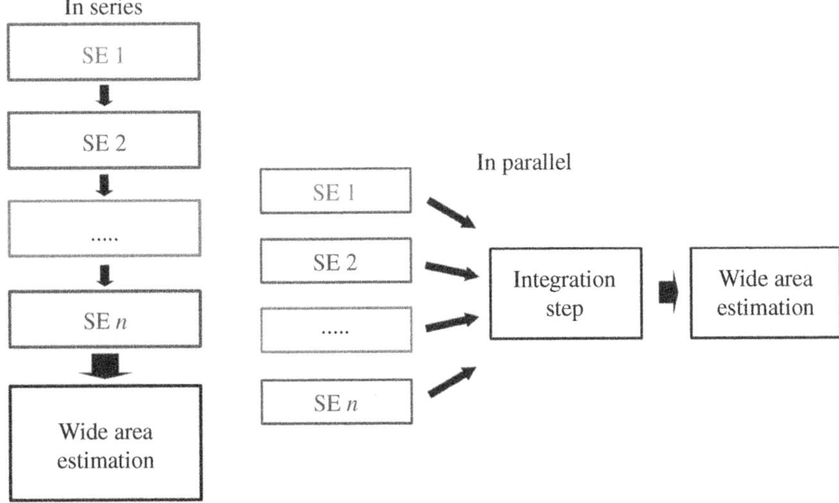

Figure 11.2 Multi-area state estimation: in-series and in-parallel execution.

lines of the adjoining areas. In this case, also the communication costs are quite low, since the communication can be limited only to the estimation results of the border quantities.

On the other hand, the major drawback of such approach is the execution time. Nevertheless, it is worth underlining that a sequential SE execution on smaller networks can allow saving computation time with respect to an estimation process carried out on the whole network.

11.1.8.2.2 In-Parallel Implementation The local SEs are performed simultaneously (as in [32, 33]) and are followed by suitable procedures of integration of the local estimation results. The communication costs, the storage requirements, and the accuracy results of the final estimation are strictly dependent on the particular methodology used for the integration procedure. An appropriate measurement infrastructure guaranteeing the observability of all the areas is commonly required. The main benefit provided by this approach is clearly on the execution times.

In practical cases, hybrid solutions can be also planned [34, 35]. It is possible to design procedures integrating some local SEs performed in parallel, and some other SEs enabled to run only after the acquisition of the estimation results from the upstream sub-networks.

11.1.8.3 Architecture

The architecture used to collect information and to perform the estimation is a very important element in a MASE. Possible options can be classified in two main categories, centralized and decentralized architecture, as presented in Figure 11.3, which also shows the main features for each architecture.

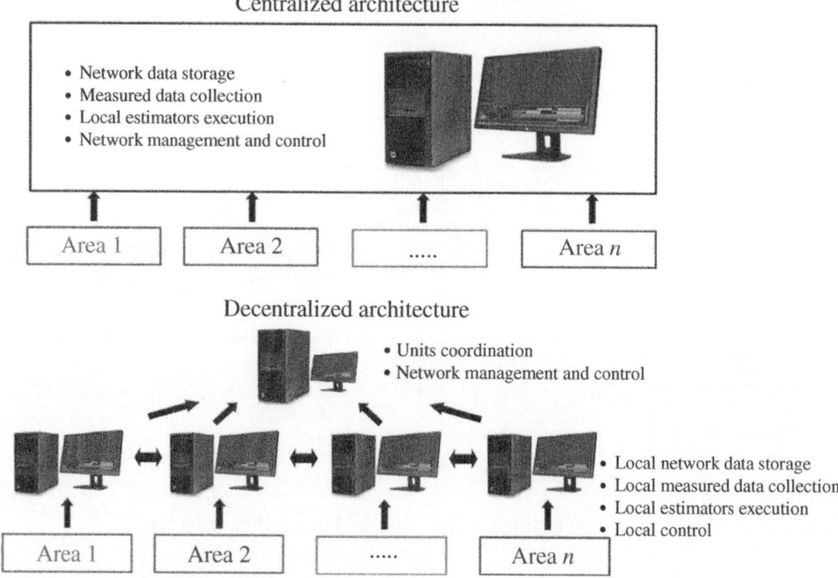

Figure 11.3 MASE computing architecture.

11.1.8.3.1 Centralized Solution In this case, there is a central computer (a mainframe installed in the control center) storing the network data of all the sub-networks, handling pseudo-measurements and performing all the main functions. Each area communicates the real-time measurements to the central unit, which collects the measured data and evaluates all the quantities of interest.

Execution and coordination of the estimators of different areas in a single computing unit can facilitate the implementation of advanced MASE techniques, with benefits for the accuracy of the estimation results. Moreover, having the available data in the same computing unit can provide additional advantages in terms of a more straightforward management and control of the network. On the other hand, high communication costs and storage requirements can derive from this solution.

11.1.8.3.2 Decentralized Architecture Each area handles its own real-time measurements, pseudo-measurements, and network data. The areas have a dedicated unit, which is in charge of the execution of the local SE and of the communication with the neighboring areas of the data needed for the refinement of the local results. This allows a reduction of the communication and storage requirements but with possible limitations for the design of the MASE technique.

Control functions can be implemented not only in the sub-network units but also in a central computing center. This center can exist in any case, possibly with functions limited to the coordination and supervision of the local units and to the monitoring of the network operation.

Hybrid solutions can also be also conceived, combining some of the benefits of both the aforementioned approaches. The architecture could be designed so that each node handles the measurements and pseudo-measurements of its area and performs the local SE. Then, the local SE results are sent to a central computing center that refines the estimations. In a hybrid solution, the advantages are in the distribution of the processing tasks and in the reduction of the data to be communicated to the central unit.

In the distribution system scenario, centralized solutions allow a good level of flexibility for the MASE design. On the other hand, given the large number of nodes and the amount of data involved in the process, a decentralized solution can provide important benefits for the communication and storage requirements.

11.1.8.4 Solution Methodology

Different specific methodologies and algorithms can be followed to perform MASE. In all the techniques, the overall network status is estimated by merging the results of local area computations with the aim to reduce the estimation uncertainty.

The algorithms belong to two main classes depending on the coordination level used in the process: MASE algorithms that perform integration at SE level and algorithms that operate at the iteration level. The first class mainly includes two-step methodologies that perform the following activities:

- Local SE for each sub-network.
- Estimation refinement by means of integration of local SE.

The second class includes algorithms that are founded on data exchange between adjacent areas. At each iteration of the estimation process, each local area estimator includes information coming from neighboring zones, which is typically given by border voltages or inter-area power flows estimated by local estimators at previous iteration. Each estimator works on its network models by iteratively refining boundary conditions to improve the knowledge on its state.

It is clear that the algorithms exploiting continuous data exchange on a per iteration basis are much more communication consuming and ask for a synchronization of all the local area processes. Such algorithms are usually much slower because the SE bottleneck is given by data waiting from the neighboring areas. Such methodologies can achieve optimal solutions of the estimation process, that is, solutions that are equivalent to the SE performed on the whole network without partitioning, but many iterations can be required. For these reasons, the first methodology class seems more promising in a distribution system multi-area perspective, since it aims at computational efficiency with low communication constraints. Two-step algorithms try to keep accuracy targets, even if optimality can only be achieved under special circumstances, when the network topology and practical constraints (management, distances, etc.) allow a custom design of the measurement system and of the division into areas. It is interesting to recall that, relaxing the speed constraints and using the hybrid approach, it is possible to repeat the two-step algorithm more than once, so that accuracy can be further refined if needed.

In the following, a novel strategy for multi-area estimation in distribution systems is illustrated, and examples are given to show its potentialities.

11.2 DISTRIBUTION SYSTEM MULTI-AREA STATE ESTIMATION (DS-MASE) APPROACH

As presented in the previous sections, multi-area techniques can be helpful, or even essential in some cases, to design an effective and efficient SE algorithm for the distribution system. In this section, a recently proposed MASE approach for distribution systems, namely, DS-MASE, is presented, which aims at combining the requirements of high accuracy and robustness together with the needs of low execution times, distributed computation and storage, and limited communication costs [36]. The DS-MASE solution relies on a two-step algorithm: the first step provides the local estimation for each sub-area, whereas, at the second step, estimations are refined through an ad hoc designed harmonization procedure that allows enhancing the final accuracies of the estimates.

11.2.1 Multi-Area Partition and Architecture

The starting point for the design of any multi-area technique is to identify the sub-areas composing the overall grid under analysis. The multi-area approach here presented does not put specific constraints for the partitioning of the grid, but some general criteria should be followed, whenever possible, to maximize the efficiency of the multi-area scheme. Main guidelines or considerations for the network partition are as follows:

- Sub-areas should be possibly composed of a similar number of nodes; this allows having a good distribution of computation and storage requirements in each sub-area and allows maximizing the efficiency of the MASE algorithm, lowering execution times, and avoiding the presence of local estimations acting as a bottleneck for the whole architecture.
- Distribution systems can be composed by parts of the grid operated at different voltage levels or with different technical characteristics; a partitioning made according to the technical features of the system is usually convenient, for example, to reduce potential issues of ill-conditioning in the SE algorithm.
- Portions of the distribution system could be in some cases handled by different DSOs: in this scenario, the partition of the grid is automatically constrained by the areas under control of each DSO; the proposed multi-area framework fosters the exchange of information among different DSOs, unlocking an enhanced situational awareness on the local grids.
- Network division should ensure the observability for each individual area (pseudo-measurements are also used to this purpose at distribution level);

this is essential not only for the applicability of the first-step estimations on the single areas but also to guarantee robustness against possible failures in the communication triggering the second step (in this case, first-step estimations with minimum accuracy requirements are anyhow available and can be used for the local management or control).

Once defined the different sub-areas composing the overall distribution grid, the DS-MASE scheme builds upon the following design characteristics:

- **Area partition**: adjacent areas have to share an overlapping node; the overlapping node can be also shared by more than two sub-areas, for example, in the case in which several feeders depart from the overlapping node. This minimum overlapping solution allows an important simplification in the design of the harmonization process performed at the second step of the MASE approach. Moreover, the placement of meters on this node can also help in reaching the observability requirements on the different sub-areas by deploying a minimum amount of measurement devices. As shown in the following, in general, the approach does not require the mandatory presence of measurements on the shared node. However, if the bus voltage and the powers/currents on the branches converging to that node are monitored, significant benefits can be achieved in terms of final estimation accuracy; in addition, if pseudo-measurements are available for each load or generation node, the presence of measurement points at the shared nodes is enough to guarantee the observability of the sub-grid.
- **Computing architecture:** a distributed, decentralized architecture is considered. In the described scheme each sub-area is endowed with a local mini-control center, which is responsible for (i) collection of the measurements from the meters deployed within the sub-grid, (ii) management/update of the pseudo-measurements (if any) required to reach the local area observability, (iii) storage of all the sub-network data necessary as input to the local SE algorithm, (iv) execution of the first- and second-step local estimation processes, (v) communication of the relevant data with the adjacent sub-areas, (vi) storage of the local SE results for possible activation of local automation routines or for possible interrogation by an upper-level supervision and monitoring unit. In such an architecture, each local mini-control center is fully autonomous, and the tasks of computation, storage, and communication are distributed among the different areas, thus relaxing the overall requirements otherwise present for the whole distribution grid.
- **Estimations execution:** both the first and the second step of the estimation process are executed in parallel among the different sub-areas. This brings the obvious advantage to minimize the overall execution times, making also possible to find a trade-off between the desired reporting rate of the SE process and the level of partitioning of the grid.
- **Coordination of local estimations:** the coordination of the local SE processes and the harmonization of the results is at SE level. Consequently, each

area performs a complete SE run at local level and, only afterward, the results are exchanged between neighboring zones to enable the refinement of the estimations taking place during the second step. A single communication step is thus required during the overall SE process between first and second step. Differently from the MASE solutions coordinated at iteration level (exchange of data carried out after each iteration of the local SE algorithms), this solution allows limiting the requirements in terms of communication bandwidth and synchronization of the local estimation processes and considerably reduces the risks associated with possible communication failures.

11.2.2 DS-MASE Procedure

The MASE procedure here presented relies upon two estimation steps performed locally by each sub-area. Following subsections provide a high-level overview of the tasks performed at the different steps and the set of inputs and outputs used/generated by each step. More details on the harmonization process implemented at the second step of the MASE and on the mathematical foundations for its design are provided in Section 11.2.3.

11.2.2.1 First Step

In the first step of the MASE procedure, a common WLS algorithm is adopted to locally perform SE in each sub-area. The inputs needed for the SE algorithm are the topology of the sub-network, the impedance characteristics of the lines, and the measurements available in the considered area. In particular, the input measurement vector $\boldsymbol{y}$ for the SE algorithm is composed of the real measurements $\boldsymbol{y}_{\mathbf{rm}}$ collected from the instrumentation deployed in the considered sub-area, the pseudo-measurements $\boldsymbol{y}_{\mathbf{pm}}$ for the forecasts of power consumption or generation at the different nodes of the sub-network, and the virtual measurements $\boldsymbol{y}_{\mathbf{vm}}$ associated with the zero injections of the sub-grid. Regarding the real measurements, a particular note applies to the case of a measurement point installed in the overlapping node shared among different areas. In this case, the possible measurements of bus voltage and of powers (or currents) flowing from the node into the area (inner branches) are automatically included in the vector $\boldsymbol{y}_{\mathbf{rm}}$. In addition to these, if power (or current) measurements are also available at the branches that are external to the area (outer branches) but connected to the shared node, then these measurements are aggregated to create an equivalent power (or current) injection measurement at the node. Figure 11.4 depicts a simple example where three sub-areas share the node s: it is possible to see how the measurements on outer branches converging to the shared node are taken into account to create the equivalent injection measurement $P_{\mathrm{eq},s}$.

According to the WLS method, given the input measurements $\boldsymbol{y}$, the vector $\boldsymbol{x}$ of the state variables representing the sub-area (a minimum set of electrical quantities describing the portion of the network) are obtained by the iterative solution of the so-called normal equations:

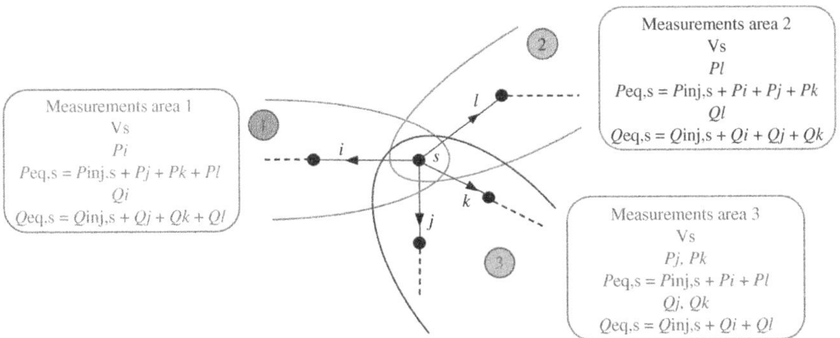

Figure 11.4 Example of equivalent power injection creation at a shared bus.

$$\Delta \boldsymbol{x} = \boldsymbol{G}^{-1}\boldsymbol{H}^T\boldsymbol{W}[\boldsymbol{y} - \boldsymbol{h}(\boldsymbol{x})] \tag{11.1}$$

where $\Delta \boldsymbol{x}$ is the updating vector summed at each iteration to the state vector $\boldsymbol{x}$ to update its result, $\boldsymbol{h}(\boldsymbol{x})$ is the vector of measurement functions expressing the same quantities measured in $\boldsymbol{y}$ in terms of the state variables in $\boldsymbol{x}$, $\boldsymbol{W}$ is the weighting matrix obtained including in the ith term of the diagonal the weight associated with the ith measurement in $\boldsymbol{y}$ (calculated as the inverse of the measurement variance), $\boldsymbol{H}$ is the Jacobian of the measurement functions $\boldsymbol{h}(\boldsymbol{x})$, and finally, $\boldsymbol{G} = \boldsymbol{H}^T\boldsymbol{W}\boldsymbol{H}$ is the so-called gain matrix. The iterative solution of the equation system in (11.1) stops when the largest term of $\Delta \boldsymbol{x}$ (in absolute value) is smaller than a chosen threshold.

The variables included in the state vector $\boldsymbol{x}$ can vary depending on the formulation chosen for the WLS DSSE algorithm. Regardless of the specific choice, all the electrical quantities of the sub-grid (thus voltages, currents, and powers both at the branches and injected at the nodes) can be retrieved, starting from the estimated variables, through corresponding linking functions $\boldsymbol{m}(\boldsymbol{x})$. Moreover, the WLS method offers the possibility to compute all the uncertainties associated with both the estimated variables $\boldsymbol{x}$ and to the indirectly computed quantities $\boldsymbol{m}(\boldsymbol{x})$. The uncertainties of the state estimates $\boldsymbol{x}$ can be obtained in the form of covariance matrix $\boldsymbol{\Sigma}_x$, as the inverse of the gain matrix:

$$\boldsymbol{\Sigma}_x = \boldsymbol{G}^{-1} \tag{11.2}$$

From the covariance matrix $\boldsymbol{\Sigma}_x$, the covariance matrix $\boldsymbol{\Sigma}_m$ of any set of indirectly estimated quantities $\boldsymbol{m}(\boldsymbol{x})$ can be calculated through the following matrix multiplication, which represents the law of propagation of the uncertainties (first order expansion, see [37]):

$$\boldsymbol{\Sigma}_m = \boldsymbol{M}\,\boldsymbol{\Sigma}_x\,\boldsymbol{M}^T \tag{11.3}$$

where $\boldsymbol{M} = \partial \boldsymbol{m}/\partial \boldsymbol{x}^T$ is the Jacobian of the functions $\boldsymbol{m}(\boldsymbol{x})$.

As a result, the local SE performed in each area provides as output the estimation of the different electrical quantities internal to the sub-grid together with the related information on the associated uncertainty. The flowchart in Figure 11.5

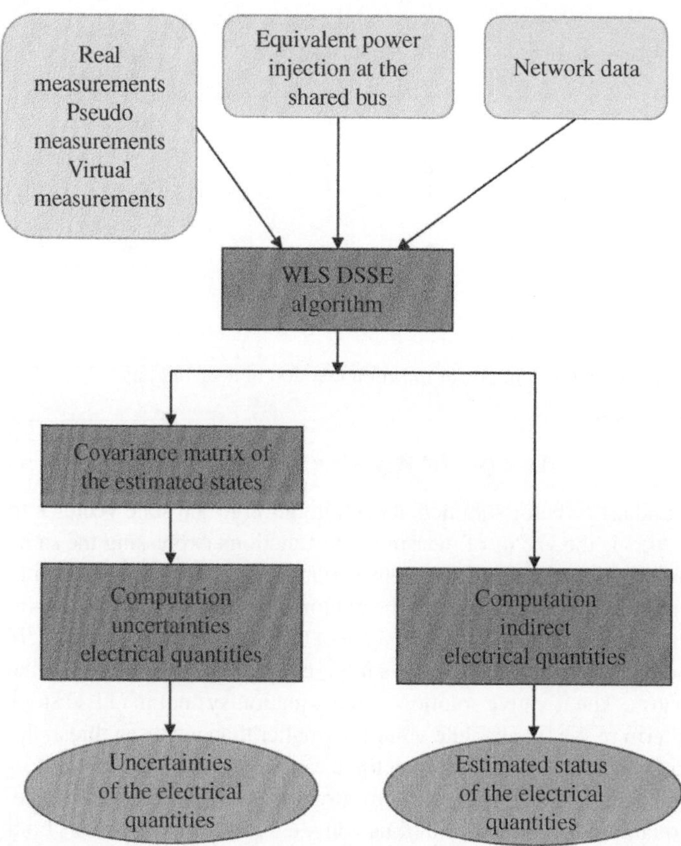

Figure 11.5 Flowchart of the MASE first step.

shows schematically the different steps involved in the first stage of the MASE procedure.

11.2.2.2 Second Step

The goal of the second step is to exploit the information available in the adjacent zones and to suitably integrate it with the local results in order to improve the accuracy of the final estimation. In particular, the focus of the designed second step is on the refinement of the voltage profile. As seen in Section 11.2.1, in the devised MASE scheme, neighboring areas share an overlapping node. Therefore, different voltage estimations are available on the shared node, provided by the different areas sharing it. The idea behind the harmonization performed in the second step is to integrate in each sub-network the additional information given by the voltage estimated on the shared nodes by the adjacent areas. To this purpose, in the DS-MASE architecture, the mini-control centers of each area need to communicate and exchange data with the control centers of the neighboring areas. In particular, the necessary data is limited to the voltage estimate on the overlapping node and the associated estimation uncertainty. The information retrieved from the adjacent areas is then post-processed, together with the local voltage estimate on a reference node, in order to obtain an improved estimation of the voltage status. The obtained voltage is then used to update the whole voltage profile of the sub-grid. The details on the mathematical steps carried out during the harmonization process used in the second step are provided in the following section.

11.2.3 Harmonization/Integration Method

The goal of the second step of the DS-MASE procedure is to exploit the availability of multiple voltage estimates on the overlapping nodes and to suitably merge them with the local results to enhance the final estimation. In Section 11.2.3.1, a simplified WLS formulation to obtain the integration of these data is shown, which is equivalent to the execution of a second SE process on the whole area, but requires a considerably lower computational effort. The proposed procedure has general validity but can suffer a degradation of the accuracy performance when measurement points are installed at the overlapping nodes. In this case, indeed, measurements are shared by the bordering areas, and this creates correlations among the voltage estimates of these areas. As it will be shown in Section 11.2.3.2, if not duly taken into account, these correlations can bring important errors during the execution of the MASE second step. Based on the mathematical analysis presented in Section 11.2.3.1, Section 11.2.3.3 presents an "adjustment" of the simplified WLS procedure for the second step, which accounts for the arising correlations and allows compensating for the impact brought by the shared measurements. Finally, Section 11.2.3.4 summarizes the harmonization procedure, recalling the mathematical steps required to integrate the data coming from the neighboring areas and to obtain the final update of the estimation results.

11.2.3.1 Simplified WLS Procedure

As recalled in previous sections, the second step aims at integrating in each sub-area the voltage estimates on the shared nodes obtained in the adjacent zones with the local results. A possible solution to this problem is to carry out a second WLS estimation, where the new voltage information is added to the locally estimated states to obtain the input vector (equivalent measurement vector) for the WLS algorithm. Even though this procedure works and allows successfully integrating the additional information contained into the voltage estimates provided by the neighboring areas, gained experience shows that achievable improvements are limited to the voltage estimation profile, whereas branch current estimates essentially remain the same as in the first step. Consequently, a simplified algorithm can be conceived, where the goal is to only refine the voltage magnitude estimation at a reference node of the sub-area, exploiting both the local estimate and the inputs arriving from the adjacent areas. The refinement of the voltage estimation at a single reference node is sufficient to improve the whole voltage profile of the grid, since the remaining bus voltages can be computed using the first-step current estimates by means of a simple computation of the voltage drops along the lines.

To design the second-step WLS procedure, an understanding of the impact brought by the additional voltage inputs to the sub-grid state estimates is necessary. To this purpose, let us refer to a WLS formulation where the branch currents of the grid are used as state variables of the system. It is worth noting that formulations based on branch current variables have been often advanced for use in distribution systems (see, for instance, [6] and [7]); moreover, [38] and [39] proved that different WLS formulations provide same estimation results regardless of the state variables chosen to translate the measurement information into estimates. Consequently, the state vector here used is helpful to simplify the mathematical analysis but found results have general validity and do apply to any WLS formulation with any choice of the state vector. For the purposes of our analysis, let us refer to a formulation where the state variables of the considered sub-grid are the complex currents at the different branches. To have a full monitoring of all the electrical quantities of the grid, a node voltage has also to be included to the state vector [7]. The set of considered state variables therefore is

$$\boldsymbol{x} = \left[v_{\text{ref}}, i_1, i_2, \ldots, i_{N_{\text{br}}}\right]^T \tag{11.4}$$

where v_{ref} is the complex voltage at the chosen reference bus of the grid, i_j is the complex current at the generic jth branch, and N_{br} is the total number of branches of the grid.

If a second SE process on the whole sub-grid were applied, using as input the previous state estimates and the voltage at the overlapping nodes as estimated by the neighboring areas, the following measurement vector would be used:

$$\boldsymbol{y}^{(2)} = \left[\hat{v}_{s1}^{(1)A_1}, \ldots, \hat{v}_{sn}^{(1)A_n}, \hat{v}_{\text{ref}}^{(1)}, \hat{i}_1^{(1)}, \hat{i}_2^{(1)}, \ldots, \hat{i}_{N_{\text{br}}}^{(1)}\right]^T \tag{11.5}$$

where $\hat{v}_{sk}^{(1)A_k}$ indicates the first-step voltage estimate at the shared node provided by the adjacent area A_k while $\hat{v}_{\text{ref}}^{(1)}$ and the branch currents $\hat{i}_j^{(1)}$ are the local estimates of the state variables of the grid.

Under the considered assumptions, the local first-step estimates are directly mapped to the state vector in (11.4), whereas the generic voltage at the shared node coming from the kth neighboring area A_k can be expressed in terms of the state vector through the following linear measurement function:

$$v_{sk}^{A_k} = v_{\text{ref}} - \sum_{j \in \Gamma_k} z_j i_j \tag{11.6}$$

where z_j is the complex impedance of the jth branch and Γ_k is the set of branches in the path between the reference node of the grid and the node shared with area A_k.

Due to the linearity of all the measurement functions linking the vector $\boldsymbol{y}^{(2)}$ to the state variables $\boldsymbol{x}$, the WLS problem can be solved through a linear equation system, which directly gives the state vector estimation:

$$\boldsymbol{G}^{(2)} \cdot \hat{\boldsymbol{x}} = \boldsymbol{H}^{(2)T} \boldsymbol{W}^{(2)} \cdot \boldsymbol{y}^{(2)} \tag{11.7}$$

where $\boldsymbol{G}^{(2)}$, $\boldsymbol{H}^{(2)}$, and $\boldsymbol{W}^{(2)}$ are the gain, Jacobian, and weighting matrix, respectively, analogously to what is already described in Section 11.2.2.1.

To assess the estimation of the reference bus voltage, let us focus on the first equation resulting from the equation system in (11.7). Such equation involves the multiplication between the first row of the gain matrix $\boldsymbol{G}^{(2)}$ and the state vector $\boldsymbol{x}$ at the left member, whereas the first row of the transpose of $\boldsymbol{H}^{(2)}$ multiplies the weighted measurements $(\boldsymbol{W}^{(2)} \cdot \boldsymbol{y}^{(2)})$ at the right member. Considering the form of the state vector $\boldsymbol{x}$ given in (11.4), the resulting equation can be rewritten as

$$G_{11}^{(2)} \cdot v_{\text{ref}} + \sum_{j=2}^{N_{\text{br}}+1} G_{1j}^{(2)} \cdot i_{j-1} = \sum_{h=1}^{M} H_{h1}^{(2)} \cdot W_{hh}^{(2)} \cdot y_h^{(2)} \tag{11.8}$$

where the subscripts adopted for $\boldsymbol{G}^{(2)}$, $\boldsymbol{H}^{(2)}$, and $\boldsymbol{W}^{(2)}$ elements are used to indicate the position of a specific term within the matrix. To better understand the relationship between input measurements and final estimation of our target voltage v_{ref}, a deeper analysis of the gain matrix and the Jacobian terms involved in (11.8) is necessary. The building of the Jacobian in the proposed framework is

$$\boldsymbol{H}^{(2)} = \begin{bmatrix} \dfrac{\partial \boldsymbol{h}_{v_s}}{\partial v_{\text{ref}}} & \dfrac{\partial \boldsymbol{h}_{v_s}}{\partial \boldsymbol{i}^T} \\ \dfrac{\partial h_{v_{\text{ref}}}}{\partial v_{\text{ref}}} & \dfrac{\partial h_{v_{\text{ref}}}}{\partial \boldsymbol{i}^T} \\ \dfrac{\partial \boldsymbol{h}_i}{\partial v_{\text{ref}}} & \dfrac{\partial \boldsymbol{h}_i}{\partial \boldsymbol{i}^T} \end{bmatrix} \tag{11.9}$$

where subscripts v_s and i indicate the shared voltages and branch current measurements, respectively, and $\boldsymbol{i}$ is the sub-vector of $\boldsymbol{x}$ relating to branch currents. In (11.9), the derivatives of all the voltage measurements with respect to v_{ref} are equal to 1, whereas the derivatives of all the branch currents with respect to the same state variable are always equal to 0:

$$\frac{\partial h_{v_{sk}}}{\partial v_{\text{ref}}} = 1 \qquad \forall k \tag{11.10}$$

$$\frac{\partial h_{v_{\text{ref}}}}{\partial v_{\text{ref}}} = 1 \tag{11.11}$$

$$\frac{\partial h_{i_h}}{\partial v_{\text{ref}}} = 0 \qquad \forall h \tag{11.12}$$

For the derivatives with respect to the branch current state variables, instead, the following holds:

$$\frac{\partial h_{v_{sk}}}{\partial i_j} = \begin{cases} -z_j & \text{if } j \in \Gamma_k \\ 0 & \text{if } j \notin \Gamma_k \end{cases} \tag{11.13}$$

$$\frac{\partial h_{v_{\text{ref}}}}{\partial i_j} = 0 \qquad \forall j \tag{11.14}$$

$$\frac{\partial h_{i_h}}{\partial i_j} = \begin{cases} 1 & \text{if } h = j \\ 0 & \text{if } h \neq j \end{cases} \tag{11.15}$$

In (11.13), the derivative with respect to the jth branch current is equal to the impedance of the branch changed in sign, only if that branch is included in the path Γ_k between the reference node and the considered shared voltage; otherwise the derivative is null. The derivatives of the reference voltage v_{ref} are always null, and the same holds for the derivatives of the branch currents i_h, except when measurement and state variable refer to the same branch ($h = j$) so that the derivative is equal to 1.

The Jacobian matrix can be thus represented in compact form with the following submatrices:

$$H^{(2)} = \begin{bmatrix} \vdots & & \vdots & \\ 1 & \cdots & \begin{cases} -z_j & \text{if } j \in \Gamma_k \\ 0 & \text{if } j \notin \Gamma_k \end{cases} & \cdots \\ \vdots & & \vdots & \\ 1 & & 0 & \\ \mathbf{0}_{N_{\text{br}} \times 1} & & \boldsymbol{I}_{N_{\text{br}}} & \end{bmatrix} \tag{11.16}$$

Using the results obtained in (11.16), the right member of the equation in (11.8) can be calculated, and it is

$$\sum_{h=1}^{M} H_{h1}^{(2)} \cdot W_{hh}^{(2)} \cdot y_h = \sum_{h=1}^{M_V} w_{V_h} \cdot \hat{v}_h^{(1)} \tag{11.17}$$

where M_V is the total number of voltage measurements included in the input vector (considering both the first-step voltage estimates collected from the adjacent areas and the local estimate at the reference node), w_{V_h} is the weight associated with the hth voltage measurement, and $\hat{v}_h^{(1)}$ is the hth first-step voltage estimate used as input in this second step of the MASE.

In addition, from (11.16), the terms of the gain matrix included in the left member of (11.8) can also be computed:

$$G_{11}^{(2)} = \sum_{h=1}^{M_V} w_{V_h} \tag{11.18}$$

$$G_{1j}^{(2)} = -z_{j-1} \cdot \sum_{k \in S} w_{V_k} \tag{11.19}$$

In (11.18), $G_{11}^{(2)}$ is equal to the sum of all the weights associated with the M_V voltage measurements included in the input vector $\mathbf{y}^{(2)}$; in (11.19), each jth term of the gain matrix row is equal to the impedance of the associated branch $j-1$ (changed in sign), which multiplies the sum of the weights for all the k voltages that have the branch $j-1$ in the path S between the reference bus and the node corresponding to the voltage k.

Using the results obtained in (11.17), (11.18), and (11.19), Eq. (11.8) can be rearranged to obtain the following equivalent relationship:

$$\sum_{h=1}^{M_V} w_{V_h} \cdot v_{\text{ref}} = \sum_{h=1}^{M_V} w_{V_h} \cdot \left(\hat{v}_h^{(1)} + \sum_{j \in \Gamma_h} z_j \cdot i_j \right) \tag{11.20}$$

Let us focus now on the right member of (11.20). The terms between parenthesis express, for each voltage measurement h considered in the second step, the sum between the first-step voltage estimate and all the voltage drops in the path between reference node and bus associated with the considered voltage input. This also corresponds to the calculation of the resulting voltage at the reference bus, obtained by using the starting voltage $\hat{v}_h^{(1)}$ as input and accounting for the different voltage drops along the grid. Under the assumption that the new current estimates i_j are approximatively equal to the first-step estimates $\hat{i}_j^{(1)}$ (the validity of this assumption will be shown via simulation results in Section 11.3.2), the term between parentheses can be calculated in advance and is

$$\hat{v}_h^{(1)} + \sum_{j \in \Gamma_h} z_j \cdot \hat{i}_j^{(1)} = \hat{v}_{\text{ref},h}^{(1)} \tag{11.21}$$

where $\hat{v}_{\text{ref},h}^{(1)}$ is the equivalent voltage at the reference node derived from the hth voltage input in $\mathbf{y}^{(2)}$.

Introducing the result in (11.21) into (11.20), the following holds true:

$$v_{\text{ref}} = \frac{\sum_{h=1}^{M_V} w_{V_h} \cdot \hat{v}_{\text{ref},h}^{(1)}}{\sum_{h=1}^{M_V} w_{V_h}} \tag{11.22}$$

Equation (11.22) clearly shows that the second-step estimation of v_{ref} can be directly calculated as a weighted average, thus avoiding the use of a complete SE procedure involving multiple iterations and the computation of the whole vector of state variables. The computation in (11.22) can be shown to be also equal to the application of a simplified WLS procedure where the only state variable is v_{ref} and the input measurements are the voltages $\hat{v}_{\text{ref},h}^{(1)}$. In this case, indeed, in the simplified solution, (11.7) would translate into

$$[1 \;\ldots\; 1]\begin{bmatrix} w_{V_1} & 0 & 0 \\ 0 & \ddots & 0 \\ 0 & 0 & w_{V_{M_V}} \end{bmatrix}\begin{bmatrix} 1 \\ \vdots \\ 1 \end{bmatrix}\cdot v_{\text{ref}} = [1 \;\ldots\; 1]\begin{bmatrix} w_{V_1} & 0 & 0 \\ 0 & \ddots & 0 \\ 0 & 0 & w_{V_{M_V}} \end{bmatrix}\cdot\begin{bmatrix} \hat{v}^{(1)}_{\text{ref},1} \\ \vdots \\ \hat{v}^{(1)}_{\text{ref},M_V} \end{bmatrix} \tag{11.23}$$

which can be easily proved to deliver the same result as in (11.22).

In the second step of the DS-MASE procedure, the weighted average in (11.22) (or, equivalently, the simplified WLS procedure in (11.23)) is considered to refine the voltage estimation at the reference bus of the area. The whole voltage profile of the sub-grid is then obtained by updating the remaining bus voltages through the computation of the voltage drops given by the first-step current estimates.

11.2.3.2 Impact of Shared Measurements

The analysis performed in the previous section is based on the assumption that all the inputs to the second step are uncorrelated and that, consequently, the weighting matrix can be built as a diagonal matrix. However, one of the characteristics of the DS-MASE architecture is that the partitioning of the distribution system has overlapping zones and that adjacent areas share an overlapping node. When a voltage measurement is installed at such node, the measured voltage is taken as input by all the sub-areas that share that node: in this case, correlations arise among the voltage estimations carried out in the involved sub-areas.

Table 11.1 shows, as an example, the correlation factors arising between the voltage estimations at the shared node performed by two adjacent areas when different measurement configurations are considered (the number of measurements between areas *A* and *B* can be switched leading to the same results). It is possible to observe that, if a shared voltage measurement exists, the lower the total number of voltage measurements available in the areas, the higher the correlation factor.

TABLE 11.1 Correlation factors between voltage estimates with a shared voltage measurement.

Number of measurements			
Area *A*	Area *B*	Shared measurements	Voltage estimate correlation factor
Any	Any	0	0
1	1	1	1
2	1	1	0.69
3	1	1	0.60
2	2	1	0.50
3	2	1	0.42
4	1	1	0.53
4	2	1	0.38

Correlations can be accounted for in the WLS formulation by introducing the corresponding covariance terms in the weighting matrix. Neglecting correlations has been proved to have the potential to degrade the accuracy performance of SE algorithms, in particular when the correlation factors are high [40]. In the DS-MASE scheme, since the used partitioning criterion is likely to introduce correlations, countermeasures have to be found to avoid possible accuracy degradations. To this purpose, the results obtained in (11.22) and (11.23) will be used to assess how shared measurements affect the estimation results. A simple evaluation can be done by considering two sub-areas sharing an overlapping node and comparing the estimation results achievable when a voltage measurement is placed, or not, at the shared node. Firstly, let us refer to the scenario shown in Figure 11.6.

In this case, both area A and area B have two local voltage measurements, and they do not share any meter. Assuming to have the a priori knowledge of the currents in the grid, according to (11.22), the voltage in any bus of the grid can be computed as a weighted average of the voltage measurement input. Therefore, the following voltage estimates can be derived for area A and area B, respectively, at the shared node s:

$$\hat{v}_s^{(1)A} = \frac{w_{V_1} \cdot v_{s,1} + w_{V_2} \cdot v_{s,2}}{w_{V_1} + w_{V_2}} \tag{11.24}$$

$$\hat{v}_s^{(1)B} = \frac{w_{V_3} \cdot v_{s,3} + w_{V_4} \cdot v_{s,4}}{w_{V_3} + w_{V_4}} \tag{11.25}$$

where $v_{s,\,h}$ is the equivalent voltage measurement obtainable through (11.21) at the shared node s starting from the voltage measurement h. Considering the simplified WLS model in (11.23), it is possible to calculate the variance associated with the estimated voltages (as inverse of the gain matrix), which results

$$\sigma^2\left(\hat{v}_s^{(1)A}\right) = \frac{1}{w_{V_1} + w_{V_2}} \tag{11.26}$$

$$\sigma^2\left(\hat{v}_s^{(1)B}\right) = \frac{1}{w_{V_3} + w_{V_4}} \tag{11.27}$$

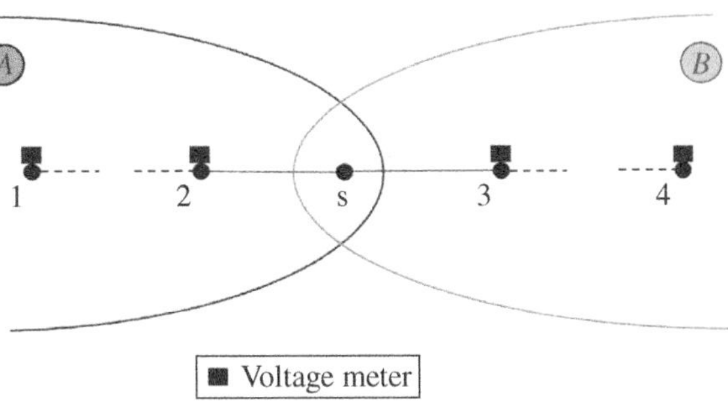

Figure 11.6 Sub-areas without measurement points at the shared node.

From the findings in (11.24)–(11.27), it is possible to apply the harmonization process designed for the second step of the MASE in order to refine the voltage at the shared node. Integrating the results obtained in area A and area B through (11.22), and choosing the shared node s as reference node for both the areas, the following updated voltage $\hat{v}_s^{(2)}$ results:

$$\hat{v}_s^{(2)} = \frac{w_{\hat{v}_s^{(1)A}} \cdot \hat{v}_s^{(1)A} + w_{\hat{v}_s^{(1)B}} \cdot \hat{v}_s^{(1)B}}{w_{\hat{v}_s^{(1)A}} + w_{\hat{v}_s^{(1)B}}} \tag{11.28}$$

which, after the mathematical calculations, is

$$\hat{v}_s^{(2)} = \frac{w_{V_1} \cdot v_{s,1} + w_{V_2} \cdot v_{s,2} + w_{V_3} \cdot v_{s,3} + w_{V_4} \cdot v_{s,4}}{w_{V_1} + w_{V_2} + w_{V_3} + w_{V_4}} \tag{11.29}$$

From (11.29), it is possible to see that the final voltage estimation is equal to the weighted average of all the measurements available in the network, reported to the shared node s through (11.21). By applying (11.22), this can be also proved to be the same result achievable if the overall grid was considered as a whole, thus applying SE on the full grid without any multi-area partition.

Let us now consider as second scenario the one given in Figure 11.7, where areas A and B have both two voltage measurements, but one of them is placed at the overlapping node and is shared between the areas. In this scenario, therefore, the three voltage measurements are totally present in the overall grid. Repeating the same steps done for the first scenario, in this context, the first-step local estimates will result in

$$\hat{v}_s^{(1)A} = \frac{w_{V_1} \cdot v_{s,1} + w_{V_2} \cdot v_{s,2}}{w_{V_1} + w_{V_2}} \tag{11.30}$$

$$\hat{v}_s^{(1)B} = \frac{w_{V_2} \cdot v_{s,2} + w_{V_3} \cdot v_{s,3}}{w_{V_2} + w_{V_3}} \tag{11.31}$$

The associated variances are analogous to those found in (11.26) and (11.27):

$$\sigma^2\left(\hat{v}_s^{(1)A}\right) = \frac{1}{w_{V_1} + w_{V_2}} \tag{11.32}$$

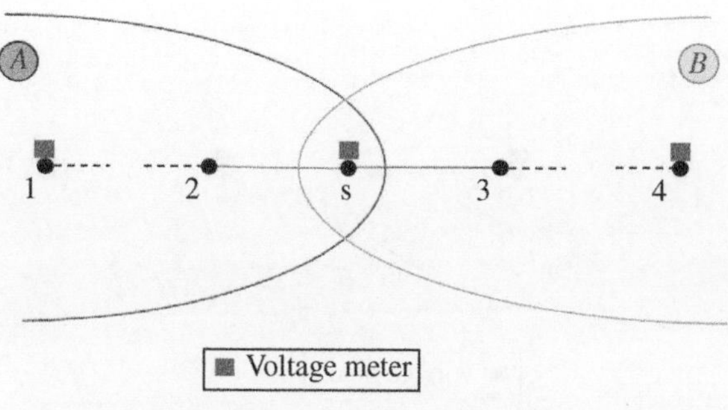

Figure 11.7 Sub-areas with measurement point installed at the shared node.

$$\sigma^2\left(\hat{v}_s^{(1)B}\right) = \frac{1}{w_{V_2} + w_{V_3}} \tag{11.33}$$

Applying again (11.22) at the second step of the MASE, the updated voltage $\hat{v}_s^{(2)}$ will be

$$\hat{v}_s^{(2)} = \frac{w_{V_1}\cdot v_{s,1} + 2w_{V_2}\cdot v_{s,2} + w_{V_3}\cdot v_{s,3}}{w_{V_1} + 2w_{V_2} + w_{V_3}} \tag{11.34}$$

As it can be observed, the proposed analysis allows clearly highlighting the negative impact brought by the shared measurement. In particular, it is possible to see that in this case the final voltage estimate is not the expected weighted average of the voltages (reported to the shared node) available in the grid, because the voltage at the shared node is weighted more than the other ones. Generalizing the result achieved in (11.34), it is possible to demonstrate that voltage measurements at the shared node always have weight n times larger than expected, where n is the number of sub-areas sharing the overlapping node. This reflects the fact that such measurement is used in the local estimation of each area and then is integrated at the MASE second step without accounting for the correlations arising among the local area voltage estimates.

11.2.3.3 Modified WLS Procedure

To deal with the issue highlighted in (11.34), the second step relies upon a WLS procedure derived from (11.23), which includes a small modification able to compensate the overweighting of the shared measurement. In particular, the proposed modification consists in including the shared measurement in the vector $\boldsymbol{y}^{(2)}$ used as input to the second step and in providing to it a weight equal to $-(n-1)\cdot w_{V_s}$, where n is the number of sub-areas sharing the overlapping bus.

Mapping this solution to the scenario described in Figure 11.7, it is possible to find

$$\begin{bmatrix}1 & 1 & 1\end{bmatrix}\begin{bmatrix} w_{\hat{v}_s^{(1)A}} & 0 & 0 \\ 0 & w_{\hat{v}_s^{(1)B}} & 0 \\ 0 & 0 & -w_{V_{s2}} \end{bmatrix}\begin{bmatrix}1\\1\\1\end{bmatrix}\cdot\hat{v}_s^{(2)}$$
$$= \begin{bmatrix}1 & 1 & 1\end{bmatrix}\begin{bmatrix} w_{\hat{v}_s^{(1)A}} & 0 & 0 \\ 0 & w_{\hat{v}_s^{(1)B}} & 0 \\ 0 & 0 & -w_{V_{s2}} \end{bmatrix}\cdot\begin{bmatrix}\hat{v}_s^{(1)A}\\ \hat{v}_s^{(1)B}\\ v_{s,2}\end{bmatrix} \tag{11.35}$$

Processing the matrix multiplications, and considering the first voltage estimates in (11.30) and (11.31) and the weights arising from the variances in (11.32) and (11.33), it is possible to find the following updated second-step voltage at the shared node:

$$\hat{v}_s^{(2)} = \frac{w_{V_1}\cdot v_{s,1} + w_{V_2}\cdot v_{s,2} + w_{V_3}\cdot v_{s,3}}{w_{V_1} + w_{V_2} + w_{V_3}} \tag{11.36}$$

The solution in (11.36) is coherent with the result obtainable by applying (11.22) or (11.23) to the overall grid (without any multi-area partition) and clearly allows compensating the overweighting of the shared voltage measurement in the DS-MASE scheme. It is also worth noting that all the sub-areas already have the information about the measurement at the shared node and its uncertainty since the first-step estimation, thus no additional communication or computational effort is required to include these data in the modified WLS procedure.

11.2.3.4 Summary of the Second-Step Process

Previous sections showed the theory and the mathematical foundation behind the design of the harmonization process conceived for the second step of the DS-MASE. Some simplifications were adopted for the sake of clarity in the presentation of the main concepts, like the use of complex quantities in the state vector, which is not the case in the real implementation of the MASE. Consequently, in the following, the main steps of the harmonization process are summarized referring to the real implementation of the algorithm and removing possible simplifications used during the mathematical analysis:

1. Choose a bus to be used as reference node, in each area, for the following calculation. The second step will update the voltage estimate at this bus, whose result can be later transferred to all the other nodes of the sub-grid in order to update the whole voltage profile.
2. Collect as input data the local voltage estimate at the reference bus, the voltage estimate at the shared node as calculated by all the neighboring areas (outer estimations), and the starting voltage measurements at the shared nodes, if any; all these data have to be accompanied by their associated uncertainty. Note that in the case of conventional measurement system (non-synchronized measurements), only voltage magnitudes have to be considered as input; in the case of measurement systems based on PMUs, both voltage magnitude and phase angle have to be acquired.
3. Referring to (11.21), all the voltage inputs have to be transformed in equivalent voltages at the reference bus of the sub-area. In the case of conventional measurement systems, each voltage magnitude is transformed in an equivalent voltage phasor by using the voltage phase angle estimated at the shared node by the considered sub-area. Then (11.21) is applied to obtain the equivalent voltage phasor at the reference bus, and, finally, the corresponding voltage magnitude is extracted for being used as input in the following WLS. If a measurement infrastructure composed of PMU measurements is instead available, the voltage phasor at the shared node can be directly obtained, and an equivalent voltage phasor can be obtained at the reference bus by direct application of (11.21). Both the voltage magnitude and phase angle are then considered as input to the following WLS procedure.
4. Apply the modified WLS procedure presented in Section 11.2.3.3 to refine the voltage estimate at the reference bus. In the case of conventional

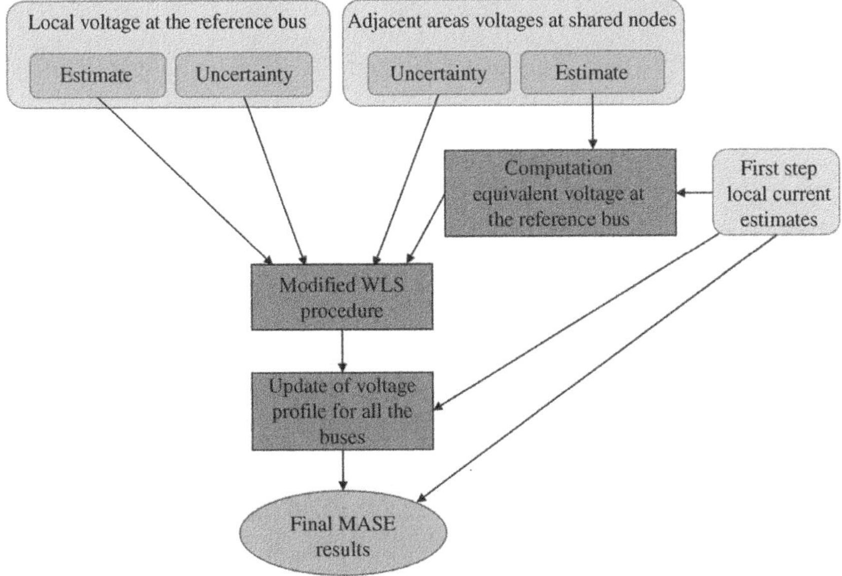

Figure 11.8 Flowchart of the MASE second step.

measurement system, only the voltage magnitude has to be updated using as input the equivalent voltage magnitude inputs calculated in the previous steps. If a PMU-based measurement system is considered, the WLS procedure is extended to refine both voltage magnitude and phase angle at the reference bus, using as inputs the voltage magnitudes and phase angles of the equivalent phasors calculated in the previous step.

5. Starting from the updated voltage estimate at the reference bus, compute the new voltage state of all the remaining nodes in each area, considering the voltage drops along the branches given by the first-step local current estimates.

Figure 11.8 shows schematically the flowchart of the abovementioned steps.

11.3 APPLICATION OF THE DS-MASE APPROACH

In this section, the results achievable through the application of the DS-MASE approach will be described by referring to different possible scenarios. At the beginning, the generic framework and assumptions used to run the presented test cases and the metrics adopted to assess the results will be introduced. Then, simulation results achieved in four different scenarios, characterized by a different setup of the measurement infrastructure, will be presented to show the behavior of the MASE in different conceivable conditions. Finally, a discussion not only summarizing the benefits and strengths of the DS-MASE method but also underlining the boundaries of validity and applicability of the novel MASE approach will be given to provide an overall overview of the proposed solution and highlight the open challenges associated with real implementation of the MASE technique on the field.

11.3.1 Description of the Scenario and Assumptions

The following results refer to the application of the MASE approach on the 95-bus network depicted in Figure 11.9. As it is often the case for distribution systems, the grid has a radial topology, and it is composed by some main feeders that could drive the design of the multi-area partition. For the sake of simplicity, in the following, an equivalent single-phase model of the network will be considered, but the DS-MASE approach also applies to the case of three-phase unbalanced grids.

Proposed scenarios differ due to the measurement configuration considered in the grid. Results and implications associated with the use of two different measurement systems will be presented, referring to the following:

- Conventional measurement system, composed of voltage magnitude and branch power measurements; for the purposes of the simulations, accuracies equal to 1 and 3% (for the voltage and the power, respectively) are taken into account.
- PMU-based measurement system, which provides voltage and branch current synchrophasor measurements; for the purposes of the simulations, accuracies of 0.7% and 0.7 crad are used for the magnitude and phase-angle measurements, respectively: the combination of these values gives a TVE equal to 1%, which is the maximum uncertainty allowed under steady-state conditions for the compliance of the PMU to the standards IEEE C37.118.1 [10, 12] and IEC/IEEE 60255-118-1 [13].

In addition to the abovementioned measurement devices, pseudo-measurements of power consumption/injection at each load/generation node will be considered to guarantee the observability of the grid. A normally distributed uncertainty equal to 50% will be taken into account for this type of forecasts.

Presented results will mainly refer to the Monte Carlo characterization of the accuracy performance of the MASE approach. To this purpose, reference values of the electrical quantities of the grid are generated through a power flow computation and considered as the true operating conditions of the system. Starting from this reference, measurements and pseudo-measurements are extracted by adding to the "true" quantities a random noise according to the abovementioned uncertainties; such measurement values are then used as input for the DSSE processes. The results obtained through 25 000 Monte Carlo trials are then processed to obtain the statistical characteristics of the output. In particular, accuracy performance will be evaluated by referring to the root-mean-square error (RMSE) of the estimated electrical quantities, which is a typical index used to quantify estimators' accuracy performance.

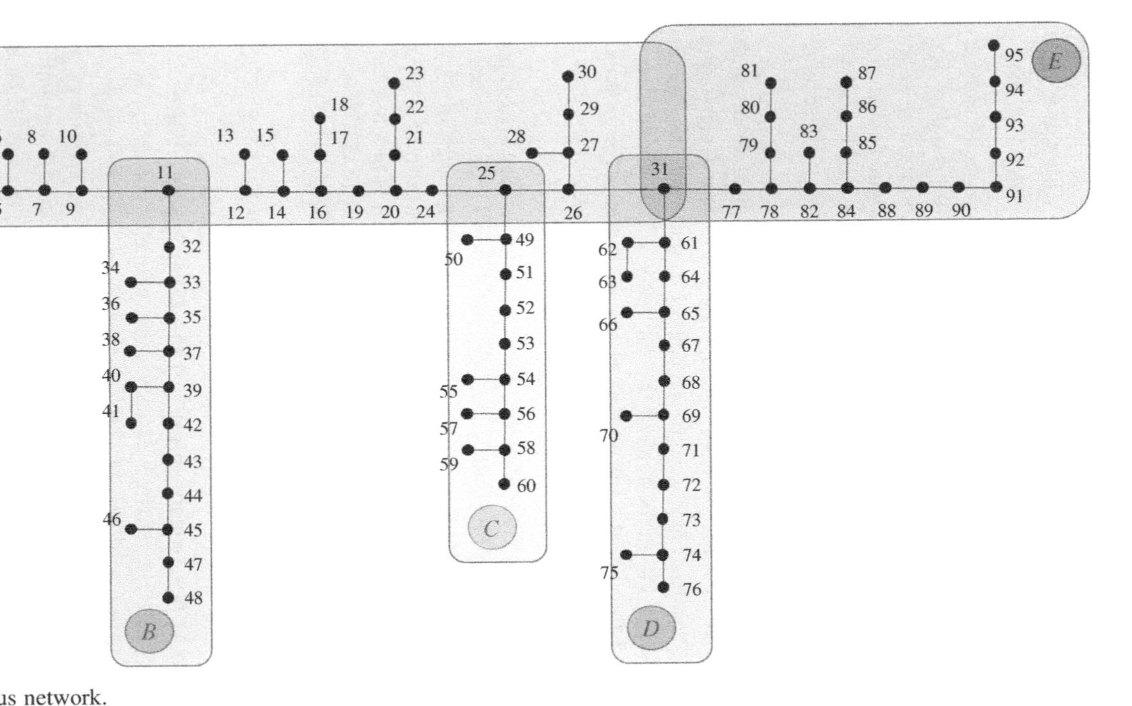

Figure 11.9 95-bus network.

In order to better highlight the characteristics and the accuracy performance of the DS-MASE approach (indicated in the following as MASE WLS_M to remark the use of the modified WLS procedure described in Section 11.2.3.3), the results obtained through other SE processes will be also used as terms of comparison in the presentation of the results. In particular, the following SEs will be considered:

- **Integrated state estimation (ISE):** it refers to the results achievable by executing SE on the whole grid without applying any multi-area decomposition of the network. This is the main term of comparison for any MASE approach, since multi-area schemes usually lead to a degradation of the accuracy performance with respect to the ISE. The closer the accuracy of MASE to the one of the ISE, the better the MASE approach from an accuracy standpoint.
- **Local state estimation (LSE):** it refers to the results obtained locally at the first step of the MASE procedure, without applying any second-step harmonization of the results. These results are helpful to underline the benefits achievable through the refinement of the estimates at the second step of the MASE and through the integration of the measurement information exchanged among neighboring areas. They also give an idea of the estimation accuracy achievable if inter-area communication fails.
- **MASE with simplified WLS (MASE WLS_S):** it refers to the MASE performed by using the simplified WLS described in Section 11.2.3.1 as second step, but without including the WLS modifications described in Section 11.2.3.3. The associated results are used here as a term of comparison to emphasize the importance of a proper consideration of possible measurements/estimates correlations during the MASE procedure and to prove the advantages given by the modification introduced in Section 11.2.3.3.

11.3.2 Example of MASE Applications

11.3.2.1 Scenario 1: No Measurements in the Overlapping Nodes

The first test case considers a general scenario where the meter placement is not coordinated with the design of the MASE scheme. Measurement devices are installed in nodes and branches that are not within the overlapping zones between neighboring areas indicated in Figure 11.9. Such a situation can occur, for example, when a measurement system is already available on the field and the MASE-based monitoring of the grid is developed a posteriori. It is worth to notice that, in this case, the information about the equivalent power injection at the overlapping buses seen from each sub-area (which also includes the power going to those branches connected to the overlapping bus but external to the sub-area) is unknown. Since the observability of all the sub-grids is needed for the applicability of the DS-MASE approach, this can bring some limitations for the design of the multi-area partition or can call for the deployment of additional meters (if the grid partitioning is constrained). Alternative solutions based on the combination of in-series and in-

TABLE 11.2 Measurement location for Scenario 1.

Type of measurement	Measurement location (nodes)				
	Area *A*	Area *B*	Area *C*	Area *D*	Area *E*
Voltage magnitude	1, 19, 26	43, 48	52	72	84, 95
Active and reactive power	1 → 3	43 → 44	52 → 53	72 → 73	84 → 82
	19 → 20	48 → 47			84 → 88
	26 → 31				95 → 94

parallel executions of the first-step local estimations could be also devised to overcome this limitation, but this would have an impact on some of the features presented in Section 11.2.1. Table 11.2 provides the details about the considered measurement location for this scenario.

One of the main assumptions in the design of the DS-MASE approach is that branch currents already have an acceptable estimation at the first step and consequently the second step can focus only on the refinement of the voltage profile. Figure 11.10 shows the results obtained in this test case for the different branch currents of the grid, by comparing the estimates given by the ISE and the MASE (since the final current estimates of the MASE are those obtained at the first step, no difference exists among the two MASE approaches and the LSE). From the plot, it is possible to notice a very fluctuating uncertainty of the estimates. In general, RMSE results above (or close) to 15% are associated with the lateral branches, while the results with RMSE below 5% are associated with the branches in the

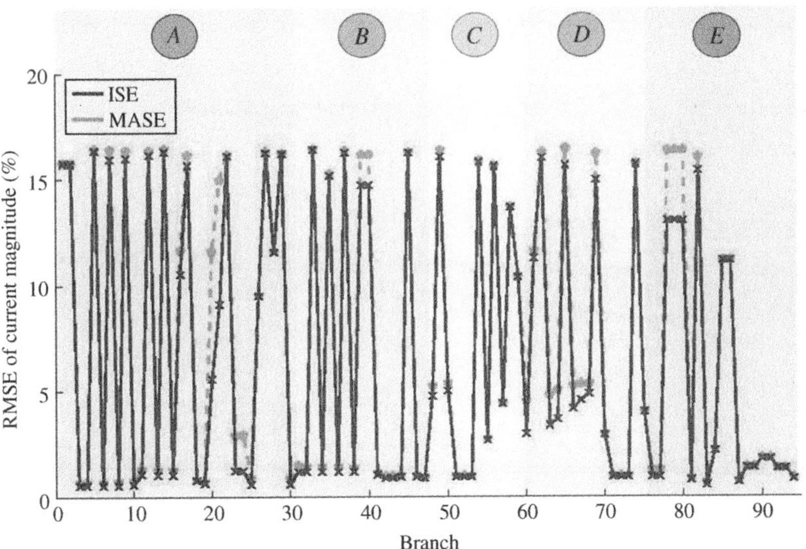

Figure 11.10 Current magnitude estimation in Scenario 1.

main feeders. For most of the branches, the same estimation accuracy as provided by the ISE can be obtained. It is worth highlighting that an accuracy degradation is in general expected when using MASE approaches, due to the fact that not all the measurements available in the grid can be exploited in the local SEs. In the presented result, accuracy degradation is present in some lateral branches where, with the used network partition and measurement configuration, it is not possible to improve the information given by the pseudo-measurements; in the feeders only few branches, close to the sub-area borders, present a slight degradation. As it will be shown in the following test cases, the scenario here presented is a worst case for the estimation of the branch currents: in fact, the placement of branch power measurements at the borders of the sub-areas allows avoiding this performance degradation.

Regarding the voltage magnitude estimation, Figure 11.11 shows the results achievable in the presented scenario. An interesting conclusion can be immediately drawn: the accuracy performance is strictly dependent on the number of voltage measurements being processed. This is clearly visible looking at the LSEs in the different sub-areas: area A, where three voltage measurements are available, has the best accuracy result, while areas *C* and *D* (only one voltage measurement) have the worst performance. Due to the same reason, the ISE (where nine measurements are present in total) can obtain significantly better results. Looking at the MASE results, the same kind of reasoning still applies, and the accuracy level of the voltage estimates depends on the total number of voltage measurements being processed, here including also those that are integrated through the harmonization process at the second step. As a consequence, area *A* is able with achieve the same accuracy performance as the ISE, since the information associated with

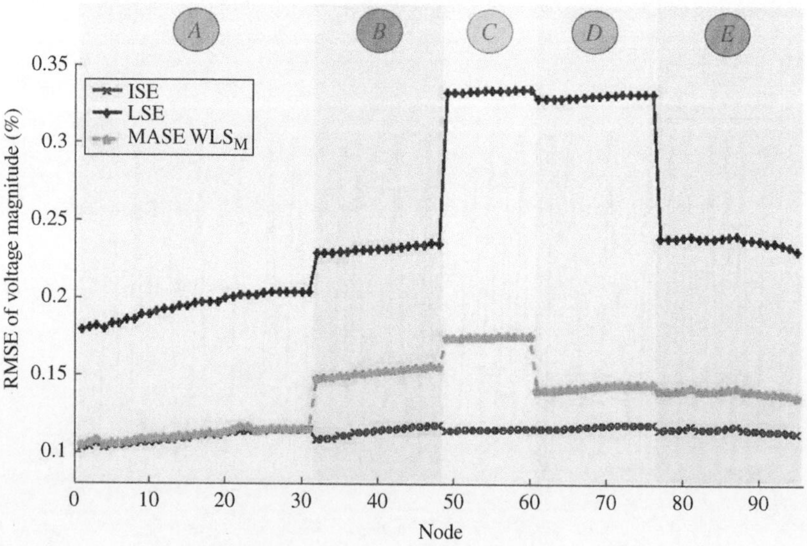

Figure 11.11 Voltage magnitude estimation in Scenario 1.

the six voltage measurements located in the adjacent areas is acquired during the second step and integrated with the three voltage measurements already processed at the first step. Area *B*, instead, can only rely on two local measurements and on the information associated with the three measurements arriving from area *A*: for this reason, an evident improvement is present with respect to the outcome of the LSE, but the same results as the ISE cannot be reached. Following the same concepts, areas *D* and *E* have slightly better performance than area *B* (they overall include six voltage measurements), whereas area *C* has the worst estimation accuracy (four measurements are processed in total). In this scenario, given the lack of meters at the shared nodes, no correlations among the different first-step voltage estimates arise (coherently with the values shown in Table 11.1). Therefore, the modifications implemented in the second-step WLS are not playing any role, and the results of the DS-MASE approach are exactly the same as those given by the MASE WLS_S (not shown in Figure 11.11 for the sake of simplicity).

11.3.2.2 Scenario 2: Coordinated Network Partition and Measurement System

The second scenario aims at showing the potential of the DS-MASE procedure when the partition of the network and design of the measurement configuration can be suitably coordinated. Given a chosen multi-area partition, an optimal meter placement involves a measurement point at all the overlapping nodes between adjacent areas and is composed of a voltage measurement at the shared buses and the monitoring of the flows at all the branches converging to those nodes. To prove the benefits deriving from such a measurement infrastructure design, this scenario considers a minimal measurement configuration where only the above-mentioned measurements at the overlapping nodes are placed. Table 11.3 shows the detail of the voltage and power measurements' location with reference to the node numbering provided in Figure 11.9. With respect to the meter placement in Scenario 1, no observability issues arise in this case. Indeed, if pseudo-measurements about the power consumptions (injections) at the load (generation)

TABLE 11.3 Measurement location for Scenario 2.

	Measurement location (nodes)				
Type of measurement	Area *A*	Area *B*	Area *C*	Area *D*	Area *E*
Voltage magnitude	1, 11, 25, 31	11	25	31	31
Active and reactive power	1 → 3 9 → 11 11 → 12 24 → 25 25 → 26 26 → 31	11 → 32	25 → 49	31 → 61	31 → 77

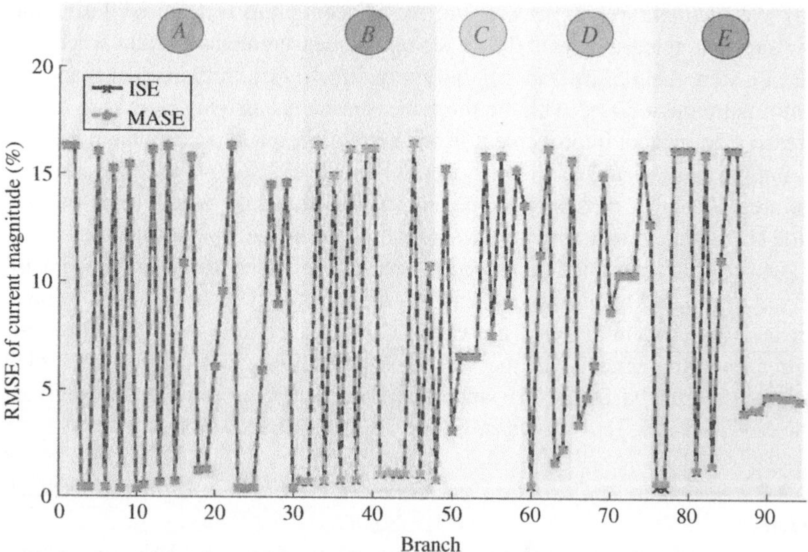

Figure 11.12 Current magnitude estimation in Scenario 2.

nodes are available, the placement of the abovementioned measurements at the overlapping buses is sufficient to ensure the observability of each sub-area.

Figure 11.12 shows the results achievable with such a measurement configuration for the branch current magnitude estimation. It is possible to observe that, in this case, the accuracy performance obtained at the first step of the MASE process is almost identical to the one provided by the ISE. In fact, the full monitoring of the flows at the branches converging to the overlapping buses allows having an accurate information of the conditions at the boundaries of the sub-areas; in this way, the accuracy degradation given by the multi-area partition (already seen in Scenario 1) is minimized, and the MASE can achieve estimation results extremely close to those of the ISE. In such a scenario, therefore, the choice made in (11.21) of substituting the current values with the first-step estimates is fully supported and does not imply any approximation.

The use of the considered measurement configuration has a significant impact also on the voltage magnitude estimation results. Moreover, in this case, given the presence of voltage measurements at the shared nodes, the importance of applying the modifications described in Section 11.2.3.3 to the WLS formulation used at the second step also arises. Figure 11.13 shows the accuracy performance in terms of RMSE for the different sub-areas. Two main results can be observed: first, the DS-MASE obtains the same accuracy performance as the ISE; second, the modifications introduced into the WLS formulation are essential to reach the best possible performance. Similarly to the previous scenario, when assessing the voltage estimation performance, the main factor affecting the accuracy level is the number of processed voltage measurements. In area *A*, since it includes all the voltage measurements available on the grid, the LSE has the same

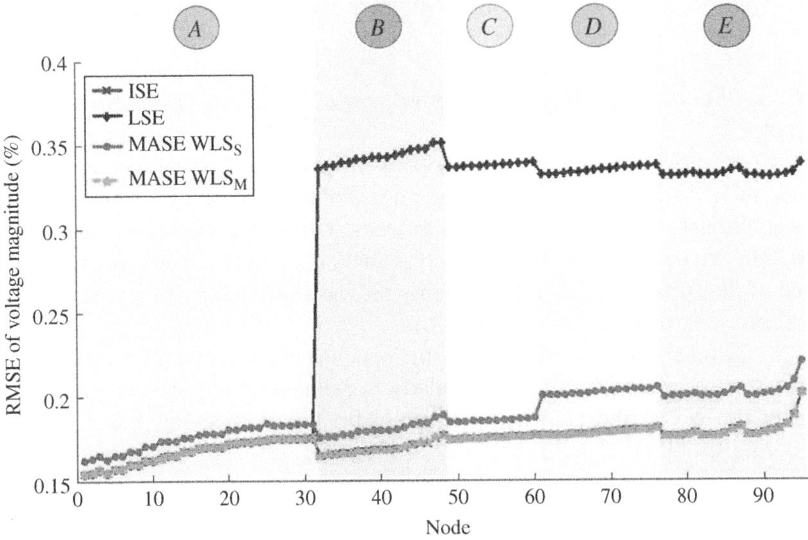

Figure 11.13 Voltage magnitude estimation in Scenario 2.

performance as the ISE. When performing the second step, it is possible to observe that a degradation of the accuracy appears if the classical WLS model is applied. As described in Section 11.2.3.2, the reason is due to the presence of correlations among the voltage estimates used as input to the second step that are not duly taken into account into the WLS model. The DS-MASE approach is able to properly fix this issue, and no degradation of the estimation accuracy occurs with the modified version of the WLS model. Looking at the other sub-areas, all of them are characterized by a quite large RMSE at the first-step LSE, since their estimation only relies upon one voltage measurement in input. Since all these areas are adjacent with area A, the second step allows integrating the information associated with the remaining measurements of the grid, and in this way an accuracy performance equal to the one achievable through the ISE can be reached. This result emphasizes the importance to apply a suitably designed second step to refine local state estimations when using multi-area solutions for the monitoring of large distribution grids. Moreover, similarly to the case of area A, the improvements obtained through the DS-MASE formulation (accounting for the effects brought by shared measurements) are evident, in particular in areas D and E where multiple correlated estimates are used as input.

As final outcome for this scenario, it is possible to remark that when measurements are installed at the overlapping buses and all the measurements are included in one sub-area (in this sample case, area A), both current and voltage estimates can achieve the same performance as the ISE. This outcome can be important to design the partitioning of the grid and to decide the measurement placement accordingly. In particular, in the case of deployment of the measurements with an incremental strategy, found results emphasize the possibility to achieve already good results through a minimal placement of meters at the overlapping buses.

11.3.2.3 Scenario 3: Meter Placement on a Pre-existing Measurement System

The test case proposed in this scenario not only involves an ad hoc placement of the measurement devices at the overlapping buses but also considers the presence of other additional measurements in the sub-areas. Therefore, differently from Scenario 2, in this case, the measurements available on the field are not all contained in one of the sub-areas. Table 11.4 reports the indication of the buses where the considered measurements are located.

In the proposed scenario, most of the conclusions already drawn in the previous test cases can be confirmed. Similarly to Scenario 2, the presence of measurement devices at the branches converging to the overlapping buses allows monitoring the flows at the boundaries of the sub-areas and guarantees that the same accuracy performance as in the ISE can be obtained for the branch current magnitude estimates. Regarding the voltage estimations, the total number of voltage measurements integrated into the estimation process gives a rough, but reliable, indication of the expected performance (Figure 11.14) [41]. Area *A*, which is adjacent with all the other areas of the grid, is again able to reach the same performance as the ISE. The other areas instead, since they cannot access information associated with all the voltage measurements, have accuracy performance slightly worse than those of the ISE. Comparing these results with those presented in Figure 11.13 for Scenario 2, it is possible to observe that even if the performance of the ISE cannot be equalized, the presence of additional measurements provides clear benefits: beyond area *A*, also areas *B*, *D*, and *E* have a RMSE lower than 0.15%, which was instead the lowest threshold in the case of Scenario 2. Furthermore, also in this test case, it is possible to observe that the correlations introduced by the shared measurements have the potential to degrade the estimation performance, but the solution conceived in the DS-MASE approach is capable of compensating for the adverse effects.

TABLE 11.4 Measurement location for Scenario 3.

	Measurement location (nodes)				
Type of measurement	Area *A*	Area *B*	Area *C*	Area *D*	Area *E*
Voltage magnitude	1, 11, 25, 31	11, 48	25	31	31, 95
Active and reactive power	1 → 3	11 → 32	25 → 49	31 → 61	31 → 77
	9 → 11	48 → 47			95 → 94
	11 → 12				
	24 → 25				
	25 → 26				
	26 → 31				

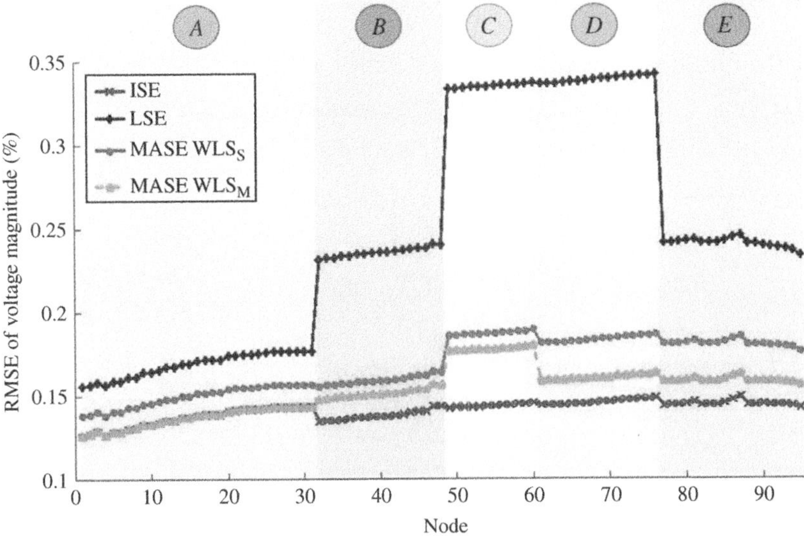

Figure 11.14 Voltage magnitude estimation in Scenario 3.

11.3.2.4 Scenario 4: PMU-Based Measurement System

In this scenario, the same meter placement used for Scenario 3 and summarized in Table 11.4 is taken into account, but conventional measurement devices are substituted with new-generation PMUs. PMUs provide phasor measurements of voltage and current synchronized with respect to the UTC time. The time synchronization is one of the main advantages given by the adoption of this measurement technology, since it allows obtaining phase-angle measurements of the electrical quantities referred to a universal reference. As a consequence, phase-angle results related to different estimation processes can be coherently compared, or integrated, exploiting the availability of the common reference.

Figures 11.15 and 11.16 show the results obtained in this test case for the voltage magnitude and phase-angle estimation. Concerning the voltage magnitude, it is possible to note that the trend of the voltage profile in the different sub-areas is exactly the same as the one presented in Figure 11.14 for Scenario 3. The only difference is for the levels of RMSE, which are lower in this case due to the starting assumption that voltage magnitude measurements given by the PMUs are more accurate than those provided by conventional meters (assumed accuracies are reported in Section 11.3.1 and are 0.7 and 1% for voltage measurements given by the PMUs and by the conventional meters, respectively). It is important to recall that when an accurate time tag is not available, the accuracy of the measurements included in DSSE degrades, and this is another point for PMU measurements.

For the voltage phase angles, as proved in [41], the assessment of the accuracy performance is based on considerations very similar to those carried out for the voltage magnitude estimation. From Figure 11.16 it is possible to notice that the shape of the voltage phase-angle estimation uncertainty profile along the different

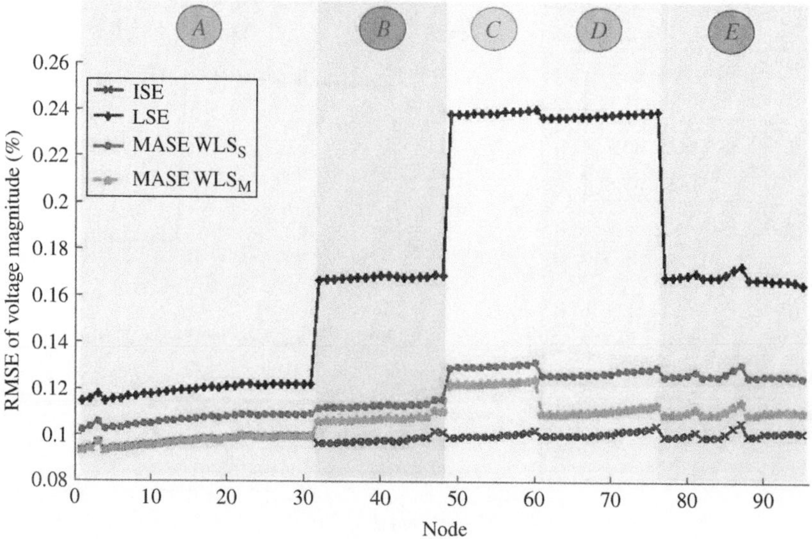

Figure 11.15 Voltage magnitude estimation in Scenario 4.

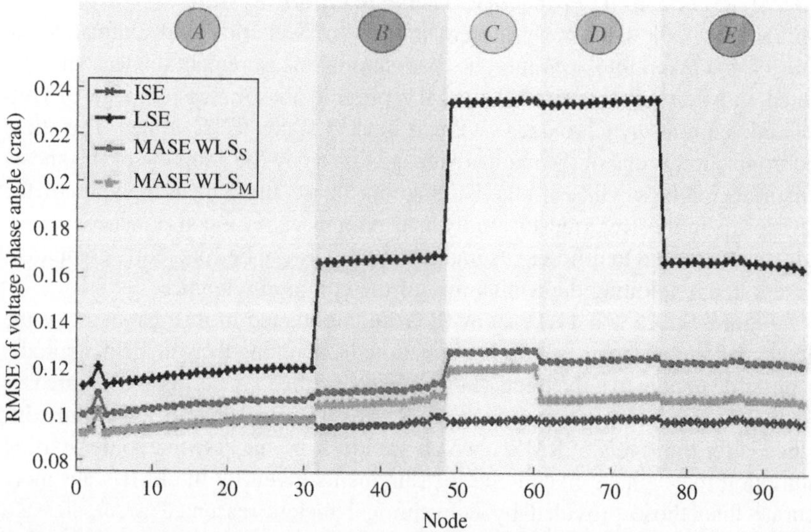

Figure 11.16 Voltage phase angle estimation in Scenario 4.

sub-areas is almost identical to the one reported in Figure 11.15 for the voltage magnitudes. As a matter of fact, also for the voltage phase angles, the main factor affecting the final estimation accuracy is the number of voltage phase-angle measurements involved in the SE process. In the case of the MASE, the measurements to be considered are therefore both those available locally and those arriving with the information acquired from the neighboring areas, which are used as input for the second step. The presence of shared voltage phase-angle measurements also leads to the same negative effects as in the case of voltage magnitudes. The proposed modification applied to the WLS approach is in this case extended to both voltage magnitude and phase-angle measurements, and it permits to suitably deal with this issue, allowing optimizing the second step and, consequently, the accuracy of the final MASE results.

As a final consideration, it is important to recall that, as shown in [7, 38, 39], PMUs allow an implementation of the DSSE that is computationally very efficient. In this regard, considering the above examples, the local estimation speed is improved for every area when PMUs are employed, thus making the whole MASE estimation more attractive for in-field implementation. Great efforts are ongoing to design PMUs that are low cost and more approriate for distribution systems, but it is likely that the whole monitoring system will be composed of heterogenous devices and there will be a trade-off in terms of costs and accuracy.

11.4 VALIDITY AND APPLICABILITY OF DS-MASE APPROACH

As shown through the test cases carried out in simulation environment, the DS-MASE approach allows a partition of large distribution grids in multiple sub-networks and permits distributing the SE task among the different areas, without sacrificing the accuracy performance of the monitoring. Such an approach, by means of simultaneous and distributed computation with minimal data exchange, drastically reduces the computational effort and, therefore, the computation time of DSSE.

The applicability of the DS-MASE procedure can be generally extended to any distribution grid, but some criteria have to be met for the proper operation of the presented scheme:

- Sub-areas need to have local observability; as it is often the case for distribution grids, observability can also be reached by means of pseudo-measurements at the load or generation buses of the network. While some constraints on observability could be relaxed by using an in-series execution of the local estimates (thus exploiting the estimation outputs of adjacent zones as measurements to achieve the local observability), this solution here is not contemplated because it would affect some of the specific features of the proposed scheme. In particular, the illustrated MASE has to guarantee the robustness to possible communication failures and the consequent possibility to have local estimates with minimum accuracy performances also in case of loss of the communication with some of the adjacent sub-networks.
- Even if never mentioned before, local area estimations need to have bad data detection capabilities in order to prevent the use of any misleading result for grid management or control purposes. This usually represents a significant

problem at distribution system level, due to the very low redundancy of measurements and the consequent difficulties to properly identify the erroneous data. In this chapter, the problem of bad data detection and identification is out of the scope and is not deeply discussed, but the existence of this issue should be always kept in mind in order to design ad hoc meter placements and/or to identify proper routines aimed at disregarding the estimation results when possible bad data are detected.

- As highlighted in the simulation results, possible placement of measurement infrastructure on the shared nodes allows achieving important benefits from an accuracy performance standpoint. In addition, the monitoring of the voltage and of the power/currents at the converging branches also ensures the achievement of the observability requirements when pseudo-measurements are available. While the placement of meters at the shared node between two areas is not mandatory, it is highly recommended, whenever possible, in order to maximize the benefits achievable through the presented MASE architecture.
- Last but not least, the DS-MASE approach relies upon the use of a coordinated data exchange between adjacent areas. The underlying assumption is that each area has its own mini-control center, which has communication capabilities and suitable time synchronization. While the DS-MASE procedure is designed to minimize the communication requirements and the amount of data to be exchanged, still some coordination in the SE execution is needed in order to perform the second step and to manage possible communication issues. Moreover, model and format of the data to be exchanged have to be suitably agreed among the different sub-networks.

All the abovementioned points are necessary for the proper implementation on the field of the DS-MASE scheme. At the same time, it can be recognized that similar issues also occur for the normal implementation of any SE scheme (need to have observability, bad data detection capabilities, ad hoc meter placement to ensure minimum redundancy requirements, and proper time synchronization and communication infrastructure to collect the measurements from the field). As a consequence, the DS-MASE procedure does not bring a significant increase of complexity or particular issues in the deployment of the monitoring tool. On the other hand, the following important advantages can be achieved:

- Distribution of the computation, communication, and storage requirements among different computing units that are able to operate autonomously without any need of upper-level coordination.
- Possibility to decouple a large distribution grid, possibly also characterized by different technical features or managed by different DSOs, in smaller areas while including some important information on the boundary conditions and having the possibility to significantly refine the knowledge of the local operating conditions through the exchange of data with the neighboring zones.

- Possibility to have accuracy performances equal, or very close, to those guaranteed by the execution of SE on the whole grid, just introducing the communication of few data among adjacent areas.
- Significant reduction of the execution times, thanks to the parallel execution of the local SE processes; this also allows increasing the reporting rate at which SE can be performed, making possible a monitoring at a fine-grained resolution that can be important to better understand the dynamic behavior of distribution grids. In a future PMU-based monitoring system, high refresh rates in SE perfectly fit the high reporting rates of new measurement devices.
- Robustness to possible failures in the communication with neighboring areas and to possible issues in the local SE performed by different zones. Accuracy of local estimation is kept as much as possible, without relying on continuous communication and synchronization activities at the boundaries.
- Easy integration of PMU measurements. The DS-MASE allows exploiting the most interesting PMU-based algorithms for DSSE presented in the literature. DS-MASE is also completely open to different estimation techniques for local SE, since it allows each local control center to define its own procedures for first-step computation.
- Minimization, with respect to other MASE approaches, of the communication requirements and of the complexity in the coordination of the processes performed by different areas. DS-MASE also maximizes the accuracy performance thanks to a proper consideration of the measurement aspects and, in particular, of the possible correlation arising due to the presence of measurements shared by multiple areas.
- Estimation of the accuracy as additional outcome. DS-MASE computes its accuracy based on the measurement accuracy information after each step, thus allowing integration with upper-level application routines based on simple thresholds or more complex algorithms, which become aware of the network status along with its uncertainty ranges.

REFERENCES

1. European Technology Platform SmartGrids. Strategic research agenda for Europe's electricity networks of the future. http://www.smartgrids.eu/documents/sra2035.pdf (accessed 13 March 2020).
2. Heydt, G.T. (Dec. 2010). The next generation of power distribution systems. *IEEE Transactions on Smart Grid* 3 (3): 225–235.
3. Farhangi, H. (Jan.–Feb. 2010). The path of the smart grid. *IEEE Power and Energy Magazine* 8 (1): 18–28.
4. Muscas, C., Sulis, S., Angioni, A. et al. (Sep. 2014). Impact of different uncertainty sources on three-phase state estimator for distribution networks. *IEEE Transactions on Instrumentation and Measurement* 63 (9): 2200–2209.

5. Baran, M.E. and Kelley, A.W. (Aug. 1994). State estimation for real-time monitoring of distribution systems. *IEEE Transactions on Power Systems* 9: 1601–1609.
6. Wang, H. and Schulz, N. (Feb. 2004). A revised branch current-based distribution system state estimation algorithm and meter placement impact. *IEEE Transactions on Power Systems* 19: 207–213.
7. Pau, M., Pegoraro, P.A., and Sulis, S. (Sep. 2013). Efficient branch current-based distribution system state estimation including synchronized measurements. *IEEE Transactions on Instrumentation and Measurement* 62 (9): 2419–2429.
8. Pegoraro, P.A., Angioni, A., Pau, M. et al. (Nov. 2017). Bayesian approach for distribution system state estimation with non-Gaussian uncertainty models. *IEEE Transactions on Instrumentation and Measurement* 66 (11): 2957–2966.
9. Zanni, L., Sarri, S., Pignati, M., et al. (2014). Probabilistic assessment of the process-noise covariance matrix of discrete Kalman filter state estimation of active distribution networks. *Proceedings of the International Conference on Probabilistic Methods Applied to Power Systems*, Durham, UK (July 2014), pp. 1–6.
10. IEEE Standard for Synchrophasor Measurements for Power Systems. IEEE Std. C37.118.1-2011 (Revis. IEEE Std C37.118-2005), pp. 1–61, Dec. 2011.
11. IEEE Standard for Synchrophasor Data Transfer for Power Systems. IEEE Std C37.118.2-2011 (Revision of IEEE Std C37.118-2005), Dec. 2011.
12. IEEE Standard for Synchrophasor Measurements for Power Systems – Amendment 1: Modification of Selected Performance Requirements. IEEE Std. C37.118.1a-2014 Amend. IEEE Std. C37.118.1-2011, pp. 1–25, Apr. 2014.
13. IEEE/IEC International Standard – Measuring relays and protection equipment – Part 118-1: Synchrophasor for power systems – Measurements. IEC/IEEE 60255-118-1:2018, pp. 1–78, 19 Dec. 2018.
14. Castello, P., Muscas, C., Pegoraro, P. A., and Sulis, S. (2016). Analysis of PMU response under voltage fluctuations in distribution grids. *Proceedings of IEEE International Workshop on Applied Measurements for Power Systems, AMPS 2016*, Aachen, Germany (September 2016).
15. Castello, P., Junqi, L., Muscas, C. et al. (Dec. 2014). A fast and accurate PMU algorithm for P+M class measurement of synchrophasor and frequency. *IEEE Transactions on Instrumentation and Measurement* 63 (12): 2837–2845.
16. de la O Serna, J.A. (Oct. 2007). Dynamic phasor estimates for power system oscillations. *IEEE Transactions on Instrumentation and Measurement* 56 (5): 1648–1657.
17. Premerlani, W., Kasztenny, B., and Adamiak, M. (Jan. 2008). Development and implementation of a synchrophasor estimator capable of measurements under dynamic conditions. *IEEE Transactions on Power Delivery* 23 (1): 109–123.
18. Petri, D., Fontanelli, D., Macii, D., and Belega, D. (2013). A DFT-based synchrophasor, frequency and ROCOF estimation algorithm. *Proceedings of the IEEE International Workshop AMPS*, Aachen, Germany (September 2013), pp. 85–90.
19. Castello, P., Muscas, C., Pegoraro, P.A., and Sulis, S. (2018). Active phasor data concentrator performing adaptive management of latency. *Sustainable Energy, Grids and Networks* 16: 270–277. https://doi.org/10.1016/j.segan.2018.09.004.
20. European Parliament. Directive 2014/32/EU of the European Parliament and of the Council of 26 February 2014 on the harmonisation of the laws of the Member States relating to the making available on the market of measuring instruments. http://eur-lex.europa.eu/legal-content/EN/TXT/?uri=CELEX:32014L0032 (accessed 13 March 2020).
21. Gungor, V., Sahin, D., Kocak, T. et al. (Feb. 2013). A survey on smart grid potential applications and communication requirements. *IEEE Transactions on Industrial Informatics* 9 (1): 28–42.

22. Ma, R., Chen, H.H., Huang, Y.R., and Meng, W. (Mar. 2013). Smart grid communication: Its challenges and opportunities. *IEEE Transactions on Smart Grid* 4 (1): 36–46.
23. Atzori, L., Iera, A., and Morabito, G. (Oct. 2010). The internet of things: a survey. *Computer Networks* 54 (15): 2787–2805.
24. Nitti, M., Pilloni, V., Colistra, G., and Atzori, L. (Nov. 2015). The virtual object as a major element of the internet of things: a survey. *IEEE Communication Surveys and Tutorials* 18: 1228–1240.
25. Meloni, A., Pegoraro, P.A., Atzori, L. et al. (Jan. 2018). Cloud-based IoT solution for state estimation in smart grids: exploiting virtualization and edge-intelligence technologies. *Computer Networks* 130: 156–165.
26. Pegoraro, P.A., Meloni, A., Atzori, L. et al. (Apr. 2017). PMU-based distribution system state estimation with adaptive accuracy exploiting local decision metrics and IoT paradigm. *IEEE Transactions on Instrumentation and Measurement* 66 (4): 704–714.
27. Gomez-Exposito, A., de la Villa Jaen, A., Gomez-Quiles, C. et al. (Apr. 2011). A taxonomy of multi-area state estimation methods. *Electric Power Systems Research* 81 (4): 1060–1069.
28. Falcao, D., Wu, F., and Murphy, L. (May 1995). Parallel and distributed state estimation. *IEEE Transactions on Power Systems* 10 (2): 724–730.
29. Zhao, M. and Abur, A. (May 2005). Multi area state estimation using synchronized phasor measurements. *IEEE Transactions on Power Systems* 20 (2): 611–617.
30. Conejo, A., De La Torre, S., and Canas, M. (Feb. 2007). An optimization approach to multiarea state estimation. *IEEE Transactions on Power Systems* 22 (1): 213–221.
31. Van Cutsem T., Horward J. L. and Ribbens-Pavella M., "A Two-Level Static State estimator for electric power systems," *IEEE Transactions on Power Apparatus and Systems*, vol. PAS-100, no. 8, pp. 3722–3732, Aug. 1981.
32. Muscas, C., Pau, M., Pegoraro, P.A. et al. (May 2015). Multiarea distribution system state estimation. *IEEE Transactions on Instrumentation and Measurement* 64 (5): 1140–1148.
33. Nusrat, N., Lopatka, P., Irving, M.R. et al. (Jul. 2015). An overlapping zone-based state estimation method for distribution systems. *IEEE Transactions on Smart Grid* 6 (4): 2126–2133.
34. Nusrat, N., Irving, M., and Taylor, G. (2011). Development of distributed state estimation methods to enable smart distribution management systems. *2011 IEEE International Symposium on Industrial Electronics (ISIE)*, Gdansk, Poland (Jun. 2011), pp. 1691–1696.
35. De Alvaro Garcia, L. and Grenard, S. (2011). Scalable distribution state estimation approach for distribution management systems. *2011 2nd IEEE PES International Conference and Exhibition on Innovative Smart Grid Technologies (ISGT Europe)*, Manchester, UK (Dec. 2011), pp. 1–6.
36. Pau, M., Ponci, F., Monti, A. et al. (May 2017). An efficient and accurate solution for distribution system state estimation with multiarea architecture. *IEEE Transactions on Instrumentation and Measurement* 66 (5): 910–919.
37. ISO/IEC 98-3:2008 (2008). Uncertainty of measurement - Part 3: guide to the expression of uncertainty in measurement (reissue of the 1995 version of the GUM).
38. Pau, M., Pegoraro, P.A., and Sulis, S. (2013). WLS distribution system state estimator based on voltages or branch-currents: accuracy and performance comparison. *2013 IEEE International Instrumentation and Measurement Technology Conference (I2MTC)*, Minneapolis, MN (May 2013), pp. 493–498.

39. Pau, M., Pegoraro, P.A., and Sulis, S. (2015). Performance of three-phase WLS distribution system state estimation approaches. *2015 IEEE International Workshop on Applied Measurements for Power Systems (AMPS)*, Aachen, Germany (Sep. 2015), pp. 138–143.
40. Muscas, C., Pau, M., Pegoraro, P.A., and Sulis, S. (Dec. 2014). Effects of measurements and pseudomeasurements correlation in distribution system state estimation. *IEEE Transactions on Instrumentation and Measurement* 63 (12): 2813–2823.
41. Muscas, C., Pau, M., Pegoraro, P.A., and Sulis, S. (May 2016). Uncertainty of voltage profile in PMU-based distribution system state estimation. *IEEE Transactions on Instrumentation and Measurement* 65 (5): 988–998.

PART IV

PARALLEL/ DISTRIBUTED PROCESSING

CHAPTER 12

HIERARCHICAL MULTI-AREA STATE ESTIMATION

Ye Guo[1], Lang Tong[2], Boming Zhang[3], Wenchuan Wu[3], and Hongbin Sun[3]

[1]*Tsinghua-Berkeley Shenzhen Institute, Shenzhen, China*
[2]*Cornell University, Ithaca, USA*
[3]*Tsinghua University, Beijing, China*

12.1 INTRODUCTION

12.1.1 Problem Description

Large electric power systems are typically interconnected. This is the case of North America, Europe, and China. In North America, for instance, much of the power system is operated by independent system operators (ISOs) as shown in Figure 12.1. Each ISO is responsible for the administration of its regional operation and market. To this end, each ISO relies on a state estimator (SE) to monitor the real-time operating state, which is represented by nodal voltage magnitudes and phase angles. Essentially, an SE extracts the real-time power flow model from raw measurements. It is a cornerstone of today's energy management systems (EMS) whose results are widely used in other advanced applications including static and transient security analysis and reactive power optimization [2].

Traditionally, each ISO separately collects raw measurements through its internal transmission owners and then performs its regional state estimation. Such a technique, however, is generally suboptimal when it is compared with a centralized SE that collects data from and state estimate for all areas. The main reason of suboptimality is twofold: One is that each ISO may have no access to measurements at the remote side of tie-lines [3], which indicates a loss of the measurement redundancy; the other is that each ISO does not consider the optimal condition

Advances in Electric Power and Energy: Static State Estimation, First Edition.
Edited by Mohamed E. El-Hawary.

Published 2021 by John Wiley & Sons, Inc.

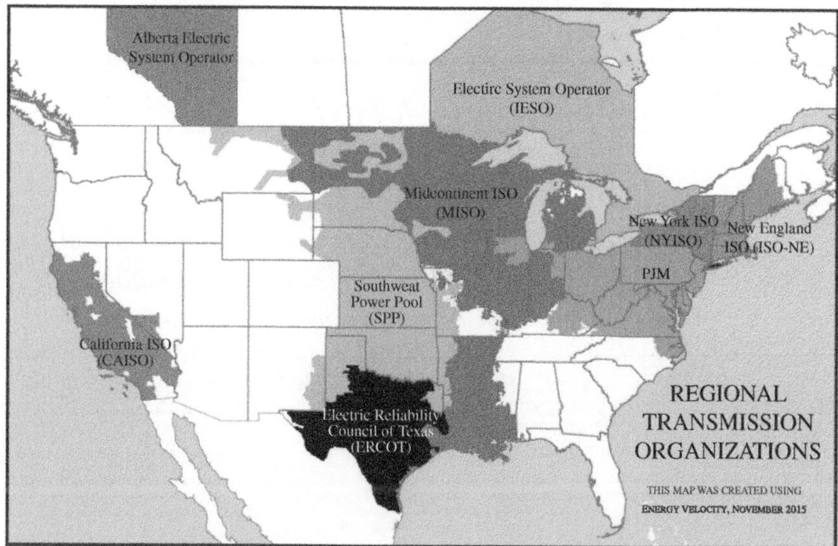

Figure 12.1 The interconnection of the power system in North America [1].

for the entire system. In particular, separate and asynchronous state estimations in different ISOs may lead to discrepancies in their state estimates, especially for boundary buses.

For the emerging smart distribution system in the presence of significant penetration of distributed energy resources with multiple interfaces with neighboring operating regions and microgrids, efficient multi-area state estimation is expected to play a crucial role. In this context, the challenges arise from potentially more volatile voltage profiles and tremendous computation costs associated with a large number of buses, which places a premium on high performance and computationally efficient distributed state estimation techniques [4].

Therefore, a coordinated SE for multi-area power systems, which provides the same state estimate as a centralized estimator but is solved in a distributed manner, is of high importance in today's interconnected transmission systems and active distribution networks. Taking into account that state estimation is a basic function in the EMS that is periodically invoked in online operations, a rapid convergence and moderate computation and communication costs are highly favorable properties [5, 6].

12.1.2 Classification of Existing Methods

There are many existing methods to solve the problem of multi-area state estimation. Taking survey papers [7] and [8] as references, we list in this subsection some important attributes of multi-area SEs.

12.1.2.1 Architecture: Hierarchical Versus Fully Decentralized

Hierarchical multi-area SEs usually involve a coordinator who exchanges information with all system operators and syncretizes boundary state estimates from different areas. In these methods, there is a common information pool shared by all system operators and managed by the coordinator. When there is no physical coordinator available, an equivalent scenario is that any shared information is allowed to be accessed, directly or indirectly, by all system operators. Fully decentralized methods, on the contrary, only allow communications between areas connected physically with tie-lines. Moreover, each area is usually prohibited to forward data received from one neighboring area to another neighbor. Hierarchical and fully decentralized architectures are illustrated in Figure 12.2.

Hierarchical estimators, by and large, are more efficient than fully decentralized ones because they assume the absence of any hurdle in data sharing and communications and allow a global viewpoint at the coordinator with exchanged

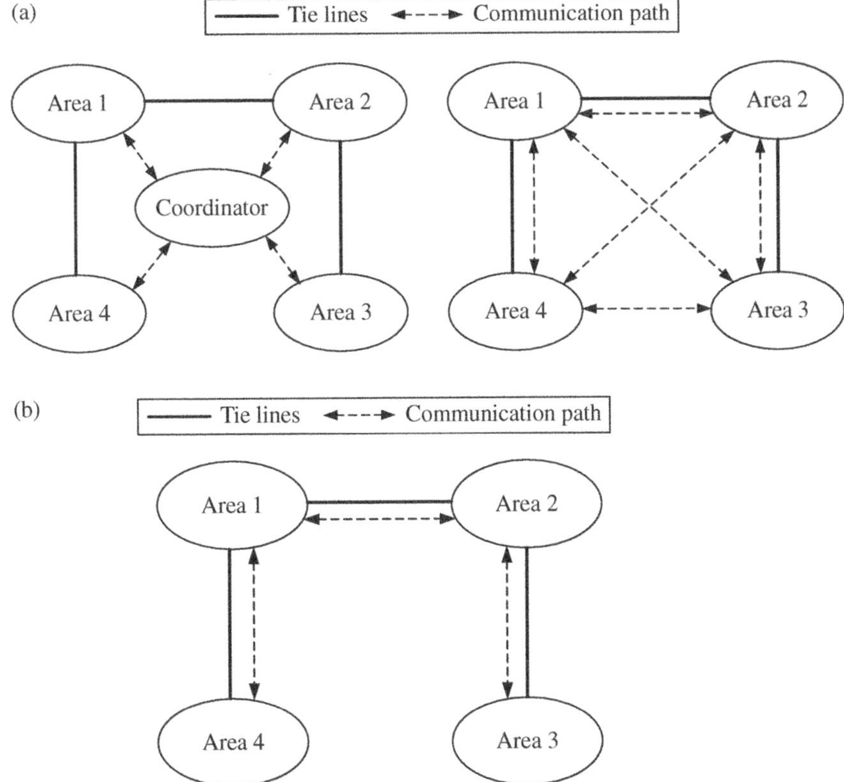

Figure 12.2 Architectures of multi-area state estimators: hierarchical versus fully decentralized. (a) Hierarchical architecture with/without a physical coordinator. (b) Fully decentralized architecture (each area cannot forward data from one neighbor to another).

information. Their limitation, on the other hand, is that they require stronger synchronizations and communication agreements across operating regions. In this chapter, we mainly focus on hierarchical multi-area SEs, with brief discussions on fully decentralized methods.

12.1.2.2 Optimality: Optimal Versus Suboptimal

A multi-area SE is *optimal* if its state estimate converges to an arbitrarily small neighborhood of the optimal estimate of the centralized problem, and it is *suboptimal* otherwise. While optimal multi-area SE achieves the same performance as the optimal centralized estimator, it comes with additional computation and communication costs. Because achieving optimality typically requires iterations among neighboring control centers or between a coordinator with regional control centers, the cost of iterations may be substantial. Therefore, suboptimal estimators are also studied as alternatives to obtain a unique boundary state estimate that is in general different from the centralized estimator. Many suboptimal estimators are non-iterative with limited computation and communication burdens. Some argue that the optimality is not the highest priority of multi-area state estimation as long as the accuracy is in an acceptable range. However, few suboptimal estimators can clearly define what "an acceptable accuracy" is and prove that accuracies of their state estimates are indeed "acceptable." Therefore, choosing between optimal and suboptimal estimators remains an open question, which essentially is to find a reasonable trade-off between accuracy and efficiency.

12.1.2.3 Communication Frequency: Inter-area and Intra-area Gauss–Newton Iterations

When optimal estimators are of concern, interregional iterations are almost inevitable. There are two types of iterations: One performs its intra-area Gauss–Newton iteration until convergence and then exchanges data with the coordinator or its neighbors; the other solves in each step the inter-area normal equation in a distributed way. In general, intra-area Gauss–Newton iterations have advantages of lower communication costs and less reliant to possible communication delays or failures. Their convergences and computation efficiencies depend on updating rules of boundary states. Inter-area Gauss–Newton iterations may have similar convergence as the centralized estimator. However, these methods may suffer from heavy communication burdens and vulnerabilities to communication delays or failures.

There are other factors such as the partition method of the multi-area power system, the synchronization of measurements, the compatibility for bad data identification (BDI), and observability analysis. In this chapter, we highlight architecture, optimality, and communication mechanism as three most representative attributes for characteristics of multi-area SEs. The selection of architectures depends on the deregulation structure of the market. In United States, for example, recent implementations of the coordinated tie-line scheduling [9] indicate that it is possible to set up a coordinator. For the other two attributes, in general, an optimal intra-area Gauss–Newton estimator with rapid convergence and limited computation burden is the most ideal option.

12.1.3 Organization of this Chapter

In this chapter, we present a general model of multi-area state estimation and elaborate some representative solutions. In Section 12.3, we first introduce the model and the standard solution of the general power system state estimation problem. Integral add-on functions of observability analysis and BDI are also introduced. In Section 12.4, we describe the model of multi-area state estimation and formulate the centralized state estimation. In Section 12.5, we review existing researches on this topic and give detailed descriptions of selected methods.

Thereafter, we present a recent technique that achieves the same rate of convergence as general Gauss–Newton iterations with moderate computation and communication costs [10]. Specifically, in Section 12.6, we review key components of the method, namely, the derivation of sensitivity functions, the formulation and solution of the coordinator's problem, and the complete iteration scheme. In Section 12.7, we explicate approaches for observability analysis and BDI for method in [10]. The established properties in [10] are reviewed in Section 12.8. Simulation results are presented in Section 12.9.

12.2 PRELIMINARIES

12.2.1 Measurement Model

We present here a measurement model based on the steady-state power flow equation. See [11] for more details and basic concepts about electric power systems.

In power systems, nodal voltage magnitudes V and phase angles θ are usually selected as state variables that fully describe the static power flow model. Specifically, active and reactive power flow from bus i to bus j are calculated by

$$\begin{aligned} P_{ij} &= V_i^2 g_{ij} - V_i V_j \left(g_{ij} \cos\left(\theta_i - \theta_j\right) + b_{ij} \sin\left(\theta_i - \theta_j\right) \right) \\ Q_{ij} &= -V_i^2 \left(b_{ij} + b_{ijs} \right) - V_i V_j \left(g_{ij} \sin\left(\theta_i - \theta_j\right) - b_{ij} \cos\left(\theta_i - \theta_j\right) \right) \end{aligned} \tag{12.1}$$

where g_{ij} and b_{ij} are line conductance and susceptance between buses i and j, respectively, and b_{ijs} is the half susceptance of the shunt branch associated with the branch between buses i and j.

Active and reactive power injections at bus i are calculated by

$$\begin{aligned} P_i &= V_i \sum_{j \in \mathbb{N}_i} V_j \left(\operatorname{Re}\left(Y_{ij}\right) \cos\left(\theta_i - \theta_j\right) + \operatorname{Im}\left(Y_{ij}\right) \sin\left(\theta_i - \theta_j\right) \right) \\ Q_i &= V_i \sum_{j \in \mathbb{N}_i} V_j \left(\operatorname{Re}\left(Y_{ij}\right) \sin\left(\theta_i - \theta_j\right) - \operatorname{Im}\left(Y_{ij}\right) \cos\left(\theta_i - \theta_j\right) \right) \end{aligned} \tag{12.2}$$

where Y_{ij} is the (i, j) entry of the bus admittance matrix and $\mathbb{N}_i$ is the set of bus indices that are directly connected to bus i including bus i itself.

Traditionally, system operators collect measurements of branch power flow, nodal injection power, and nodal voltage magnitudes through the supervisory control and data acquisition (SCADA) system. Recently, the deployment of phasor measurement units (PMU) makes it possible to measure phasors of nodal voltages and branch currents at a much higher frequency. Anyway, at a specific snapshot, all real-time measurements can be summarized as follows:

$$z = h(x) + e \tag{12.3}$$

where x is the state vector composed of V and θ. The vector z represents values of all collected measurements. The mapping $h(\bullet)$ denotes measurement functions. Specifically, for measurement m, there is $h_m(x) = V_i$ if it is a nodal voltage magnitude measurement, $h_m(x) = \theta_i$ if it is a nodal phase angle measurement, and $h_m(x)$ as (12.1) or (12.2) if it is a branch power or power injection measurement. The vector e represents measurement errors, which is usually assumed to be Gaussian.

12.2.2 Weighted Least Square (WLS) Estimator

The standard formulation of power system state estimation is the weighted least square estimator as the following:

$$\tilde{x} = \arg\min_{x} \quad f(x) = \frac{1}{2} r(x)^T W r(x) \tag{12.4}$$

where $\tilde{x}$ is the optimal state estimate, W is the diagonal measurement weight matrix, and $r(x) = z - h(x)$ is the residual vector. The superscript T denotes transposition. For brevity, we consider in this chapter an unconstrained optimization model. See [12] for the effect of zero injection constraints.

Problem (12.4) is usually solved by the Gauss–Newton method. At each iteration, the Gauss–Newton method linearizes the measurement function at the given point of x and solves the following quadratic programming to obtain the update Δx:

$$\begin{aligned} &\min_{\Delta x} \frac{1}{2} r_\Delta(\Delta x)^T W r_\Delta(\Delta x) \\ &r_\Delta(\Delta x) := z - h(x) - H(x)\Delta x \end{aligned} \tag{12.5}$$

where $H(x)$ is the measurement Jacobian at x.

The solution to (12.5) is obtained from

$$G(x)\Delta x + g(x) = 0 \tag{12.6}$$

where

$$g(x) = -H(x)^T W r(x), \quad G(x) = H(x) W H(x) \tag{12.7}$$

Here the vector $g(x) = \nabla_x f$ is the gradient of the objective function of model (12.4), and the matrix $G(x)$ is referred to as the gain matrix. Equation (12.6) is referred to as the *normal equation*.

Empirically, values of entries in $H(x)$ and $G(x)$ change slightly with x. Setting the stopping rule as $|\Delta x| < \varepsilon$ with ε as a small positive constant, the Gauss–Newton iteration stops at $\hat{x}$ that is in a neighborhood of the optimal state estimate $\tilde{x}$.

Next, we consider the convergence rate of Gauss–Newton iteration. For an infinite sequence $\{x_k\}$ that converges to L, if there is

$$\lim_{k\to\infty}\frac{|x_{k+1}-L|}{|x_k-L|^q}>0,\quad q>1 \tag{12.8}$$

then the sequence $\{x_k\}$ is said to have a superliner rate of convergence [13]. In particular, if $q = 2$, then the sequence $\{x_k\}$ is said to have a quadratic rate of convergence. For brevity, we also say an algorithm has a linear/superliner/quadratic rate of convergence if it generates a sequence with the same rate of convergence.

For the Gauss–Newton iteration, its convergence rate depends on the optimal value of (4): Assuming that the measurement function $h(x)$ is Fréchet differentiable in the neighborhood of the optimal state estimate $\tilde{x}$, then the Gauss–Newton iteration has a quadratic rate of convergence when the optimal residual $r(\tilde{x}) = 0$, a superliner rate of convergence with a relatively small $r(\tilde{x})$, and a linear rate of convergence in the worst scenario [13]. Measurement functions in power systems are smooth, and the optimal residual $r(\tilde{x})$ is reasonably small. Therefore, the Gauss–Newton iteration usually has a superliner rate of convergence for power system state estimation problems. This is consistent with our experience that the Gauss–Newton iteration typically converges rapidly in power system state estimation problems.

In the rest of this chapter, for brevity, we will simplify $r(\tilde{x})$ as $\tilde{r}$ if it does not cause confusion. The same for the measurement Jacobian H, the gain matrix G, and the gradient g.

12.2.3 Add-On Functions

Besides the core module that solves the WLS estimator (12.4) by iteratively solving the normal equation (12.6), real SEs in power systems contain several add-on functions. In this chapter, we review two essential add-on functions: observability analysis and BDI.

Observability analysis determines, prior to the state estimation process, the uniqueness of the optimal state estimate $\tilde{x}$ in its small neighborhood. Identically, it is to judge if the gain matrix $G(x)$ in (12.7) is full rank at $x = \tilde{x}$. Due to the high computation cost to calculate the rank of a large matrix, we usually employ heuristic methods, namely, topological or numerical methods. These methods typically assume that active and reactive power measurements always appear in pairs and that the update for x does not change the rank of $G(x)$.

Illustrated with the example in Figure 12.3, topological methods sequentially perform two phases: observable area exploration and the fusion of observable areas.

First, we explore observable areas using branch power measurements and power injection measurements that can be converted to branch power measurements. The basic rule here is the fact that for any branch, if we know the voltage magnitude and phase angle of one of its terminal bus and active and reactive power flow measurements at either side of the branch, then we can estimate the state of the other terminal bus. In Figure 12.3a, assume that we know the voltage magnitude and phase angle at bus 1, then we are able to estimate the state of bus 2 because

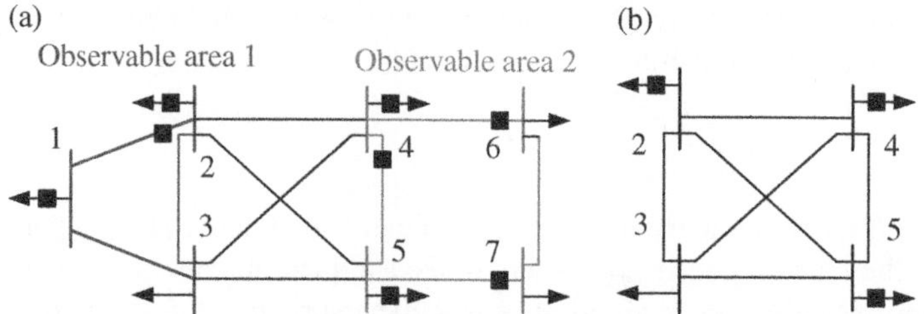

Figure 12.3 Illustration of topological methods for observability analysis. (a) System configuration. (b) Merging of observable areas. ■ Denotes power measurements.

there is a power measurement on branches 1–2. Subsequently, the power flow from bus 1 to bus 2 can be estimated, and the power injection measurement at bus 1 can be converted to the power flow measurement from bus 1 to bus 3, and the state of bus 3 can also be estimated. Therefore, buses 1–3 form the first observable area. Similarly, buses 4–7 form the second observable area.

Second, we try to merge observable areas using the rest power injection measurements. The basic rule here is the fact that if we know the voltage magnitude and phase angle at one bus and power injections at other buses, then we can calculate the state of all buses via the power flow calculation. In Figure 12.3b, we eliminate internal buses in all observable areas, and the network remaining satisfies the condition above. Assuming there is at least one voltage magnitude measurement, we conclude that the power system in Figure 12.3a is observable.

Numerical methods check if there is any nonzero entry in the null space of the measurement Jacobian $H(x)$. To simplify calculation, these methods typically adopt a decoupled state estimation model in which voltage phase angles are solely determined by active power measurements and voltage magnitudes are solely determined by reactive power measurements. They may also assume flat state (all voltage magnitudes are 1.0 p.u., and all phase angles are zero) and equal weights.

An open question is defining the largest observable island when the entire power system is unobservable, especially in distribution networks where decoupled model cannot be directly used and active and reactive power measurements may not always appear in pairs [14].

BDI is an integral part of a real SE. It removes suspicious measurements out of the original measurement set. The performance of BDI is crucial for real SEs, due to the vulnerability of the WLS estimator (12.4) to gross errors. The standard technique is based on the largest normalized residual principle. The normalized residual of measurement m is defined as

$$\hat{r}_{Nm} := \frac{|\hat{r}_m|}{\sqrt{\hat{\Omega}_{mm}}} \tag{12.9}$$

where $\hat{\Omega}$ is the residual covariance matrix calculated with $\hat{x}$:

$$\hat{\Omega} := W^{-1} - \hat{H}\hat{G}^{-1}\hat{H}^T \tag{12.10}$$

After the Gauss–Newton iteration (12.6) converges, the state estimate $\hat{x}$ is delivered to the BDI who calculates normalized residuals for all measurements (12.9). If the largest normalized residual is less than a given threshold, we consider there is no gross error and $\hat{x}$ is the final state estimate. Otherwise, we remove the measurement with the largest normalized residual out of the measurement set and re-estimate the system state. A complete SE scheme needs to iterate the bad data removal-state estimation procedure until the maximum normalized residual in the measurement set is small enough.

BDI is another active research field with many remarkable advances, including the removal of multiple bad data in a single iteration [15], the restoration of once removed measurements [16], and robust estimators [17, 18]. In this chapter, we use the simplest approach that removes only one bad data per iteration as an example to explain the idea of BDI.

12.3 MODELING AND PROBLEM FORMULATION

12.3.1 Multi-area Power System Modeling

We describe in this section a model of multi-area state estimation. For simplicity, we use a three-area power system in Figure 12.4 as an example. Similar methods can be derived when there are more than three areas. Areas 1, 2, and 3 are connected by tie-lines. Terminal buses of these tie-lines form the boundary bus sets, and the other buses are internal buses. Note that between different areas, there may be multiple tie-lines and boundary buses on each side.

All measurements $z = [z_1, z_2, z_3]$ are partitioned according to the three areas, where z_i is the measurement vector in area i ($i = 1, 2, 3$). Specifically, vector z_i includes values of internal measurements, boundary power injection measurements, and power flow measurements on tie-lines on the side of area i. The boundary power measurements in area 1, which are entries in vector z_1, are highlighted with arrows in Figure 12.4. We assume that the local SE of area i has access to z_i but not z_j, ($j \neq i$).

All state variables are partitioned into internal state variables x_i ($i = 1, 2, 3$) and boundary state variables x_{ij} ($ij = 12, 13, 23$). When considering the local state

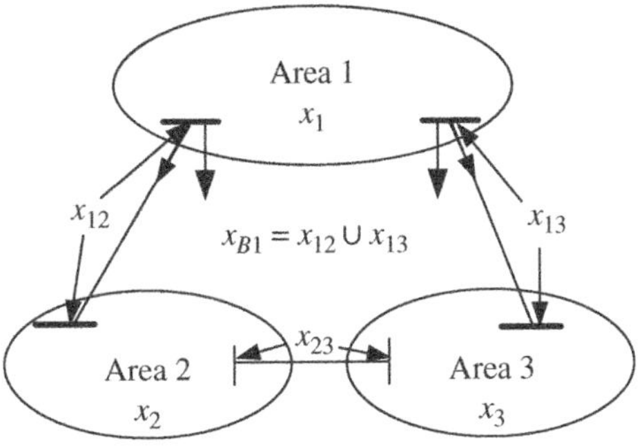

Figure 12.4 A three-area power system.

estimation problem of area i, we summarize all boundary state variables associated with area i as x_{Bi}. In Figure 12.4, for instance, $x_{B1} = x_{12} \cup x_{13}$ and corresponding buses are thickened. Note that x_{B1} includes terminal buses on both sides of tie-lines. When considering the coordinator's problem, we still use x_{ij} as a non-overlapping description of boundary state variables. One can conveniently convert between x_{Bi} and x_{ij} by

$$x_{Bi} = \bigcup_j x_{ij}, \quad x_{ij} = x_{Bi} \cap x_{Bj} \tag{12.11}$$

The measurement model, under the system partition described above, is given by

$$z_i = h_i(x_i, x_{Bi}) + e_i, \forall i \in \{1, 2, 3\} \tag{12.12}$$

where $h_i(x_i, x_{Bi})$ represents measurement functions in area i. They are functions of all internal and boundary state variables associated with area i. Vector e_i represents measurement errors in area i.

12.3.2 Centralized State Estimation

Taking the three-area power system in Figure 12.4 as a whole, a centralized SE is formulated as

$$\min_{\{x_i, x_{ij}\}} f = \sum_{i=1}^{3} f_i(x_i, x_{Bi}) \tag{12.13}$$

where $f_i(x_i, x_{Bi})$ is the objective function of area i defined by

$$\begin{aligned} f_i(x_i, x_{Bi}) &= \frac{1}{2}[r_i(x_i, x_{Bi})]^T W_i r_i(x_i, x_{Bi}) \\ r_i(x_i, x_{Bi}) &= z_i - h_i(x_i, x_{Bi}) \end{aligned} \tag{12.14}$$

The first-order optimal condition for (12.13) is given by

$$\begin{cases} \nabla_{x_i} f = g_{i,i}(x_i, x_{Bi}) = 0 & (12.15) \\ \nabla_{x_{ij}} f = g_{i,ij}(x_i, x_{Bi}) + g_{j,ij}(x_j, x_{Bj}) = 0 & (12.16) \end{cases}$$

The centralized state estimation (12.13) can be solved via the Gauss–Newton iteration (12.6), where the gain matrix G, update for state variables Δx, and the gradient vector g take the following block form:

$$G = \begin{bmatrix} G_{1,1-1} & & & G_{1,1-12} & G_{1,1-13} & \\ & G_{2,2-2} & & G_{2,2-12} & & G_{2,2-23} \\ & & G_{3,3-3} & & G_{3,3-13} & G_{3,3-23} \\ G_{1,12-1} & G_{2,12-2} & & G_{1,12-12} + G_{2,12-12} & G_{1,12-13} & G_{2,12-23} \\ G_{1,13-1} & & G_{3,13-3} & G_{1,13-12} & G_{1,13-13} + G_{3,13-13} & G_{3,13-23} \\ & G_{2,2-23} & G_{3,23-3} & G_{2,23-12} & G_{3,23-13} & G_{2,23-23} + G_{3,23-23} \end{bmatrix} \tag{12.17}$$

$$\Delta x = [\Delta x_1 \quad \Delta x_2 \quad \Delta x_3 \quad \Delta x_{12} \quad \Delta x_{13} \quad \Delta x_{23}]^T \tag{12.18}$$

$$g = \left[g_{1,1} \quad g_{2,2} \quad g_{3,3} \quad g_{1,12} + g_{2,12} \quad g_{1,13} + g_{3,13} \quad g_{2,23} + g_{3,23}\right]^T \tag{12.19}$$

where

$$\begin{aligned} &g_{i,i} = -H_{i,i}^T W_i r_i, g_{i,ij} = -H_{i,ij}^T W_i r_i \\ &G_{i,i-i} = H_{i,i}^T W_i H_{i,i}, G_{i,i-ij} = H_{i,i}^T W_i H_{i,ij}, G_{i,ip-ij} \quad i \in \{1,2,3\}, ij, ip \in \{12, 13, 23\} \\ &= H_{i,ip}^T W_i H_{i,ij} \end{aligned} \tag{12.20}$$

For the measurement Jacobian H, gradient vectors g, and gain matrices G, their first and second subscripts denote, respectively, area indices for measurements and states. For instance, measurement Jacobian $H_{i,ij}$ describes derivatives of measurements in area i (z_i) with respect to the boundary state variables between areas i and j (x_{ij}). In gain matrix $G_{i,i-i}$, for another instance, the first subscript i means it is computed only with measurements in area i (z_i), and the second subscript $i-i$ means its row and column are both associated with internal states in area i (x_i).

A centralized SE iterates its Gauss–Newton update (12.6) until convergence. In the next section, we review existing approaches to solve problem (12.13).

12.4 A BRIEF SURVEY OF SOLUTION TECHNIQUES

12.4.1 Overview

The idea of multi-area state estimation was proposed in the third part of Schweppe's milestone papers about power system state estimation [19]. The first systematic study, to our best knowledge, was published in 1972 [20]. Since then, the popularity of this topic has waxed and waned: it was very attractive in the 1970s and 1980s, became less popular afterward, and then has been reinvigorating rapidly in recent years. However, it is interesting to notice that early works and recent ones were motivated quite differently.

Initially, Schweppe raised the idea of multi-area state estimation in 1970 as a possible technique to implement state estimations in large-scale power systems more efficiently. By decomposing the power system into smaller subareas, one was easier to estimate operating states of power systems with limited computation and storage capabilities at that time. Since the 1990s, speed and storage of computers had been hugely improved. Most system operators had enough capabilities to solve their local state estimations, and the idea of using multi-area SEs to reduce computation burden became less attractive. However, with the deregulation of the electricity market in North America at the end of the twentieth century and the development of smart grid technology, an effective multi-area SE has been increasingly desirable, which can eliminate discrepancies on boundary state estimates,

TABLE 12.1 Attributes of selected works on multi-area state estimation.

First author/paper/year	Architecture	Optimality	Communication
Clements/[20]/1972	Hierarchical	Suboptimal	Non-iterative
Cutsem/[21]/1981	Hierarchical	Suboptimal	Non-iterative
Lin/[22]/1992	Fully decentralized	Optimal	Inter-area Gauss–Newton
Falcao/[23]/1995	Hierarchical	Suboptimal	Inter-area Gauss–Newton
Habiballah/[24]/1996	Hierarchical	Suboptimal	Non-iterative
Ebrahimian/[25]/2000	Hierarchical	Optimal	Inter-area Gauss–Newton
Zhao/[26]/2005	Hierarchical	Suboptimal	Non-iterative
Sun/[27]/2005	Hierarchical	Optimal	Intra-area Gauss–Newton
Conejo/[28]/2007	Fully decentralized	Optimal	Intra-area Gauss–Newton
Jiang/[29]/2008	Hierarchical	Optimal	Inter-area Gauss–Newton
Gómez-Expósito/[30]/2009	Hierarchical	Optimal	Intra-area Gauss–Newton
Gómez-Expósito/[6]/2011	Hierarchical	Optimal	Intra-area Gauss–Newton
Korres/[31]/2011	Hierarchical	Optimal	Inter-area Gauss–Newton
Xie/[32]/2012	Fully decentralized	Optimal	Intra-area Gauss–Newton
Kekatos/[33]/2013	Fully decentralized	Optimal	Intra-area Gauss–Newton
Marelli/[34]/2015	Fully decentralized	Optimal	Intra-area Gauss–Newton
Minot/[35]/2016	Fully decentralized	Optimal	Inter-area Gauss–Newton
Zheng/[36]/2017	Fully decentralized	Optimal	Intra-area Gauss–Newton
Guo/[10]/2017	Hierarchical	Optimal	Intra-area Gauss–Newton

monitor interregional power flows, and estimate operating states of active distribution networks.

We select some representative papers out of all relevant works and summarize their attributes on architecture, optimality, and communication frequency as Table 12.1:

In Table 12.1, papers [20, 21, 23, 24, 26] are referred to as two-level single-iteration SEs. Methods in this category, many are early works, have priorities for higher computation efficiencies. Therefore, they adopt non-iterative structures and achieve only suboptimal performance.

Papers [22, 25, 29, 31, 35] are inter-area Gauss–Newton methods. Many methods in this category are optimal and computational efficient. However, as mentioned in Section 12.2.2, these methods may have higher communication costs. Latent communication delay or failure may substantially deteriorate the performance of these estimators.

Papers [6, 10, 27, 28, 30, 32–34, 36] are intra-area Gauss–Newton methods. In these methods, interregional coordination happens only after local Gauss–Newton iterations converge. Therefore, they in general have lower communication costs and are less reliant to the quality of communication channels.

In following subsections, we provide further descriptions on the three categories of methods by selecting some representative works as examples.

12.4.2 Two-Level Single-Iteration State Estimators

These methods typically proceed three non-iterative steps between the local system operator level and the coordinator level. First, each local system operator i estimates its internal state x_i and boundary state x_{Bi}, note that x_{Bi} in different areas are overlapped and may be inconsistent. Second, the coordinator reconciles initial boundary state estimates x_{Bi} from all areas and communicates consistent x'_{Bi} to corresponding areas. Third, each local system operator i estimates its internal state x'_i with fixed x'_{Bi}. The resulting x'_i and x'_{Bi} form the state estimate in area i.

Such a structure was proposed in [20]. Authors of paper [37] suggested iterating this process until convergence. In what follows, we describe paper [21] as an example to illustrate the underlying idea of two-level single-iteration estimators.

As the first step of [21], each local system operator i estimates its internal state x_i and boundary state variables on area i side $\bar{x}_i$, which is a subset of x_{Bi}. Only internal measurements $\bar{z}_i$ are used here, which are the result of removing boundary power injections and tie-line power flows from the original measurement set z_i. The resulting measurement set $\bar{z}_i$ only relates to x_i and $\bar{x}_i$. The model of local state estimation is

$$[x_i, \bar{x}_i] = \arg\min \left\{ \frac{1}{2} [\bar{r}_i(x_i, \bar{x}_i)]^T \overline{W}_i \bar{r}_i(x_i, \bar{x}_i) \right\} \tag{12.21}$$

where $\bar{r}_i(x_i, \bar{x}_i) = \bar{z}_i - \bar{h}_i(x_i, \bar{x}_i)$ is the residual of $\bar{z}_i$ and $\overline{W}_i$ is the diagonal weight matrix for $\bar{z}_i$. System operator i sends the estimated $\bar{x}_i$ to the coordinator.

Second, the coordinator takes $\bar{x}_i$ estimated by individual areas as pseudo-measurements and re-estimate them together with tie-line power flow measurements by solving

$$[\bar{x}'_1 \ldots \bar{x}'_n] = \arg\min \left\{ \frac{1}{2} [r_C(\bar{x}')]^T W_C r_C(\bar{x}') \right\} \tag{12.22}$$

where $r_C(\bar{x}') = z_C - h_C(\bar{x}')$ is the residual vector for the measurement set z_C composed of boundary state pseudo-measurements and tie-line power flow measurements. For boundary state pseudo-measurements, there is $r_C(\bar{x}') = z_C - h_C(\bar{x}') = \bar{x} - \bar{x}'$. For tie-line power flow measurements, z_C is the vector of measurement values, and $h_C(\bar{x}')$ is given by the branch power flow function (12.1). The weight matrix W_C for pseudo-measurements should be gain matrices of (12.21). For simplicity, the authors suggested to empirically assign a diagonal matrix for W_C offline. Boundary injection measurements are difficult to be incorporated in this work.

Third, the coordinator sends the updated $\bar{x}'_i$ back to local system operators who re-estimate their internal state x_i via (12.21) with boundary state $\bar{x}_i = \bar{x}'_i$ fixed. Results from this step are final state estimates for local system operators.

The method above is a highly representative early research on multi-area state estimation. Its primary goal is to improve the computation efficiency, and it is successful in doing so by decomposing the original state estimation problem into smaller ones and adopting a non-iterative structure. Its shortcoming, on the other hand, is the loss of optimality. First, the authors claimed that it was not easy to incorporate boundary injection measurements into the method. Second, even if there is no boundary injection in original measurement sets z_i, this method is still suboptimal because its boundary state estimates $\overline{x}_i^t$ may not satisfy the optimal condition (12.16). Third, the problem of bad data detection under this setting becomes complicated. In particular, the results of the maximum normalized residual test may be different from a centralized method due to the suboptimality in residuals and the approximation of residual covariance matrices.

To alleviate the suboptimality problem, paper [24] discussed the issue of boundary injections and [26] incorporated phasor measurements into the coordinator's problem.

12.4.3 Inter-area Gauss–Newton State Estimators

Another important direction is a hierarchical solution of the centralized normal Eq. (12.6). These methods first approximate the centralized gain matrix (12.17) with a block-diagonal matrix so that each area can solve for its update for x_i and x_{Bi} locally. Subsequently, they exchange off-diagonal entries and correct their updates. By exploiting the sparse structure of the gain matrix, such corrections can be very efficient. An important theoretical basis is the matrix inverse lemma as follows:

$$\left(G_I + H_B^T W_B H_B\right)^{-1} = G_I^{-1} - G_I^{-1} H_B^T \left(W_B^{-1} + H_B G_I^{-1} H_B^T\right)^{-1} H_B G_I^{-1} \tag{12.23}$$

where G_I represents contributions in the gain matrix (12.17) from all internal measurements. It is a block-diagonal matrix. Terms H_B and W_B represent, respectively, Jacobian and weight matrix of boundary measurements.

An early study along this direction was in 1985 [38], and an efficient diakoptics-based approach was proposed in [29]. Here, we use paper [31] as an example to describe how (12.23) is used in the hierarchical solution of (12.5). The original paper used a constrained optimization model to derive the method. In this chapter, we interpret this method from a different perspective.

We partition all measurements into internal ones with subscript I and boundary ones with subscript B. Namely, boundary measurements include tie-line power flow measurements and boundary bus injection measurements. Accordingly, the normal equation (12.5) changes to

$$\left(G_I + H_B^T W_B H_B\right)\Delta x = H_I^T W_I r_I + H_B^T W_B r_B \tag{12.24}$$

where $G_I = H_I^T W_I H_I$ is the contribution of internal measurements to the gain matrix. It is a block-diagonal matrix. Applying the matrix inverse lemma, we have

$$\begin{aligned}
\Delta x &= \left(G_I + H_B^T W_B H_B\right)^{-1}\left(H_I^T W_I r_I + H_B^T W_B r_B\right)\\
&= G_I^{-1} H_I^T W_I r_I + G_I^{-1} H_B^T W_B r_B - G_I^{-1} H_B^T K^{-1} H_B G_I^{-1}\left(H_I^T W_I r_I + H_B^T W_B r_B\right)\\
&= \Delta y + G_I^{-1} H_B^T K^{-1}\left[K W_B r_B - H_B \Delta y - H_B G_I^{-1} H_B^T W_B r_B\right]\\
&= \Delta y + G_I^{-1} H_B^T K^{-1}\left(r_B - H_B \Delta y\right)
\end{aligned} \tag{12.25}$$

where

$$K = W_B^{-1} + H_B G_I^{-1} H_B^T, \quad \Delta y = G_I^{-1} H_I^T W_I r_I \tag{12.26}$$

Assume that the matrix G_I is composed of blocks G_i, the term Δy can be solved locally. Furthermore, this paper defines $\lambda = K^{-1}(r_B - H_B \Delta y)$. Then the update of state in area i can be calculated by

$$\Delta x_i = \Delta y_i + G_i^{-1} H_{Bi}^T \lambda^1 \tag{12.27}$$

All terms in (12.27) are based on local information except for $\lambda = K^{-1}(r_B - H_B \Delta y)$, which is calculated at the coordinator based on information from all areas. The matrix K is the sum of contributions from boundary measurements in all areas.

In principle, the method iterates the following phases to solve for the update of (12.27):

1. Each area separately solves for Δy_i using internal measurements and computes its contribution to matrix K, denoted by K_i, then communicates K_i and Δy_i to the coordinator.
2. The coordinator calculates $\lambda = \left(\sum_i K_i\right)^{-1} (r_B - H_B \Delta y)$ and distributes λ to individual areas. Here for brevity, we assume the coordinator knows residuals of boundary measurements in each iteration.
3. Each area obtains the Gauss–Newton update via (12.27).

This method achieves the optimal state estimate because it iteratively solves the centralized normal equation (12.5) until convergence. In each iteration, the solution of (12.5) is in a hierarchical and non-iterative manner. The convergence of the interregional iteration is the same as centralized Gauss–Newton. Therefore, this method has satisfactory convergence and computation efficiency. Its shortage, on the other hand, is that its communication cost is high, in terms of both total amount and frequency. An implementation research of this work was published in [39].

12.4.4 Intra-area Gauss–Newton State Estimators

We elaborate in this subsection the method in [6] as an example of intra-area Gauss–Newton SEs. This work proposed a multi-level state estimation framework that involves state estimations at the distribution feeder level, substation level, local system operator level, and the coordinator level. In this chapter, we focus on the coordination between local SEs in transmission system operators and the coordinator. This method introduced intermediate state variables y, and the measurement model (12.12) was written as

[1] Here the update Δx_i is for both internal and boundary state variables in area i.

$$\begin{aligned} z_i &= h_y(y_i, y_{Bi}) + e \\ y_i &= x_i + e_{yi}, \quad y_{Bi} = x_{Bi} + e_{yB} \end{aligned} \tag{12.28}$$

where y is taken as state variables in the measurement function $z_i = h_y(y_i, y_{Bi}) + e$ and as pseudo-measurements when estimating the true state vector x. The difference between x and y is that, boundary state $x_{ij} = x_{Bi} \cap x_{Bj}$ must have same values in x_{Bi} and x_{Bj}, whereas values in y_{Bi} and y_{Bj} are in general inconsistent in their common entries.

The first stage of the method is that, each local system operator i estimates the intermediate state y locally by iteratively solving its normal equations until convergence:

$$\begin{bmatrix} G_{i,i-i} & G_{i,i-Bi} \\ G_{i,Bi-i} & G_{i,Bi-Bi} \end{bmatrix} \begin{bmatrix} \Delta y_i \\ \Delta y_{Bi} \end{bmatrix} + \begin{bmatrix} g_{i,i} \\ g_{i,Bi} \end{bmatrix} = 0 \tag{12.29}$$

The expressions of the coefficient matrix G and the vector g are given in (12.17) and (12.19), respectively.

Each local system operator i communicates its estimate of y_{Bi} and the corresponding gain matrix that is the coefficient matrix in (12.29), to the coordinator.

As the second stage, the method in [6] estimates x using y as pseudo-measurements. Here we use the non-overlapping notation x_{ij} to represent boundary state variables. For the three-area system in Figure 12.4, the measurement function at this stage is

$$\begin{bmatrix} y_1 \\ y_2 \\ y_3 \\ y_{B1} \\ y_{B2} \\ y_{B3} \end{bmatrix} = \begin{bmatrix} I & & & & & \\ & I & & & & \\ & & I & & & \\ & & & B_{B1-12} & B_{B1-13} & \\ & & & B_{B2-12} & & B_{B2-23} \\ & & & & B_{B3-13} & B_{B3-23} \end{bmatrix} \begin{bmatrix} x_1 \\ x_2 \\ x_3 \\ x_{12} \\ x_{13} \\ x_{23} \end{bmatrix} + \begin{bmatrix} e_1 \\ e_2 \\ e_3 \\ e_{12} \\ e_{13} \\ e_{23} \end{bmatrix} \tag{12.30}$$

where B is the incidence matrix composed of 0 and 1 with $x_{Bi} = \sum_j B_{Bi-ij} x_{ij}$. The weight matrix is given by (12.17) with each sub-matrix computed in the corresponding local system operator via (12.29). Because the measurement functions are linear, state variables x_i and x_{ij} can be estimated by solving the following Gauss–Newton equation:

$$\begin{bmatrix} B^T G_{BB} B & B^T G_{BI} \\ G_{IB} B & G_{II} \end{bmatrix} \begin{bmatrix} x_B \\ x_I \end{bmatrix} = \begin{bmatrix} B^T G_{BB} & B^T G_{BI} \\ G_{IB} & G_{II} \end{bmatrix} \begin{bmatrix} y_B \\ y_I \end{bmatrix} \tag{12.31}$$

where

$$x_I = [x_1 \quad x_2 \quad x_3]^T, \quad x_B = [x_{12} \quad x_{13} \quad x_{23}]^T \tag{12.32}$$

$$y_I = [y_1 \quad y_2 \quad y_3]^T, \quad y_B = [y_{12} \quad y_{13} \quad y_{23}]^T \tag{12.33}$$

$$B = \begin{bmatrix} B_{B1-12} & B_{B1-13} & \\ B_{B2-12} & & B_{B2-23} \\ & B_{B3-13} & B_{B3-23} \end{bmatrix}, \quad G_{BI} = G_{IB}^T = \begin{bmatrix} G_{1,12-1} & G_{2,12-2} & \\ G_{1,13-1} & & G_{3,13-3} \\ & G_{2,2-23} & G_{3,23-3} \end{bmatrix} \tag{12.34}$$

$$G_{II} = \begin{bmatrix} G_{1,1-1} & & \\ & G_{2,2-2} & \\ & & G_{3,3-3} \end{bmatrix},$$

$$G_{BB} = \begin{bmatrix} G_{1,12-12} + G_{2,12-12} & G_{1,12-13} & G_{2,12-23} \\ G_{1,13-12} & G_{1,13-13} + G_{3,13-13} & G_{3,13-23} \\ G_{2,23-12} & G_{3,23-13} & G_{2,23-23} + G_{3,23-23} \end{bmatrix} \tag{12.35}$$

The Gauss–Newton Eq. (12.31) should be solved in a distributed way. Paper [6] provided two options: One is solving for x_B at the coordinator by eliminating x_I from (12.31) and then solving for x_I in local SEs; the other is Gauss–Seidel iteration between the two rows in (12.31).

With the newest estimate for x_I and x_B, local system operators update their gain matrices in (12.29) and estimate their y_i and y_{Bi} by iterating (12.29) with a constant gain matrix until convergence. Subsequently, the estimates for x_I and x_B are updated via (12.31). Such iterations converge when the estimated x are close enough in two consecutive iterations.

In this method, local system operators iterate their local Gauss–Newton updates until convergence then exchange information with the coordinator. Therefore, the communication cost is expected to be moderate.

Fully decentralized methods in this category are proposed in [32, 33]. Note that in intra-area Gauss–Newton methods, each interregional iteration requires heavier computation burden. Therefore, a solid commitment on the rate of convergence is extremely valuable. Hereinafter, we present a recent work in this field with theoretical commitment on its convergence [10].

12.5 HIERARCHICAL STATE ESTIMATOR VIA SENSITIVITY FUNCTION EXCHANGES

12.5.1 Outline

The method in [10] is a hierarchical and optimal multi-area SE with intra-area Gauss–Newton iterations. Its architecture is illustrated in Figure 12.5.

An underlying feature of the estimator in [10] is that it uses sensitivity functions to represent optimality conditions of local state estimations. This is fundamentally different from exchanging intermediate boundary state estimates or

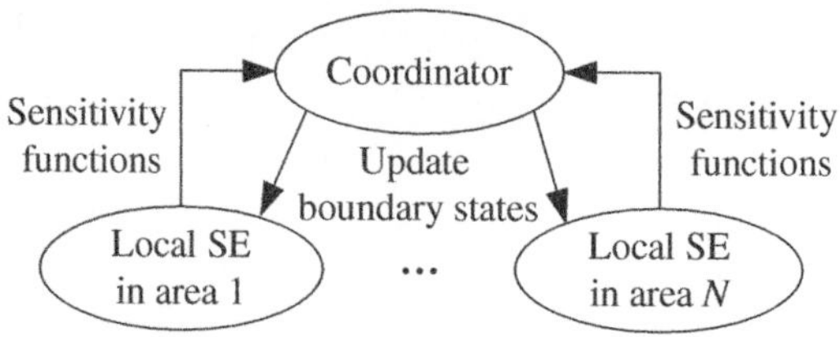

Figure 12.5 Architecture of the method in [10]. Source: Reproduced with permission of IEEE.

pseudo-measurements, which only contain information about a few buses. From this perspective, this method is referred to as *Sensitivity Function-based Hierarchical State Estimator (SFHSE)*.

SFHSE involves interactions between local SEs and a coordinator. Specifically, it iterates the following two phases:

1. In each area, with fixed boundary state x_{Bi}, the local SE solves for the internal state estimate $\hat{x}_i$ and derives its sensitivity function of the local objective function f_i in (12.14) with respect to boundary states x_{Bi}.
2. The coordinator collects sensitivity functions from all areas, optimizes the boundary state for the next iteration, and communicates the newest x_{Bi} to individual areas.

The idea of exchanging sensitivities in distributed optimization has been considered in the literature, and its benefit in the rate of convergence has been estimated for special types of distributed convex optimization problems (see [40, 41]). Applying these ideas in MASE, however, is not trivial due to the fact that the power system state estimation model is not convex.

12.5.2 Initialization

Prior to iterations between the coordinator and local SEs, the coordinator needs to initialize the boundary state $x_{Bi}^{(0)}$. A trivial initialization is the flat start, i.e. set all voltage magnitudes as one and all phase angles as zero. In this paper, for the sake of fewer iteration times and higher efficiencies, the boundary state $x_{Bi}^{(0)}$ are initialized based on initial local state estimates.

The coordinator initiates the multi-area state estimation by sending a starting signal to local SEs upon which each area solves an initial state estimation:

$$\min_{\{x_i, x_{Bi}\}} f_i(x_i, x_{Bi}) = \frac{1}{2}[r_i(x_i, x_{Bi})]^T W_i r_i(x_i, x_{Bi}) \tag{12.36}$$

where decision variables are internal and boundary state variables associate with area i. Note that an artificial slack bus is required to solve (12.36) in each area. Each area communicates its boundary state estimate $\hat{x}_{Bi}^{(0)}$ in (12.36) to the coordinator.

Subsequently, the coordinator initializes $x_{Bi}^{(0)}$ according to their local estimates $\hat{x}_{Bi}^{(0)}$. Note that $\hat{x}_{Bi}^{(0)}$ from different areas are inconsistent at their common

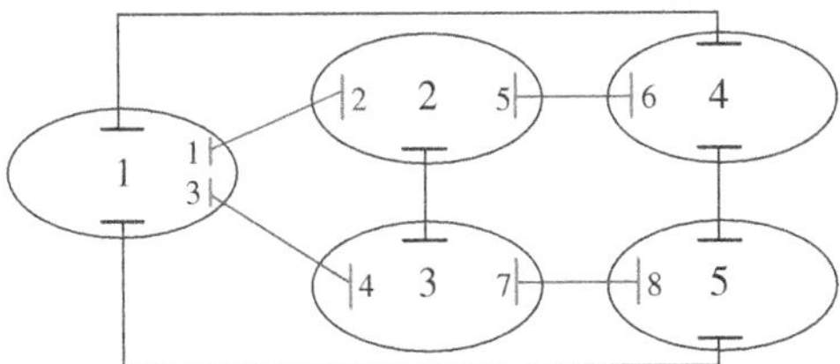

Figure 12.6 Illustration of the phase angle rotation method.

entries. Moreover, phase angle estimates in $\hat{x}_{Bi}^{(0)}$ are not referring to the same bus. To address this issue, the coordinator rotates $\hat{x}_{Bi}^{(0)}$ with respect to a unique phase angle reference. Taking the multi-area system in Figure 12.6 as an example, the phase angle rotation technique is elucidated as the following:

First, a spanning tree was arbitrarily constructed in which each node represents one area. In Figure 12.6, for example, area 1 is selected as the root node, and the spanning tree is constructed via the breadth-first search, as grey lines in Figure 12.6. Without loss of generality, the global phase angle reference is selected as bus 1 in the root area.

Second, initial boundary phase angle estimates $\hat{x}_{Bi}^{(0)}$ are rotated to match with its parent node at one boundary bus. Take area 2 as an example, its phase angle estimates $\hat{\theta}_{B2}^{(0)}$ should be rotated to match with area 1 at bus 1 by

$$\left[\hat{\theta}_{B2}^{(0)}\right]_R = \hat{\theta}_{B2}^{(0)} - \left[\hat{\theta}_{B2}^{(0)}(1) - \hat{\theta}_{B1}^{(0)}(1)\right] \tag{12.37}$$

where $\hat{\theta}_{B2}^{(0)}$ and $\left[\hat{\theta}_{B2}^{(0)}\right]_R$ represent, respectively, the initial boundary phase angle estimates in area 2 before and after the phase angle rotation. Terms $\hat{\theta}_{B1}^{(0)}(1)$ and $\hat{\theta}_{B2}^{(0)}(1)$ represent, respectively, the initial phase angle estimates for bus 1 in areas 1 and 2. Equation (12.37) rotates every entry in $\hat{\theta}_{B2}^{(0)}$ by $\left[\hat{\theta}_{B2}^{(0)}(1) - \hat{\theta}_{B1}^{(0)}(1)\right]$ such that the $\left[\hat{\theta}_{B2}^{(0)}\right]_R$ after rotation takes bus 1 as its reference. Similarly, the initial boundary phase angle estimates in areas 3, 4, and 5 are rotated to match with their parent nodes (areas 1, 2, and 3, respectively) at one boundary bus (can be selected as buses 3, 5, and 7, respectively).

After the phase angle rotation, the coordinator can initialize $x_{Bi}^{(0)}$ as the mean of the rotated $\left[\hat{x}_{Bi}^{(0)}\right]_R$ from corresponding areas. The initial value of $x_{Bi}^{(0)}$ may vary with different constructions of the spanning tree. This does not matter in the initialization step since subsequent SFHSE iterations will update the boundary state until it reaches the neighborhood of its optimum estimate.

With $x_{Bi}^{(0)}$ initialized at the coordinator, each system operator solve its local problem and derive its sensitivity function, as elaborated in the following subsection.

12.5.3 Local State Estimator

In iteration t, the local SE of area i estimates the internal state x_i with given boundary state $x_{Bi} = x_{Bi}^{(t)}$. It solves the following WLS problem:

$$\min_{x_i} f_i(x_i, x_{Bi}) = \frac{1}{2}[r_i(x_i, x_{Bi})]^T W_i r_i(x_i, x_{Bi}) \tag{12.38}$$

In (12.38) decision variables are x_i only, and x_{Bi} are parameters of the optimization. The Gauss–Newton iteration for problem (12.38) is

$$G_{i,i-i}\Delta x_i + g_{i,i} = 0 \tag{12.39}$$

where $G_{i,\,i-i}$, Δx_i, and $g_{i,\,i}$ are sub-matrices and sub-vectors given in (12.17)–(12.19).

The optimal solution $\tilde{x}_i$ to (12.38) and the value $f_i(\tilde{x}_i, x_{Bi})$ can be written as functions of x_{Bi}:

$$\tilde{x}_i = \phi_i(x_{Bi}) \tag{12.40}$$

$$\widetilde{J}_i(x_{Bi}) := f_i(\tilde{x}_i, x_{Bi}) = f_i(\phi_i(x_{Bi}), x_{Bi}) \tag{12.41}$$

Like the idea of Gauss–Newton method, the measurement function $\widetilde{J}_i(x_{Bi})$ is locally linearized. Consequently, the first- and second-order derivatives of $\widetilde{J}_i(x_{Bi})$ are calculated by

$$\begin{aligned} \tilde{g}_{Bi} &= \nabla_{x_{Bi}}\widetilde{J}_i(x_{Bi}) = \nabla_{x_{Bi}} f_i(\tilde{x}_i, x_{Bi}) + (\nabla_{x_{Bi}}\tilde{x}_i)^T \nabla_{x_i} f_i(\tilde{x}_i, x_{Bi}) \\ &= \nabla_{x_{Bi}} f_i(\tilde{x}_i, x_{Bi}) = -\widetilde{H}_{i,Bi}^T W_i \tilde{r}_i \end{aligned} \tag{12.42}$$

$$\widetilde{G}_{Bi} = \nabla_{x_{Bi}}^2 \widetilde{J}_i = -\widetilde{H}_{i,Bi}^T W(\nabla_{x_{Bi}} r_i + \nabla_{x_i} r_i \nabla_{x_{Bi}} \tilde{x}_i) = \widetilde{H}_{i,Bi}^T W\left(\widetilde{H}_{i,Bi} - \widetilde{H}_{i,i}\nabla_{x_{Bi}}\tilde{x}_i\right) \tag{12.43}$$

The term $(\nabla_{x_{Bi}}\tilde{x}_i)^T$ can be obtained from the local normal equation associated with area i, which is part of the normal equation of the centralized SE:

$$\begin{bmatrix} G_{i,i-i} & G_{i,i-Bi} \\ G_{i,Bi-i} & G_{i,Bi-Bi} \end{bmatrix}\begin{bmatrix} \Delta x_i \\ \Delta x_{Bi} \end{bmatrix} + \begin{bmatrix} g_{i,i} \\ g_{i,Bi} \end{bmatrix} = 0 \tag{12.44}$$

The internal optimal state estimate $\tilde{x}_i$ ensures $\tilde{g}_{i,i} = 0$. From the first row of (12.44), there is

$$\nabla_{x_{Bi}}\tilde{x}_i = -\widetilde{G}_{i,i-i}^{-1}\widetilde{G}_{i,i-Bi} \tag{12.45}$$

By substituting (12.45) to (12.43), there is

$$\widetilde{G}_{Bi} = \nabla_{x_{Bi}}^2 \widetilde{J}_i = \widetilde{H}_{i,Bi}^T W\left(\widetilde{H}_{i,Bi} - \widetilde{H}_{i,i}\widetilde{G}_{i,i-i}^{-1}\widetilde{G}_{i,i-Bi}\right) = \widetilde{G}_{i,Bi-Bi} - \widetilde{G}_{i,Bi-i}\widetilde{G}_{i,i-i}^{-1}\widetilde{G}_{i,i-Bi} \tag{12.46}$$

The matrix $\widetilde{G}_{Bi}$ is calculated by selecting the sub-matrices associated with area i from the global gain matrix in (12.17), as shown in (12.44), then eliminating all internal state variables of area i.

With the local linearization of measurement functions at $\tilde{x}_i$ and the given x_{Bi}, the relation between the optimal objective function of area i and the given x_{Bi} is represented by the quadratic function $\widetilde{J}_i(x_{Bi})$ with gradient $\widetilde{g}_{Bi}$ as (12.42) and Hessian $\widetilde{G}_{Bi}$ as (12.46). Assuming that the given x_{Bi} deviates for Δx_{Bi}, then the change in $\nabla_{x_{Bi}}\widetilde{J}_i$ is estimated by

$$\nabla_{x_{Bi}}\widetilde{J}_i(x_{Bi} + \Delta x_{Bi}) = \widetilde{g}_{Bi}(x_{Bi}) + \widetilde{G}_{Bi}(x_{Bi})\Delta x_{Bi} \tag{12.47}$$

In practice, however, local SE i terminates at $\hat{x}_i$ that is in a neighborhood of the optimal point $\tilde{x}_i$. Their distance is bounded by the convergence criterion $\|\hat{g}_{i,i}\|_2 \le \varepsilon_i$, with ε_i as the convergence threshold for the local SE in area i. The gradient and Hessian computed with point $\hat{x}_i$ are respectively denoted by $\hat{g}_{Bi}$ and $\hat{G}_{Bi}$ and let $\hat{J}_i(x_{Bi}) := f_i(\hat{x}_i, x_{Bi})$. Qualitative performance of SFHSE does not become invalid with reasonably small convergence tolerances. The system operator of area i communicates $\hat{g}_{Bi}$ and $\hat{G}_{Bi}$ to the coordinator. The formulation and solution of the coordinator's problem are presented in the next subsection.

12.5.4 Coordinator's Problem

The coordinator's objective is to solve for the update for boundary state that best satisfies the other optimal condition (12.16). The basis for this step is the sensitivity functions (12.47) communicated from local SEs.

By substituting (12.47) into the other optimal condition (12.16) we obtain

$$\begin{aligned}\nabla_{x_{ij}} f\left(\hat{x}_i, x_{ij} + \Delta x_{ij}\right) &\approx \nabla_{x_{ij}} f\left(\tilde{x}_i, x_{ij} + \Delta x_{ij}\right) = \sum_{i=1}^{3} \nabla_{x_{Bi}}\widetilde{J}_i(x_{Bi} + \Delta x_{Bi}) \\ &\approx \sum_{i=1}^{3} \nabla_{x_{Bi}}\hat{J}_i(x_{Bi} + \Delta x_{Bi}) = 0 \\ \hat{G}_{B\Sigma}\left(x_{ij}\right)\Delta x_{ij} + \hat{g}_{B\Sigma}\left(x_{ij}\right) &= 0\end{aligned} \tag{12.48}$$

where $\hat{G}_{B\Sigma}$ and $\hat{g}_{B\Sigma}$ are the sums of $\hat{G}_{Bi}$ and $\hat{g}_{Bi}$ according to the connections among different areas. For example, in the power system in Figure 12.4, the matrix $\hat{G}_{B\Sigma}$ is calculated by

$$\hat{G}_{B\Sigma} = \begin{bmatrix} \hat{G}_{B1,12-12} + \hat{G}_{B2,12-12} & \hat{G}_{B1,12-13} & \hat{G}_{B2,12-23} \\ \hat{G}_{B1,13-12} & \hat{G}_{B1,13-13} + \hat{G}_{B3,13-13} & \hat{G}_{B3,13-23} \\ \hat{G}_{B2,23-12} & \hat{G}_{B3,23-13} & \hat{G}_{B2,23-23} + \hat{G}_{B3,23-23} \end{bmatrix} \tag{12.49}$$

where $\hat{G}_{B1,12-12}$ represents the sub-matrix of $\hat{G}_{B1}$ associated with boundary buses between area 1 and 2, similar for other sub-matrices.

If $\|\hat{g}_{B\Sigma}\|_2 \leq \varepsilon$, then the current system-wide state estimate $(x_{ij}, \hat{x}_i)$ satisfies both (12.15) and (12.16), respectively, with suboptimal level ε_i and ε. Note that in the coordinator's problem, $\hat{x}_i$ is used to approximate the optimal $\tilde{x}_i$. To make such an approximation valid, it is suggested to set ε_i smaller than ε, e.g. set $\varepsilon_i = 0.1\varepsilon$.

If current x_{ij} does not satisfy the optimal condition (12.16), the solution to the coordinator's problem (12.48) provides an update for x_{ij} such that Eq. (12.16) is satisfied under the local linearization of measurements. The coordinator communicates the updated $x_{Bi}^{(t+1)}$ to area i.

12.5.5 Complete Scheme

The complete scheme of SFHSE is illustrated in Figure 12.7.

Each step in Figure 12.7 is explained as follows:

Step 0: The coordinator sends a starting signal to each area and set iteration time $t = 0$.

Step 1: Each area i performs its initial local state estimation (12.36).

Step 2: The coordinator initializes boundary state $x_{ij}^{(0)}$ according to $\hat{x}_{Bi}^{(0)}$ like Section 12.6.2.

Step 3: The coordinator converts $x_{ij}^{(t)}$ to $x_{Bi}^{(t)}$ via (12.11) and broadcasts $x_{Bi}^{(t)}$ to area i.

Step 4: Given $x_{Bi}^{(t)}$, each area solves its local state estimation (12.38). Due to its stopping rule, the local SE terminates at $\hat{x}_i$ that is in a neighborhood of $\tilde{x}_i$. Note that each area should not select extra angle reference bus here because the specified $x_{Bi}^{(t)}$ already provides a phase angle reference.

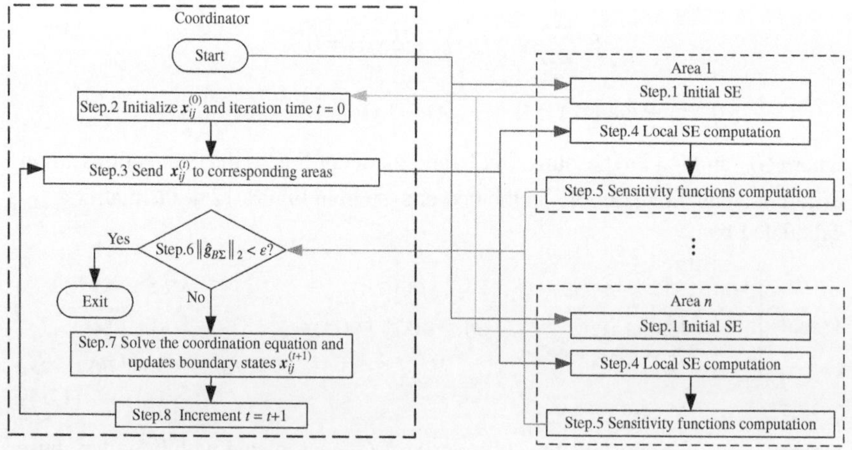

Figure 12.7 Complete scheme of SFHSE.

Step 5: Each area computes its sensitivity function represented by $(\hat{g}_{Bi}, \hat{G}_{Bi})$ using the local state estimate $\hat{x}_i$ and communicates them to the coordinator. As an alternative, because entries in the gain matrix change slightly with state variables, $\hat{G}_{Bi}$ can be transferred together with $\hat{g}_{Bi}$ only once in the first iteration. This will reduce communication cost with the price of potentially more iterations. In subsequent iterations, only $\hat{g}_{Bi}$ vectors are communicated.

Step 6: The coordinator computes the sum $\hat{G}_{B\Sigma}$ and $\hat{g}_{B\Sigma}$. If $\|\hat{g}_{B\Sigma}\|_2 < \varepsilon$, SFHSE converges, exit. Otherwise, go to step 7.

Step 7: The coordinator solves Eq. (12.48) to update boundary state $x_{ij}^{(t+1)} = x_{ij}^{(t)} + \Delta x_{ij}^{(t)}$.

Step 8: Increment iteration time $t = t + 1$, turn to step 3.

12.6 ADD-ON FUNCTIONS IN MULTI-AREA STATE ESTIMATION

12.6.1 Observability Analysis

Assuming that the global power system is always observable, i.e. the gain matrix $G(x)$ in (12.17) has full rank at whatever x, paper [10] proved that the SFHSE was also observable, in the sense that it had a unique optimal state estimate in its small neighborhood.

For the global power system, without loss of generality, the global phase angle reference bus can be selected to be one on the boundary. By removing the corresponding row and column associated with the global reference bus, the remaining gain matrix in (12.17) should be non-singular and positive definite.

For the local SE, the determinant of its gain matrix in (12.39) is one of the leading principal minors of the global gain matrix in (12.17), which is always positive. Therefore, the gain matrix $G_{i,\, i-i}$ in (12.39) is non-singular, the local SE (12.38) is always observable. To better interpret this, note that in the local SE (12.38), boundary state variables are fixed, and only the internal state variables are estimated.

For the coordinator's problem (12.48), the coefficient matrix $\hat{G}_{B\Sigma}$ is the result of eliminating all internal buses from the global gain matrix (12.10). Assuming that the global system is always observable, then the coordinator's problem also has a unique solution for Δx_{ij}. Hence in summary, SFHSE is observable as long as the centralized estimator (12.13) is observable.

However, the state estimation (12.36) in the initialization step may be unobservable. When this occurs, $\hat{x}_{Bi}^{(0)}$ can be empirically initialized as flat start or one of its historical estimates.

12.6.2 Bad Data Identification

A distributed BDI for SFHSE was also presented in [10], in which assisted by the coordinator, each area calculates the normalized residual for its own measurements z_i and removes bad data independently. The method is described as follows:

Step I: Perform SFHSE as Figure 12.7 until convergence.

Step II: For area i, the coordinator eliminates all other boundary state variables beside x_{Bi} from $\hat{G}_{B\Sigma}$. For example, in the three-area example with $\hat{G}_{B\Sigma}$ shown in (12.49), the resulting matrix for area 1, denoted by $\hat{G}_{B\Sigma,1}$, is calculated by

$$\hat{G}_{B\Sigma,1} = \begin{bmatrix} \hat{G}_{B1,12-12} + \hat{G}_{B2,12-12} & \hat{G}_{B1,12-13} \\ \hat{G}_{B1,13-12} & \hat{G}_{B1,13-13} + \hat{G}_{B3,13-13} \end{bmatrix} - \begin{bmatrix} \hat{G}_{B2,12-23} \\ \hat{G}_{B3,13-23} \end{bmatrix} \left(\hat{G}_{B2,23-23} + \hat{G}_{B3,23-23}\right)^{-1} \begin{bmatrix} \hat{G}_{B2,23-12} & \hat{G}_{B3,23-13} \end{bmatrix} \tag{12.50}$$

The coordinator communicates $\hat{G}_{B\Sigma,i}$ to area i.

Step III: Each area computes the normalized residuals of local measurements by (12.9). In particular, the residual variance matrix $\hat{\Omega}_i$ is calculated by

$$\hat{\Omega}_i = W_i^{-1} - \begin{bmatrix} \hat{H}_{i,i} \\ \hat{H}_{i,Bi} \end{bmatrix} \begin{bmatrix} G_{i,i-i} & G_{i,i-Bi} \\ G_{i,Bi-i} & G_{i,Bi-Bi} + \hat{G}_{B\Sigma,i} \end{bmatrix}^{-1} \begin{bmatrix} \hat{H}_{i,i} & \hat{H}_{i,Bi} \end{bmatrix} \tag{12.51}$$

It can be seen that $\hat{G}_{B\Sigma,i}$ communicated from the coordinator is incorporated. Area i communicates its largest normalized residual to the coordinator.

Step IV: The coordinator finds the largest normalized residual for all areas. If it is smaller than a certain threshold, which is set as 3.0 in this paper, exit. Otherwise, the coordinator informs the corresponding area to eliminate its most suspicious measurement then turn to Step I.

By ignoring the stopping tolerances in SFHSE and the centralized state estimator (CSE), it can be concluded that these two approaches obtain same normalized residuals. The reasons are as follows:

1. SFHSE is proved to be an optimal estimator. Therefore, residuals calculated in individual areas in SFHSE are equal to those in the centralized estimator.
2. To compute the denominator $\hat{\Omega}_i$, the centralized estimator calculates the inverse of the global gain matrix in (12.17) then takes the sub-matrix associated with internal and boundary buses in area i. The decentralized approach, on the other hand, preserves internal and boundary buses in area i, eliminates other buses from the global gain matrix in (12.17), and then calculates the inverse of the resulting sub-matrix. From the property of Gauss elimination,

these two operations are equivalent. Therefore, the residual variance matrix $\hat{\Omega}_i$ in (12.51) is equal to that of a centralized approach.

In summary, the distributed BDI is equivalent to the centralized state estimation and BDI. For approaches that identify multiple bad data per iteration, their distributed versions can be derived similarly.

12.7 PROPERTIES

12.7.1 Assumptions

Paper [10] established properties on optimality and convergence for SFHSE. These properties are considered locally within a neighborhood of the optimal solution. The considered neighborhoods for boundary state associated with area i, internal state in area i, and all state variables are respectively denoted as S_{Bi}, S_i, and S. Throughout, the following assumptions are made:

Assumption 12.1 The Hessian matrix of the overall SE objective, $\nabla_x^2 f$, is strong convex within region S, i.e. there exists a positive constant m such that

$$\nabla_x^2 f \succeq mI, \quad \forall x \in S \tag{12.52}$$

Assumption 12.2 For area i, the derivatives of its measurement function $h_i(x_i, x_{Bi})$ and the mapping $\tilde{x}_i = \phi_i(x_{Bi})$ are bounded, i.e. there exist constants K_{hi} and $K_{\phi i}$ such that

$$\|\nabla_{x_i} h_i\| \leq K_{hi}, \forall x_i \in S_i, x_{Bi} \in S_{Bi}; \quad \|\nabla_{x_{Bi}} \phi_i\| \leq K_{\phi i}, \forall x_{Bi} \in S_{Bi} \tag{12.53}$$

Assumption 12.3 For area i, the measurement function $h_i(x_i, x_{Bi})$ and the mapping $\tilde{x}_i = \phi_i(x_{Bi})$ are Fréchet differentiable in the neighborhood of the optimal state estimate, i.e. there exist constants K_{hi1} and $K_{\phi i1}$, such that [13]

$$\begin{gathered}
\|h_i(x_i + \Delta x_i, x_{Bi} + \Delta x_{Bi}) - h_i(x_i, x_{Bi}) - H_{i,i}\Delta x_i - H_{i,Bi}\Delta x_{Bi}\| \leq K_{hi1} \left\| \begin{bmatrix} \Delta x_i \\ \Delta x_{Bi} \end{bmatrix} \right\|^2, \\
\forall x_i \in S_i, x_{Bi} \in S_{Bi} \\
\|\phi_i(x_{Bi} + \Delta x_{Bi}) - \phi_i(x_{Bi}) - \nabla_{x_{Bi}}\phi_i \Delta x_{Bi}\| \leq K_{\phi i1} \|\Delta x_{Bi}\|^2, \forall x_{ij} \in S_{Bi}
\end{gathered} \tag{12.54}$$

For Assumption 12.1, the gain matrix $G = H^T WH$ is always strong convex. It is not the Hessian of the objective function, but their difference is usually small. Therefore, Assumption 12.1 holds in most cases. For Assumptions 12.2 and 12.3, the commonly used measurement functions are smooth and well behaved. The mapping ϕ in (12.40) is also continuous and Fréchet differentiable according to the parametric programming theory [42]. Thus Assumptions 12.2 and 12.3 also hold in real power systems.

12.7.2 Optimality

On the issue of optimality, taking into account that stopping rules used by local SEs and the coordinator may affect the accuracy, the following theorem is established:

Theorem 12.1 Under Assumption 12.1 and assuming that SFHSE converges, then

$$\lim_{\varepsilon, \varepsilon_i \to 0} [f(\hat{x}) - f(\tilde{x})] = 0 \tag{12.55}$$

where $\hat{x}$ is the vector of all state variables estimated by SFHSE and $\tilde{x}$ is the optimal state estimate.

Proof: When SFHSE converges in step 6, according to its stopping criterion, there is $\|\hat{g}_{B\Sigma}\|_2 < \varepsilon$. Note that local SEs have just converged in step 4, and from the stopping criterion there is $\|\hat{g}_{i,i}\|_2 < \varepsilon_i$ for all i. Therefore, at the state estimate from SFHSE $\hat{x}$, the norm of the gradient vector of the centralized SE (12.13) is bounded by

$$\|\nabla_x f(\hat{x})\|_2 \le \|\hat{g}_{i,i}\|_2 + \|\hat{g}_{B\Sigma}\|_2 < \varepsilon + \varepsilon_i \tag{12.56}$$

With Assumption 12.1, there is [43]:

$$f(x') \ge f(x) - \frac{1}{2m}\|\nabla_x f(x)\|_2^2, \quad \forall x, x' \in S \tag{12.57}$$

Thus there is

$$\begin{aligned} & f(\hat{x}) - f(\tilde{x}) \le \frac{1}{2m}\|\nabla_x f(\hat{x})\|_2^2 < \frac{1}{2m}(\varepsilon + \varepsilon_i)^2 \\ & \lim_{\varepsilon, \varepsilon_i \to 0} [f(\hat{x}) - f(\tilde{x})] = 0 \end{aligned} \tag{12.58}$$

Inequality (12.58) proves Theorem 12.1.

Remark: Theorem 12.1 shows that SFHSE can terminate in an arbitrarily small neighborhood of the optimal state estimate $\tilde{x}$, i.e. it is an optimal multi-area SE. Theorem 12.1 holds for all possible paths for $(\varepsilon, \varepsilon_i) \to 0$. □

12.7.3 Convergence

On the convergence rate of SFHSE, the following theorem was proved in [10]:

Theorem 12.2 Under Assumptions 12.1, 12.2, and 12.3, SFHSE has the same convergence rate as the Gauss–Newton method for the centralized power system state estimation.

Proof: The theorem was proved in two steps in [10]. First, SFHSE is proved to be equivalent to the Gauss–Newton iteration to solve a special state estimation problem. Second, the special state estimation problem is proved to have Fréchet differentiable measurement functions. Note that the rate of convergence describes how fast an algorithm converges within a small neighborhood of the optimum. Thus, in this proof, it is assumed that one can set local stopping tolerances ε_i small enough such that $\hat{x}_i \approx \tilde{x}_i$.

Recall that in area i, the optimal internal state estimate can be represented by $\tilde{x}_i = \phi_i(x_{Bi})$. Therefore, measurements in area i are written as functions of x_{Bi} only:

$$v_i(x_{Bi}) := h_i(\tilde{x}_i, x_{Bi}) = h_i(\phi_i(x_{Bi}), x_{Bi}) \tag{12.59}$$

Then the centralized state estimation model (12.13) changes to

$$\min_{x_{Bi}} J_\Sigma = \frac{1}{2}\sum_i (z_i - v_i(x_{Bi}))^T W_i (z_i - v_i(x_{Bi})) \tag{12.60}$$

Model (12.60) is a state estimation problem with state variables x_{Bi} and measurement functions $v_i(x_{Bi})$. Given current value of x_{Bi}, the Gauss–Newton update to solve (12.60) is determined by the following optimization:

$$\min_{\Delta x_{ij}} J_\Delta(\Delta x_{ij}) = \frac{1}{2}\sum_i r_{\Delta i}^T W_i r_{\Delta i} \tag{12.61}$$

where

$$r_{\Delta i}(\Delta x_{Bi}) = z_i - v_i(x_{Bi}) - \nabla_{x_{Bi}} v_i \Delta x_{Bi} \tag{12.62}$$

The normal equation for Gauss–Newton iteration is in the form of

$$\left[\sum_i (\nabla_{x_{Bi}} v_i)^T W_i (\nabla_{x_{Bi}} v_i)\right] \Delta x_{ij} = \sum_i (\nabla_{x_{Bi}} v_i)^T W_i (z_i - v_i(x_{Bi})) \tag{12.63}$$

The gradient of $v_i(x_{Bi})$ is calculated by

$$\nabla_{x_{Bi}} v_i = \nabla_{x_{Bi}} h_i + \nabla_{x_i} h_i \nabla_{x_{Bi}} \phi_i = \widetilde{H}_{i,Bi} - \widetilde{H}_{i,i} \widetilde{G}_{i,i-i}^{-1} \widetilde{G}_{i,i-Bi} \tag{12.64}$$

Substitute (12.64) to (12.63):

$$\begin{aligned}\sum_i (\nabla_{x_{Bi}} v_i)^T W_i (\nabla_{x_{Bi}} v_i) &= \sum_i (\widetilde{H}_{i,Bi}^T W_i \widetilde{H}_{i,Bi} - \widetilde{G}_{i,Bi-i} \widetilde{G}_{i,i-i}^{-1} \widetilde{H}_{i,i}^T W_i \widetilde{H}_{i,Bi} - \\ &\quad \widetilde{H}_{i,Bi}^T W_i \widetilde{H}_{i,i} \widetilde{G}_{i,i-i}^{-1} \widetilde{G}_{i,i-Bi} + \widetilde{G}_{i,Bi-i} \widetilde{G}_{i,i-i}^{-1} \widetilde{H}_{i,i}^T W_i \widetilde{H}_{i,i} \widetilde{G}_{i,i-i}^{-1} \widetilde{G}_{i,i-Bi}) \\ &= \sum_i \left(\widetilde{G}_{i,Bi-Bi} - \widetilde{G}_{i,Bi-i} \widetilde{G}_{i,i-i}^{-1} \widetilde{G}_{i,i-Bi}\right) = G_{B\Sigma}\end{aligned} \tag{12.65}$$

$$\begin{aligned}\sum_i (\nabla_{x_{Bi}} v_i)^T W_i (z_i - v_i(x_{Bi})) &= \sum_i \widetilde{H}_{i,Bi}^T W_i \widetilde{r}_i - \sum_i \widetilde{G}_{i,iB-i} \widetilde{G}_{i,i-i}^{-1} \widetilde{H}_{i,i}^T W_i \widetilde{r}_i \\ &= \sum_i \widetilde{H}_{i,Bi}^T W_i \widetilde{r}_i = -g_{B\Sigma}\end{aligned} \tag{12.66}$$

Equations (12.63)–(12.66) show that the Gauss–Newton update in (12.63) is equivalent to the coordinator's problem (12.48) of SFHSE. By exploiting sensitivity functions of local SEs, SFHSE is equivalent to the Gauss–Newton method solving the centralized state estimation problem (12.60) that takes x_{Bi} as state variables and $v_i(x_{Bi})$ as measurement functions.

Second, measurement function $v_i(x_{Bi})$ is proved to be Fréchet differentiable. With Assumptions 12.2 and 12.3, there is

$$\begin{aligned}
&\|v_i(x_{Bi} + \Delta x_{Bi}) - v_i(x_{Bi}) - \nabla_{x_{Bi}} v_i \Delta x_{Bi}\| \\
&= \|h_i(x_i + \Delta x_i, x_{Bi} + \Delta x_{Bi}) - h_i(x_i, x_{Bi}) - \nabla_{x_{Bi}} h_i \Delta x_{Bi} - \nabla_{x_i} h_i \nabla_{x_{Bi}} \phi_i \Delta x_{Bi}\| \\
&\leq \|h_i(x_i + \Delta x_i, x_{Bi} + \Delta x_{Bi}) - h_i(x_i, x_{Bi} + \Delta x_{Bi}) - \nabla_{x_i} h_i \Delta x_i\| \\
&+ \|h_i(x_i, x_{Bi} + \Delta x_{Bi}) - h_i(x_i, x_{Bi}) - \nabla_{x_{Bi}} h_i \Delta x_{Bi}\| + \|\nabla_{x_i} h_i\| \|\Delta x_i - \nabla_{x_{Bi}} \phi_i \Delta x_{Bi}\| \\
&\leq K_{hi1} \|\Delta x_i\|^2 + K_{hi1} \|\Delta x_{Bi}\|^2 + K_{\phi i1} \|\nabla_{x_i} h_i\| \|\Delta x_i\|^2 \\
&\leq \left(K_{hi1} K_{\phi i}^2 + K_{hi1} + K_{\phi i1} K_{hi} K_{\phi i}^2\right) \|\Delta x_{Bi}\|^2 \triangleq K_{vi} \|\Delta x_{Bi}\|^2
\end{aligned} \tag{12.67}$$

which shows that measurement functions $v_i(x_{Bi})$ defined in (12.59) are Fréchet differentiable.

As proved in Theorem 12.1, SFHSE is an optimal multi-area SE. Let $(\varepsilon, \varepsilon_i) \to 0$, there is

$$z_i - v_i(\tilde{x}_{Bi}) = z_i - h_i(\tilde{x}_i, \tilde{x}_{Bi}) = r(\tilde{x}_i, \tilde{x}_{Bi}) \tag{12.68}$$

In summary, SFHSE is equivalent to use Gauss–Newton iteration to solve another version of centralized state estimation in which x_{Bi} are state variables and $v_i(x_{Bi})$ are Fréchet differentiable measurement functions. The centralized Gauss–Newton iteration has a quadratic rate of convergence when $r(\tilde{x}_i, \tilde{x}_{Bi}) = 0$. At this time, SFHSE also has a quadratic rate of convergence with $z_i - v_i(\tilde{x}_{Bi}) = 0$. When the centralized Gauss–Newton has a superliner rate of convergence, SFHSE also has the same convergence rate with the same residual. □

Remark: Theorem 12.2 describes how fast the iteration error decays in a small neighborhood of the optimal state estimate. It may not reflect the behavior at the early stage of the iteration. Simulations in [10] showed that SFHSE converged rapidly during the entire process.

12.7.4 Computation and Communication Costs

The computational burden of SFHSE mainly includes the following:

1. The local SE in each area
2. The computation of the coefficients in the sensitivity function via (12.42) and (12.46)
3. The solution to the coordinator's problem (12.48)

The first term is a standard state estimation procedure. The computation of the sensitivity function is in fact the results of the Gaussian elimination in each area, which can be naturally obtained from the results of local state estimations. The coordination Eq. (12.48) is a linear equation with a relatively low dimension (due to the limited number of boundary buses). The computation cost of SFHSE is moderate.

SFHSE coordinates after the local state estimation converges, which requires a lower frequency of data exchange than those coordinate per iteration. Specifically, the information exchange includes the following:

1. Each area uploads the upper triangular part of the matrix $\hat{G}_{Bi}$ to the coordinator at the first iteration.
2. Each area uploads the vector $\hat{g}_{Bi}$ to the coordinator.
3. The coordinator sends the newest boundary state x_{Bi} to corresponding areas.
4. The data exchange for the BDI.

Terms 1 and 4 only proceed once per SFHSE computation (multiple SFHSE may be solved when there is bad data), and terms 2 and 3 only exchange vectors whose sizes are equal to the number of boundary state variables. None of them require excessive data exchange. Communicating matrix $\hat{G}_{Bi}$ in each SFHSE iteration will not add substantial communication burden due to its limited size.

In addition, Theorem 12.2 indicates that SFHSE converges rapidly, which is also beneficial in limiting its total computation and communication costs.

12.8 SIMULATIONS

12.8.1 Tests Setup

Paper [10] tested the performance of SFHSE on the IEEE 14-bus, 118-bus, and real power systems. The program was developed based on a C++ decoupled SE program. Networked computers were used to emulate an actual implementation [44, 45]. The convergence criterion was set as $\|\hat{g}_{B\Sigma}\|_2 < 0.01$ for SFHSE and $\|\hat{g}_{i,i}\|_2 \leq 0.001$ for local SEs. See [10] for more details in the configuration of simulations.

12.8.2 Tests on the IEEE 14-Bus System with Two Areas

The IEEE 14-bus system was first used to illustrate the scheme of SFHSE. The system was divided into two areas: Area 1 included buses 1–5, and the rest of buses belonged to area 2. Branches 4–7, 4–9, and 5–6 were tie-lines. No bad data presented in this test. The scheme of SFHSE was illustrated step by step for this test.

Step 1: Areas 1 and 2 performed initial SEs by iteratively solving (12.44) and selecting buses 1 and 14 as phase angle references. Both areas needed four iterations to converge from flat start. Their boundary state estimates

TABLE 12.2 Initialization for boundary state variables.

Bus indices		4	5	6	7	9
Area 1	V (p.u.)	1.0181	1.0195	1.0688	1.0601	1.0605
	θ (rad)	−0.1812	−0.1539	−0.2499	−0.2323	−0.2678
	θ rotated	0	0.0273	−0.0687	−0.0511	−0.0866
Area 2	V (p.u.)	1.0185	1.0226	1.0699	1.0626	1.0568
	θ (rad)	0.1019	0.127	0.0320	0.0468	0.0190
	θ rotated	0	0.0251	−0.0699	−0.0551	−0.0829
Initial value	V (p.u.)	1.0183	1.0211	1.0694	1.0614	1.0587
	θ (rad)	0	0.0262	−0.0693	−0.0531	−0.0848

were given in Table 12.2. Areas 1 and 2 sent these boundary states (rows 2, 3, 5, and 6), totally 20 floating point data, to the coordinator.

Step 2: The coordinator sets bus 4 as the global angle reference bus, and the former estimated phase angles were rotated to this reference (rows 4 and 7). Then boundary states were initialized by the mean values of estimated voltage magnitudes (rows 2 and 5) and rotated phase angles (rows 4 and 7), shown in the last two rows of Table 12.2.

Step 3: The coordinator sent initial boundary states to areas 1 and 2, totally 20 floating point data was sent in this step.

Step 4: Areas 1 and 2 performed local SEs by iteratively solving (12.39). Here local SEs started with the newest states, both areas only needed one iteration to converge. The previous local angle reference buses (1 and 14) were not used here.

Step 5: Areas 1 and 2 computed their $\hat{g}_{Bi}$ vectors and $\hat{G}_{Bi}$ matrices by (12.42) and (12.46) and then sent them to the coordinator. The two $\hat{G}_{Bi}$ (active and reactive power) matrices were in sizes of 5×5, and $\hat{g}_{Bi}$ vectors had lengths of five. Each area needed to send 40 floating point data to the coordinator in this step.

Step 6: The norm $\left\|\hat{g}_{B\Sigma}^{(0)}\right\|_2 = 0.0302 > 0.01$. Continue step 7.

Step 7: The coordinator solved Eq. (12.48). Two equations with sizes of five were solved.

Step 8: Increment the iteration time index t to one, turn to step 3, and start another iteration.

Steps 3–6 for a new iteration: In step 4, the local SEs in the two areas started from the newest states and only needed one iteration to converge. In step 5, each area only communicated its vector $\hat{g}_{Bi}$ to the coordinator. In step 6, $\left\|\hat{g}_{B\Sigma}^{(1)}\right\|_2 = 0.00257 < 0.01$, indicating SFHSE converged within single iteration. Forty floating point data was transferred in this iteration.

TABLE 12.3 Performance of SFHSE and the comparison with the centralized state estimator.

	CSE	SFHSE
Iteration times	—	1
Number of solutions of local normal equations	6	12
Exchanged floating point data	—	160
Computation time cost	0.079 s	0.053 s
Objective function values	170.6345	170.7636

The comparison between a CSE and SFHSE was given in Table 12.3.

1. Iteration times: In this test, SFHSE needed a single iteration to converge. It solved local normal equations for 12 times. Note that local SE iterations in SFHSE had smaller size than the CSE and can be solved in parallel.
2. Data exchange: SFHSE only needed to exchange 160 floating point data between the coordinator and two areas.
3. Computation time: The computation time of SFHSE was about 67% of that of CSE.
4. Objective function values: The objective function values for SFHSE was very close to the CSE. According to Theorem 12.1, the objective function of SFHSE will converge to its optimal value when its stopping tolerances converge to zero. In practice, however, proper stopping tolerances must be assigned for both SFHSE and CSE.

In this test, SFHSE demonstrated satisfactory performance in accuracy, rate of convergence, and communication and computation efficiencies.

12.8.3 Tests on the IEEE 118-Bus System with Three Areas

The IEEE 118-bus system was divided into three areas as shown in Figure 12.8. When no bad data exists, SFHSE needed only one iteration. Table 12.4 compared the performances of SFHSE and the CSE, along with the results from method [31].

The method of [31] updates the boundary state with inter-area Gauss–Newton iterations. Its objective function was close to that of the centralized estimator, and its computation time cost was less than that of the centralized estimator. However, its communication cost was heavier than that of SFHSE.

SFHSE needed one coordination iteration, and it solved local normal equations for 12 times. Its computation time was about 53% of CSE and 69% of the method in [31], which was the fastest among these methods. The objective function value of SFHSE was also very close to CSE.

To test the performance of distributed BDI, three internal measurements and three boundary measurements were set as bad data. Their residuals r_i, the values of Ω_{ii}, and the normalized residuals r_{Ni} were given in Table 12.5.

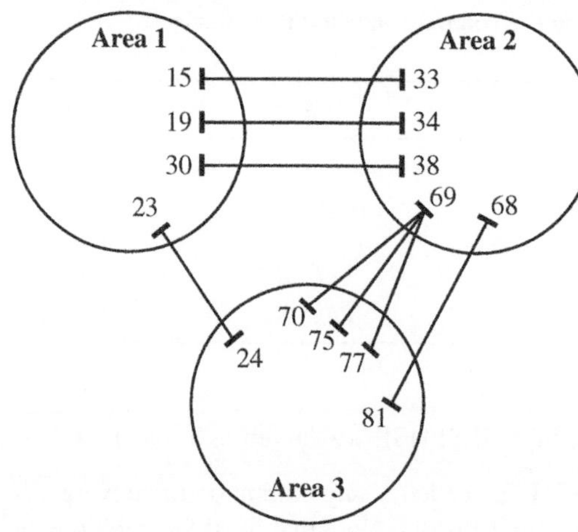

Figure 12.8 IEEE 118-bus three area system [46]. Source: Reproduced with permission of IEEE.

TABLE 12.4 Comparison between SFHSE, the CSE, and another MASE method for IEEE 118-bus system test.

Methods	CSE	SFHSE	Method in [31]
Iteration times	—	1	—
Number of solutions of local normal equations	5	12	5
Exchanged floating point data	—	580	5696
Computation time cost	0.211 s	0.112 s	0.163
Objective function values	1325.30	1326.99	1327.03

TABLE 12.5 Distributed bad data identification in the IEEE 118-bus system test.

	Centralized SE and BDI			SFHSE and distributed BDI		
Measure	r_i (p.u.)	Ω_{ii}	r_{Ni}	r_i (p.u.)	Ω_{ii}	r_{Ni}
P_{18}	−0.2373	3.46×10^{-4}	12.76	−0.2368	3.45×10^{-4}	12.75
Q_{34-36}	0.0916	3.75×10^{-4}	4.73	0.0922	3.75×10^{-4}	4.76
P_{61-62}	0.3897	3.84×10^{-4}	19.88	0.3894	3.83×10^{-4}	19.89
P_{15}	0.2455	3.26×10^{-4}	13.58	0.2450	3.25×10^{-4}	13.59
P_{23-24}	0.0845	3.68×10^{-4}	4.40	0.0849	3.68×10^{-4}	4.43
Q_{69-77}	0.2078	3.93×10^{-4}	10.47	0.2073	3.93×10^{-4}	10.46

Table 12.5 showed that no matter for internal or boundary bad data, SFHSE and hierarchical bad data identification could get the same results with the centralized approach.

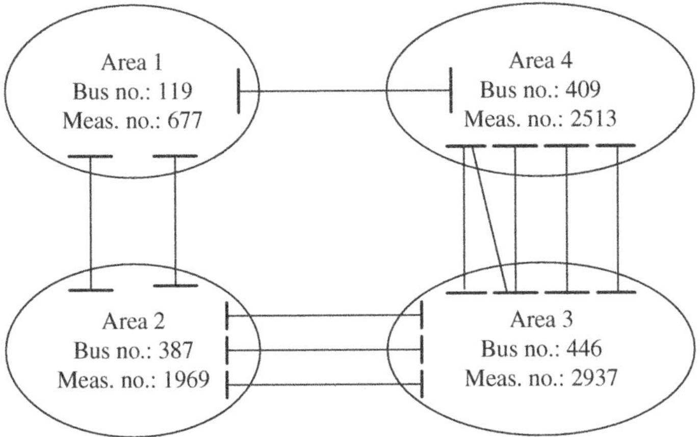

Figure 12.9 Four-area real power system.

TABLE 12.6 Comparison between SFHSE and the CSE for real system test.

	CSE	SFHSE
Data exchange amount	—	7878 float data
Computation time	1.012 s	0.452 s
Objective function values	4598.36	4599.70

12.8.4 Tests on Real Power System with Four Areas

This real power system has 1361 computation buses and 8096 measurements and was partitioned into four areas. Its schematic was shown in Figure 12.9.

For this real system test, BDI was necessary, and both SFHSE and CSE needed iterative computation. CSE needed six iterative SEs. SFHSE needed the same number of SEs, as the distributed BDI was equivalent to the centralized approach. During these six times of SFHSE iterations, the total iteration number was 9. Other performances of SFHSE and CSE were compared in Table 12.6.

In this real power system test, SFHSE also had outstanding performance. It needed few iterations to converge, and its entire computation time cost was 0.452 seconds, which was less than 45% of CSE. The total data exchange amount was 7878 floating point data, including the communication cost for the BDI. Such a communication cost is acceptable in practice. The objective function value for SFHSE is very close to CSE, which gave practical supports of the optimality claim.

12.9 CONCLUSIONS

The idea of multi-area state estimation was first proposed to reduce the computation burden of state estimations in large-scale power systems. Today, it has been endowed with new values. It is an essential tool in the future's smart grid, especially in the monitorization of interregional power flows and the modeling of distribution networks. In this chapter, we review relevant works in this field, including two-level single-iteration estimators, inter-area Gauss–Newton methods, and intra-area Gauss–Newton methods. In particular, we elaborate a recently published work where local system operators communicate their sensitivity functions to the coordinator. These sensitivity functions fully represent local optimal conditions, and consequently, this method has a theoretical commitment on its rate of convergence.

In the future, research on this topic would be increasingly motivated by several factors. First, real system operators have been attaching more and more importance to the problem of interregional coordination in which multi-area state estimation is a basic problem. Second, the deployment of phasor measurement units has hugely increased the amount of measurement data, and the idea of distributed data center has been developed. Third, at the distribution level, the rapid development of distributed energy resources and their growing impact to the bulk system also call for effective coordination between transmission and distribution systems. In summary, we expect the topic of multi-area state estimation to be extremely attractive in the future.

ACKNOWLEDGMENTS

The work of Ye Guo was supported in part of the National Science Foundation of China under Award 51977115. The work of Lang Tong was supported in part by the National Science Foundation under Awards 1932501 and 1816397.

REFERENCES

1. Regional Transmission Organizations (RTO)/Independent System Operators (ISO). https://www.ferc.gov/industries-data/electric/power-sales-and-markets/rtos-and-isos (accessed September 2020).
2. Liacco, T.D. (Apr. 1990). The role and implementation of state estimation in an energy management system. *International Journal of Electrical Power & Energy Systems* 12 (2): 75–79.
3. Control Center Requirements Manual. NYISO, June 2019. https://www.nyiso.com/documents/20142/2923231/M-21-CCRM-v4.0-final.pdf (accessed September 2020).
4. Giustina, D.D., Pau, M., Pegoraro, P.A. et al. (Dec.2014). Electrical distribution system state estimation: measurement issues and challenges. *IEEE Instrumentation and Measurement Magazine* 16 (6): 36–42.
5. Ipakchi, A. and Albuyeh, F. (Mar. 2009). Grid of the future. *IEEE Power and Energy Magazine* 7 (2): 52–62.
6. Gómez-Expósito, A., Abur, A., Jaén, A.d.l.V., and Gómez-Quiles, C. (Jun. 2011). A multilevel state estimation paradigm for smart grids. *Proceedings of the IEEE* 99 (6): 952–976.
7. Cutsem, T.V. and Ribbens-Pavella, M. (Oct. 1983). Critical survey of hierarchical methods for state estimation of electrical power systems. *IEEE Transactions on Power Apparatus and Systems* 102 (10): 3415–3424.

8. Gómez-Expósito, A., Jaén, A.d.l.V., Gómez-Quiles, C. et al. (Apr. 2011). A taxonomy of multi-area state estimation methods. *Electrical Power System Research* 81 (4): 1060–1069.
9. White, M. and Pike, R. (Jan. 2011). ISO New England and New York ISO interregional interchange scheduling: analysis and options, ISO white paper, https://www.iso-ne.com/static-assets/documents/pubs/whtpprs/iris_white_paper.pdf (accessed September 2020).
10. Guo, Y., Tong, L., Wu, W. et al. (Jan. 2017). Hierarchical multi-area state estimation via sensitivity function exchanges. *IEEE Transactions on Power Systems* 32 (1): 442–453.
11. Abur, A. and Gómez-Expósito, A. (2004). *Power System State Estimation: Theory and Implementation*. New York: Marcel Dekker.
12. Guo, Y., Wu, W., Zhang, B., and Sun, H. (Aug. 2013). An efficient state estimation algorithm considering zero injection constraints. *IEEE Transactions on Power Systems* 28 (3): 2651–2659.
13. Schaback, R. (1985). *Numerische Mathematik*, 281–309. New York: Springer-Verlag.
14. Magnago, F., Zhang, L., and Celik, M.K. (2016). Multiphase observability analysis in distribution systems state estimation. *2016 Power Systems Computation Conference (PSCC)* (June 2016). Genoa, Italy: IEEE, pp. 1–7.
15. Monticelli, A., Wu, F.F., and Yen, M. (Jul. 1986). Multiple bad data identification for state estimation by combinatorial optimization. *IEEE Transactions on Power Delivery* 1 (3): 361–369.
16. Zhang, B. and Lo, K. (Feb. 1991). A recursive measurement error estimation identification method for bad data analysis in power system state estimation. *IEEE Transactions on Power Systems* 6 (1): 191–198.
17. Mili, L., Cheniae, M., Vichare, N., and Rousseeuw, P. (May 1996). Robust state estimation based on projection statistics. *IEEE Transactions on Power Systems* 11 (2): 1118–1127.
18. Wu, W., Guo, Y., Zhang, B. et al. (Nov. 2011). Robust state estimation method based on maximum exponential square. *IET Generation, Transmission and Distribution* 5 (11): 1165–1172.
19. Schweppe, F.C. (Jan. 1970). Power system static-state estimation, part III: implementation. *IEEE Transactions on Power Apparatus and Systems* 89 (1): 130–135.
20. Clements, K., Denison, O., and Ringlee, R. (1972). A multi-area approach to state estimation in power system networks. *1972 IEEE PES Summer Meeting* (July 1972). San Francisco, CA: IEEE, pp. 1–5.
21. Cutsem, T.V., Horward, J., and Ribbens-Pavella, M. (Aug. 1981). A two-level static state estimator for electric power systems. *IEEE Transactions on Power Apparatus and Systems* 100 (8): 3722–3732.
22. Lin, S. (May 1992). A distributed state estimator for electric power systems. *IEEE Transactions on Power Systems* 7 (2): 551–557.
23. Falcao, D., Wu, F., and Murphy, L. (May 1995). Parallel and distributed state estimation. *IEEE Transactions on Power Systems* 10 (2): 724–730.
24. Habiballah, I. (Mar. 1996). Modified two-level state estimation approach. *IEE Proceedings. Generation, Transmission and Distribution* 143 (2): 193–199.
25. Ebrahimian, R. and Baldick, R. (Nov. 2000). State estimation distributed processing. *IEEE Transactions on Power Systems* 15 (4): 1240–1246.
26. Zhao, L. and Abur, A. (May 2005). Multiarea state estimation using synchronized phasor measurements. *IEEE Transactions on Power Systems* 20 (2): 611–617.
27. Sun, H. and Zhang, B. (May 2005). Global state estimation for whole transmission and distribution networks. *Electric Power Systems Research* 74 (2): 187–195.

28. Conejo, A., Torre, S.d.l., and Canas, M. (Feb. 2007). An optimization approach to multiarea state estimation. *IEEE Transactions on Power Systems* 22 (1): 213–221.
29. Jiang, W., Vittal, V., and Heydt, G.T. (Nov. 2008). Diakoptic state estimation using phasor measurement units. *IEEE Transactions on Power Systems* 23 (4): 1580–1589.
30. Gómez-Expósito, A. and Jaén, A.d.l.V. (May 2009). Two-level state estimation with local measurement pre-processing. *IEEE Transactions on Power Systems* 24 (2): 676–684.
31. Korres, G. (Feb. 2011). A distributed multiarea state estimation. *IEEE Transactions on Power Systems* 26 (1): 73–84.
32. Xie, L., Choi, D., Kar, S., and Poor, H. (Sep. 2012). Fully distributed state estimation for wide-area monitoring systems. *IEEE Transactions on Smart Grid* 3 (3): 1154–1169.
33. Kekatos, V. and Giannakis, G. (May 2013). Distributed robust power system state estimation. *IEEE Transactions on Power Systems* 28 (2): 1617–1626.
34. Marelli, D., Ninness, B., and Fu, M. (2015). Distributed weighted least-squares estimation for power networks. *IFAC-PapersOnLine* 48 (28): 562–567.
35. Minot, A., Lu, Y.M., and Li, N. (Sep. 2016). A distributed Gauss–Newton method for power system state estimation. *IEEE Transactions on Power Systems* 31 (5): 3804–3815.
36. Zheng, W., Wu, W., Gómez-Expósito, A. et al. (Jan. 2017). Distributed robust bilinear state estimation for power systems with nonlinear measurements. *IEEE Transactions on Power Systems* 32 (1): 499–509.
37. Kobayashi, H., Narita, S., and Hamman, M. (1974). Model coordination method applied to power system control and estimation problems. In: *4th IFAC/IFIP International Conference on Digital Computer Applications to Process Control, Zürich, Switzerland*. Springer, Berlin, Heidelberg, pp. 114–128.
38. Seidu, K. and Mukai, H. (May 1985). Parallel multi-area state estimation. *IEEE Transactions on Power Apparatus and Systems* 104 (5): 1025–1034.
39. Korres, G.N., Tzavellas, A., and Galinas, E. (Sep. 2013). A distributed implementation of multi-area power system state estimation on a cluster of computers. *Electric Power Systems Research* 102: 20–32.
40. Wei, E., Ozdaglar, A., and Jadbabaie, A. (Sep. 2013). A distributed newton method for network utility maximization–I: algorithm. *IEEE Transactions on Automatic Control* 58 (9): 2162–2175.
41. Wei, E., Ozdaglar, A., and Jadbabaie, A. (Sep. 2013). A distributed newton method for network utility maximization–part II: convergence. *IEEE Transactions on Automatic Control* 58 (9): 2176–2188.
42. Pistikopoulos, E., Georgiadis, M., and Dua, V. (2007). *Multi-Parametric Programming: Theory, Algorithms and Applications*. Weinheim: Wiley-VCH.
43. Boyd, S. and Vandenberghe, L. (2004). *Convex Optimization*. Cambridge: Cambridge University Press.
44. Chen, S. and Chen, J. (May 2000). Fast load flow using multiprocessors. *International Journal of Electrical Power & Energy Systems* 22 (4): 231–236.
45. Tomim, M., Martí, J., and Wang, L. (Oct. 2009). Parallel solution of large power system networks using the Multi-Area Thévenin Equivalents (MATE) algorithm. *International Journal of Electrical Power & Energy Systems* 31 (9): 497–503.
46. Min, L. and Abur, A. (Aug. 2006). Total transfer capability computation for multiarea power systems. *IEEE Transactions on Power Systems* 21 (3): 1141–1147.

CHAPTER 13

PARALLEL DOMAIN-DECOMPOSITION-BASED DISTRIBUTED STATE ESTIMATION FOR LARGE-SCALE POWER SYSTEMS

Hadis Karimipour[1] *and Venkata Dinavahi*[2]

[1]*School of Engineering, University of Guelph, Canada*

[2]*Department of Electrical and Computer Engineering, University of Alberta, Canada*

13.1 INTRODUCTION

Continuous growth in electricity demand and complexity of the power systems brings up new challenges in online monitoring and state estimation of large-scale power system. Therefore, power engineers are always exploring methods for fast and efficient solutions for these problems. In addition, the evolution of power systems toward the new smart grid era is increasing the size and complexity of the power grids.

Compared with the traditional state estimation, state estimation of smart grids has a lot of key differences. One of the main differences is that the power grid can be distributed in a variety of environments. The traditional centralized state estimation is not scalable enough to process the huge amount of data generated all over the grid. In addition, the traditional state estimation updates much slower than the measurement cycle, so it is not fast enough to predict the real-time behavior of power grids and respond to emergencies [1], such as the 2003 US–Canada blackout [2].

Advances in Electric Power and Energy: Static State Estimation, First Edition.
Edited by Mohamed E. El-Hawary.

Published 2021 by John Wiley & Sons, Inc.

State estimation is a fundamental problem in monitoring and control of large-scale power systems. A state estimator provides accurate estimate of the state variables, which usually includes static states like bus voltages and phase angles or dynamic states such as generator rotor angles and speed. There are several key functions in power networks that are developed based on state estimation such as contingency analysis, optimal power flow, and economic dispatch [3]. Therefore a fast and accurate state estimation is invaluable for secure and economical operation of complex power systems.

The block diagram on Figure 13.1 shows the components of a state estimator.

Remote terminal unit (RTU) updates the system regarding the current status of the power system and encodes measurement transducer outputs into digital signals. A central master station, located at the control center, gathers information through the supervisory control and data acquisition (SCADA) system. SCADA systems are designed to collect field information, transfer it to a central computer facility, and display the information to the operator graphically or textually, thereby allowing the operator to monitor or control an entire system from a central location in real time [4]. Typical measurements include power flows (both active and reactive), power injections, voltage magnitude, phase angles, and current magnitude.

In addition, direct measurement of voltage phase angle is available through phasor measurement units (PMUs). A PMU provides measurements by sampling the AC voltage and current waveforms collected at secondary of instrument transformers (CT and PT) while synchronizing with a global positioning system (GPS) clock with the accuracy of one second. The voltage and current phasors are then transmitted to the SCADA server via phasor data concentrators (PDCs). Compared with SCADA measurements, which are usually updated every three to five seconds, PMU measurements are more accurate and can deliver up to 50 measurements per second [5].

The network topology processor determines the topology of the network from the telemetered status of circuit breakers. The updated electrical model of the power transmission system is sent to the state estimator program together with the analog measurements. The state estimator processes all data before being used by other programs, except the analog measurements of generator outputs, which are used directly, by the automatic generation control (AGC) program.

The output of the state estimator consists of all bus voltage magnitudes, phase angles, power injections, and power flows. The bad data is also identified, detected, and, if possible, eliminated by the estimator. The output data together with the electrical model, developed by the network topology program, provides the basis for the economic dispatch program and contingency analysis program.

There are two main types of state estimation: static state estimation (SSE) and dynamic state estimation (DSE). SSE relies on the fact that under normal operation condition, the power system is considered as a quasi-static system [6]. It mainly tries to estimate static parameters of the power system, such as voltage magnitude and phase angle, using the measurers provided by PMUs and SCADA system. Unlike the former method, DSE employs the previous states to predict the state one step ahead of the time and then correct it using current measurements [7]. The term DSE also implies to the estimation of dynamic states of the synchronous generator.

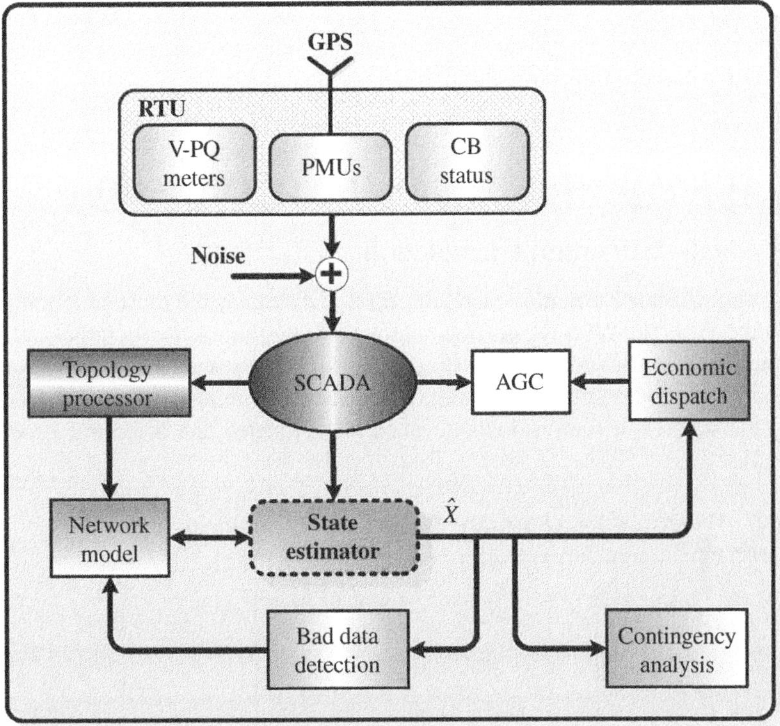

Figure 13.1 State estimation process block diagram.

Conventional methods to improve the cycle time for state estimation mainly include the following strategies:

- Complexity reduction
- Hierarchical state estimation
- Distributed state estimation
- Parallel state estimation

Although above approaches tried to improve the process of state estimation, they all have their own drawbacks. Performance degradation may occur in complexity reduction approaches due to the constraints in number of measurements or neglecting the model details. The main issue with hierarchical methods is the communication overhead between subsystems and the delay caused by coordination stage. In addition, in both distributed and hierarchical methods, the observability of the subsystems is the underlying assumption that may not be feasible in practice. Both distributed and parallel approaches accelerate the computational process; however, they randomly decompose the original system neglecting interaction of subsystems, which may lead to an inaccurate estimation.

Overall, there is an increasing demand for fast and real-time simulations that can be incorporated within the energy management system to determine the critical system limits based on the current conditions of the system.

13.2 FUNDAMENTAL THEORY AND FORMULATION

13.2.1 State Estimation Formulation

Among various state estimation methods, WLS algorithm is the most commonly used method [6]. The WLS is fundamental for other algorithms but for large-scale state estimation can be prohibitively slow. The weighted least squares method is a commonly used method for state estimation that tries to minimize the weighted sum of the squares of the residuals between the estimated and actual measurements [3].

13.2.1.1 Weighted Least Square Static State Estimation

Consider the measurement set vector m as follows:

$$m = \begin{bmatrix} m_1 \\ m_2 \\ \vdots \\ m_{2(1+n)+1} \end{bmatrix} = \begin{bmatrix} h_1(x_1, \ldots, x_n) \\ h_2(x_1, \ldots, x_n) \\ \vdots \\ h_{2(1+n)+1}(x_1, \ldots, x_n) \end{bmatrix} + \begin{bmatrix} \varepsilon_1 \\ \varepsilon_2 \\ \vdots \\ \varepsilon_{2(1+n)+1} \end{bmatrix} \tag{13.1}$$
$$= h(x) + \varepsilon$$

where m, $h(x)$, and ε are the vectors of measurements, nonlinear measurement functions, and measurement errors, respectively. For a system with n buses and l lines, there are $2l + 2n + 1$ elements in each vector: $2l$ power flows, $2n$ power injections, and slack bus measurements. x is a vector of system states comprising of voltage magnitudes and phase angles. Since the phase angle in slack bus is considered 0, there are $2n - 1$ states to be estimated.

For simplicity, it is assumed that

- $\text{Ex}[\varepsilon_i] = 0 \quad i = 1, 2, \ldots, 2(l+n)+1$
- $\text{Ex}[\varepsilon_i \varepsilon_j] = 0$

Therefore $\text{Cov}[\varepsilon] = R = \text{Ex}[\varepsilon\varepsilon^T] = \text{diag}\left(\sigma_1^2, \ldots, \sigma_{2(1+n)+1}^2\right)$, where σ_i is the standard deviation of measurement i.

Substituting the first-order Taylor's expansion of $h(x)$ around x_0 in (13.1), we obtain:

$$m = H\Delta(x) + \varepsilon \tag{13.2}$$

where $\Delta(x) = x - x_0$ is the $(2n - 1) \times 1$ state mismatch vector and H is the $(2l + 2n + 1) \times (2n - 1)$ Jacobian matrix defined as follows:

$$H(x) = \begin{bmatrix} \frac{\partial h_1(x)}{\partial x_1} & \cdots & \frac{\partial h_1(x)}{\partial x_n} \\ \vdots & \ddots & \vdots \\ \frac{\partial h_n(x)}{\partial x_1} & \cdots & \frac{\partial h_n(x)}{\partial x_n} \end{bmatrix} \tag{13.3}$$

The objective function $J(x)$ to be minimized by the WLS formulation can be expressed as follows:

$$J(x) = \sum_{k=1}^{2l+2n+1} (m_k - h_k(x))^2 R_{kk}^{-1} = [m - h(x)]^T R^{-1} [m - h(x)] \tag{13.4}$$

where R is the $(2l + 2n + 1) \times (2l + 2n + 1)$ covariance matrix. Index k refers to the kth measurement. The following equation satisfies the first-order optimality condition at the minimum of $J(x)$:

$$g(x) = \frac{\partial J(x)}{\partial x} = H^T R^{-1} [m - h(x)] = 0 \tag{13.5}$$

where $g(x)$ is the $(2n - 1) \times 1$ matrix of gradient of the objective function. Substituting the first-order Taylor's expansion of $g(x)$ in (13.4), the following equation is solved iteratively to find the solution that minimizes $J(x)$:

$$G(x)\Delta(x) = H^T R^{-1} [m - h(x)] \tag{13.6}$$

where $G(x) = \frac{\partial g(x)}{\partial x}$ is $(2n - 1) \times (2n - 1)$ gain matrix. The WLS state estimation algorithm given by (13.1)–(13.5) can be solved iteratively until convergence of $\Delta(x)$. Figure 13.2 shows the block diagram of state estimation process.

13.2.1.2 Extended Kalman Filter-Based Dynamic State Estimation

Using the present and previous states of the network, DSE predicts the state vector one step ahead of the time. The generic power system for DSE can be described by

$$x_{k+1} = f(x_k) + w_k \tag{13.7}$$

$$m_{k+1} = h(x_{k+1}) + \varepsilon_{k+1}, \quad \varepsilon_k \sim N(0, R_k) \tag{13.8}$$

where x is a vector of system states comprising of voltage magnitudes and phase angles at all buses except the slack bus where $V_1 = 1\angle 0°$ p.u. $f(x)$, m, and $h(x)$ are vectors of nonlinear system transition function, unified measurements, and nonlinear measurement functions, respectively. ε and w are measurements and system noises assuming normal distribution with zero mean, and covariance R. Equation (13.7) can be linearized as follows if the time frame is small enough:

$$x_{k+1} = F_k x_k + a_k + \omega_k, \quad \omega_k \sim N(0, Q_k) \tag{13.9}$$

where F_k represents the $(2n - 1) \times (2n - 1)$ state transition matrix between two time frames, a_k is the vector of associated behavior of the state trajectory, and ω_k is the Gaussian noise vector with zero mean and covariance matrix Q_k. Generally, EKF is

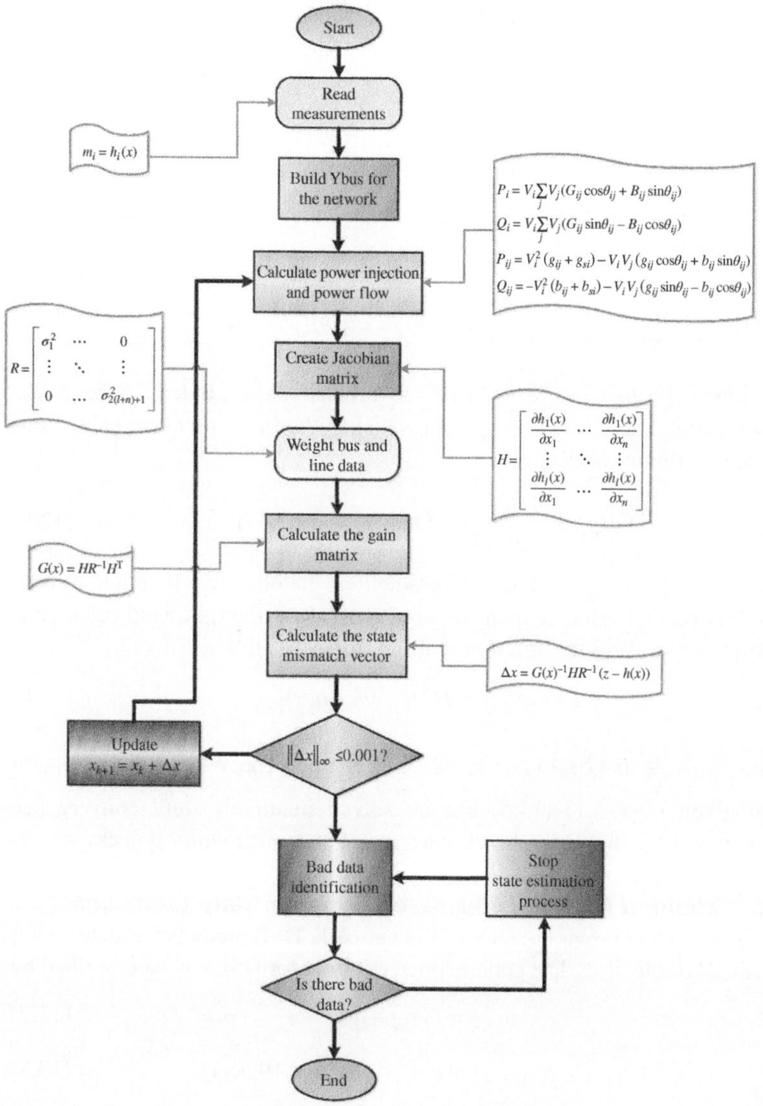

Figure 13.2 State estimation flowchart.

composed of three major steps: identification, prediction, and filtering, which are explained in details as follows:

13.2.1.2.1 Parameter Identification To evaluate the dynamic model, unknown parameters need to be calculated online. Holt's exponential smoothing technique [8] was used for identification of F_k and a_k. Based on this method F_k and a_k can be described as follows if $\tilde{x}$ and $\hat{x}$ represent the predicted and estimated value of the states, respectively:

$$\begin{aligned}
F_k &= \alpha(1+\beta)I_{\text{idn}}, \quad 0<\alpha,\beta<1 \\
a_k &= (1-\alpha)(1+\beta)\tilde{x}_k - \beta\gamma_{k-1} + (1+\beta)\xi_{k-1} \\
\gamma_k &= \alpha\hat{x}_k + (1+\alpha)\tilde{x}_k \\
\xi_k &= \beta(\gamma_k - \gamma_{k-1}) + (1-\beta)\xi_{k-1}
\end{aligned} \tag{13.10}$$

where α and β are smoothing parameters. Under normal operation conditions it is possible to adjust F_k and a_k such that Q_k remains constant. I_{idn} is the $(2n-1)\times(2n-1)$ identity matrix. However, considering the dynamic behavior of the network, neglecting the changes in Q_k may result in inaccurate prediction. Online estimation of Q_k can be formulated as [9]

$$\hat{Q}_{k+1} = Q_k\sqrt{\frac{\text{trace}\left\{H_{k+1}\left(F_k\rho_k F_k^T + \hat{Q}_k\right)H_{k+1}^T\right\}}{\text{trace}\left\{H_{k+1}\left(F_k\rho_k F_k^T + Q_k\right)H_{k+1}^T\right\}}} \tag{13.11}$$

13.2.1.2.2 State Prediction Using the measurement and estimated states at the time instant k, the predicted value $\tilde{x}_{k+1}$ can be formulated as

$$\begin{aligned}
\tilde{x}_{k+1} &= F_k\hat{x}_k + a_k, \quad (x_k - \hat{x}_k) \sim N(0,\rho_k) \\
\tilde{\rho}_{k+1} &= F_k\rho_k F_k^T + \hat{Q}_k, \quad (x_k - \tilde{x}_k) \sim N(0,\tilde{\rho}_k)
\end{aligned} \tag{13.12}$$

where ρ and $\tilde{\rho}$ are $(2n-1)\times(2n-1)$ error covariance matrices for estimated and predicted values, respectively. The objective function $J(x)$ was chosen to minimize both estimation and prediction errors:

$$J(x) = \min_x \, [m-h(x)]^T R^{-1}[m-h(x)] + [x_k - \tilde{x}_k]^T \tilde{\rho}^{-1}[x_k - \tilde{x}_k] \tag{13.13}$$

The following equation satisfies the first-order optimality condition at the minimum of $J(x)$:

$$g(x) = H^T R^{-1}[m-h(x)] - \tilde{\rho}^{-1}[x_k - \tilde{x}_k] = 0 \tag{13.14}$$

where $g(x)$ is the $(2n-1)\times 1$ vector of gradient of the objective function and $H = \dfrac{\partial h(x)}{\partial x}$ is the $(2m+2n+1)\times(2n-1)$ Jacobian matrix. Using Taylor's expansion of $h(x)$ around $\tilde{x}_0$, (13.14) can be expressed as follows:

$$\begin{aligned}
G(x)\Delta(x) &= H^T(\tilde{x})R^{-1}[m-h(\tilde{x})] \\
G(x) &= H^T(\tilde{x})R^{-1}H(\tilde{x}) + \tilde{\rho}^{-1}
\end{aligned} \tag{13.15}$$

where $\Delta(x) = \hat{x} - \tilde{x}_0$, is the $(2n-1)\times 1$ state mismatch vector and $G(x) = \dfrac{\partial g(x)}{\partial x}$ is the $(2n-1)\times(2n-1)$ gain matrix. The state estimation algorithm given by (13.13)–(13.15) can be solved iteratively until convergence of $\Delta(x)$ to a specified threshold.

13.2.1.2.3 State Filtering This step updates the predicted values using the next set of measurements at the time instant $k + 1$. The updated state through EKF can be written as

$$\begin{aligned}\hat{x}_{k+1} &= \tilde{x}_{k+1} + K_{k+1}(m_{k+1} - h(\tilde{x}_{k+1}))\\ K_{k+1} &= \tilde{\rho}_{k+1}H_{k+1}^T\left[H_{k+1}\tilde{\rho}_{k+1}H_{k+1}^T + R\right]^{-1}\\ \rho_{k+1} &= \tilde{\rho}_{k+1} - K_{k+1}H_{k+1}\tilde{\rho}_{k+1}\end{aligned} \tag{13.16}$$

where K is the $(2n-1)\times(2n-1)$ Kalman gain matrix. For the same reasons mentioned earlier, this step is also a good candidate for parallelization.

13.2.2 Measurement and Component Modeling

An individual transmission line is typically modeled as a single-phase π circuit equivalent. An equivalent π model of a two-bus (i, j) system is shown in Figure 13.3. The measurements in an AC system are mainly of three types, bus power injection, line power flows, and bus voltage magnitudes.

These quantities can be expressed using the state variables.

The real and reactive power injection at a bus can be expressed as

$$\begin{aligned}P_i &= V_i\sum_{j=1}^{2l+2n+1} V_j\left(G_{ij}\cos\theta_{ij} + B_{ij}\sin\theta_{ij}\right)\\ Q_i &= V_i\sum_{j=1}^{2l+2n+1} V_j\left(G_{ij}\sin\theta_{ij} - B_{ij}\cos\theta_{ij}\right)\end{aligned} \tag{13.17}$$

where P_i and Q_i are the real and reactive bus power injection at bus i, respectively. V_i and θ_i are the voltage magnitude and phase angle at bus i and $\theta_{ij} = \theta_i - \theta_j$.

$(G_{ij} + jB_{ij})$ is the ijth element of the complex bus admittance matrix.

The real and reactive power flows from bus i to bus j are expressed as follows:

$$\begin{aligned}P_{ij} &= V_i^2\left(g_{si} + g_{ij}\right) - V_iV_j\left(b_{ij}\sin\theta_{ij} + g_{ij}\cos\theta_{ij}\right)\\ Q_{ij} &= V_i^2\left(b_{si} + b_{ij}\right) - V_iV_j\left(g_{ij}\sin\theta_{ij} - b_{ij}\cos\theta_{ij}\right)\end{aligned} \tag{13.18}$$

where P_{ij} and Q_{ij} are the real and reactive bus power flows from bus i to bus j, respectively.

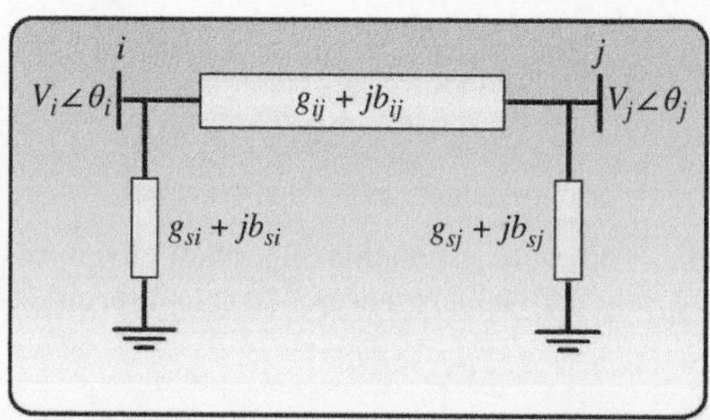

Figure 13.3 Standard transmission line π model.

$(g_{ij} + jb_{ij})$ is the admittance of the series branch connecting buses i and j.
$(g_{si} + jb_{si})$ is the admittance of the shunt branch connected at bus i.
Jacobian of the real power injection measurement can be calculated as

$$\begin{aligned}
\frac{\partial P_i}{\partial \theta_i} &= \sum_{j=1}^{n} V_i V_j \left(-G_{ij}\sin\theta_{ij} + B_{ij}\cos\theta_{ij}\right) - V_i^2 B_{ii} \\
\frac{\partial P_i}{\partial \theta_j} &= V_i V_j \left(G_{ij}\sin\theta_{ij} - B_{ij}\cos\theta_{ij}\right) \\
\frac{\partial P_i}{\partial V_i} &= \sum_{j=1}^{n} V_j \left(G_{ij}\cos\theta_{ij} + B_{ij}\sin\theta_{ij}\right) - V_i^2 G_{ii} \\
\frac{\partial P_i}{\partial V_j} &= V_i \left(G_{ij}\cos\theta_{ij} + B_{ij}\sin\theta_{ij}\right)
\end{aligned} \tag{13.19}$$

Jacobian of the reactive power injection measurement is given in following equations:

$$\begin{aligned}
\frac{\partial Q_i}{\partial \theta_i} &= \sum_{j=1}^{n} V_i V_j \left(G_{ij}\cos\theta_{ij} + B_{ij}\sin\theta_{ij}\right) - V_i^2 G_{ii} \\
\frac{\partial Q_i}{\partial \theta_j} &= V_i V_j \left(-G_{ij}\cos\theta_{ij} - B_{ij}\sin\theta_{ij}\right) \\
\frac{\partial Q_i}{\partial V_i} &= \sum_{j=1}^{n} V_j \left(G_{ij}\sin\theta_{ij} - B_{ij}\cos\theta_{ij}\right) - V_i^2 B_{ii} \\
\frac{\partial Q_i}{\partial V_j} &= V_i \left(G_{ij}\sin\theta_{ij} - B_{ij}\cos\theta_{ij}\right)
\end{aligned} \tag{13.20}$$

Jacobian of real line power flow measurement can be derived as follows:

$$\begin{aligned}
\frac{\partial P_{ij}}{\partial \theta_i} &= V_i V_j \left(g_{ij}\sin\theta_{ij} - b_{ij}\cos\theta_{ij}\right) \\
\frac{\partial P_{ij}}{\partial \theta_j} &= -V_i V_j \left(g_{ij}\sin\theta_{ij} - b_{ij}\cos\theta_{ij}\right) \\
\frac{\partial P_{ij}}{\partial V_i} &= -V_i V_j \left(g_{ij}\cos\theta_{ij} - b_{ij}\sin\theta_{ij}\right) - 2V_i \left(g_{ij} + g_{si}\right) \\
\frac{\partial P_{ij}}{\partial V_j} &= -V_i V_j \left(g_{ij}\cos\theta_{ij} + b_{ij}\sin\theta_{ij}\right)
\end{aligned} \tag{13.21}$$

Jacobian of reactive line power flow measurement can be derived as follows:

$$\begin{aligned}
\frac{\partial Q_{ij}}{\partial \theta_i} &= -V_i V_j \left(g_{ij}\cos\theta_{ij} + b_{ij}\sin\theta_{ij}\right) \\
\frac{\partial Q_{ij}}{\partial \theta_j} &= V_i V_j \left(g_{ij}\cos\theta_{ij} + b_{ij}\sin\theta_{ij}\right) \\
\frac{\partial Q_{ij}}{\partial V_i} &= -V_i \left(g_{ij}\sin\theta_{ij} - b_{ij}\cos\theta_{ij}\right) - 2V_i \left(b_{ij} + b_{si}\right) \\
\frac{\partial Q_{ij}}{\partial V_j} &= -V_i \left(g_{ij}\sin\theta_{ij} - b_{ij}\cos\theta_{ij}\right)
\end{aligned} \tag{13.22}$$

To construct the Jacobian matrix (H), the partial derivative of line flows and bus power with respect to θ_i, θ_j, V_i, and V_j should be computed using Eqs. (13.19)–(13.22):

$$H(x) = \left[\begin{array}{ccc|ccc}
\frac{\partial V_1}{\partial V_1} & \cdots & \frac{\partial V_1}{\partial V_n} & \frac{\partial V_1}{\partial \theta_2} & \cdots & \frac{\partial V_1}{\partial \theta_n} \\
\vdots & \ddots & \vdots & \vdots & \ddots & \vdots \\
\frac{\partial V_n}{\partial V_1} & \cdots & \frac{\partial V_n}{\partial V_n} & \frac{\partial V_n}{\partial \theta_2} & \cdots & \frac{\partial V_n}{\partial \theta_n} \\
 & \vdots & & & \vdots & \\
\cdots & \frac{\partial P_{ij}}{\partial V_i}, \frac{\partial P_{ij}}{\partial V_j} & \cdots & \cdots & \frac{\partial P_{ij}}{\partial \theta_i}, \frac{\partial P_{ij}}{\partial \theta_j} & \cdots \\
 & \vdots & & & \vdots & \\
 & \vdots & & & \vdots & \\
\cdots & \frac{\partial Q_{ij}}{\partial V_i}, \frac{\partial Q_{ij}}{\partial V_j} & \cdots & \cdots & \frac{\partial Q_{ij}}{\partial \theta_i}, \frac{\partial Q_{ij}}{\partial \theta_j} & \cdots \\
 & \vdots & & & \vdots & \\
\frac{\partial P_1}{\partial V_1} & \cdots & \frac{\partial P_1}{\partial V_n} & \frac{\partial P_1}{\partial \theta_1} & \cdots & \frac{\partial P_1}{\partial \theta_n} \\
\vdots & \ddots & \vdots & \vdots & \ddots & \vdots \\
\frac{\partial P_n}{\partial V_1} & \cdots & \frac{\partial P_n}{\partial V_n} & \frac{\partial P_n}{\partial \theta_1} & \cdots & \frac{\partial P_n}{\partial \theta_n} \\
\frac{\partial Q_1}{\partial V_1} & \cdots & \frac{\partial Q_1}{\partial V_n} & \frac{\partial Q_1}{\partial \theta_1} & \cdots & \frac{\partial Q_1}{\partial \theta_n} \\
\vdots & \ddots & \vdots & \vdots & \ddots & \vdots \\
\frac{\partial Q_n}{\partial V_1} & \cdots & \frac{\partial Q_n}{\partial V_n} & \frac{\partial P_n}{\partial \theta_1} & \cdots & \frac{\partial Q_n}{\partial \theta_n}
\end{array}\right] \tag{13.23}$$

13.2.3 Parallel Processing

Parallel processing is an information processing technique in which two or more processors work simultaneously on the solution of a problem. Parallel processing techniques attracted considerable research interest in different types of power

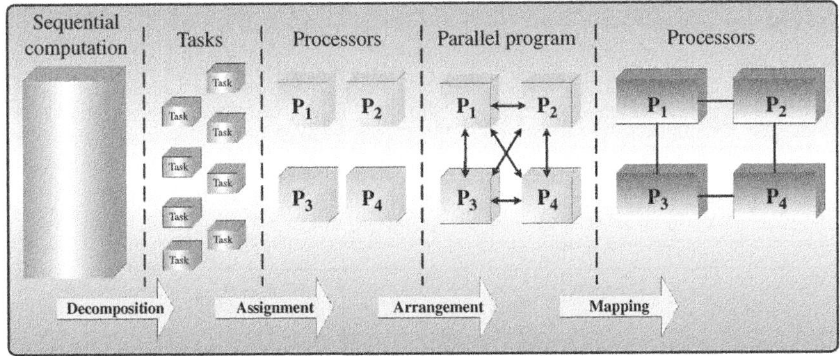

Figure 13.4 Steps of parallel algorithm generation.

system computations such as state estimation [10–15], power flow analysis [16–18], transient stability [19, 20], and electromagnetic simulations [21–23].

Parallel processing is a type of computation where many calculations are performed simultaneously. This method of computation is based on the fact that large problems can be divided into smaller pieces and then solved concurrently. In parallel processing the single CPU is replaced by multiple CPUs whose overall parallel performance accelerates the simulation.

Figure 13.4 shows the four main steps in creating a parallel program that includes (i) decomposition of computation in tasks; (ii) assigning tasks to processors; (iii) arrangement of data access, communication, and synchronization; and (iv) mapping processes to processors.

The application of parallel processing in power system analysis is motivated by the desire for faster computation and not because of the structure of problems. Except for those analytical procedures that require repeated solutions, like contingency analysis, there is no obvious parallelism inherent in the mathematical structure of power system problems. Thus, for a particular problem a parallel (or near-parallel) formulation needs to be found that is amenable to formulation as a parallel algorithm.

13.2.3.1 CPU and GPU Architecture

There exist two major processor architectures: the multicore central processing unit (CPU) and the many-core graphics processing unit (GPU). The CPU is composed of only a few cores that can handle a few software threads at a time. In contrast, the GPU that is an energy-efficient processor on the market is composed of hundreds of cores known as stream processors (SP) that can simultaneously handle thousands of threads [24].

The modern GPU consists of multiprocessors that map the data elements to the parallel processing threads. A multicore CPU has six to seven times larger cache than the GPU's cache system. On the other hand, the GPU has many more cores than the CPU. Owing to the fact that more transistors are devoted to data processing than caching and flow control, the GPU has significantly larger

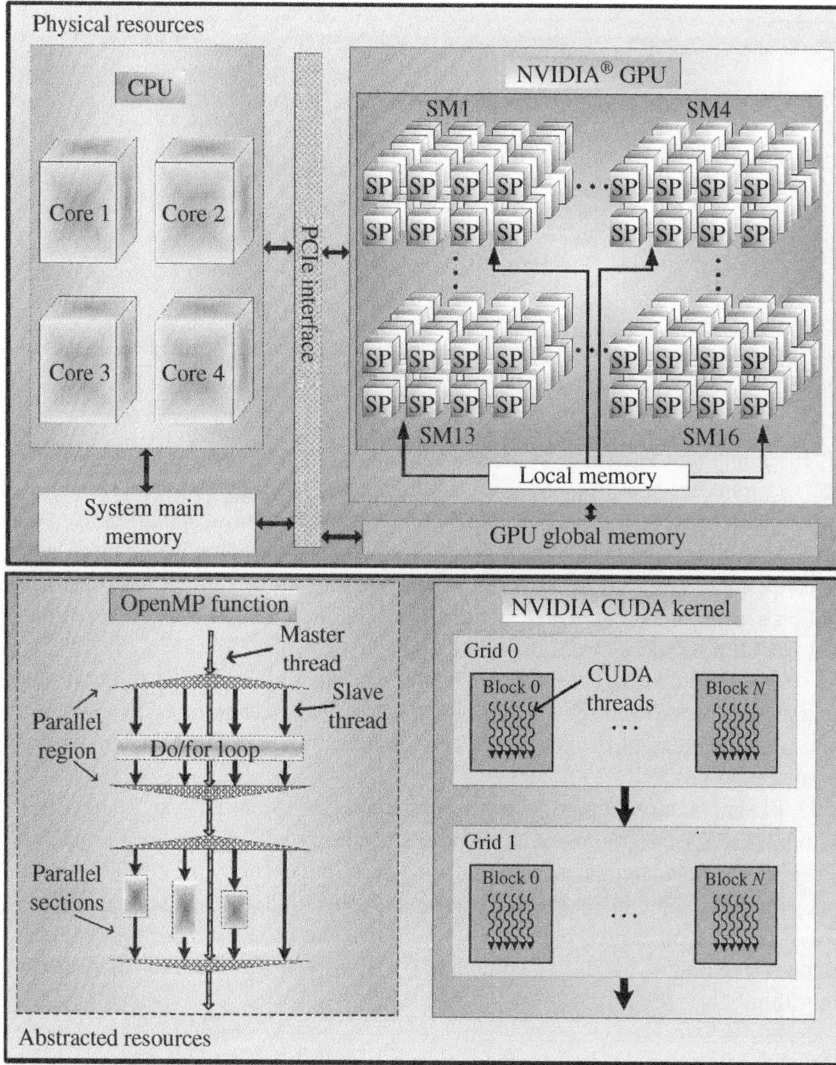

Figure 13.5 CPU, GPU, CUDATM, and OpenMP resources.

computational power compared with a multicore CPU [25]. In order to highlight the differences between CPU and GPU, Figure 13.5 is provided, which shows the physical and abstracted resources in a CPU and a GPU.

13.2.3.2 OpenMP

OpenMP is a standard application programming interface (API) for multicore CPUs, which does not require major code reformation for parallelization [26]. It supports shared memory; however, it is limited to a couple hundred cores due to thread management overheads and cache coherence hardware requirements.

OpenMP also includes directives, library routines, and environment variables to facilitate scheduling and parallelism at runtime with high level of portability. It provides support for sharing and privatizing data using extension of the C and C++ languages with tasking constructs, device constructs, work sharing constructs, and synchronization constructs. The program begins as a single process called a master thread, which executes sequentially. The master thread creates a group of slave threads within the parallel construct (Figure 13.5). At the end of the construct, only the master thread remains, while the rest of the slave threads synchronize and terminate.

13.2.3.2.1 Simple Example An example is provided here to illustrate how a simple loop may be parallelized in a shared memory programming model using OpenMP. Suppose we have a function that takes two $N \times 1$ vectors A and B and adds them up in a third vector C. On the CPU, a *for* loop can be used over all array elements as follows:

```
for (i = 0 : N)
C[i] = A[i] + B[i]
i = i + 1
end
```

There are two levels of active computation: outside the loop and inside the loop. Outside the loop, the loop counter i is increasing and compared with the length of vectors N, while inside the loop, the actual computation is performed on arrays at a fixed position determined by the loop counter. The calculations performed on each data element in the vectors are independent of each other.

The OpenMP version of the same function can be written as

```
#pragma omp parallel num_thread(n)
#pragma omp parallel for private (i), shared (--)
#pragma omp parallel reduction (-- : --)

for (i = 1 : i < N, i ++)
C[i] = A[i] + B[i]
end
```

Using parallel construct a team of n threads that are defined by *num_thread(n)* is formed to start parallel execution of the corresponding code. *num_thread(n)* is an internal control variable that defined the maximum number of threads in the team executing the parallel region. Appropriate iterations are assigned to the individual threads by compiler.

The data environment consists of the following:

- Private variable – which determines private variables in the thread executing region.
- Shared variable – which determines variables that are shared among the team of threads executing the parallel region.

- Reduction variable – which identifies which shared value and operation will be used.

This model of execution is referred to master/slave model where multiple threads of execution perform tasks defined implicitly or explicitly by OpenMP directives. Given a sequential program, the master/slave makes it easy to get loop level parallelism in an incremental fashion that takes advantages of a multiprocessor system.

13.2.3.3 Many-Core GPU

A data-parallel application consists of large streams of data elements in the form of matrices and vectors that run the same computation code. The first general-purpose programming model for the GPU hardware was compute unified device architecture (CUDA) that provides a C-like syntax to execute and manage computations on the GPU as a data-parallel computing device.

The executable code that runs on the device is known as a CUDA kernel. Kernels are functions designed to run in parallel on multiple streaming multiprocessors (SMs) of the GPU. The GPU runs its own kernel independently under the CPU's control. The execution starts with the host and moves to the device after a kernel function is invoked. All the kernels have their own unique coordinates to distinguish themselves and to identify the specific portion of the data to process [24].

13.2.3.3.1 Simple GPU Kernel To illustrate how the CUDA works, a simple GPU kernel is described here. Consider the same example provided in Section 13.2.3.2. The computation on the GPU is performed by separating the outer loop from the inner calculations. First of all, enough memory space on device memory should be allocated for each vector using *CudaMalloc* commands:

```
CudaMalloc ((void**) &dA, (sizeof(float) *N);

CudaMalloc((void**) &dB, (sizeof(float) *N);
```

d_A and d_B specify the location of the vectors *A* and *B* in device memory. The size of allocated memory is defined by measuring the size of the variable (float in our case) and multiplying it by the size of vector *A*. The next step is to transfer data to the GPU by executing the following command:

```
CudaMemcpy
(dA; hA; (sizeof(float) *N); cudaMemcpyHostToDevice);

CudaMemcpy
(dB; hB; (sizeof(float) N); cudaMemcpyHostToDevice);
```

h_A and h_B specify the location of vector on host memory. The kernel code to perform the operation can be written as

```
_global_void Add(float*A, float*B, int N)
{
int id.x = blockIdx.x*blockDim.x+ threadIdx.x;
if (id.x< N)
C[id.x] = A[id.x] +B[id.x];
__synchthread();
}
```

The global qualifier specifies that the kernel is callable from the CPU and will be executed on the GPU. There is not any for loop, and instead of that a new parameter, called *id.x*, is defined to control the execution of the kernel. Each thread will perform on the vector elements specified by *id.x*; *blockDim.x* returns the number of threads in each block.

Every thread in a block and every block in a grid have a unique index, which is accessible through the *threadIdx* and *blockIdx*, respectively. The next line performs the addition while each element is calculated by a different thread. The number of threads has to be equal or more than the number of elements in vectors to ensure that calculation is performed on all elements. The *syncthreads()* call ensures that all threads are synchronized. To invoke this kernel from a CPU-based code, we need to add a syntax as shown below:

```
Add< < <grid,block> > > (dA,dB,N)
```

The grid dimensions and the block dimension in execution configuration (<<< >>>) are defined by grid and block, respectively. At the end the result can be transferred to host memory using the *CudaMemcpy* command.

13.2.4 Numerical Methods for Solving Linear Systems

Solution of a linear system is usually the most computationally expensive step in various power system analyses such as power flow and state estimation. In general, a linear system can be written as

$$Ax = b \tag{13.24}$$

where A is a $n \times n$ square matrix known as coefficient matrix, b is an $n \times 1$ vector, and x is a $n \times 1$ vector.

Methods of solving linear systems fall into two general categories: direct methods and iterative methods. Direct methods obtain the exact solution of the system in a definite number of operations, whereas iterative methods calculate sequences of approximations that may or may not converge to the solution. While direct methods obtain an exact solution of the linear system, they require significantly more computations than iterative methods.

13.2.4.1 *Direct Method*

To solve the linear system $Ax = b$, several different algorithms can be used. One is to explicitly calculate the inverse of the coefficient matrix and multiply it by vector b. This method is computationally very expensive especially for large, sparse matrices, such as those encountered in power systems. Thus, various methods have been developed to solve a system of linear equations without explicitly calculating the inverse of the linear system. The most famous direct methods are LU decomposition and Cholesky decomposition [27].

Depending on the properties of the matrix A, different factorizations are used:

- For a $n \times n$ symmetric positive definite matrix, the Cholesky factorization $A = LL^T$ is usually computed, where L is a lower triangular matrix.
- For a $n \times n$ asymmetric matrix, its LU decomposition $A = LU$ is computed where L is a unit lower triangular matrix and U is an upper triangular matrix.

13.2.4.1.1 LU Decomposition The idea is to factor $A = LU$ where L is lower triangular and U is upper triangular. Then you solve the pair of equations as follows:

$$\underbrace{\begin{bmatrix} a_{1,1} & a_{1,2} & \cdots & \cdots & a_{1,n} \\ a_{2,1} & a_{2,2} & \cdots & \cdots & a_{2,n} \\ \vdots & \vdots & \vdots & \ddots & \vdots \\ a_{n,1} & a_{n,2} & \cdots & \cdots & a_{n,n} \end{bmatrix}}_{A} = \underbrace{\begin{bmatrix} l_{1,1} & l_{1,2} & \cdots & \cdots & l_{1,n} \\ l_{2,1} & l_{2,2} & \cdots & \cdots & l_{2,n} \\ \vdots & \vdots & \vdots & \ddots & \vdots \\ l_{n,1} & l_{n,2} & \cdots & \cdots & l_{n,n} \end{bmatrix}}_{L} \underbrace{\begin{bmatrix} u_{1,1} & u_{1,2} & \cdots & \cdots & u_{1,n} \\ u_{2,1} & u_{2,2} & \cdots & \cdots & u_{2,n} \\ \vdots & \vdots & \vdots & \ddots & \vdots \\ u_{n,1} & u_{n,2} & \cdots & \cdots & u_{n,n} \end{bmatrix}}_{U} \tag{13.25}$$

Then the linear system of (13.24) becomes

$$L\underbrace{Ux}_{Z} = b \Rightarrow LZ = b \tag{13.26}$$

13.2.4.1.2 Cholesky Decomposition The Cholesky factorization is a special LU factorization technique that decomposes the coefficient matrix into LL^T, where L is a lower triangular matrix with real and positive diagonal entries and L^T denotes the transpose of L. For symmetric positive definite matrices, Cholesky factorization needs less computation and memory space, since only the elements $a_{i,j}$, $i = j, \ldots, n; j = 1, \ldots, n$ should be stored in memory.

13.2.4.2 Iterative Method

In contrast with direct methods, iterative methods construct a series of solution approximations such that it converges to the exact solution of a system. It starts with an approximation to the solution of (13.24) and improves this approximate solution in each iteration. The approximate solution may converge to the exact solution in an infinite or infinite number of iterations. The iterative method can be stopped whenever the desired accuracy in the solution is obtained. Some of the famous methods in this category include conjugate gradient (CG) and Gauss–Jacobi method. Detailed description of the properties of iterative methods and their algorithms is provided in [28].

13.2.4.2.1 Preconditioned Conjugate Gradient Method Starting from (13.24) and assuming A is a symmetric matrix, the CG method was originally developed to minimize following quadratic function [29]:

$$f_g(x) \cong \left[\frac{\partial f(x)}{\partial x_1}\ \frac{\partial f(x)}{\partial x_2} \cdots \frac{\partial f(x)}{\partial x_n}\right]^T \tag{13.27}$$

$$f_g(x) = \frac{1}{2}\left(A^T X + AX\right) - b \tag{13.28}$$

Since A is symmetric, then $A = A^T$ and (13.28) can be rewritten as

$$f_g(x) = AX - b \tag{13.29}$$

Therefore, setting the gradient vector to zero and finding the critical point of $f(x)$ is equal to solving the linear system $Ax = b$. The process starts from a gauss about the solution, and the method tries to reduce the residual in each iteration and get close to the exact solution as much as possible. The residual vectored is defined as $r = b - Ax$. If r is less than the threshold or the iteration has exceeded the allowed maximum iterations, the algorithm will stop.

Iterative solvers are mainly less robust compared with direct solvers. To combat this problem, preconditioned CG are developed to improve the performance by speeding the convergence rate, which leads to less number of iterations and thus less runtime. Preconditioning transforms the original linear system into one that has the same solution but with a better condition number [30]. The condition number of the matrix is often used to quantify the eigenvalue spread of a matrix, and it is defined as

$$\text{Cond}_A = \frac{\lambda_{\max}(A)}{\lambda_{\min}(A)} \tag{13.30}$$

where Cond_A denotes the condition number, $\lambda_{\max}$ is the maximum eigenvalue, and $\lambda_{\min}$ is the minimum eigenvalue. The system is said to be ill-conditioned if the condition number is considerably more than unity.

Consider (13.24), which A is symmetric and positive definite; let M be a preconditioner, which approximates A in preserving the same solution. It is assumed that M is also symmetric positive definite. Then, the following preconditioned system could be solved:

$$M^{-1}Ax = M^{-1}b \tag{13.31}$$

To preserve symmetry one can decompose M in its Cholesky factorization and split the preconditioner between left and right:

$$L^{-1}A\underbrace{L^{-T}u}_{X}=L^{-1}b \tag{13.32}$$

13.2.4.2.2 Gauss–Jacobi Methods One of the most famous iterative methods used for relaxation-based solutions is the Gauss–Jacobi (G-J) methods. Consider a set of equations with n unknowns as follows:

$$\begin{gathered} a_{11}X_1 + a_{12}X_2 + \ldots + a_{1n}X_n = b_1 \\ a_{21}X_1 + a_{22}X_2 + \ldots + a_{2n}X_n = b_2 \\ \vdots \\ a_{n1}X_1 + a_{n2}X_2 + \ldots + a_{nn}X_n = b_2 \end{gathered} \tag{13.33}$$

In order to calculate unknown values, (13.33) can be rewritten as

$$\begin{gathered} X_1 = \frac{1}{a_{11}}(b_1 - a_{12}X_2 - \ldots - a_{1n}X_n) \\ X_2 = \frac{1}{a_{22}}(b_2 - a_{22}X_2 - \ldots - a_{2n}X_n) \\ \vdots \\ X_n = \frac{1}{a_{nn}}(b_{n1} - a_{n2}X_2 - \ldots - a_{nn}X_n) \end{gathered} \tag{13.34}$$

In general, for any unknown value in (13.33), following equation can be written:

$$X_i = \frac{1}{a_{ii}}\left[b_i - \sum_{j=1,i\neq j}^{n} a_{ij}X_j\right], \quad i = 1, 2, \ldots, n \tag{13.35}$$

The algorithm starts with an initial guess for the solution. At each iteration it updates x. The iterations stop when the absolute relative approximate error is less than a prespecified tolerance for all unknowns. If in each iteration the algorithm uses the results of previous iteration, it is called G-J (13.36) method. G-J iteration sequence for two subsystems is described in Figure 13.6:

$$X_i^{k+1} = \frac{1}{a_{ii}}\left[b_i - \sum_{j=1}^{i-1} a_{ij}X_j^k - \sum_{j=i+1}^{n} a_{ij}X_j^k\right], \quad i = 1, 2, \ldots, n \tag{13.36}$$

In the relaxation-based G-J (RG-J) method, the ith subsystem uses the current iterate value from subsystems $(1, \ldots, i-1)$ and the previous iterate value from subsystems $(i+1, \ldots, n)$ as inputs. Consider the following differential equation that can be used for RG-J:

$$\dot{X}^{\mathrm{L}} = f\left(X^{\mathrm{L}}, u^{\mathrm{L}}, X^{\mathrm{G}}, u^{\mathrm{G}}, t\right) \tag{13.37}$$

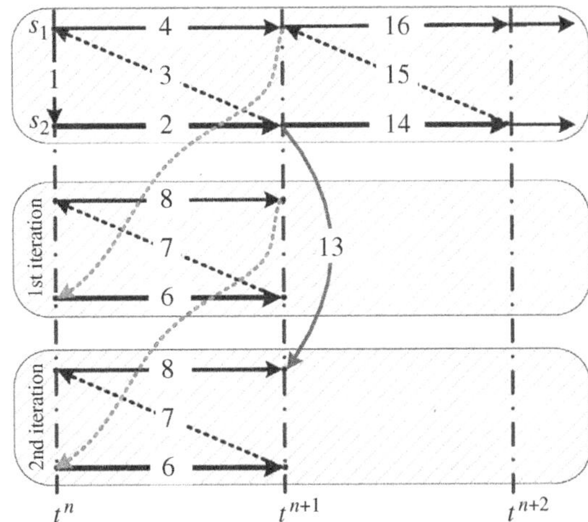

Figure 13.6 Gauss–Jacobi iterative method for two subsystems.

where x^L and u^L are local variables that define the dynamic behavior of subsystem i and x^G and u^G are global variables that define all subsystems excluding subsystem i.

In the RG-J the solution of subsystem i for the current time step is fully independent from the solution of other subsystems. To solve each subsystem (13.37) should follow all three steps explained earlier in Section 13.4.2. Following the convergence of iterative solutions in all subsystems, the state and algebraic variables calculated from the last time step are updated. Figure 13.7 shows the flowchart of RG-J method for duration of [0, T].

13.2.5 Additive Schwarz Method (ASM)

Additive Schwarz Method (ASM) is a relaxation-based approach that partitions the system into a number of subsystems based on either the system equations or component connectivity, which reduces the complexity [31, 32]. It facilitates the parallel solution of the small subsystems and the exchange of computational data between them. This method can be used at different levels of equations by breaking the system into subsystems in a way that the variables inside of each subsystem (local variables) are strongly interdependent while the dependency between variables in two different subsystems (local and global variables) is weak enough to ignore their interconnection. The relaxation method was the first attempt to exploit both space and time parallelism [33].

Consider a decomposition of the domain into M nonoverlapping subdomains:

$$\Omega = \bigcup_{i=1}^{M} \Omega_i, \quad \Omega_i \cap \Omega = \{\emptyset\}_{i \neq j} \tag{13.38}$$

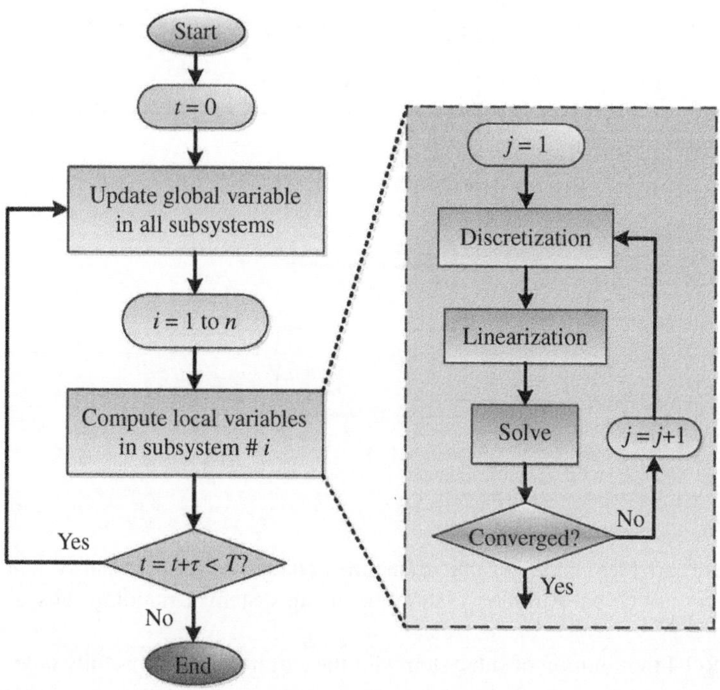

Figure 13.7 Flowchart of ASM method with time stem τ. i, current subsystem; n, total number of subsystems; j, iteration number.

Let $\Delta(x)_i^{(0)}$ denote the initial condition in subdomain i. A general RG-J algorithm for state estimation can be written as

$$\begin{cases} G(x)_i^{(k+1)}\Delta(x)_i^{(0)} = g(x)_i^{(k+1)} & \text{in } \Omega_i \\ x_i^{(k+1)} = x_i^{(k)} & \text{on } \partial\Omega_i \cap \Omega_j \\ Z_i^{(k+1)} = Z_i^{(k)} & \text{on } \partial\Omega_i \cap \Omega_j \end{cases} \tag{13.39}$$

In other word, the ASM is a type of domain decomposition method that approximately solves a boundary value problem by splitting it into subproblems on smaller domains [25, 34]. The process starts from an initial condition for all the subsystems. To achieve convergence several iterations may be required, where each of the subsystems exchanges boundary information and is then solved with updated data collected from other subsystems. This process is repeated using a Jacobi method to iterate among subsystems until all variables converge with the necessary accuracy.

The flowchart of the relaxation-based Jacobi implementation of WLS SSE on CPU for a time interval of [0, T] is depicted in Figure 13.8. The same as other parallel computation algorithms, the preliminary step in utilizing ASM method is to decompose the problem into smaller tasks that can be distributed among several processors. After decomposition, each subsystem will be solved independently.

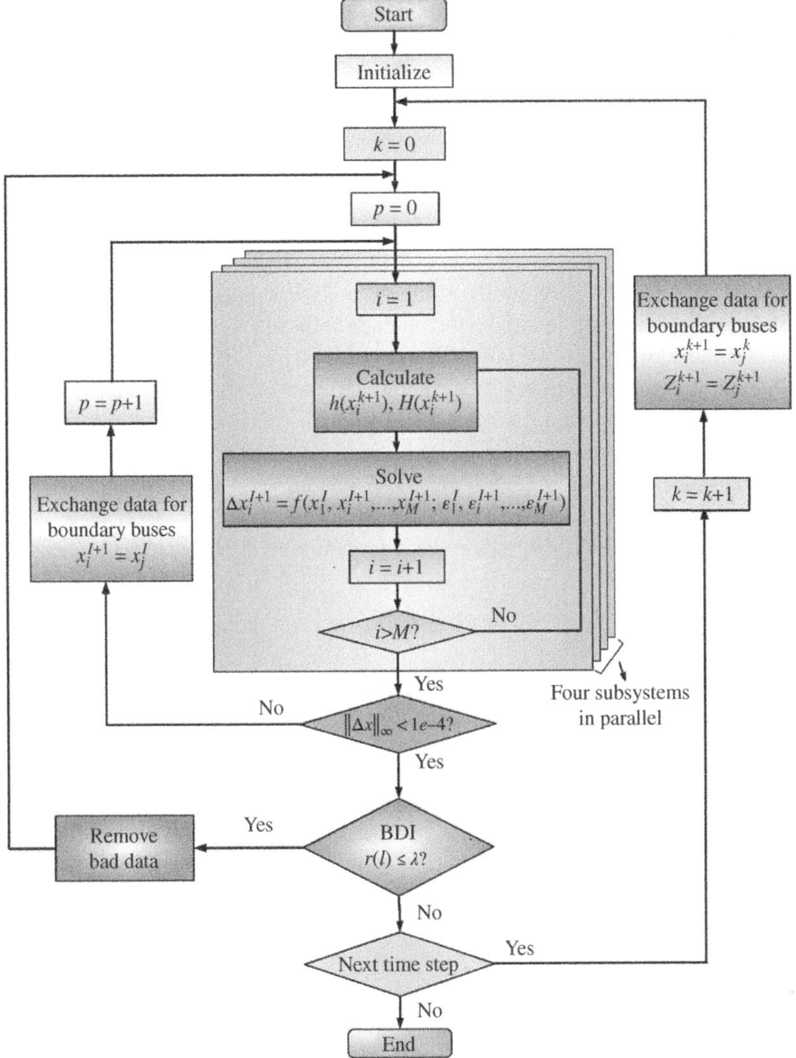

Figure 13.8 The ASM-based Jacobi WLS algorithm with BDD. k, time step; i, the number of subsystems; p, iteration counter; l, index of component in residual vector.

In every iteration each subsystem is solved for its local variables, while other subsystems are considered constant or relaxed during the time step. Each subsystem uses the previous value of other subsystems as a guess for its new iteration. At the end of each iteration, the global variables of all subsystems are exchanged for the next time step, and this process is repeated until convergence is gained.

The parallelism inherent in the ASM method offers a coarse-grained parallelization as a top level algorithm that should be implemented before using a numerical method for solving the system of equations.

13.2.5.1 Domain Decomposition

Any technique that divides a system of equations into several subsets that can be solved individually can be classified as domain decomposition method. Domain decomposition techniques are primarily partitioning methods that try to split a system into several subsystems that can be solved individually [35, 36]. The main advantage of decomposition techniques is that they are suitable for parallel application on multiprocessors since independent subsystems can be solved simultaneously. This feature makes them an excellent candidate for distributed state estimation.

Generally, from a power system point of view, domain decomposition methods reduce the problem size by dividing the network into several subsystems, which results in less computation effort in each individual subnetwork. The subsystem is a subset of system variables. Since subsystems can be solved simultaneously, decomposition techniques are suitable for parallel hardware architectures.

There are two main approaches for decomposing a domain: overlapping and nonoverlapping subdomains. In this case study nonoverlapping decomposition considering interfaces between subdomains is utilized. Consider a power network that is decomposed into two parts as shown in Figure 13.9.

As an example, the objective function for static WLS state estimation can be formulated as

$$J(x) = \min_{x_1, x_2} \begin{cases} [m_1 - h_1(x_1)]^T R_1^{-1} [m_1 - h_1(x_1)] + \\ [m_2 - h_2(x_2)]^T R_2^{-1} [m_2 - h_2(x_2)] + \\ [m_B - h_B(x_B)]^T R_B^{-1} [m_B - h_B(x_B)] \end{cases} \tag{13.40}$$

subject to $(x_{B_1} - x_{B_2}) = 0$

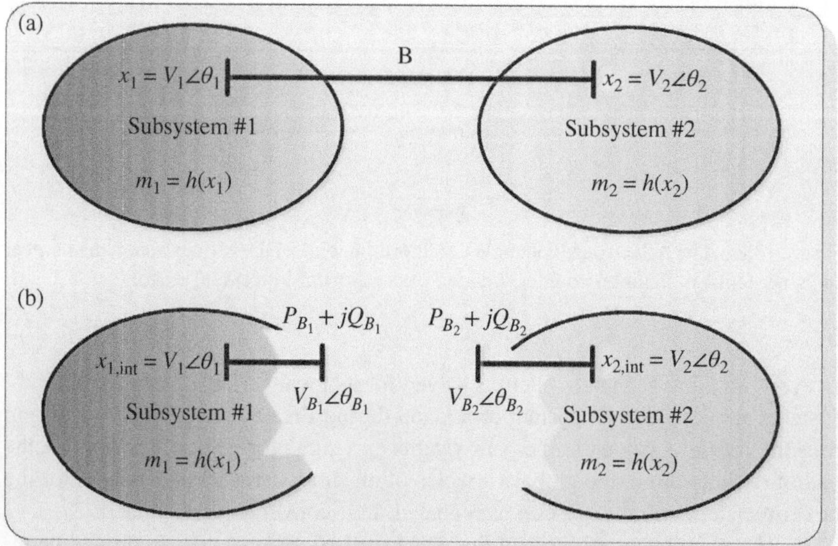

Figure 13.9 Domain decomposition: (a) interconnection of two subsystems and (b) split of two subsystems.

where $h_i(x_i)$; $i = 1, 2$ is the nonlinear measurement function for each subsystem i while $\frac{\partial h_1(x)}{\partial x_{B_1}} = \frac{\partial h_2(x)}{\partial x_{B_2}} = 0$, and the complete vectors are as follows:

- $x_1 = [x_{1,\text{int}}, x_{B_1}]^T$
- $x_2 = [x_{2,\text{int}}, x_{B_2}]^T$

It is already shown that the rate of convergence in ASM method is highly dependent on the method of partitioning [37]. There are different approaches to decompose a power system for parallel processing:

- Decomposition based on geographical distance: however, this approach stymied by the computational load balancing problem and may result in inaccuracy due to neglecting the effect of subsystems.
- Distributing equal numbers of generators and buses among subsystems: however, this is not an efficient method since the generator models vary in both size and complexity and network buses have different connectivity.
- Splitting the computation burden among processors based on the total number of equations: however, this method will increase the programming complexity.

The best way to decompose a system for parallel processing is to distribute equal amount of work among processors. The best result guaranteed when subsystems are independent of each other or in another word tightly coupled variables are gathered in the same subsystem.

13.2.6 Coherency Analysis

From power system point of view, determination of tightly coupled variables can find a physical meaning. Following a disturbance in the system, some generators lose their synchronism with the network that causes sudden changes in the buses connected to those generators and naturally partition the system into several areas. Generators in each of these areas are said to be coherent. In this situation, state estimation will take more iterations for some area since it should be repeated after clearing the disturbance.

Coherent generators can be grouped in the same subsystem, which can be solved independently from other subsystems. Partitioning the system into several areas in which generators are in step together or are coherent will increase the accuracy of the state estimation and save time by localizing the effect of disturbances. The partitioning achieved using the coherency property is independent of the size of disturbance and the level of detail used in the generators, which makes it suitable for our case studies [38].

In a network a pair of generators is called coherent if the difference between their rotor angles remains constant over time (13.42):

$$\delta_i(t) = \delta_j(t) = \Delta\delta_{ij} \pm \epsilon \tag{13.41}$$

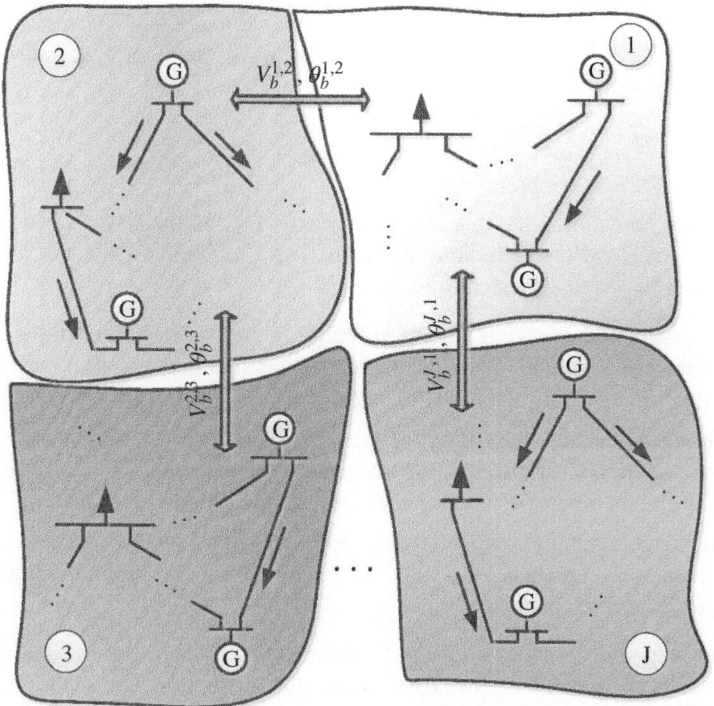

Figure 13.10 Original power system decomposed into J subsystems for RJDSE implementation.

where $\Delta\delta_{ij}$ is a constant value and ϵ is a small positive number. For efficient parallelization, in our work load balancing is also considered in domain decomposition. Equal load distribution among subsystems reduces the complexity resulting in faster computations, which also makes it efficient for implementation in a multi-GPU architecture. Figure 13.10 shows a power system decomposed into J subsystems. Decomposition based on the coherency approach and equal load work criteria divides the full set of equation into several independent functions running in separate GPUs.

13.3 EXPERIMENTAL RESULTS

Large-scale systems were constructed by duplicating the IEEE 39-bus system shown in Figure 13.11. The uniform set of measurements that are the input to the state estimation algorithms are obtained by corrupting online power flow results of the test power systems with Gaussian noise of zero mean and covariance R. Therefore, to assess the accuracy of the state estimator, the results are verified using bus voltage magnitudes and phase angles for all test cases modeled in PSS/E®.

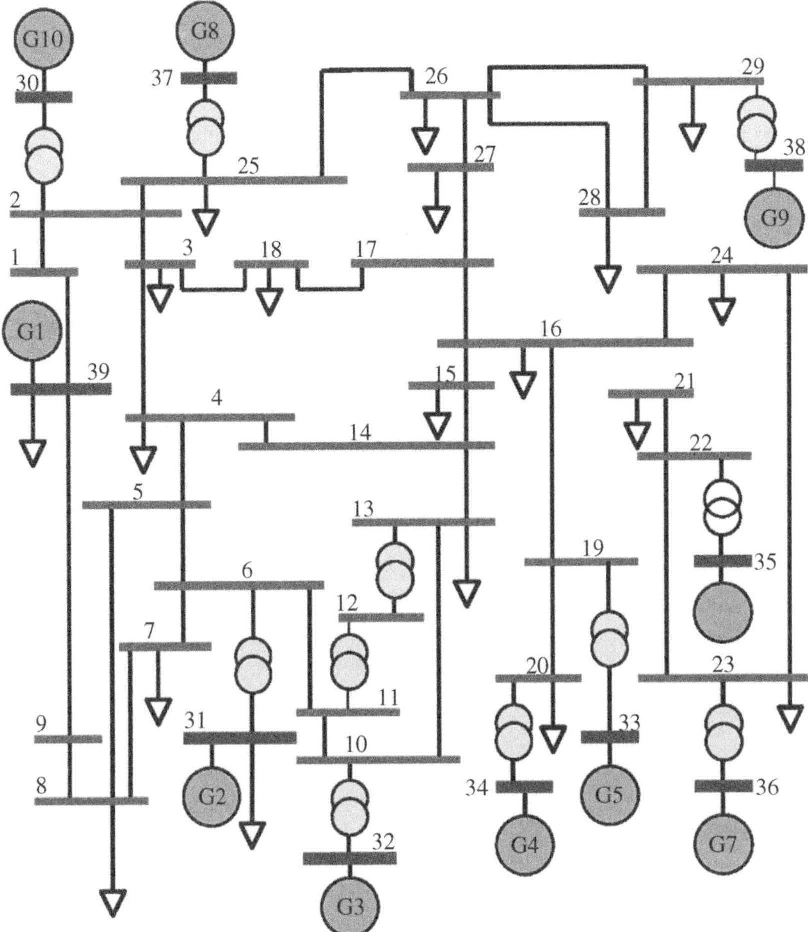

Figure 13.11 IEEE 39-bus power system used to build large-scale test cases.

To evaluate accuracy and efficiency of the parallel DSE algorithms, experiments were conducted based on two separate simulation codes: multithread CPU-based code in C++ and a massive-thread GPU-based code written in C++ and CUDA. Since the matrices are highly sparse in state estimation, all matrices and vectors are stored in compressed sparse row format to reduce the computational burden.

13.3.1 Hardware Setup

The hardware used in this work is the hexa-core Intel Xeon™ E5-2620v2 with 2.1 GHz core clock and 32 GB memory with 51.2 GB/s memory bandwidth, running the 64-bit Windows 8.1r operating system. The GPU used in this work is one unit of Tesla™ S2050 GPU from NVIDIA® with 148 GB/s memory bandwidth. Each

Fermi™ GPU inside Tesla™ S2050 has 448 cores, which deliver up to 515 gigaflops of double-precision peak performance. This device contains 16 SMs, each with 32 streaming processors (SPs), an instruction unit and on-chip memory. Each SM has 4 special function units (SFUs) that execute transcendental instructions such as sin, cosine, and 16 load/store units, allowing source and destination addresses to be calculated for 16 threads per clock [24].

Figure 13.12 shows the architecture of the Fermi™ GPU.

13.3.2 Extraction of Parallelism

In this work several aspects of parallelism are combined to utilize the full capability of CPU and GPU. The following types of parallelism are used in this work:

- Task parallelism – In this level, the traditional serial algorithm is converted into various smaller and independent tasks, which can be solved in parallel. All of the independent tasks in the process of state estimation are calculated in parallel to accelerate the algorithm.
- Data parallelism – This level employs the fine-grained type of parallelism that can be used on the SIMD-based architectures such as GPUs for the basic computations in the algorithm. Generally, matrix–vector and matrix–matrix products are time-consuming for large data sets. There are several independent *for* loops in the implementation of each matrix–matrix and matrix–vector products, which make them the best candidates for parallelization utilizing GPU threads. Assigning each iteration in a loop to individual threads, the task can be executed in parallel by converting into a kernel.
- Parallelism in linear solver – Solution of $\Delta(x)$ by inversion of $G(x)$ is considerably expensive due to the sheer size of the inverted matrix. Two alternatives, LU decomposition as a direct method and preconditioned conjugate gradient (PCG) as an iterative method, were used in this work. For iterative solvers, the preconditioner is the most challenging part to parallelize.

13.3.3 Parallel ASM-Based WLS Static State Estimation

Proposed method was implemented on CPU architecture to evaluate the efficiency for further implementation in multi-GPU architecture. After decomposition, each subdomain is stored and computed by a processor core. All processor cores solve the subdomains in parallel.

The process starts from an initial condition for all the subsystems. To achieve convergence several iterations may be required, where each of the subsystems exchanges boundary information and is then solved with updated data collected from other subsystems. If the stopping criteria are not satisfied, new iteration is performed. This process is repeated using a Jacobi method to iterate among subsystems until all variables converge with the necessary accuracy. In case of bad measurements, state estimation will repeat only for the subdomain affected by bad measurement.

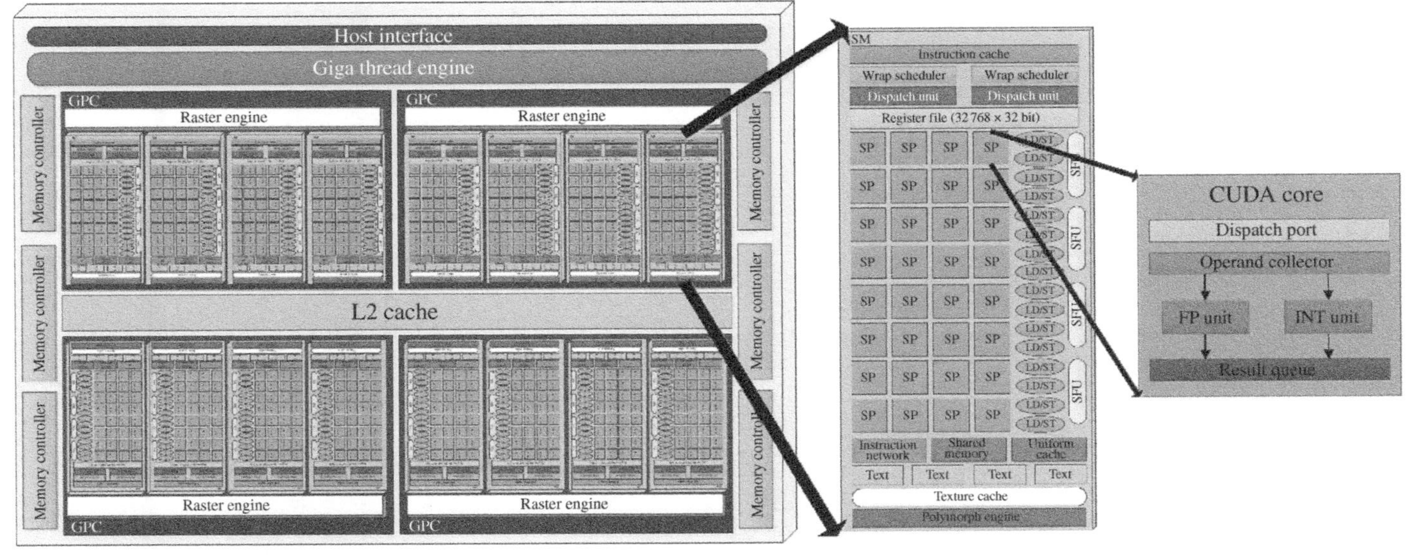

Figure 13.12 Fermi GPU architecture.

Programming was done in Matlab® using its Parallel Computing Toolbox™. After decomposition, each subdomain is stored and computed by a processor core. All processor cores solve the subdomains in parallel.

After each iteration, estimated states of boundary nodes are transferred among processors to update each subdomain's boundary information. If the stopping criteria are not satisfied, new iteration is performed. In case of bad measurements, state estimation will repeat only for the subdomain affected by bad measurement.

One of the main problems associated with parallel ASM is the communication overhead among subsystems. In our work, since subsystems are fully independent of each other, only necessary boundary bus values are needed to be exchanged. Also, PMU measurements are considered to be installed in the boundary buses to accelerate data transfer among subsystems.

13.3.3.1 Accuracy Analysis

To evaluate accuracy of the proposed algorithms, the results were compared with traditional centralized state estimation. It is assumed that PMUs are installed at the boundary buses. To assess the accuracy of the state estimator, the results were also compared with the original power flow results from PSS/E. The role of bad data detection in state estimation is necessary since bad measurements easily affect the accuracy of the results. One of the popular methods for BDD that is used in this paper is based on normalized residual test [39]:

$$r_{\mathrm{N}}(l) = \frac{|r(l)|}{|\sigma_{\mathrm{r}}(l)|} \leq \tau \tag{13.42}$$

where $r_{\mathrm{N}}(l)$ is the largest residual among all and $\sigma_{\mathrm{r}}(l)$ is the standard deviation of the lth component of the residual vector. In this work the measurements having the largest normalized residual and larger than 3 were considered as bad data, with a 99.7% confidence level. After removing bad data, state estimation was repeated starting from the most recent estimate.

The results of parallel distributed state estimation are compared with traditional centralized state estimation method. The simulations were done using the test data sets listed in Table 13.1, with a tolerance of 0.0001 for convergence of the estimated parameters. Performance of the proposed method was evaluated for different case studies. The estimated states for Case 1 are shown in Figures 13.13 and 13.14. It is clear from the results that the proposed approach can accurately estimate the voltage magnitude and phase angle.

The normalized Euclidian norm of the state estimation is also defined as a factor to evaluate the accuracy using

$$E_x = \frac{\|x - \dot{x}\|}{\sqrt{\dim(x)}} \leq \tau \tag{13.43}$$

where E_x is the normalized Euclidian norm of the estimation error and $\dim(x)$ is the dimension of vector x. x and $\dot{x}$ are vectors of true states and estimated states, respectively. Table 13.1 shows the accuracy index for both voltage magnitude (E_V) and

TABLE 13.1 Results for comparison of parallel ASM WLS with centralized WLS.

Case	No. of buses	No. of measurements	E_V^{Cen}	E_ϕ^{Cen}	E_V^{Dec}	E_ϕ^{Dec}	$T_{\text{Ex}}^{\text{Cen}}$	$T_{\text{Ex}}^{\text{Dec}}$	S_{p}
1	39	171	0.009	0.85	0.006	0.46	0.08s	0.11s	0.72
2	78	347	0.004	0.6	0.003	0.43	0.26s	0.33s	0.78
3	156	699	0.0032	0.55	0.001	0.47	0.51s	0.61s	0.83
4	312	1 421	0.0041	0.6	0.0014	0.48	1.49s	0.91s	1.6
5	624	2 865	0.0033	0.7	0.0012	0.49	5.2s	1.84s	2.8
6	1 248	5 825	0.004	0.65	0.0011	0.5	24.5s	8.1s	3.1
7	2 496	11 553	0.0044	0.6	0.0013	0.49	68.3s	15.5s	4.4
8	4 992	23 151	0.005	0.65	0.0017	0.49	364.5s	56.3s	6.5

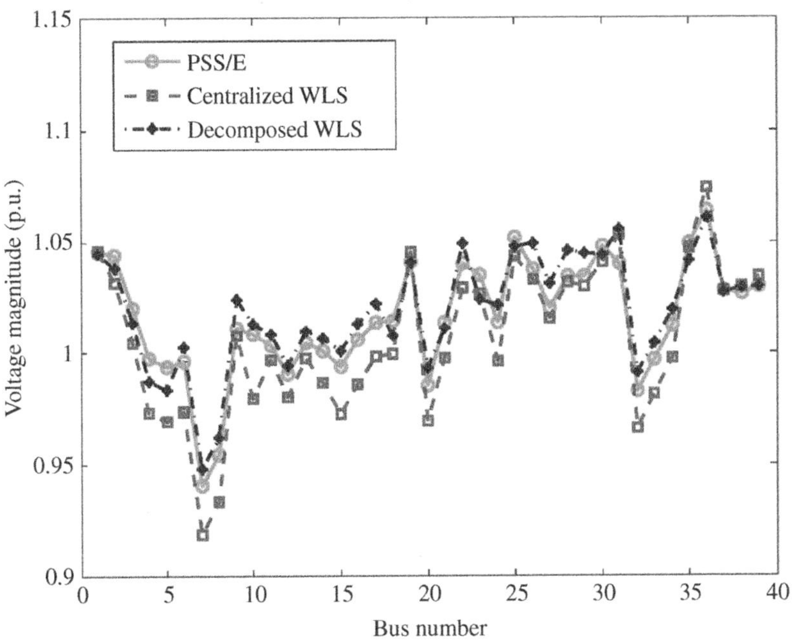

Figure 13.13 Voltage magnitudes for Case 1 with respect to system size.

phase angle (E_ϕ) for all case studies that clarifies the performance of the proposed method for large-scale systems.

13.3.3.2 Efficiency Analysis

Using the partitioning pattern mentioned in [40] Case 1, IEEE 39-bus system has been divided into three subsystems: {1, 8, 9}, {2, 3, 4, 5, 6, 7}, and {10}. For computational load balancing other criteria were considered to have almost equal number of buses in each subdomain. Satisfying both conditions simultaneously resulted in following four domains, which are shown in Figure 13.15.

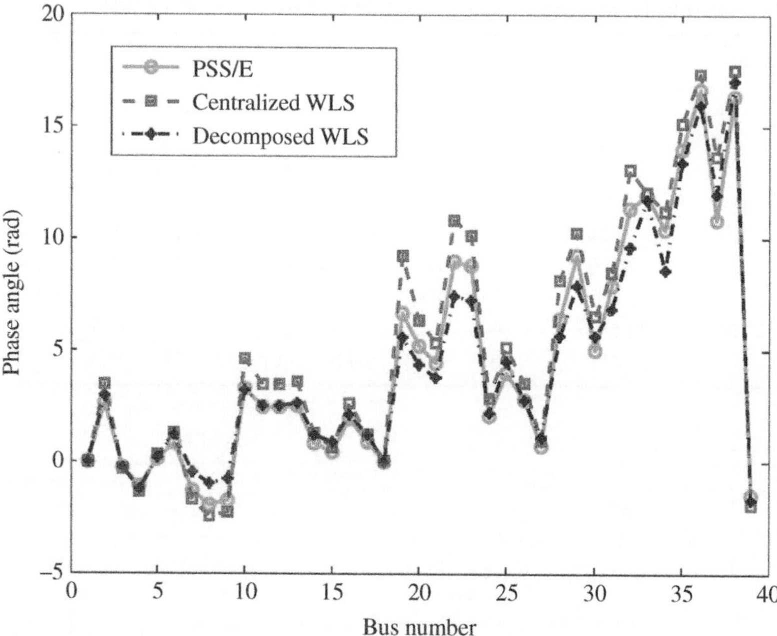

Figure 13.14 Phase angles for Case 1 with respect to system size.

To demonstrate the efficiency of the proposed approach, execution time using decomposed SE ($T_{\rm Ex}^{\rm Dec}$) is compared with traditional centralized ($T_{\rm Ex}^{\rm Cen}$) SE method. As can be seen from Figure 13.16, the percentage of required execution time for the centralized WLS method increases very fast, which shows the higher complexity of this method. In contrast, the effect of domain decomposition and relaxation approach on SE process results in slower increase in execution time, which is close to a linear behavior.

Generally, when a system with α nodes is partitioned into M subsystems, each subdomain has approximately $\alpha/_M$ nodes. Assume that solving a linear system with iterative method has complexity of $O(\alpha^\beta)$ where $\beta > 1$.

Using domain decomposition technique, the complexity of solving each subsystem is $O(\alpha/M)^\beta$, which results in the complexity of $O(\alpha)^\beta/(M)^\beta$ for the entire system. It should be noted that this only occurs in the ideal case. However, it still justifies the speedup reported in Table 13.1 and Figure 13.16. As can be seen from Figure 13.16, the percentage of required execution time for the centralized WLS method increases very fast, which shows the higher complexity of this method.

13.3.4 Parallel ASM-Based Dynamic State Estimation on GPU

In this section a parallel ASM-based DSE is implemented on 4 GPUs each with a massively data-parallel architecture. All equations are expressed based on the unique SIMD-based architecture of GPU. Instead of using single element values,

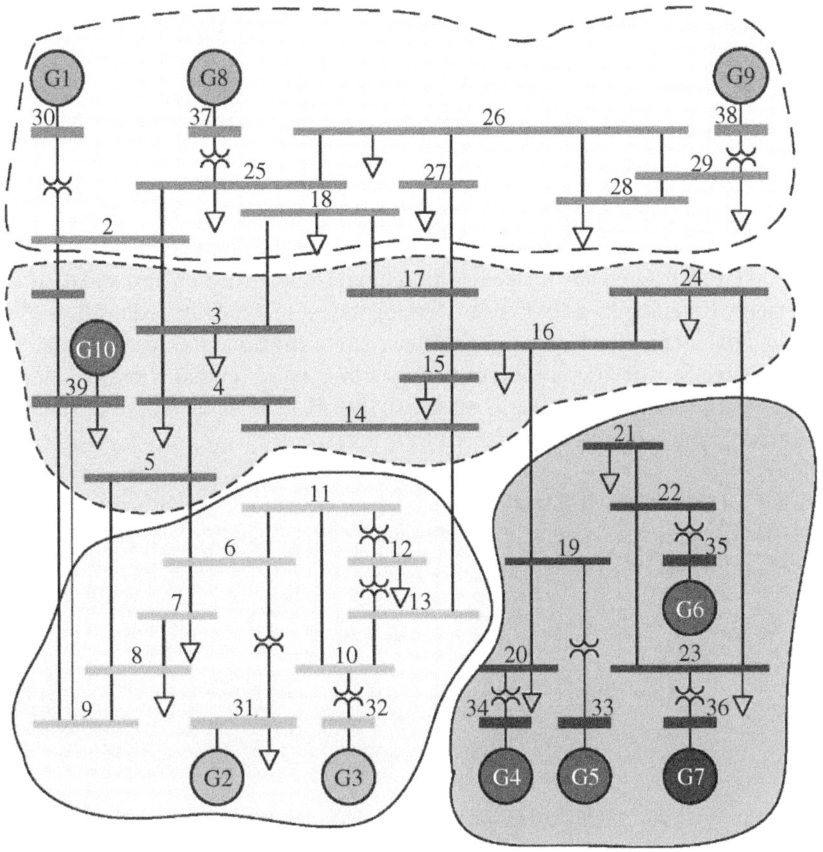

Figure 13.15 Decomposing a Case 1 into four subsystems to apply the ASM algorithm.

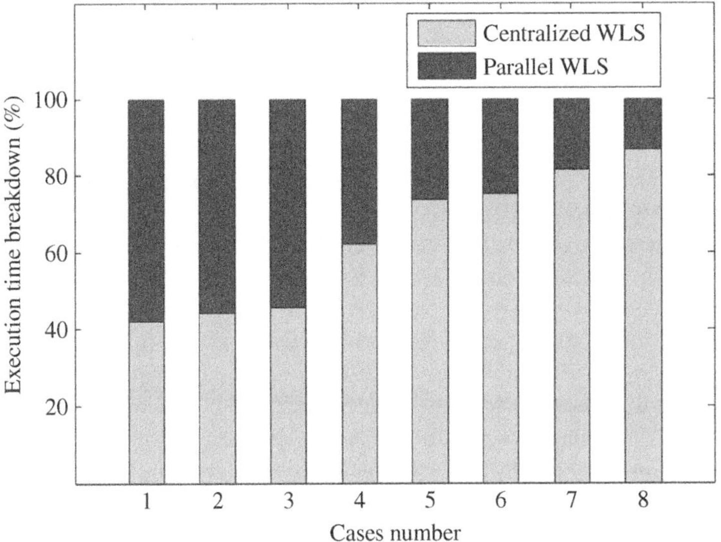

Figure 13.16 Percentage of execution time breakdown with respect to system size.

vectors or matrices of them are used. In addition, all matrix–matrix and matrix–vector that include many independent *for* loops are implemented in a fully parallel manner.

CUDA that is the general-purpose programming model for the GPU hardware was used in this work.

In the proposed GPU implementation, most of the computational steps are moved to GPU to avoid any unnecessary communication between host and device. The OpenMP standard was utilized to develop the multicore DSE code. All the *for* loops and parallel sections were assigned to separate threads and cores. Each thread is responsible for specific portion of the tasks to execute on that core. The entire DSE task was divided into several subsets to distribute an equal workload among the threads, which equals the number of the cores.

13.3.4.1 *Hierarchy of Parallelism*

After partitioning the system, the following set of equations can describe the dynamics of each subsystem:

$$\begin{cases} f\left(x_1^t, \ldots, x_i^{t+\tau}, \ldots, x_J^t, \dot{x}_1^t, \ldots, \dot{x}_i^{t+\tau}, \ldots, \dot{x}_J^t, t, u_1^t, \ldots, u_i^{t+\tau}, \ldots, u_j^t\right) = 0 \\ g\left(x_1^t, \ldots, x_i^{t+\tau}, \ldots, x_J^t, t, u_1^t, \ldots, u_i^{t+\tau}, \ldots, u_j^t\right) = 0 \\ h\left(x_1^t, \ldots, x_i^{t+\tau}, \ldots, x_J^t, t, Z_1^t, \ldots, Z_i^{t+\tau}, \ldots, Z_j^t\right) = 0 \end{cases} \tag{13.44}$$

where indices L and G stand for local and global variables, respectively.

Equation (3.44) is solved in parallel and iteratively for all subsystems. After each iteration, the global state variables are exchanged and updated between all interconnected subsystems. In order to calibrate the results, an overall loop based on the Gauss–Jacobi algorithm is applied outside the solutions of subsystems. Since the Gauss–Jacobi algorithm only uses the previously computed values for the solution of each subsystem, all computations of subsystems can be processed in parallel.

The algorithm starts at top level with Gauss–Jacobi iteration. By functional parallelism, almost equal tasks are assigned to each GPU. The iteration starts at the same time and in parallel inside all GPUs. Inside each iteration fine-grained parallelism is used to accelerate the process.

After each iteration, only estimated states of boundary buses are exchanged. If the Gauss–Jacobi algorithm convergence is not satisfied, new iterations will be performed.

In summary, fine-grained parallelism is performed inside the functional parallelism, and functional parallelism is a subset of coarse-grained parallelism (Figure 13.17). In the best-case scenario, coarse-grained parallelism by dividing the system into J subsystems reduces the execution time to $\tau/_J$, using functional parallelism and J_1 independent tasks results in execution time of $\tau/_{J.J_1}$, and finally utilizing J_0 independent matrix–matrix, matrix–vector, and other type of fine-grained parallelism execution time can be further reduced to $\tau/_{J.J_1.J_0}$. However,

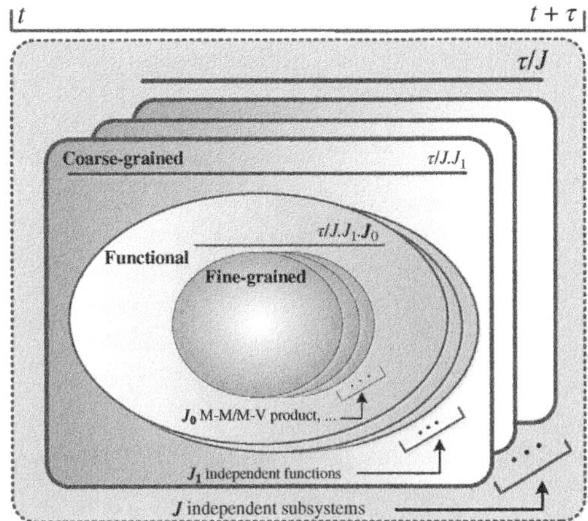

Figure 13.17 Hierarchy of parallelism. τ, integration time step; t, simulation time.

TABLE 13.2 Data sets for simulation.

Case	No. of buses	No. of measurements	Jacobian matrix $H(x)$	Gain matrix $G(x)$
1	39	171	171 × 77	77 × 77
2	78	347	347 × 155	155 × 155
3	156	699	699 × 311	311 × 311
4	312	1 421	1 421 × 623	623 × 623
5	624	2 865	2 865 × 1 247	1 247 × 1 247
6	1 248	5 825	5 825 × 2 495	2 495 × 2 495
7	2 496	11 553	11 553 × 4 991	4 991 × 4 991
8	4 992	23 151	23 151 × 9 983	9 983 × 9 983

in reality the speedup is less than this considering the different costs of parallelization and data transfer to GPU.

13.3.4.2 Accuracy Evaluation

The simulations were done using the test data sets listed in Table 13.2, with a tolerance of 0.0001 for convergence of the estimated parameters. The performance of the proposed method was evaluated under both normal and contingency conditions. The estimation error for Case 1 under a temporary fault at bus 10 at t = 10 minutes for a duration of 0.1 second is shown in Figure 13.18. It is clear that the proposed method can accurately track the system dynamics. The results show that in a long-term simulation, the average estimation error for voltage magnitudes and phase angles was 0.002 p.u. and 0.05 rad, respectively. A snapshot of the estimation error at bus numbers 10, 11, 13, and 32 is provided in Figure 13.19.

The simulation results were followed by bad data detection. The largest normalized residuals with threshold of $\chi = 3$ were considered as bad data.

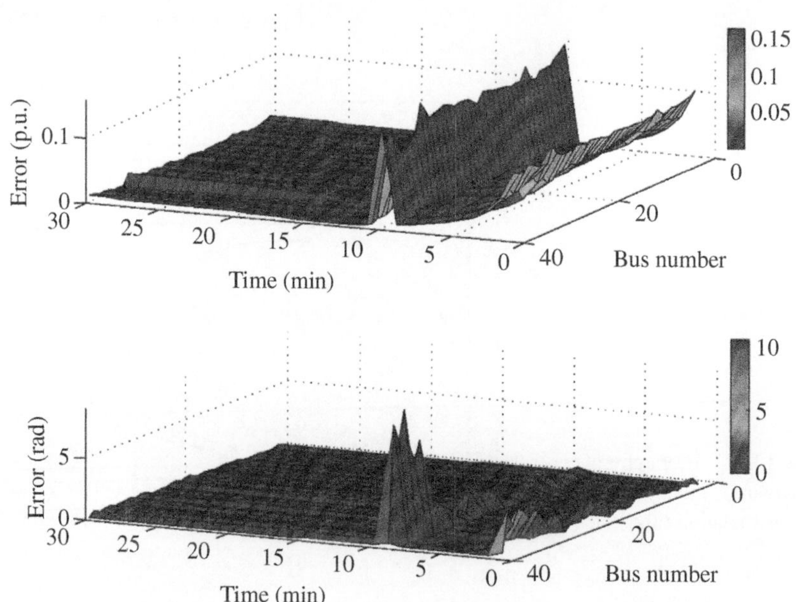

Figure 13.18 Estimation errors in GPU-based ASM for Case 1 compared with PSS/E under fault conditions.

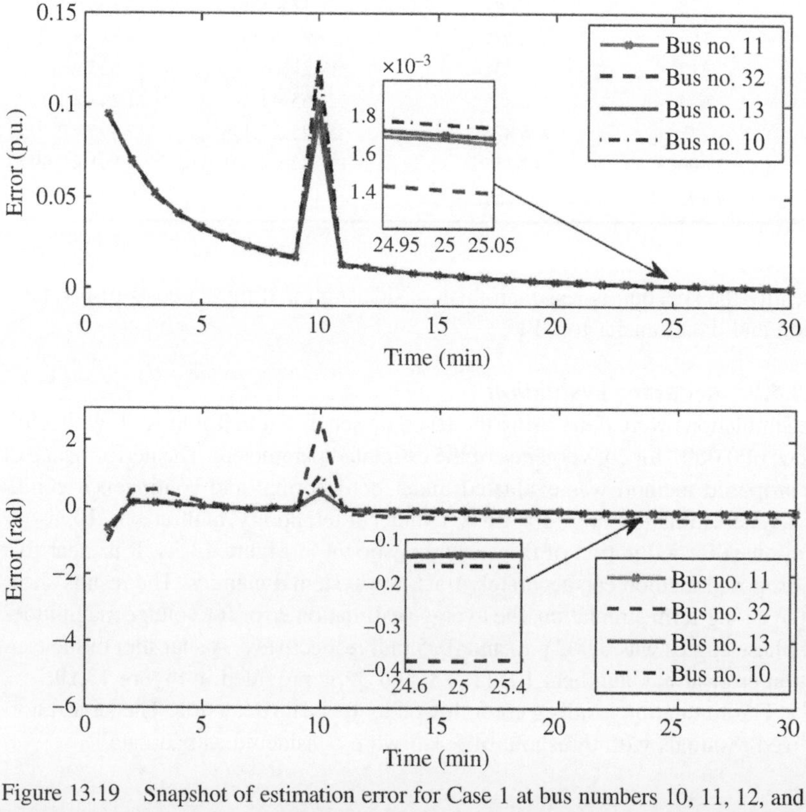

Figure 13.19 Snapshot of estimation error for Case 1 at bus numbers 10, 11, 12, and 13.

TABLE 13.3 Estimation error for different percentage of PMU installation.

	Normalized error norm in voltage magnitude			Normalized error norm in phase angle		
Case	$N_P = 10\% N_T$	$N_P = 20\% N_T$	$N_P = 40\% N_T$	$N_P = 10\% N_T$	$N_P = 20\% N_T$	$N_P = 40\% N_T$
1	0.004	0.0027	0.0015	0.59	0.37	0.23
2	0.0038	0.0028	0.0013	0.57	0.39	0.25
3	0.0036	0.0027	0.0015	0.55	0.32	0.29
4	0.0041	0.0029	0.0016	0.56	0.34	0.24
5	0.0033	0.0028	0.0017	0.58	0.36	0.27
6	0.0039	0.003	0.0017	0.59	0.37	0.28
7	0.0041	0.0026	0.0016	0.57	0.34	0.21
8	0.0036	0.0029	0.0015	0.56	0.38	0.28

Here the bad data refers to measurements with gross errors. Once bad data is identified, corresponding measurement was updated by deducting gross error $\frac{R_{ii}}{\sigma_{ii}} r_i$ from bad data. Using the updated measurements, state estimation was repeated only for the subsystems that were affected by bad data. For large-scale systems, this localization of bad data can save lots of time, which can in turn accelerate the state estimation process.

In addition, the performance of the proposed state estimation method is tested for various numbers of PMU installations. The normalized Euclidian norm of the state estimation is defined as a factor to evaluate the accuracy.

Table 13.3 shows the accuracy index for both voltage magnitude and phase angle using PMU installation (N_P) at 10, 20, and 40% of the total system buses (N_T). N_P is rounded to the next larger number. As it is shown the results are accurate for all of the case studies, and the accuracy is increased as the percentage of PMU measurements increased.

13.3.4.3 *Efficiency Analysis*

Case 1 that is the IEEE 39-bus system has been partitioned into four subdomains satisfying both computational load balancing and coherency characteristic of the generators. Similarly, all the large test cases are partitioned. The simulation starts by initialization on the CPU. After that, the measurement set corresponding to each subsystem was transferred to GPU assigned for that specific subsystem. All the subsystems start the simulation at the same time. After each iteration boundary data was exchanged among subsystems.

The total execution time under contingency scenario is reported in Table 13.4. The results obtained for both CPU ($T_{\text{Ex}}^{\text{CPU}}$) and GPU ($T_{\text{Ex}}^{\text{GPU}}$) codes as the system size increased. An incomplete Cholesky PCG iterative algorithm was used to condition the gain matrix, which reduced the number of iterations in each solution. The condition numbers before and after preconditioning are

TABLE 13.4 Execution time for CPU-based and GPU-based DSE.

Case	Cond. no. CG	Cond. no. PCG	$T_{\text{Ex}}^{\text{CPU}}$ LU	$T_{\text{Ex}}^{\text{GPU}}$ LU	$T_{\text{Ex}}^{\text{CPU}}$ PCG	$T_{\text{Ex}}^{\text{GPU}}$ PCG	S_{p} LU	S_{p} PCG
1	2.4E+02	1.6E+01	0.49s	0.29s	0.38s	0.19s	1.69	2
2	9.7E+03	6.3E+02	1.16s	0.59s	0.83s	0.39s	1.96	2.12
3	3.9E+04	1.1E+03	4.5s	2.04s	3.1s	1.1s	2.2	2.81
4	8.9E+04	1.8E+03	19.2s	6.8s	13.4s	4.3s	2.8	3.02
5	2.8E+05	4.9E+04	45.3s	12.5s	38.8s	9.8s	3.6	3.9
6	3.6E+06	2.6E+05	146s	26s	109.1s	20s	5.6	5.4
7	2.4E+07	1.3E+06	369s	40s	290.4s	33s	9.2	8.8
8	4.5E+08	9.4E+06	932s	59s	722.5s	48s	15.9	15.05

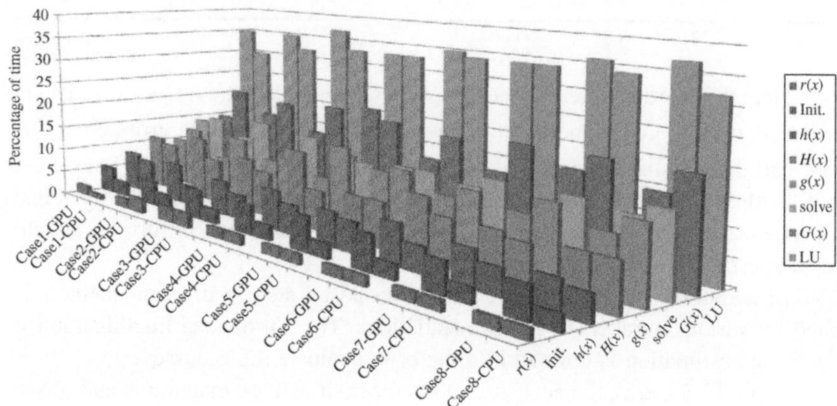

Figure 13.20 Percentage of time used for various steps in GPU-based ASM.

shown in Table 13.4. The results also show that the execution time reduced by approximately 20–50% using PCG compared with the LU decomposition method. The reason is that direct methods calculate an exact solution on a single iteration, so that they deal with large-size matrices that take more time to be solved; however, the iterative method starts from an initial guess and improves the solution in each iteration.

Figure 13.20 illustrates the percentage of time taken by various steps for all case studies.

Figure 13.21 shows the results of comparison between the CPU and GPU simulations along with the speedup of the parallel code for both iterative and direct solvers.

As can be seen, PCG converged faster than the LU decomposition algorithm, but the speedup ($S_{\text{p}} = T_{\text{Ex}}^{\text{CPU}}/T_{\text{Ex}}^{\text{GPU}}$) using GPU is almost the same for both methods. The reason is that the execution times for both CPU-based PCG and GPU-based PCG experience a similar drop in each case study. Therefore, the overall speedup is

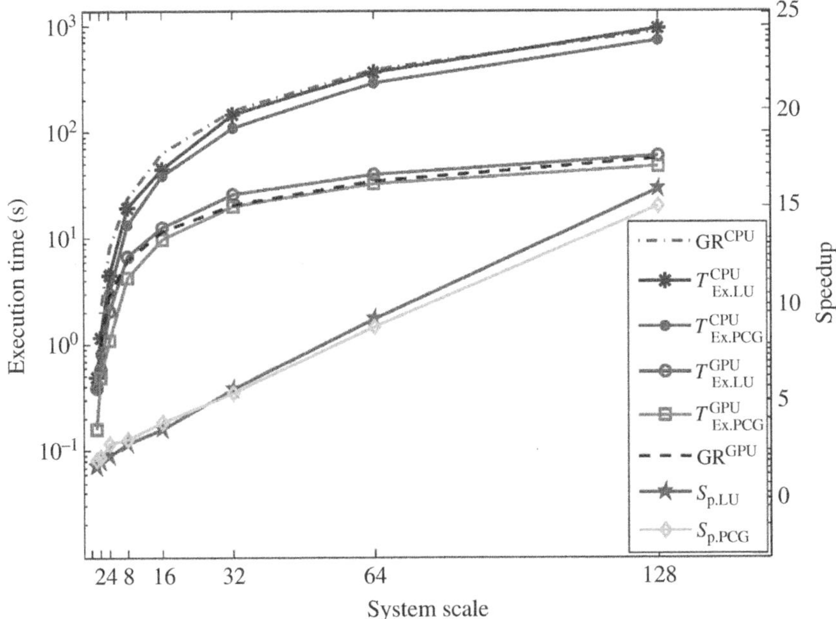

Figure 13.21 Execution time (T_{Ex}) and speedup (S_p) comparisons of multithread and massive-thread DSE along with growth rate functions.

almost the same. The accuracy was mostly the same for both linear solvers, so the details are omitted from Table 13.4.

To analyze the algorithm complexity, their efficiencies are expressed as functions of the problem size. Considering S_c as the system scale, by curve fitting the closest growth rate function ($GR = T_{Ex}(S_c)$) for the CPU-based and the GPU-based algorithms is calculated as $S_c \times \log_2(S_c)$ and $\sqrt{S_c} \times \log_2(\sqrt{S_c})$, respectively. The growth rate functions for CPU (GR^{CPU}) and GPU (GR^{GPU}) are plotted in Figure 13.21. As the size of the problem increases, the required time for execution increased proportional to these functions, which proves that the speedup will increase with growing system sizes. It can be seen that the CPU-based algorithm follows a higher-order complexity compared with the GPU-based algorithm.

13.4 CONCLUSION

State estimation is a major requirement for safe operation and control of power systems. It is the core of EMS that is computationally very demanding for large-scale power system operation and control. The state estimation problem is rich in parallelism that makes it very suitable for utilizing massively parallel processing techniques.

The application of parallel processing for static/dynamic state estimation is motivated by the desire for faster computation for online monitoring of the system behavior. The approach proposed in this chapter investigates the process of accelerating the static/dynamic for large-scale networks.

In the first part, using an ASM the solution of each subsystem was carried out by using the conventional numerical techniques and exchanging the boundary data among subsystems. To increase the accuracy slow coherency method was used to decide the domain decomposition. In addition, load balancing by distributing equal workload among processors was utilized to minimize interprocessor communication.

The proposed approach has following advantages over existing approaches:

- Reduces the execution time by splitting equal amount of work among several processors.
- Minimizes the effect of boundary buses in accuracy by exchanging data.
- Localizes the effect of bad data on the state estimation result.
- It is applicable for large-scale power systems.
- It does not require major changes in existing power system state estimation paradigm.

In the second part, the proposed method is implemented in massively parallel architecture of GPU. Overall, GPU have the following advantages over CPU clusters:

- *Parallelism* – Parallelization using GPU is fine-grained parallelization, which is a lot different from coarse-grained parallelization on CPU. In contrast to the CPU with a limited number of arithmetic cores, the GPU is composed of hundreds of cores known as SPs that can simultaneously handle thousands of threads.
- *Extensibility* – Unlike CPU with limited achievable speedup, the maximum achievable speedup by massive parallelism in GPU is proportional to the number of cores.
- *Cost* – A GPU with hundreds of core is a lot cheaper than a system with hundred CPU cores. Basically, the GPU has enormous cost advantage, gigaflops per dollar, in comparison with CPUs.

Although GPUs are well suited for large-scale data-parallel processing, writing efficient code for them is not without its difficulties. One of the most important bottlenecks in parallel GPU programming is the overhead in data transfer between host and device. The size and frequency of data transfers can create a bottleneck that significantly impacts the execution time of an algorithm. This fact is considered in both proposed methods by moving all the time-consuming steps to GPU to reduce the data transfer that resulted in significant speedup for both SSE and DSE applications.

As shown in the results, the advantage of utilizing GPU for parallelization is significant when the size of the system is increased. One explanation is that for small size of data, the communication overhead between the host and device supersedes the execution time on the CPU. With growing system sizes, the CPU is barely able to handle the computation tasks in a reasonable time. With growing system sizes less time is spent on state estimation. The reason is that as the size of the system grows, the parallel portion of the program expands faster than the serial portion, while underscoring GPU's advantages. These results are not unique due to the fact that the programming structure is one of the most important factors that affects the processing time. Therefore, a different programming style may lead to faster results; nevertheless, the speedup would still be valid for increasing system sizes although with a slightly lower numerical value.

The proposed method is general and extensible to any number of GPUs connected in a cluster. Results show that more GPUs can reduce expected computation time. Result comparisons verified the accuracy and efficiency of the proposed method. In addition, the performance of the slow coherency method as the partitioning tool was analyzed, and it was concluded that for different fault locations in the system, results derived from this method had lower amounts of error.

REFERENCES

1. Benigni, A., Junqi Liu, F., Ponci, A. et al. (May 2011). Decoupling power system state estimation by means of stochastic collocation. *IEEE Trans. Instrum. Meas.* 60 (5): 1623–1632.
2. U.S.-Canada Power System Outage Task Force (2004). *Final Report on the August 14, 2003 Blackout in the United States and Canada: Causes and Recommendations.*
3. Abur, A. and Gómez Expósito, A. (2004). *Power System State Estimation: Theory and Implementation.* Marcel Dekker.
4. Stouffer, K., Falco, J., and Kent, K. (2006). *Guide to Supervisory Control and Data Acquisition (SCADA) and Industrial Control Systems Security.* NIST, pp. 800–802.
5. De La Ree, J., Centeno, V., Thorp, J.S., and Phadke, A.G. (Jun. 2010). Synchronized phasor measurement applications in power systems. *IEEE Trans. Smart Grid* 1 (1): 20–27.
6. Jain, A. and Shivakumar, N.R. (2009). Power system tracking and dynamic state estimation. *2009 IEEE/PES Power Systems Conference and Exposition.* Seattle, WA, pp. 1–8.
7. Debs, A. and Larson, R. (Sep. 1970). A dynamic estimator for tracking the state of a power system. *IEEE Trans. Power App. Syst.* PAS-89 (7): 1670–1678.
8. Leite da Silva, A.M., Do Coutto Filho, M.B., and de Queiroz, J.F. (1983). State forecasting in electric power systems. *IET Gener. Transm. Distrib.* 130 (5): 237–244.
9. Ding, W., Wang, J., Rizos, C., and Kinlyside, D. (Sep. 2007). Improving adaptive Kalman estimation in GPS/INS integration. *J. Navig.* 60 (3): 517.
10. Seidu, K. and Mukai, H. (May 1985). Parallel multi-area state estimation. *IEEE Power Eng. Rev.* PER-5 (5): 31–32.
11. Sasaki, H., Aoki, K., and Yokoyama, R. (Aug. 1987). A parallel computation algorithm for static state estimation by means of matrix inversion lemma. *IEEE Power Eng. Rev.* PER-7 (8): 40–41.
12. Falcao, D.M., Wu, F.F., and Murphy, L. (May 1995). Parallel and distributed state estimation. *IEEE Trans. Power Syst.* 10 (2): 724–730.

13. Karimipour, H. and Dinavahi, V. (May 2015). Extended Kalman filter-based parallel dynamic state estimation. *IEEE Trans. Smart Grid* 6 (3): 1539–1549.
14. Dinavahi, V. and Karimipour, H. (Feb. 2016). Parallel relaxation-based joint dynamic state estimation of large-scale power systems. *IET Gener. Transm. Distrib.* 10 (2): 452–459.
15. Karimipour, H. and Dinavahi, V. (2015). Parallel domain decomposition based distributed state estimation for large-scale power systems. *IEEE Trans. Ind. Appl.* 52 (2): 1–1.
16. Gopal, A., Niebur, D., and Venkatasubramanian, S. (2007). DC power flow based contingency analysis using graphics processing units. *2007 IEEE Lausanne Power Tech*. Lausanne, Switzerland, pp. 731–736.
17. Garcia, N. (2010). Parallel power flow solutions using a biconjugate gradient algorithm and a Newton method: a GPU-based approach. *IEEE PES General Meeting*. Providence, RI, pp. 1–4.
18. Vilacha, C., Moreira, J.C., Miguez, E., and Otero, A.F. (2011). Massive Jacobi power flow based on SIMD-processor. *10th International Conference on Environment and Electrical Engineering, 2011*. Rome, Italy, pp. 1–4.
19. Jalili-Marandi, V. and Dinavahi, V. (Aug. 2010). SIMD-based large-scale transient stability simulation on the graphics processing unit. *IEEE Trans. Power Syst.* 25 (3): 1589–1599.
20. Jalili-Marandi, V., Zhou, Z., and Dinavahi, V. (2012). Large-scale transient stability simulation of electrical power systems on parallel GPUs. *2012 IEEE Power and Energy Society General Meeting*. San Diego, CA, pp. 1–11.
21. Zhou, Z. and Dinavahi, V. (2015). Parallel massive-thread electromagnetic transient simulation on GPU. *2015 IEEE Power & Energy Society General Meeting*. Denver, CO, pp. 1–1.
22. Zhou, Z. and Dinavahi, V. (2017). Fine-grained network decomposition for massively parallel electromagnetic transient simulation of large power systems. *IEEE Power Energy Technol. Syst. J.* 4 (3): 1–11.
23. Liu, P. and Dinavahi, V. (2017). Finite-difference relaxation for parallel computation of ionized field of HVDC lines. *IEEE Trans. Power Del.* 33 (1): 1–10.
24. Lindholm, E., Nickolls, J., Oberman, S., and Montrym, J. (Mar. 2008). NVIDIA Tesla: a unified graphics and computing architecture. *IEEE Micro* 28 (2): 39–55.
25. Li, Z., Donde, V.D., Tournier, J.-C., and Yang, F. (2011). On limitations of traditional multi-core and potential of many-core processing architectures for sparse linear solvers used in large-scale power system applications. *2011 IEEE Power and Energy Society General Meeting*. Detroit, MI, pp. 1–8.
26. Chapman, B., Jost, G., and van der Pas, R. (2007). *Using OpenMP: Portable Shared Memory Parallel Programming*. Massachusetts Institute of Technology.
27. Davis, T.A. (2006). *Direct Methods for Sparse Linear Systems*. Philadelphia, PA: SIAM.
28. Saad, Y. (2003). *Iterative Methods for Sparse Linear Systems*. SIAM.
29. Shewchuk, J.R. (1994). *An Introduction to the Conjugate Gradient Method Without the Agonizing Pain*. Carnegie Mellon University.
30. Benzi, M. (Nov. 2002). Preconditioning techniques for large linear systems: a survey. *J. Comput. Phys.* 182 (2): 418–477.
31. Ortega, J.M. and Rheinboldt, W.C. (1970). *Iterative Solution of Nonlinear Equations in Several Variables*. Academic Press.
32. La Scala, M. and Bose, A. (May 1993). Relaxation/Newton methods for concurrent time step solution of differential-algebraic equations in power system dynamic simulations. *IEEE Trans. Circuits Syst. I Fundam. Theory Appl.* 40 (5): 317–330.

33. Roosta, S.H. (2000). Computer architecture. In: Seyed H. Roosta (ed.), *Parallel Processing and Parallel Algorithms*, 1–56. New York: Springer.
34. Zhang, J., Welch, G., and Bishop, G. (2011). LoDiM: a novel power system state estimation method with dynamic measurement selection. *2011 IEEE Power and Energy Society General Meeting*. Detroit, MI, pp. 1–7.
35. Chan, T.F. and Goovaerts, D. (Apr. 1992). On the relationship between overlapping and nonoverlapping domain decomposition methods. *SIAM J. Matrix Anal. Appl.* 13 (2): 663–670.
36. Smith, B.F., Bjørstad, P.E., and Gropp, W. (2004). *Domain Decomposition: Parallel Multilevel Methods for Elliptic Partial Differential Equations*. Cambridge University Press.
37. Leondes, C.T. (1991). *Analysis and Control System Techniques for Electric Power Systems. Part 4 of 4*. Academic Press Inc.
38. You, H., Vittal, V., and Wang, X. (Feb. 2004). Slow coherency-based islanding. *IEEE Trans. Power Syst.* 19 (1): 483–491.
39. Nucera, R.R., Brandwajn, V., and Gilles, M.L. (May 1993). Observability and bad data analysis using augmented blocked matrices (power system analysis computing). *IEEE Trans. Power Syst.* 8 (2): 426–433.
40. Wang, X., Vittal, V., and Heydt, G.T. (Aug. 2005). Tracing generator coherency indices using the continuation method: a novel approach. *IEEE Trans. Power Syst.* 20 (3): 1510–1518.

CHAPTER 14

DISHONEST GAUSS–NEWTON METHOD-BASED POWER SYSTEM STATE ESTIMATION ON A GPU

Md. Ashfaqur Rahman and Ganesh Kumar Venayagamoorthy

Real-Time Power and Intelligent Systems Laboratory, Clemson University, SC, USA

14.1 INTRODUCTION

State estimation is one of the key operations in power system. Many other processes like the contingency analysis, optimal power flow, etc. are directly dependent on the result of this operation. On one side it needs to be very accurate, and on the other side, it should be very fast.

The most widely used estimator is the weighted least squares (WLS) estimator that was developed by Carl Friedrich Gauss in 1795 [1], and it was proposed in power system by Fred Schweppe in 1970 [2]. It is the most efficient estimator that can give the most accurate result with limited data that are costly to collect. However, due to its high cost of computation that includes solving large linear systems, researchers had proposed some modifications to the original WLS estimator based on the properties of power system.

One of the most time-consuming parts of the WLS estimator is the calculation of the Jacobian matrix in each iteration. As the states move around some fixed numbers and do not change much, the matrix can be kept constant. As a result, the inverse of the matrix also remains constant. This is known as the dishonest Gauss–Newton method [3]. Another popular modification is taken by the decoupling of estimation of voltage magnitude and voltage angle. As the real power is mainly

Advances in Electric Power and Energy: Static State Estimation, First Edition.
Edited by Mohamed E. El-Hawary.

Published 2021 by John Wiley & Sons, Inc.

related with the voltage angles and the reactive power with the magnitude, they can be separated from the beginning [4]. This reduces the size of the system to half. However, in both the cases, the accuracy is compromised a little bit. There are some other modifications like the Cholesky decomposition of the Jacobian matrix that gives a simpler way to solve linear equations. But nothing seems to be more effective than a constant Jacobian matrix for parallel implementation. Unfortunately, as mentioned in [5], it is not investigated much for state estimation. But this method is well established for stability analysis.

The notion of running the estimation in real time yields another advantage of starting values. The WLS estimator can be run at such a rate that the system dynamics do not change much. As a result, the estimated value from the previous time sample can be used as the starting value for the current iteration.

The main objective of this chapter is to explore the dishonest method in details and to implement it on a single graphics processing unit (GPU) platform to get the fastest estimator. A GPU is a moderate cheap parallel processing unit that can easily be implemented in a control center. It runs the program in blocks of threads that are efficient for executing medium to short operations. So, the process of state estimation is needed to be separated into small operations that will be suitable for the GPU. GPUs are successfully used for dynamic estimation in [6, 7], and their usefulness is tested for large systems.

The main contributions of this chapter are as follows:

- The accuracy of the dishonest method under different levels of noise and sampling rates is studied.
- The implementation of the method on a GPU with estimated required time is explained.

The rest of the chapter is organized as follows. In Section 14.2, the background of state estimation is described. The performance of the dishonest method is analyzed in Section 14.3. The plan for making the operations suitable for GPU is given in Section 14.4. The simulation results are shown in Section 14.5. The scalability of the proposal is discussed in Section 14.6. The distributed method-based parallelization is described in Section 14.7. The chapter is concluded with future plan in Section 14.8.

14.2 BACKGROUND

14.2.1 Nonlinear Power Flow Equations

In power systems, erroneous measurements are taken in the form of the real and reactive power flows of the transmission lines as well as the bus injections, voltage magnitudes, and sometimes current magnitudes. Nowadays, with the deployment of the phasor measurement units (PMUs), phase angles are also being measured. However, the magnitudes and angles of the voltages of the buses are taken as the states of the systems. For an N-bus system, the state vector $\mathbf{x}$ can be written in terms of the angles θ and magnitude V as

$$\mathbf{x} = [\theta_2\, \theta_3 \cdots \theta_N \;\; V_1\, V_2 \cdots V_N]^T \tag{14.1}$$

The measurements can be considered as the true value added with some errors. Combining all the measurements, the measurement set $\mathbf{z}$ is formed. If the relation between the true values and the states are defined with $h(.)$, then, mathematically,

$$\mathbf{z} = h(\mathbf{x}) + \mathbf{e} \tag{14.2}$$

The objective of the state estimation is to remove the error from the measurements as much as possible.

14.2.2 Weighted Least Squares Estimation

WLS estimation is an iterative process that has a good convergence rate. In this estimation, the measurements get different weights based on the accuracy of the measuring devices. The most complex part of the iteration is the calculation of the Jacobian matrix **H**. It represents the partial derivative of each measurement with respect to every state variable.

However, the iteration starts with the flat start value where all angles are zero and all magnitudes are one:

$$\mathbf{x} = [0\,0 \cdots 0 \;\; 1\,1 \cdots 1]^T$$

Then, the difference between the current estimate and the measured value is calculated and updated according to the following steps:

- Step 1: $\Delta\mathbf{x} = (\mathbf{H}^T(\mathbf{x})\mathbf{W}\mathbf{H}(x))^{-1}\mathbf{H}^T(\mathbf{x})\mathbf{W}(\mathbf{z} - \mathbf{h}(\mathbf{x}))$.
- Step 2: $\mathbf{x}_{k+1} = \mathbf{x}_k + \Delta\mathbf{x}$.
- Step 3: Update $\mathbf{h}(\mathbf{x})$.
- Step 4: Update $\mathbf{H}(\mathbf{x})$.

Here, **W** is the weight matrix. This is the basic WLS estimation that is also known as the honest Gauss–Newton method.

To measure the accuracy of the estimation, the L_2 norm of the residue is taken as the measure. If the final estimate is $\hat{\mathbf{x}}$, then

$$\|\mathbf{r}\| = \|\mathbf{z} - h(\hat{\mathbf{x}})\| \tag{14.3}$$

14.2.3 Dishonest Gauss–Newton Method

In dishonest Gauss–Newton method, step 4 of the WLS method is not executed [3]. **H** is calculated at the beginning and updated after a certain period. If **H** does not change throughout the whole process, the method is called very dishonest.

The constant **H** helps in reducing the computation of step 1. As **H** remains constant, $(\mathbf{H}^T\mathbf{W}\mathbf{H})^{-1}\mathbf{H}^T\mathbf{W}$ does not change. Therefore, a constant matrix can be multiplied with the vector $\mathbf{z} - \mathbf{h}(\mathbf{x})$ to complete step 1. The matrix–vector multiplication is very suitable for GPU. The three steps can be reorganized as follows:

- Before estimation: Calculate $\mathbf{M} = (\mathbf{H}^T\mathbf{W}\mathbf{H})^{-1}\mathbf{H}^T\mathbf{W}$.
- During estimation:
  - Take previous estimation, $\mathbf{x}$.
  - For each measurement set, repeat the following steps for several times:
    - Step i: Calculate residuals, $\mathbf{r} = \mathbf{z} - \mathbf{h}(\mathbf{x})$.
    - Step ii: Calculate $\Delta\mathbf{x} = \mathbf{M}\mathbf{r}$.
    - Step iii: Calculate $\mathbf{x}_{n+1} = \mathbf{x}_n + \Delta\mathbf{x}$.

Though the method described in [3] proposes the flat start values for calculating **H** before estimation, it is not mandatory from the viewpoint of computational requirements. In fact, instead of using one Jacobian, different Jacobians can be used for handling different situations of the system.

14.2.4 Difference Between Honest and Dishonest Method

The difference between the honest and the dishonest method can be seen geometrically in Figure 14.1. To make it simpler, a single variable function, $y = f(x)$, is shown.

Let the iterations start at $x = x_0$ with an objective of $y = y_f$ (Figure 14.1a). The slope at x_0 is denoted with m_0. In the honest method, m_0 is used with the difference between $f(x_0)$ and y_f to find the new position, x_1. For x_1, the slope is calculated as m_1, and the process is repeated to find the solution.

On the other hand, the dishonest method starts with a fixed slope, m. The difference is always multiplied with this constant to find the new position of x as shown in Figure 14.1b. The use of a constant slope, m, does not only eliminate the calculation of m but also changes the division operation to multiplication (m^{-1}). The contribution is not significant for a single variable system, but it becomes an important improvement for multidimensional large-scale systems.

However, the dishonest method does not ensure convergence for any slope, m. The choice of m depends on the functions, the region of operations, the target values, and the starting values. Calculating the Jacobian for the extreme target and extreme starting values can make the process slow. So, for each function, the Jacobian can be developed for a normal and some extreme conditions.

14.2.5 Convergence of Dishonest Method

Convergence is an important property of any iterative algorithm. It ensures the robustness of the algorithm to complete its job under difficult situations. Though the dishonest method is existing in the literature for a long time, the analysis of the convergence is not found. Recently, the authors have made a specific analysis on the typical functions of power system state estimation [8]. With a geometric

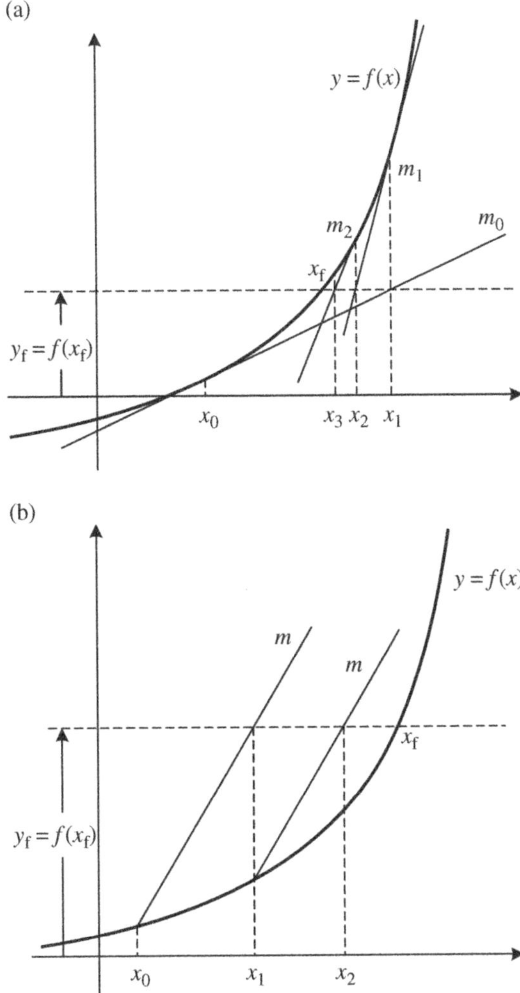

Figure 14.1 The working principle of Gauss–Newton method: (a) honest and (b) dishonest [8].

analysis on a single variable case, it is found that the range of convergence of the dishonest method can be significantly increased with the Jacobian calculated at a higher slope.

The convergence of a linear function is shown in Figure 14.2. If the slope of the search line is m and the original slope of a linear function is a_1, it can be easily shown that the process converges when

$$m > \frac{a_1}{2} \tag{14.4}$$

The analysis for the quadratic and the sinusoidal function is not shown in this chapter. However, as the states of the power system do not change much under

(a)

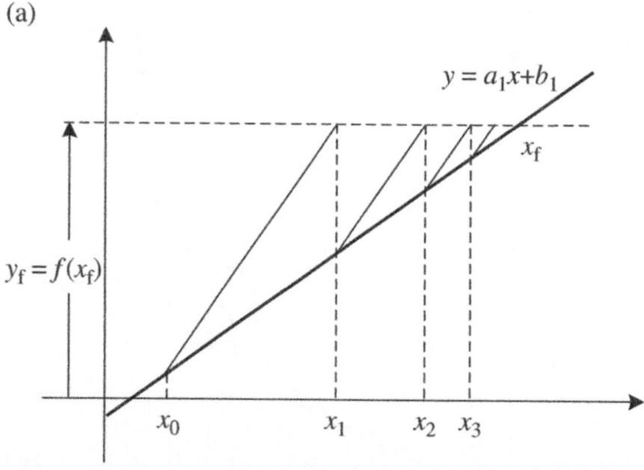

(b)

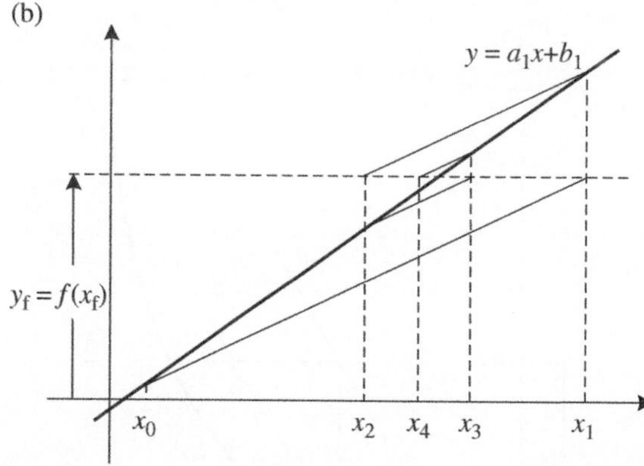

Figure 14.2 Two major ways of convergence of the dishonest method on a linear function, (a) overdamped case, (b) underdamped case [8]. Source: Reproduced with permission of IEEE.

normal operating condition, the quadratic and the sinusoidal functions can be linearized over the operating range. For this linear assumption, the analysis for the linear function can also be applied for the quadratic and the sinusoidal functions.

14.2.6 Graphics Processing Unit

A GPU is a combination of large number of processors under a specific structure. The basic structure of a GPU is shown in Figure 14.3. It contains a moderate number of streaming multiprocessors (SMs). Each SM contains a specific number of moderate powerful processors. The blocks of the kernel are assigned to the SMs. The number of blocks handled by an SM depends on the availability of the SMs. Each block contains a number of threads that are assigned to the processors.

The SMs can communicate with each other with L2 cache or global memory. Beside this, each SM has its own local memory. The local memory is combined of different levels of accessibility.

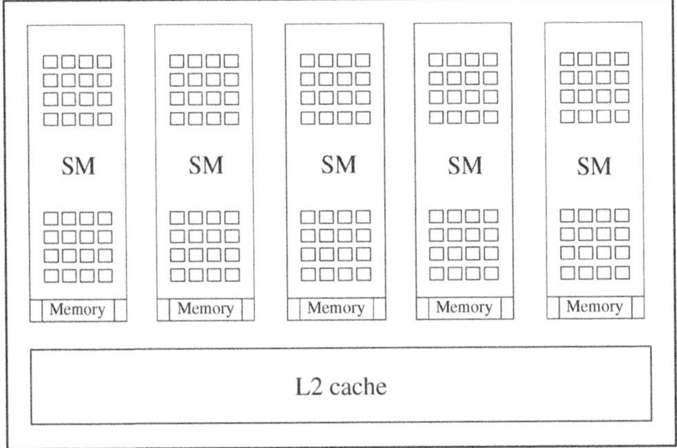

Figure 14.3 Simplified structure of a GPU.

14.3 PERFORMANCE OF DISHONEST GAUSS–NEWTON METHOD

In the dishonest method, the computations related with the Jacobian matrix are removed. This saves a lot of computations that makes the estimator fast. If the Jacobian is fixed for a very long time, it is also referred as the very dishonest Newton (VDHN) method [9]. It is mainly applied in the stability analysis.

14.3.1 Accuracy of the Estimation

Dishonest Gauss–Newton shows a high level of accuracy. Accuracy of an estimator directly depends on the level of noise. In order to measure the accuracy skipping the effects of the noise, it is measured with respect to the most accurate WLS estimator:

$$\alpha = \frac{\|\mathbf{r}_{\mathrm{WLS}}\|}{\|\mathbf{r}_{\mathrm{dis}}\|} \tag{14.5}$$

where $\mathbf{r}_{\mathrm{WLS}}$ and $\mathbf{r}_{\mathrm{dis}}$ are both measured under the same noise. Typically the value of α is in between 0.97 and 0.99. This indicates that the method is around 97–99% accurate (Figure 14.4).

Noise is not the single factor; the accuracy depends on some other factors as well. Among the other factors, measurement collection rate and the number of iterations are connected together. If the measurements are collected at a slower rate, the states change by a greater amount. So, starting from the previous estimation, it takes longer time to reach the new estimation.

However, it does not create any problem. Assume that the measurement is collected at a rate of n_{m} samples per second, and the estimator runs at n_{e} times per second with k iterations per time. Definitely, $n_{\mathrm{e}} \geq n_{\mathrm{m}}$. Now, if the rate is

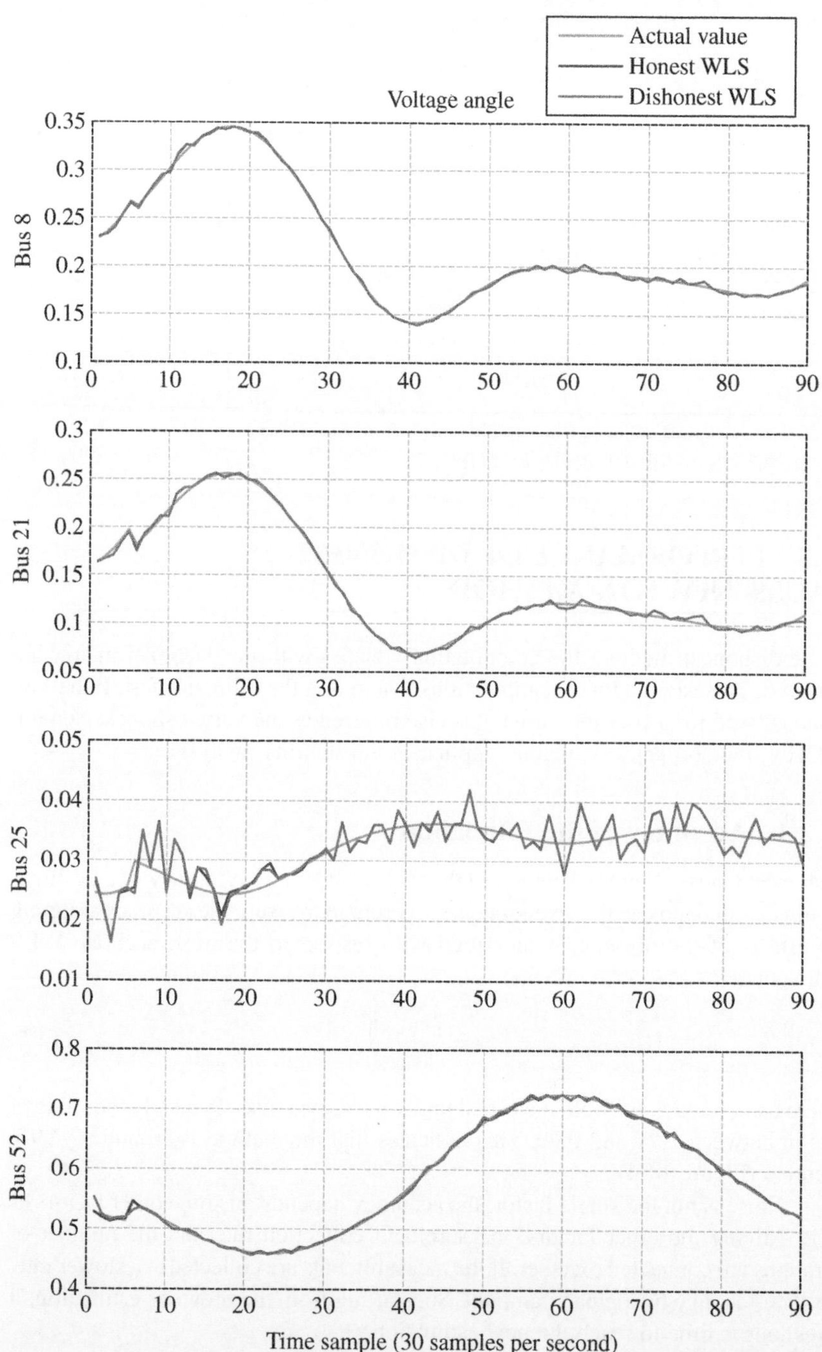

Figure 14.4 Accuracy of the dishonest Gauss–Newton method compared with the honest Gauss–Newton method. Though it takes a few samples to get accurate in the beginning due to the flat start, it performs much like the honest one afterward.

decreased by a factor of d, i.e. n_m/d, the number of iterations per sample can be increased by a factor of d. If the estimator can achieve the desired accuracy in that iterations, it will not create any problem.

In practice, the rate of collecting measurements is way too slow than the time required with dishonest method. Even with the fastest rate of the PMUs, the data is collected at every 8.33 ms where the estimator runs at 200 μs with three iterations. Even if the estimator runs for 10 iterations, it will not take more than 670 μs.

The relation between the sampling rate of the measurements and the accuracy for different number of measurements for a 68-bus system [10] is shown in Figure 14.5. It can be seen that the accuracy increases with the increase of sampling rate as well as with the number of iterations per sample. It also shows that seven iterations per sample give a good accuracy for any rate of collection over three samples per second. As the sampling rate is known, the number of iterations can be settled accordingly.

14.3.2 Fast Decoupled State Estimator

Fast decoupled state estimator is one of the most popular estimators in the industry [3]. It combines two advantages from the real power systems: one from the decoupling of the $P-\theta$ and $Q-V$ blocks of the Jacobian matrix and another from the constant Jacobian. It makes the whole process very fast.

14.3.3 Impact of Noise

In power systems, the noise is assumed to be Gaussian. It can play a big role on the accuracy. Not only the level of noise rather the pattern of noise can also change the accuracy. The same values of noise redistributed to different measurements can cause different accuracy.

The impact of noise on the accuracy is shown in Figure 14.6 for six different combinations of the noise values. The level of noise is defined as the mean noise to measurement ratio. The simulation shows that the accuracy is a bit low (around 98%) for low level of noise. However, it gets up to a certain level in between 99 and 99.5%. It does not give 100% accuracy at any time.

The effect of noise on the norm of the residue is shown in Figure 14.7. This can be referred as the absolute accuracy. It is completely linear, which means that the residue increases linearly with the increase of the level of noise. It is also shown for six different noises.

14.4 GPU IMPLEMENTATION

In order to make a suitable platform for the GPU, the process of state estimation needs to be divided in small parts for running it in different threads. Moreover, the constant values need to be calculated prior to running the estimation. As mentioned

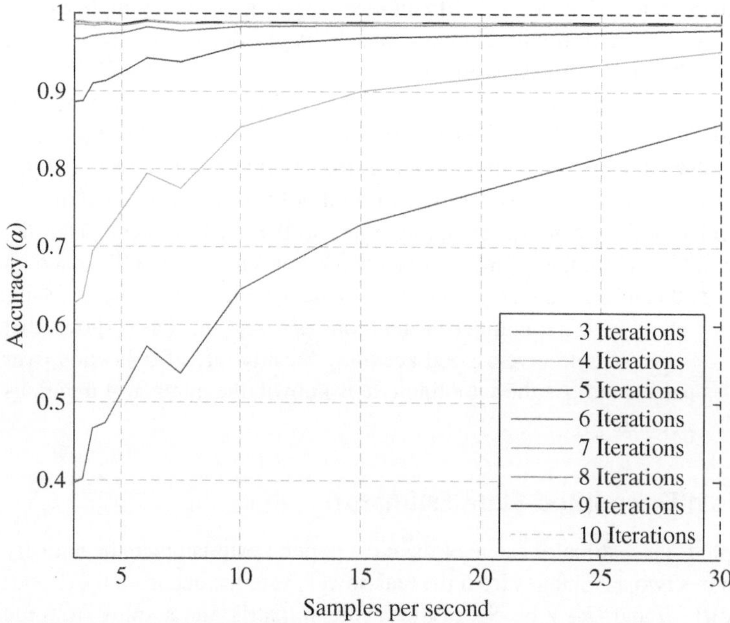

Figure 14.5 Accuracy of the dishonest Gauss–Newton method compared with the honest Gauss–Newton method for different rate of data of 68-bus system [11]. Source: Reproduced with permission of IEEE.

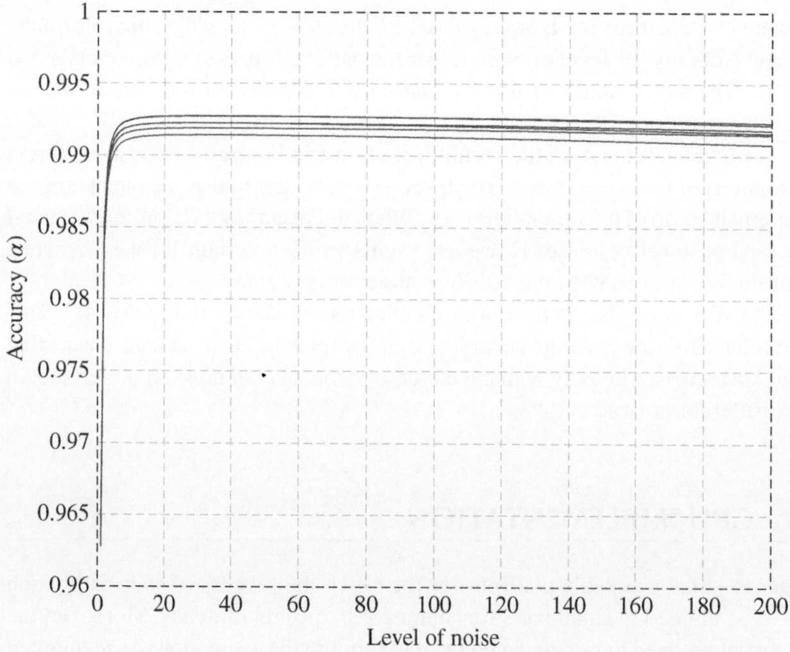

Figure 14.6 The accuracy of the estimator under different level of noise [11]. Source: Reproduced with permission of IEEE.

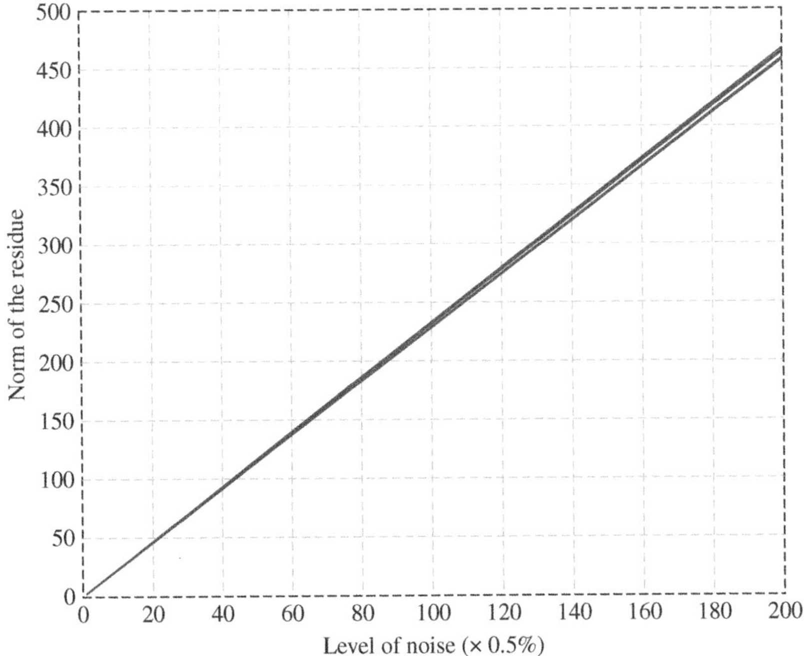

Figure 14.7 The norm of the residue of the estimated values under different level of noise [11]. Source: Reproduced with permission of IEEE.

earlier, the computation of the Jacobian matrix **H** is the most time-consuming part of the four steps. In dishonest Gauss–Newton method, step 4 is not executed.

Though it is advantageous to use the decoupling of voltage magnitudes and angles, it is not a helpful idea due to its slower convergence. It is experimented that the decoupling requires a good number of iterations to reach the same level of accuracy of the dishonest one. However, as **H** is remaining constant, it is not necessary to decompose it if there are enough processing units.

Another big advantage comes from the starting value. It reduces the number of iterations to reach the final value. Instead of starting from the flat start, the iterations can start from last estimation results.

An algorithm can be organized differently to serve different purposes. For GPU implementations, the steps of the dishonest method are organized as follows:

- Step i: Calculate residuals, $\mathbf{r} = \mathbf{z} - \mathbf{h}(\mathbf{x})$.
- Step ii: Calculate $\Delta\mathbf{x} = \mathbf{Mr}$.
- Step iii: Calculate $\mathbf{x}_{k+1} = \mathbf{x}_k + \Delta\mathbf{x}$.

The reason behind this reorder is the requirement of processors. In GPU computing, a set of threads are called to execute a kernel. The threads belong to one or more blocks. So, the number of threads needs to match the number of tiny operations. In the first step, there will be an equal number of measurements and power flow equations. So, they can be combined in a single equation and can be run in a single thread. If there are m measurements, the required number of threads becomes m. As the GPU can handle a maximum of 1024 threads per block, it will require only one block if $m < 1024$.

Step ii is a matrix–vector multiplication that can be divided into two parts – multiplication by rows to columns and addition by columns. Though it can be implemented in different ways, the best possible method is to assign one processor for every multiplication. As the size of **M** is $(2N - 1) \times m$, it can be separated by blocks and threads. Each block will be responsible for each row, and the threads of that block will take care of the columns of that row as shown in Figure 14.8. However, after completing the multiplications, the columns need to be added. Adding m elements of a row vector can be made parallel according to the method described in [12]. Instead of using the full method, a part of the method can be implemented to get good speedup. The part is shown in Figure 14.9.

In this addition, each thread adds only two elements. The process continues until all elements are added together. This simple method can add up to 2^n elements in the required time of $n + 1$ addition.

Step iii is a simple vector–vector addition. It will require $2N - 1$ threads that can be accommodated in one block for a system of 511 buses.

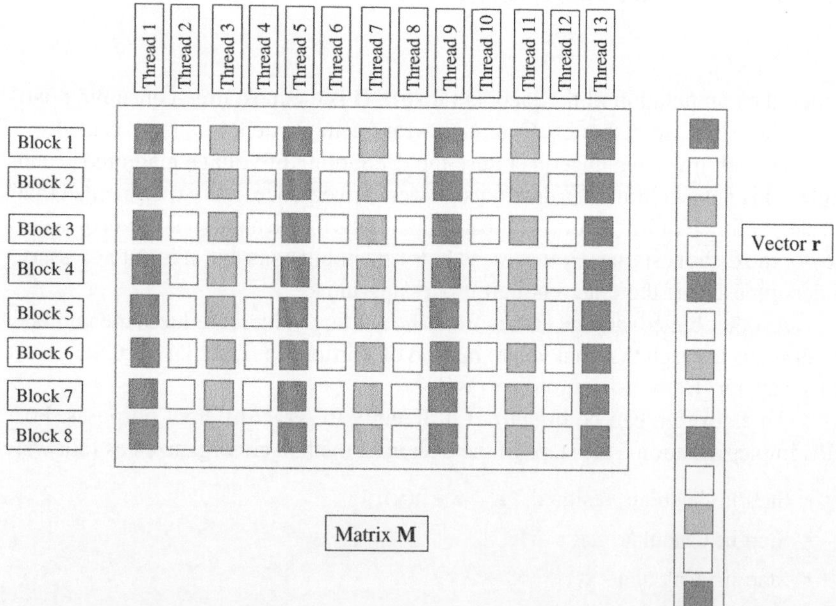

Figure 14.8 Parallel multiplication of a matrix and a vector [11]. Source: Reproduced with permission of IEEE.

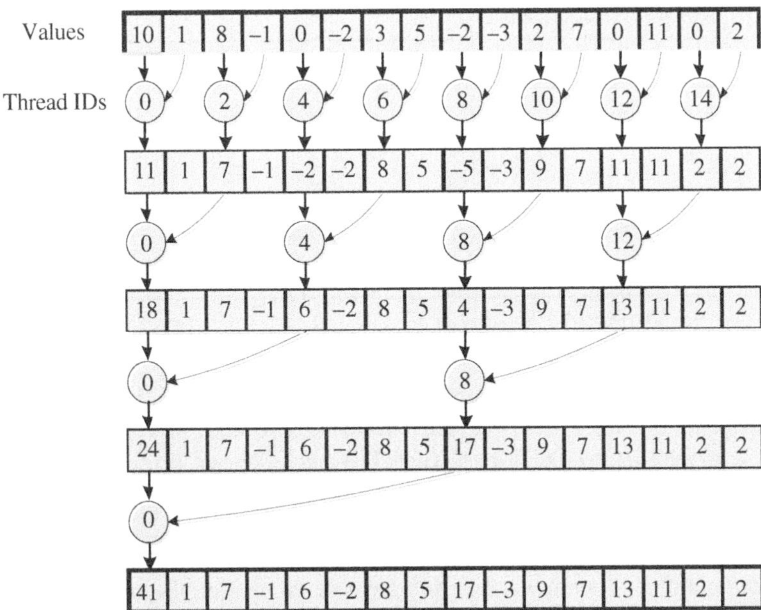

Figure 14.9 Parallel addition of 16 numbers [12].

14.5 SIMULATION RESULTS

To evaluate the actual time requirement, a 68-bus system is taken as the reference. It has 83 transmission lines with 166 line flow measurements (P and Q). For this evaluation, only line flows are taken as measurements. However, it does not matter much what measurements are taken but the number of them.

The measurements are taken from a real-time simulation on Real Time Digital Simulator (RTDS). Measurement errors are added artificially, which varies from 1 to 15% of the original value. The measurements are taken for three seconds at a rate of 30 samples per second, i.e. 33 ms per sample that is the typical sampling rate of the PMUs.

For estimating the states from the measurements, an NVIDIA Tesla K20c GPU card with compute capability 3.5 is used. Based on the three substeps, three different kernels are written that use different number of blocks and threads. The substeps are called sequentially. One of the most important advantages of the GPU card is that the time for starting the kernels takes very little time. So, stopping and starting kernels do not affect the execution time.

From the simulation, it is found that except the first estimation, the average required time for each sample with three iterations is around 124 μs. The first one requires 172 μs, which may be due to some initialization issue. The maximum and the minimum time for the subsequent estimations are 147 and 119 μs, respectively. However, this does not include the measurement collection or estimation result writing time.

TABLE 14.1 Reported required time [11].

Method	Year	Size	Required time (μs)
Distributed multi-area [13]	2011	118	3714
M-CSE [14]	2012	14	23 700
M-CSE [14]	2012	118	20 000
Distributed robust [15]	2013	118	3200
Dishonest	2015	68	124
Dishonest	2015	118	217

Source: Reproduced with permission of IEEE.

In order to understand the significance of the numbers, it can be compared with some reported times from previous works. The comparison is shown in Table 14.1. In addition to the 68-bus, IEEE 118-bus system is also estimated as most of the previous studies are done with that. The required time will be discussed more in Section 14.6.

It is important to remember that the times are taken with three iterations. With higher number of iterations, they will increase. Moreover, they also depend on the starting value, speed of the processor, stopping criteria, etc. It is discussed later.

14.6 DISCUSSIONS ON SCALABILITY

The parallel code was implemented for a system of only 68-bus. Therefore, it was not possible to simulate a larger power system on the RTDS racks at the Real-Time Power and Intelligent Systems (RTPIS) Laboratory [16] and collect the data that is necessary for the dependency of the starting values. However, it can easily be extended for larger systems with less than 1024 measurements without making any big change in the code. For example, IEEE 300-bus system has 411 transmission lines. It has 599 state variables, and if the system collects 1000 measurements to estimate the 599 state variables, the required time will be affected based on the configuration of the GPU.

14.6.1 Estimation of Time for Very Large Systems

Though it is expected that the blocks and the threads of a GPU will run simultaneously, it is not possible in practice. Like other computation devices, GPU has its own limitations.

Every block of a GPU runs a maximum number of threads at a time. In case of the k20 series, 32 threads run at a time in one block. These 32 threads are called a warp. On the other hand, the blocks are run on the SMs. These SMs are physical entity, and they are limited as well. Each SM can take care of more than one blocks, but their scheduling can differ.

For this analysis, let the maximum number of threads and blocks that can run simultaneously in a GPU be N_t and N_b, respectively. A block can hold maximum N_{max} threads. The time required for each addition/subtraction and multiplication is t_a and t_m. The number of states is n, and the number of measurements is m.

The first step has n calculations of $h(\mathbf{x})$ and n subtractions. As it is dependent on the nature of $h(\mathbf{x})$, let us take the time for $h(\mathbf{x})$ to be t_h. For m additions, it will require ceil(m/N_{max}) blocks. So, the total time for step i is

$$t_1 = t_h + \text{ceil}\left(\frac{\text{ceil}\left(\frac{m}{N_{max}}\right)}{N_b}\right) \times \text{ceil}\left(\frac{N_{max}}{N_t}\right) \times t_a \tag{14.6}$$

The most time-consuming part is step ii. There are n rows and m columns in matrix $\mathbf{M}$. If each row is assigned to one block, then there will be n blocks. As N_b blocks can run simultaneously, n blocks will be called in ceil(n/N_b) steps.

Similarly, if each column is assigned to one thread, the whole threads will be called in ceil(m/N_t) steps. For large systems, it is possible that $m \geq N_{max}$. In that case each thread has to take care of more than one multiplication. So, the total time t_{21} for the multiplication will take

$$t_{21} = \text{ceil}\left(\frac{n}{N_b}\right) \times \text{ceil}\left(\frac{N_{max}}{N_t}\right) \times \text{ceil}\left(\frac{m}{N_{max}}\right) \times t_m \tag{14.7}$$

After completing the multiplication, there will be m additions that will take the time of ceil($\log_2 m$) addition operations:

$$t_{22} = \text{ceil}(\log_2 m) \times t_a \tag{14.8}$$

In step iii, there are n additions and it will take ceil(n/N_{max}) blocks. As N_b blocks can run at a time, the total required time for step iii is

$$t_3 = \text{ceil}\left(\frac{\text{ceil}\left(\frac{n}{N_{max}}\right)}{N_b}\right) \times \text{ceil}\left(\frac{N_{max}}{N_t}\right) \times t_a \tag{14.9}$$

Now, all of them can be added together to estimate the required time of estimation t for a very large system:

$$t = t_1 + t_{21} + t_{22} + t_3 \tag{14.10}$$

It should be remembered that the launch and execution time of the threads or blocks are not always equal and there is randomness. So, a random number should be added with each term of Eq. (14.10). Moreover, each block or thread should wait for the slowest ones to complete their jobs.

There are other factors like the number of registers and size of shared memories that can limit the performance. They are excluded from the calculations for simplicity.

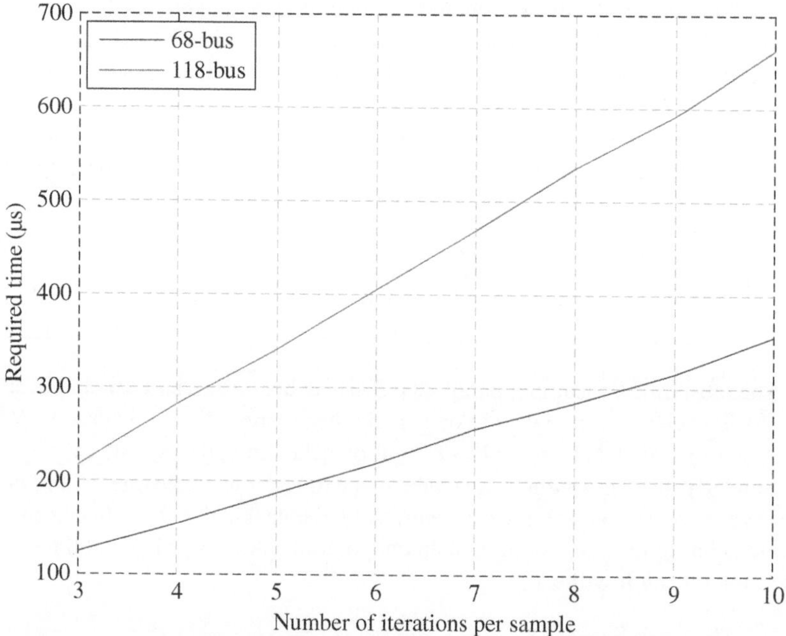

Figure 14.10 Required time for different number of iterations. Though it grows linearly, there is a fixed amount of time for transferring the data between the CPU and the GPU [11]. Source: Reproduced with permission of IEEE.

14.6.2 Communication Time

It is well known that sending the data to the GPU and taking the result back to the CPU takes a significant amount of time. The results in Table 14.1 show the total time. If the communication time is subtracted, the core computation time that includes only the time for three iterations can be found.

In Figure 14.10, the total time for 68-bus and 118-bus systems are shown for different number of iterations.

After fitting the times with a line, it is found that the fixed amounts of time required for 68-bus and 118-bus are 23 and 26 μs, respectively. This is the time for sending the measurements to the GPU and taking the estimation results out. As, in case of 118-bus, the measurements set is bigger, it takes longer time.

14.7 DISTRIBUTED METHOD OF PARALLELIZATION

Another important class of parallelization can be implemented with the distributed dishonest method. Typically the term *distributed* refers to separating the whole grid in some groups. Having different sizes for the groups do not yield much speedup for the process. The distributed process depends on the size of the largest group.

The maximum speedup can be achieved with cellular computational network (CCN) [17]. In this method, each bus is distributed to separate cells. Each cell completes its own estimation and exchanges the result with its neighbors. The neighbors update their results, and the whole process improves over exchange and update.

14.7.1 Cellular Computational Network

CCN is primarily proposed to divide a large network into small subsystems. In power systems, it forms a computational cell at each bus. The cells complete local estimations and exchange and update their result. As the cells run in parallel, it becomes a completely scalable framework. The process of exchange and update of CCN-based network is shown in Figure 14.11. However, estimation at the cellular level reveals some unique aspects that are not found at the traditional distributed estimators.

14.7.2 Challenges of Cellular Estimation

State estimation of a power system is done with respect to a reference bus. The angle of the reference bus is considered as zero. Moreover, the measured real and reactive powers as well as the voltage and current magnitudes are represented in per unit quantities. Usually the operating voltage of the reference bus is considered as the base voltage, and all measured powers are converted to per unit based on that. This is helpful for the centralized state estimation.

For the cellular estimation, it creates a problem. Each cell requires a reference bus. As the measurements are normalized using the voltage angle and magnitude of the reference bus, they are needed to be normalized with the local reference bus. But, in the beginning, the voltage magnitude of the local reference bus is unknown.

One of the solutions of the normalization can be done with the predicted values of the corresponding bus. As the estimation process is relentless, the estimated value of time t can be used to predict the value of $t + 1$. This is a single-step prediction and the accuracy can be pretty high.

However, the scaling of the measurements does not take much effort. From the standard power flow equations, it can be seen that the angles are completely relative and they do not affect the power flow. But the power flow measurements are dependent on the scaling of voltage magnitudes. As a result, every power flow

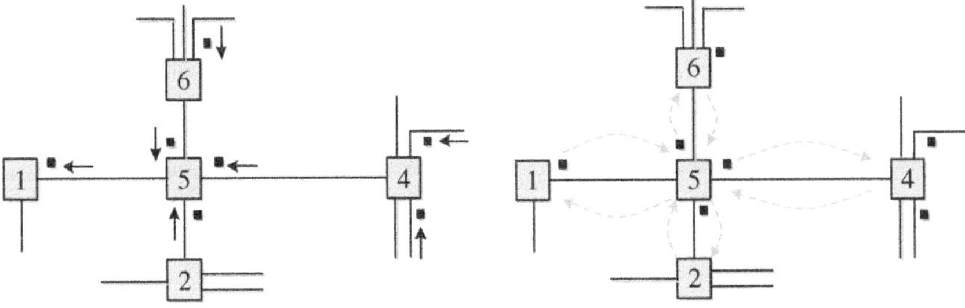

Figure 14.11 The process of exchange and update of the CCN.

and power injection measurements are needed to be scaled with the square of the voltage magnitudes of local reference bus voltage magnitude.

14.7.3 Structure of the Cellular Dishonest Estimator

The structure of the cellular dishonest method is shown in Figure 14.12. To make the whole process parallel, the predictor is also made cellular. The outputs of the predictor are fed to the cellular estimators. The full description of the cellular predictor can be found in [18].

The output of the cellular dishonest units is fed back to the predictors to predict the next sample. From the block diagram, it may look like the output of each dishonest unit is sent to each prediction unit. This is not true. Each predictor unit receives the output of its own and its neighbors only. Thus it also keeps the privacy of the cells as well.

From the structure it is realizable that each cell can be run in parallel. With this structure, the computational complexity remains the same over the size of the system.

14.7.4 Accuracy of Cellular Dishonest Method

The accuracy of the cellular dishonest method for four different buses is shown in Figure 14.13. The phase angles are shown on the left, and the voltage magnitudes are shown in the right. From the figure, it can be seen the estimated values follow the actual value quite well. Except for Bus 36, all others show an acceptable accuracy.

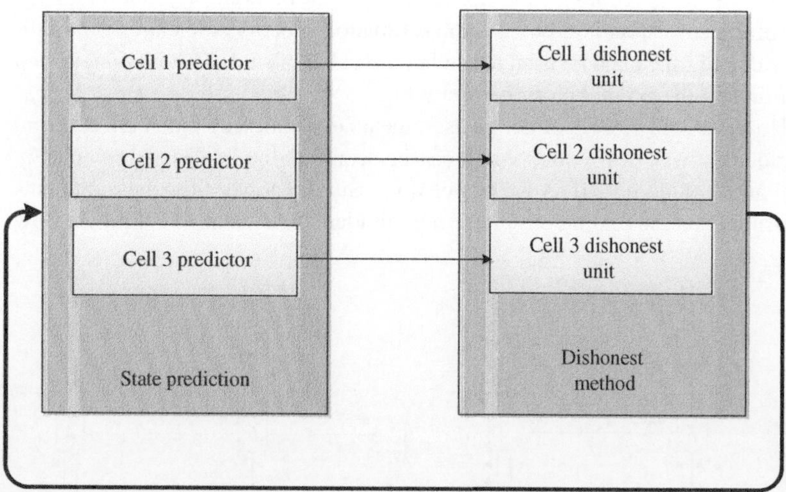

Figure 14.12 The cellular dishonest method.

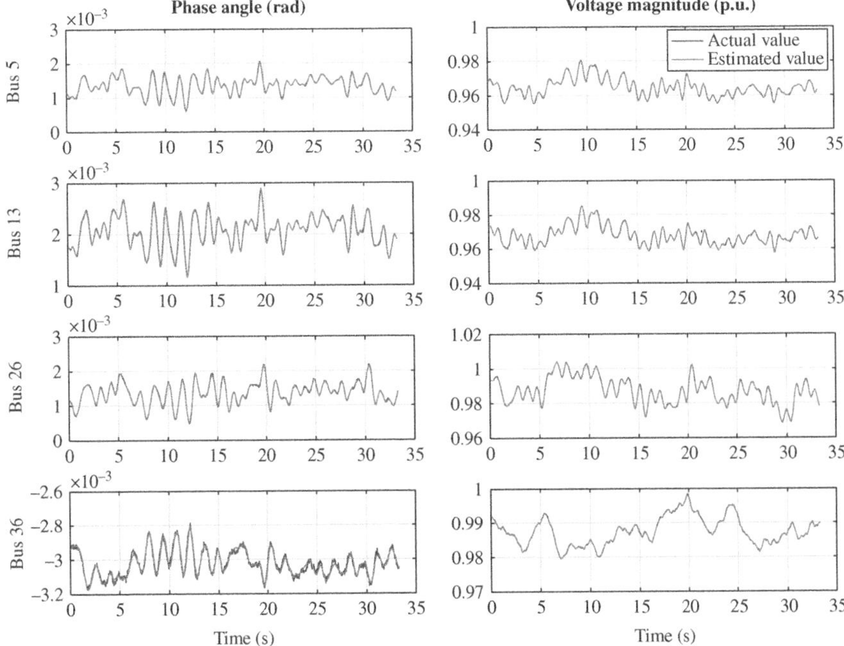

Figure 14.13 The actual and the estimated value of the cellular dishonest method.

14.7.5 Time of Cellular Dishonest Method

The cellular method is implemented in MATLAB serial mode, and the cellular time is calculated by separating the total time for each cell. The experiments are run on an Intel(R) Xeon(R) CPU (E5-2609) with 2.4 GHz core and 48GB of memory. The average required time by the whole unit is around 1.8955 ms. The dishonest unit takes around 1.7122 ms; the prediction unit takes around 0.19 ms. The communication time is excluded in all these times.

14.8 CONCLUSIONS

An algorithm can be parallelized in a number of ways for different types of architecture. This chapter presents a specific fragmentation of the calculations of the WLS estimator suitable for running on a GPU. From the studies, it is shown that the required time surpasses other distributed algorithms of recent time. It can be taken as the standard for the further experiments on fast state estimation.

The fragmentation can lead to the development of special GPUs for state estimation with high number of simultaneously running blocks for very large power systems of around tens of thousands of buses. This improvement in speed can enhance predicting the system states and avoid upcoming undesirable events.

Though the cellular dishonest method-based estimation is implemented in a stand-alone computer with limited number of cores, it can also be parallelized on a GPU. The amount of computation for each thread can be reduced with analytic solutions. This is left as a future work.

REFERENCES

1. Otto, B. (2005). *Linear Algebra with Applications*. Prentice Hall.
2. Schweppe, F.C. and Wildes, J. (1970). Power system static-state estimation, part I: exact model. *IEEE Transactions on Power Apparatus and Systems* PAS-89: 120–125.
3. Monticelli, A. (1999). *State Estimation in Electric Power Systems: A Generalized Approach*, vol. 507. Springer.
4. Abur, A. and Exposito, A. (2004). *Power System State Estimation: Theory and Implementation*. New York: Marcel Dekker Inc.
5. Semlyen, A. and De León, F. (2001). Quasi-Newton power flow using partial Jacobian updates. *IEEE Transactions on Power Systems* 16 (3): 332–339.
6. Karimipour, H. and Dinavahi, V. (2015). Extended Kalman filter-based parallel dynamic state estimation. *IEEE Transactions on Smart Grid* 6 (3): 1539–1549.
7. Karimipour, H. and Dinavahi, V. (2016). Parallel relaxation-based joint dynamic state estimation of large-scale power systems. *IET Generation, Transmission and Distribution* 10 (2): 452–459.
8. Rahman, M.A. and Venayagamoorthy, G.K. (2017). Convergence of the fast state estimation for power systems. *SAIEE Africa Research Journal* 108 (3): 117–127.
9. Chai, J.S., Zhu, N., Bose, A., and Tylavsky, D.J. (1991). Parallel Newton type methods for power system stability analysis using local and shared memory multiprocessors. *IEEE Transactions on Power Systems* 6 (4): 1539–1545.
10. Ramos, R.A., Kuiava, R., Fernandes, T.C., Pataca, L.C., and Mansour, M.R. (2014). IEEE PES task force on benchmark systems for stability controls. Pacific Northwest National Laboratory, Tech. Rep., 2012, pp. 1–49.
11. Rahman, M.A. and Venayagamoorthy, G.K. (2016). Dishonest Gauss Newton method based power system state estimation on a GPU. *Clemson University Power Systems Conference (PSC), 2016*, IEEE. SC, USA, pp. 1–6.
12. Harris, M. (2007). Optimizing parallel reduction in CUDA. *NVIDIA Developer Technology* 2 (4): 1–39.
13. Korres, G.N. (2011). A distributed multiarea state estimation. *IEEE Transactions on Power Systems* 26 (1): 73–84.
14. Xie, L., Choi, D.H., Kar, S., and Poor, H. (2012). Fully distributed state estimation for wide-area monitoring systems. *IEEE Transactions on Smart Grid* 3 (3): 1154–1169. https://doi.org/10.1109/TSG.2012.2197764.
15. Kekatos, V. and Giannakis, G. (2013). Distributed robust power system state estimation. *IEEE Transactions on Power Systems* 28 (2): 1617–1626.
16. Venayagamoorthy, G.K. (2004), Real-Time Power and Intelligent Systems Laboratory. http://rtpis.org (accessed 5 september 2020).
17. Luitel, B. and Venayagamoorthy, G.K. (2014). Cellular computational networks – a scalable architecture for learning the dynamics of large networked systems. *Neural Networks* 50: 120–123.
18. Rahman, M.A. and Venayagamoorthy, G.K. (2016). Power system distributed dynamic state prediction. *IEEE Symposium Series on Computational Intelligence (SSCI), 2016*, IEEE. Athens, Greece, pp. 1–7.

INDEX

Advances in Electric Power and Energy: Static State Estimation, First Edition.
Edited by Mohamed E. El-Hawary.

Published 2021 by John Wiley & Sons, Inc.

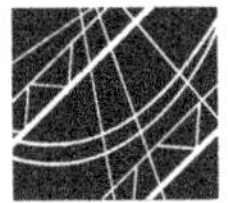

IEEE Press Series on Power Engineering

Series Editor: **M. E. El-Hawary,** Dalhousie University, Halifax, Nova Scotia, Canada

The mission of IEEE Press Series on Power Engineering is to publish leading- edge books that cover the broad spectrum of current and forward-looking technologies in this fast-moving area. The series attracts highly acclaimed authors from industry/academia to provide accessible coverage of current and emerging topics in power engineering and allied fields. Our target audience includes the power engineering professional who is interested in enhancing their knowledge and perspective in their areas of interest.

1. *Electric Power Systems: Design and Analysis, Revised Printing*
Mohamed E. El-Hawary

2. *Power System Stability*
Edward W. Kimbark

3. *Analysis of Faulted Power Systems*
Paul M. Anderson

4. *Inspection of Large Synchronous Machines: Checklists, Failure Identification, and Troubleshooting*
Isidor Kerszenbaum

5. *Electric Power Applications of Fuzzy Systems*
Mohamed E. El-Hawary

6. *Power System Protection*
Paul M. Anderson

7. *Subsynchronous Resonance in Power Systems*
Paul M. Anderson, B.L. Agrawal, and J.E. Van Ness

8. *Understanding Power Quality Problems: Voltage Sags and Interruptions*
Math H. Bollen

9. *Analysis of Electric Machinery*
Paul C. Krause, Oleg Wasynczuk, and S.D. Sudhoff

10. *Power System Control and Stability, Revised Printing*
Paul M. Anderson and A.A. Fouad

Advances in Electric Power and Energy: Static State Estimation, First Edition.
Edited by Mohamed E. El-Hawary.

Published 2021 by John Wiley & Sons, Inc.

11. *Principles of Electric Machines with Power Electronic Applications, Second Edition*
Mohamed E. El-Hawary

12. *Pulse Width Modulation for Power Converters: Principles and Practice*
D. Grahame Holmes and Thomas A. Lipo

13. *Analysis of Electric Machinery and Drive Systems, Second Edition*
Paul C. Krause, Oleg Wasynczuk, and Scott D. Sudhoff

14. *Risk Assessment for Power Systems: Models, Methods, and Applications*
Wenyuan Li

15. *Optimization Principles: Practical Applications to the Operation of Markets of the Electric Power Industry*
Narayan S. Rau

16. *Electric Economics: Regulation and Deregulation*
Geoffrey Rothwell and Tomas Gomez

17. *Electric Power Systems: Analysis and Control*
Fabio Saccomanno

18. *Electrical Insulation for Rotating Machines: Design, Evaluation, Aging, Testing, and Repair, Second Edition*
Greg C. Stone, Ian Culbert, Edward A. Boulter, and Hussein Dhirani

19. *Signal Processing of Power Quality Disturbances*
Math H. J. Bollen and Irene Y. H. Gu

20. *Instantaneous Power Theory and Applications to Power Conditioning*
Hirofumi Akagi, Edson H. Watanabe, and Mauricio Aredes

21. *Maintaining Mission Critical Systems in a 24/7 Environment*
Peter M. Curtis

22. *Elements of Tidal-Electric Engineering*
Robert H. Clark

23. *Handbook of Large Turbo-Generator Operation and Maintenance, Second Edition*
Geoff Klempner and Isidor Kerszenbaum

24. *Introduction to Electrical Power Systems*
Mohamed E. El-Hawary

25. *Modeling and Control of Fuel Cells: Distributed Generation Applications*
M. Hashem Nehrir and Caisheng Wang

26. *Power Distribution System Reliability: Practical Methods and Applications*
Ali A. Chowdhury and Don O. Koval

27. *Introduction to FACTS Controllers: Theory, Modeling, and Applications*
Kalyan K. Sen and Mey Ling Sen

28. *Economic Market Design and Planning for Electric Power Systems*
James Momoh and Lamine Mili

29. *Operation and Control of Electric Energy Processing Systems*
James Momoh and Lamine Mili

30. *Restructured Electric Power Systems: Analysis of Electricity Markets with Equilibrium Models*
Xiao-Ping Zhang

31. *An Introduction to Wavelet Modulated Inverters*
S. A. Saleh and M. Azizur Rahman

32. *Control of Electric Machine Drive Systems*
Seung-Ki Sul

33. *Probabilistic Transmission System Planning*
Wenyuan Li

34. *Electricity Power Generation: The Changing Dimensions*
Digambar M. Tagare

35. *Electric Distribution Systems*
Abdelhay A. Sallam and Om P. Malik

36. *Practical Lighting Design with LEDs*
Ron Lenk and Carol Lenk

37. *High Voltage and Electrical Insulation Engineering*
Ravindra Arora and Wolfgang Mosch

38. *Maintaining Mission Critical Systems in a 24/7 Environment, Second Edition*
Peter M. Curtis

39. *Power Conversion and Control of Wind Energy Systems*
Bin Wu, Yongqiang Lang, Navid Zargari, and Samir Kouro

40. *Integration of Distributed Generation in the Power System*
Math H. J. Bollen and Fainan Hassan

41. *Doubly Fed Induction Machine: Modeling and Control for Wind Energy Generation*
Gonzalo Abad, Jesús López, Miguel Rodrigues, Luis Marroyo, and Grzegorz Iwanski

42. *High Voltage Protection for Telecommunications*
Steven W. Blume

43. *Smart Grid: Fundamentals of Design and Analysis*
James Momoh

44. *Electromechanical Motion Devices, Second Edition*
Paul Krause, Oleg Wasynczuk, and Steven Pekarek

45. *Electrical Energy Conversion and Transport: An Interactive Computer-Based Approach, Second Edition*
George G. Karady and Keith E. Holbert

46. *ARC Flash Hazard and Analysis and Mitigation*
J. C. Das

47. *Handbook of Electrical Power System Dynamics: Modeling, Stability, and Control*
Mircea Eremia and Mohammad Shahidehpour

48. *Analysis of Electric Machinery and Drive Systems, Third Edition*
Paul Krause, Oleg Wasynczuk, Scott Sudhoff, and Steven Pekarek

49. *Extruded Cables for High-Voltage Direct-Current Transmission: Advances in Research and Development*
Giovanni Mazzanti and Massimo Marzinotto

50. *Power Magnetic Devices: A Multi-Objective Design Approach*
S. D. Sudhoff

51. *Risk Assessment of Power Systems: Models, Methods, and Applications, Second Edition*
Wenyuan Li

52. *Practical Power System Operation*
Ebrahim Vaahedi

53. *The Selection Process of Biomass Materials for the Production of Bio-Fuels and Co-Firing*
Najib Altawell

54. *Electrical Insulation for Rotating Machines: Design, Evaluation, Aging, Testing, and Repair, Second Edition*
Greg C. Stone, Ian Culbert, Edward A. Boulter, and Hussein Dhirani

55. *Principles of Electrical Safety*
Peter E. Sutherland

56. *Advanced Power Electronics Converters: PWM Converters Processing AC Voltages*
Euzeli Cipriano dos Santos Jr. and Edison Roberto Cabral da Silva

57. *Optimization of Power System Operation, Second Edition*
Jizhong Zhu

58. *Power System Harmonics and Passive Filter Designs*
J. C. Das

59. *Digital Control of High-Frequency Switched-Mode Power Converters*
Luca Corradini, Dragan Maksimovic, Paolo Mattavelli, and Regan Zane

60. *Industrial Power Distribution, Second Edition*
Ralph E. Fehr, III

61. *HVDC Grids: For Offshore and Supergrid of the Future*
Dirk Van Hertem, Oriol Gomis-Bellmunt, and Jun Liang

62. *Advanced Solutions in Power Systems: HVDC, FACTS, and Artificial Intelligence*
Mircea Eremia, Chen-Ching Liu, and Abdel-Aty Edris

63. *Operation and Maintenance of Large Turbo-Generators*
Geoff Klempner and Isidor Kerszenbaum

64. *Electrical Energy Conversion and Transport: An Interactive Computer-Based Approach*
George G. Karady and Keith E. Holbert

65. *Modeling and High-Performance Control of Electric Machines*
John Chiasson

66. *Rating of Electric Power Cables in Unfavorable Thermal Environment*
George J. Anders

67. *Electric Power System Basics for the Nonelectrical Professional*
Steven W. Blume

68. *Modern Heuristic Optimization Techniques: Theory and Applications to Power Systems*
Kwang Y. Lee and Mohamed A. El-Sharkawi

69. *Real-Time Stability Assessment in Modern Power System Control Centers*
Savu C. Savulescu

70. *Optimization of Power System Operation*
Jizhong Zhu

71. *Insulators for Icing and Polluted Environments*
Masoud Farzaneh and William A. Chisholm

72. *PID and Predictive Control of Electric Devices and Power Converters Using MATLAB®/Simulink®*
Liuping Wang, Shan Chai, Dae Yoo, Lu Gan, and Ki Ng

73. *Power Grid Operation in a Market Environment: Economic Efficiency and Risk Mitigation*
Hong Chen

74. *Electric Power System Basics for the Nonelectrical Professional, Second Edition*
Steven W. Blume

75. *Energy Production Systems Engineering*
Thomas Howard Blair

76. *Model Predictive Control of Wind Energy Conversion Systems*
Venkata Yaramasu and Bin Wu

77. *Understanding Symmetrical Components for Power System Modeling*
J. C. Das

78. *High-Power Converters and AC Drives, Second Edition*
Bin Wu and Mehdi Narimani

79. *Current Signature Analysis for Condition Monitoring of Cage Induction Motors: Industrial Application and Case Histories*
William T. Thomson and Ian Culburt

80. *Introduction to Electric Power and Drive Systems*
Paul Krause, Oleg Wasynczuk, Timothy O'Connell, and Maher Hasan

81. *Instantaneous Power Theory and Applications to Power Conditioning, Second Edition*
Hirofumi Akagi, Edson Hirokazu Watanabe, and Mauricio Aredes

82. *Practical Lighting Design with LEDs, Second Edition*
Ron Lenk and Carol Lenk

83. *Introduction to AC Machine Design*
Thomas A. Lipo

84. *Advances in Electric Power and Energy Systems: Load and Price Forecasting*
Mohamed E. El-Hawary

85. *Electricity Markets: Theories and Applications*
Jeremy Lin and Fernando H. Magnago

86. *Multiphysics Simulation by Design for Electrical Machines, Power Electronics, and Drives*
Marius Rosu, Ping Zhou, Dingsheng Lin, Dan Ionel, Mircea Popescu, Frede Blaabjerg, Vandana Rallabandi, and David Staton

87. *Modular Multilevel Converters: Analysis, Control, and Applications*
Sixing Du, Apparao Dekka, Bin Wu, and Navid Zargari

88. *Electrical Railway Transportation Systems*
Morris Brenna, Federica Foiadelli, and Dario Zaninelli

89. *Energy Processing and Smart Grid*
James A. Momoh

90. *Handbook of Large Turbo-Generator Operation and Maintenance, Third Edition*
Geoff Klempner and Isidor Kerszenbaum

91. *Advanced Control of Doubly Fed Induction Generator for Wind Power Systems*
Dehong Xu, Frede Blaabjerg, Wenjie Chen, and Nan Zhu

92. *Electric Distribution Systems, Second Edition*
Abdelhay A. Sallam and Om P. Malik

93. *Power Electronics in Renewable Energy Systems and Smart Grid: Technology and Applications*
Bimal K. Bose

94. *Distributed Fiber Optic Sensing and Dynamic Rating of Power Cables*
Sudhakar Cherukupalli and George J. Anders

95. *Power System and Control and Stability, Third Edition*
Vijay Vittal, James D. McCalley, Paul M. Anderson, and A. A. Fouad

96. *Electromechanical Motion Devices: Rotating Magnetic Field-Based Analysis and Online Animations, Third Edition*
Paul Krause, Oleg Wasynczuk, Steven D. Pekarek, and Timothy O'Connell

97. *Applications of Modern Heuristic Optimization Methods in Power and Energy Systems*
Kwang Y. Lee and Zita A. Vale

98. *Handbook of Large Hydro Generators: Operation and Maintenance*
Glenn Mottershead, Stefano Bomben, Isidor Kerszenbaum, and Geoff Klempner

99. *Advances in Electric Power and Energy: Static State Estimation*
Mohamed E. El-Hawary

Printed and bound by CPI Group (UK) Ltd, Croydon, CR0 4YY